Lecture Notes in Computer Sci

T0238499

Commenced Publication in 1973
Founding and Former Series Editors:
Gerhard Goos, Juris Hartmanis, and Jan van Leeuwen

Advanced Research in Computing and Software Science

Subline of Lectures Notes in Computer Science

Martin Charles Golumbic   Michal Stern
Avivit Levy   Gila Morgenstern (Eds.)

Graph-Theoretic Concepts in Computer Science

38th International Workshop, WG 2012
Jerusalem, Israel, June 26-28, 2012
Revised Selcted Papers

Springer

Volume Editors

Martin Charles Golumbic
Caesarea Rothschild Institute (CRI)
and Department of Computer Science
31905 Haifa, Israel
E-mail: golumbic@cs.haifa.ac.il

Michal Stern
The Academic College of Tel Aviv
Unit of Statistics and Probability Studies
64044 Yaffo, Israel
E-mail: stern@mta.ac.il

Avivit Levy
Shenkar College
52526 Ramat-Gan, Israel
and
University of Haifa
Caesarea Rothschild Institute (CRI)
31905 Haifa, Israel
E-mail: avivitlevy@shenkar.ac.il

Gila Morgenstern
Caesarea Rothschild Institute (CRI)
31905 Haifa, Israel
E-mail: gilamor@cs.bgu.ac.il

ISSN 0302-9743   e-ISSN 1611-3349
ISBN 978-3-642-34610-1   e-ISBN 978-3-642-34611-8
DOI 10.1007/978-3-642-34611-8

Springer Heidelberg Dordrecht London New York

Library of Congress Control Number: 2012950704

CR Subject Classification (1998): G.2.2, I.2.8, E.1, F.2.2, I.3.5

LNCS Sublibrary: SL 1 – Theoretical Computer Science and General Issues

Typesetting: Camera-ready by author, data conversion by Scientific Publishing Services, Chennai, India

Printed on acid-free paper

Springer is part of Springer Science+Business Media (www.springer.com)

Preface

The 38th International Workshop on Graph-Theoretic Concepts in Computer Science (WG 2012) took place in Ramat-Rachel on the outskirts of Jerusalem, Israel, during June 26–28, 2012. There were 74 participants coming from five continents, 14 different countries mostly from Europe.

The workshop continues a tradition of 37 previous WG workshops. Since 1975, WG has taken place 21 times in Germany, four times in The Netherlands, twice in Austria, France and the Czech Republic, and once in Greece, Italy, Norway, Slovakia, Switzerland, and the UK. This year, WG 2012 took place in Israel for the first time. The workshop aims to unite theory and practice by demonstrating how graph theoretic concepts can be applied to various areas in computer science, and by extracting new graph theoretic problems from applications. The goal is to present new and recent research results as well as to identify and explore directions of future research.

WG 2012 received 78 submissions, three of which were withdrawn for various reasons before finalizing the review process. Each submission was carefully reviewed by at least three members of the Program Committee. The Committee accepted 29 papers to be presented at the workshop. Unfortunately, there were several high-quality papers that had to be rejected for lack of time slots. The workshop program was enriched by three interesting invited talks by outstanding researchers: Dieter Rautenbach (Ulm, Germany), David Peleg (Rehovot, Israel), and Amitava Bhattacharya (Mumbai, India). The talk by Amitava Bhattacharya was dedicated to the memory of Uri N. Peled and was sponsored by the Caesarea Rothschild Institute at the University of Haifa.

Greetings were given by Daniel Hershkowitz, Minister of Science of the State of Israel, and a mathematician himself who has published many papers in linear algebra, matrix theory, and their relationship with combinatorics and graph theory.

In order to encourage more young scientists taking part in the workshop, for the first time in the tradition of WG, there was a Student Poster Session, where six posters were presented. The criterion for presentation of a poster was that it must be based on a research paper accepted to a refereed computer science or mathematics conference during the past year. We found the experience to be very positive and it met our expectations. The Best Student Paper Award was given to Marek Cygan, Marcin Pilipczuk, and Michał Pilipczuk for the paper "On Group Feedback Vertex Set Parameterized by the Size of the Cutset." The scientific program of the workshop was complemented by two sightseeing tours. One tour was to the Old City of Jerusalem, revealing the history of Jerusalem related to the places visited. This guided walking tour from Jaffa Gate to the Western Wall included a visit to the Church of the Holy Sepulchre, the Jewish Quarter, and a walk through the Western Wall Tunnels. For those not too tired, the tour ended with a Sound and Light Spectacular Show at the Tower of David at night. At the conclusion of the scientific program, a second (optional) tour taking participants overnight to the Dead Sea, including swimming (well, to be more accurate floating there), visiting the Botanical Gardens at Kibbutz Ein Gedi, exploring Massada, and hiking to the waterfalls and streams of Nahal David. We succeeded to exhaust everyone!

We would like to thank all who contributed to the success of WG 2012: the authors who submitted very high quality papers, the speakers, the Program Committee members for their devotion, and the referees. Special thanks to the Local Organizing Committee: first of all Danielle Friedlander, who worked tirelessly during the months of preparation and the final execution of the wonderful arrangements and coordination, and second to Hananel Hazan, who was our ever present technology assistant and guy Friday. Without their work, WG 2012 could not have been such a success. Our tour guides in Jerusalem were Donna Goldberg and Daniel Barkai. Donna (who happens to have a masters degree in computer science) continued with us to the Dead Sea, Massada, and Nachal David where she pushed us to our limits. Thanks also to Ruth Touito and Elad Cohen for their assistance.

Special thanks for the sponsoring organizations: University of Haifa, the Caesarea Rothschild Institute for Interdisciplinary Applications of Computer Science, I-Core – Israeli Center of Excellence in Algorithms, and Springer.

August 2012

Martin Golumbic
Michal Stern
Avivit Levy
Gila Morgenstern

Organization

Program Chairs

Martin Charles Golumbic	University of Haifa, Israel
Michal Stern	Academic College of Tel Aviv Yaffo, Israel
Avivit Levy	Shenkar College, Israel
Gila Morgenstern	University of Haifa, Israel

Program Committee

Threse Biedl	University of Waterloo, Canada
Hans Bodlaender	Utrecht University, The Netherlands
Andreas Brandstädt	University of Rostock, Germany
L. Sunil Chandran	Indian Institute of Science, India
Jianer Chen	Texas A&M University, USA
Lenore J. Cowen	Tufts University, USA
Celina de Figueiredo	Universidade Federal do Rio de Janeiro, Brazil
Fedor Fomin	University of Bergen, Norway
Martin Charles Golumbic	University of Haifa, Israel
Gregory Z. Gutin	Royal Holloway, University of London, UK
Magnus M. Halldorsson	Reykjavik University, Iceland
Pavol Hell	Simon Fraser University, Canada
Seok-Hee Hong	University of Sydney, Australia
Tibor Jordán	Eötvös Loránd University, Hungary
Michael Kaufmann	University of Tübingen, Germany
Dieter Kratsch	LITA, Université de Metz, France
Lap Chi Lau	The Chinese University of Hong Kong, SAR China
Avivit Levy	Shenkar College, Israel
Vincent Limouzy	Blaise Pascal University, France
Alberto Marchetti-Spaccamela	Sapienza University of Rome, Italy
Ross McConnell	Colorado State University, USA
Gila Morgenstern	University of Haifa, Israel
Rüdiger K. Reischuk	University Luebeck, Germany
Michal Stern	Academic College of Tel Aviv Yaffo, Israel
Dimitrios Thilikos	National and Kapodistrian University of Athens, Greece
Yaokun Wu	Shanghai Jiao Tong University, China
Shmuel Zaks	Technion, Israel Institute of Technology, Israel

Local Organization

Danielle Friedlander
Martin Charles Golumbic
Hananel Hazan
Avivit Levy
Gila Morgenstern
Michal Stern

Additional Reviewers

Alcón, Liliana
Ausiello, Giorgio
Bhattacharya, Amitabha
Blaeser, Markus
Bonifaci, Vincenzo
Bonomo, Flavia
Bornstein, Claudson
Bourgeois, Nicolas
Cai, Yufei
Cechlarova, Katarina
Crowston, Robert
Cygan, Marek
Del-Vecchio, Renata
Dorbec, Paul
Durán, Guillermo A.
Fleiner, Tamás
Fonseca, Guilherme D. Da
Gaspers, Serge
Giannopoulou, Archontia
Golovach, Petr
Guedes, Andre
Gusmão Pereira De Sá, Vinícius
Gutierrez, Marisa
Hanaki, Akihide
Hermelin, Danny
Herskovics, David
Hirasaka, Mitsugu
Huang, Jing
Hundt, Christian
Hunter, Paul
Hüffner, Falk
Iwata, Satoru
Jampani, Krishnam Raju
Jansen, Klaus

Jean-Claude, Bermond
Jiang, Minghui
Jin, Xian'An
Jones, Mark
Joret, Gwenaël
Kaminski, Marcin
Kang, Liying
Katz, Matthew
Kenkre, Sreyash
Kim, Eun Jung
King, Andrew
Kiraly, Tamas
Kopparty, Swastik
Korman, Amos
Kortsarz, Guy
Koutsonas, Athanassios
Krakovski, Roi
Kratochvil, Jan
Kratsch, Stefan
Krause, Philipp Klaus
Kuhn, Daniela
Kurur, Piyush
Kwok, Tsz Chiu
Le, Van Bang
Lewenstein, Moshe
Lhouari, Nourine
Lindzey, Nathan
Linn, Joseph G.
Liskiewicz, Maciej
M. S., Ramanujan
Maffray, Frederic
Makowsky, Johann
Marino, Andrea
Mary, Arnayd

Meidanis, Joao
Mertzios, George
Misra, Neeldhara
Mnich, Matthias
Moscardelli, Luca
Muciaccia, Gabriele
Muller, Haiko
Möhring, Rolf H.
N, Narayanan
Nederlof, Jesper
Obdrzalek, Jan
Paul, Christophe
Paulusma, Daniel
Peleg, David
Pemmaraju, Sriram
Pergel, Martin
Rao, Michael
Richerby, David
Ries, Bernard
Rodrigues, Rosiane De Freitas
Rossman, Benjamin
Rothvoss, Thomas
Saitoh, Toshiki

Sampaio, Rudini
Schlotter, Ildikó
Sritharan, R.
Stewart, Lorna
Szeider, Stefan
Tantau, Till
Telles, Guilherme
Thiyagarajan, Karthick
Trevisan, Vilmar
van 'T Hof, Pim
van Bevern, René
Vegh, Laszlo
Wang, Xinmao
Wang, Zengqi
Wilmer, Elizabeth
Wong, Prudence W.H.
Wu, Taoyang
Xu, Min
Yeo, Anders
Yuditsky, Yelena
Zoros, Dimitris
Zwols, Yori

Table of Contents

Account on Intervals

Dieter Rautenbach

University of Ulm, Germany

Abstract. The structural and algorithmic properties of interval graphs have received considerable attention. While many aspects of this important and classical class of graphs are well understood, some old problems are still open. One such problem is the so-called interval count problem, which asks for the minimum number of different interval lengths needed to represent a given interval graph. Whereas graphs of interval count 1 coincide with unit interval graphs, not much is known about graphs of interval count 2. In this talks we will survey some recent results and discuss several open problems related to interval count 2 graphs.

M.C. Golumbic et al. (Eds.): WG 2012, LNCS 7551, p. 1, 2012.
© Springer-Verlag Berlin Heidelberg 2012

Constructing Resilient Structures in Graphs: Rigid vs. Competitive Fault-Tolerance

David Peleg

Weizmann Institute of Science, Israel

Abstract. The setting considered in this talk is that of a structure S constructed over a given network G and intended to efficiently support some service on it (e.g., a distributed database or a query-answering oracle.) Such a structure is required to ensure certain desirable properties with respect to G. However, a failure event F might damage some of the network's vertices and edges, and cause S to malfunction. We are interested in ways of making S fault-tolerant, namely, reinforcing it so that following a failure event, its surviving part continues to satisfy the requirements. The talk will distinguish between two types of fault-tolerance, termed rigid and competitive fault tolerance, compare these two notions, and illustrate them on a number of examples.

M.C. Golumbic et al. (Eds.): WG 2012, LNCS 7551, p. 2, 2012.

Alternating Reachability and Integer Sum of Closed Alternating Trails

The 3rd Annual Uri N. Peled Memorial Lecture

Amitava Bhattacharya

Tata Institute of Fundamental Research, Mumbai, India

Abstract. We consider a graph with colored edges and study the following two problems.

(i) Suppose first that the number of colors is two, say red and blue. A nonnegative real vector on the edges is said to be *balanced* if the red sum equals the blue sum at every vertex. A *balanced subgraph* is a subgraph whose characteristic vector is balanced (i.e., red equals blue degree at every vertex). By a *sum* (respectively, *fractional sum*) of cycles we mean a nonnegative integral (respectively, nonnegative rational) combination of characteristic vectors of cycles. Similarly, we define sum and fractional sum of balanced subgraphs. We show that a balanced sum of cycles is a fractional sum of balanced subgraphs.

(ii) Next we consider the problem of finding a necessary and sufficient condition for the existence of a balanced subgraph containing a given edge. This problem is easily reduced to the alternating reachability problem, defined as follows. Given an edge colored graph (here we allow ≥ 2 colors) a trail (vertices may repeat but not edges) is called *alternating* when successive edges have different colors. Given a set of vertices called *terminals*, the *alternating reachability* problem is to find an alternating trail connecting distinct terminals, if one exists. By reduction to the classical case of searching for an augmenting path with respect to a matching we show that either there exists an alternating trail connecting distinct terminals or there exists an obstacle, called a *Tutte set*, to the existence of such trails. We also give a Gallai-Edmonds decomposition of the set of nonterminals.

This work started when Uri Peled and Murali Srinivasan met in Caesarea Edmond Benjamin de Rothschild Foundation Institute for Interdisciplinary Applications of Computer Science at the University of Haifa, Israel during May–June 2003. This led to many interesting questions and some of them are still open. In this talk we would like to discuss some of them.

M.C. Golumbic et al. (Eds.): WG 2012, LNCS 7551, p. 3, 2012.
© Springer-Verlag Berlin Heidelberg 2012

Student Poster Session

During WG 2012 there was a Student Poster Session, where the following posters were presented (alphabetically ordered by student's last name).

The Canadian Tour Operator Problem (Online Graph Exploration with Disposal)

Sabine Büttner

In the prize-collecting travelling salesman problem, we are given a weighted graph $G = (V, E)$ with edge weights $l : E \to R_+$, a special vertex $r \in V$, penalties $p : V \to R_+$ and the goal is to find a closed tour T such that $r \in V(T)$ and such that the cost $l(T) + p(V \setminus V(T))$ is minimized. We consider an online variant of the prize-collecting travelling salesman problem related to graph exploration. In the *Canadian Tour Operator Problem* (ctop) the task is to find a closed route for a tourist bus in a given network $G = (V, E)$ in which some edges are blocked by avalanches. An online algorithm learns from a blocked edge only when reaching one of its endpoints. The bus operator has the option to avoid visiting each node $v \in V$ by paying a refund of $p(v)$ to the tourists. The goal is to minimize the sum of the travel costs and the refunds. We show that no deterministic or randomized algorithm can achieve a bounded competitive ratio for the CTOP on general graphs. Further, we present a ϕ-competitive algorithm for the line and give a Ski-Rental like 3-competitive algorithm for tree networks.

Joint work with Sven O. Krumke.

Fully Dynamic Approximate Distance Oracles for Planar Graphs via Forbidden-Set Distance Labels [1]

Shiri Chechik

Distance oracle is a data structure that provides fast answers to distance queries. Recently, the problem of designing distance oracles capable of answering restricted distance queries, that is, estimating distances on a subgraph avoiding some forbidden vertices, has attracted a lot of attention. We consider forbidden set distance oracles for planar graphs. We present an efficient compact distance oracle that is capable of handing any number of failures. In addition, we consider a closely related notion of fully dynamic distance oracles. In the dynamic distance oracle problem instead of getting the failures in the query phase, we rather need to handle an adversarial online sequence of update and query operations. Each query operation involves two vertices s and t whose distance needs to be estimated. Each update operation involves inserting/deleting a vertex/edge from the graph. Our forbidden set distance oracle can be tweaked to give fully dynamic distance oracle with improved bounds compared to the previously known fully dynamic distance oracle for planar graphs.

Joint work with Ittai Abraham and Cyril Gavoille.

M.C. Golumbic et al. (Eds.): WG 2012, LNCS 7551, pp. 4–6, 2012.

Product Graphs Invariants with Applications to the Theory of Information [4]

Marcin Jurkiewicz

There are a large number of graph invariants. We consider some of them, e.g. the independence and chromatic numbers. It is well know that we cannot efficiently calculate these numbers for arbitrary graphs. We present relations between these invariants and some concepts from the theory of information. Concepts such as source coding and transmission over a noisy channel with zero probability of error are modeled using graph theoretical structures and are measured by the independence and chromatic numbers of some products of graphs, i.e. graphs arising from other graphs. It turns out that for some classes of product graphs, there exist algorithms and methods for determining these invariants. Using optimization algorithms together with some theoretical results, we can establish their values or bounds on previously mentioned invariants of product graphs.

Joint work with Marek Kubale.

Improved Approximation for Orienting Mixed Graphs [3]

Moti Medina

An instance of the maximum mixed graph orientation problem consists of a mixed graph and a collection of source-target vertex pairs. The objective is to orient the undirected edges of the graph so as to maximize the number of pairs that admit a directed source-target path. This problem has recently arisen in the study of biological networks, and it also has applications in communication networks.

In this paper, we identify an interesting local-to-global orientation property. This property enables us to modify the best known algorithms for maximum mixed graph orientation and some of its special structured instances, due to Elberfeld et al. (CPM '11), and obtain improved approximation ratios. We further proceed by developing an algorithm that achieves an even better approximation guarantee for the general setting of the problem. Finally, we study several well-motivated variants of this orientation problem.

Joint work with Iftah Gamzu.

SINR Diagram with Interference Cancellation [2]

Merav Parter

This paper studies the reception zones of a wireless network in the SINR model with receivers that employ interference cancellation (IC). IC is a recently developed technique that allows a receiver to decode interfering signals, and cancel them from the received signal in order to decode its intended message. We first derive the important topological properties of the reception zones and their relation to high-order Voronoi diagrams and other geometric objects. We then discuss the computational issues that arise when seeking an efficient description of the zones. Our main fundamental result states that although potentially there are exponentially many possible cancellation orderings, and as a result, reception zones, in fact there are much fewer nonempty such zones. We prove a linear bound (hence tight) on the number of zones and provide a polynomial time algorithm to describe the diagram. Moreover, we introduce a novel parameter, the Compactness Parameter, which influences the tightness of our bounds. We then utilize these properties to devise a logarithmic time algorithm to answer point-location queries for networks with IC.

Joint work with Chen Avin, Asaf Cohen, Yoram Haddad, Erez Kantor, Zvi Lotker, and David Peleg

Approximating the Girth [5]

Roei Tov

This paper considers the problem of computing a minimum weight cycle in weighted undirected graphs. Given a weighted undirected graph $G(V, E, w)$, let C be a minimum weight cycle of G, let $w(C)$ be the weight of C and let $w_{max}(C)$ be the weight of the maximal edge of C. We obtain three new approximation algorithms for the minimum weight cycle problem:

1. For integral weights from the range $[1, M]$ an algorithm that reports a cycle of weight at most $\frac{4}{3}w(C)$ in $O(n^2 \log n(\log n + \log M))$ time.
2. For integral weights from the range $[1, M]$ an algorithm that reports a cycle of weight at most $w(C) + w_{max}(C)$ in $O(n^2 \log n(\log n + \log M))$ time.
3. For non-negative real edge weights an algorithm that for any $\varepsilon > 0$ reports a cycle of weight at most $(\frac{4}{3} + \varepsilon)w(C)$ in $O(\frac{1}{\varepsilon}n^2 \log n(\log \log n))$ time.

Joint work with Liam Roditty.

References

1. Abraham, I., Chechik, S., Gavoille, C.: Fully dynamic approximate distance oracles for planar graphs via forbidden-set distance labels. In: Proc. the 44th Symposium on Theory of Computing Conference, STOC 2012, pp. 1199–1218 (2012)
2. Avin, C., Cohen, A., Haddad, Y., Kantor, E., Lotker, Z., Parter, M., Peleg, D.: In: Proc. the 23rd Annual ACM-SIAM Symposium on Discrete Algorithms, SODA 2012, pp. 502–515 (2012)
3. Gamzu, I., Medina, M.: Improved Approximation for Orienting Mixed Graphs. In: Even, G., Halldórsson, M.M. (eds.) SIROCCO 2012. LNCS, vol. 7355, pp. 243–253. Springer, Heidelberg (2012)
4. Jurkiewicz, M., Kubale, M.: Product Graphs Invariants with Applications to the Theory of Information. In: The 5th International Interdisciplinary Technical Conference of Young Scientists InterTech (2012)
5. Roditty, L., Tov, R.: Approximating the Girth. In: Proc. the 22nd Annual ACM-SIAM Symposium on Discrete Algorithms, SODA 2011, pp. 1446–1454 (2011)

Triangulation and Clique Separator Decomposition of Claw-Free Graphs

Anne Berry and Annegret Wagler

LIMOS UMR CNRS 6158*
Ensemble Scientifique des Cézeaux, Université Blaise Pascal,
63 173 Aubière, France
{berry,wagler}@isima.fr

Abstract. Finding minimal triangulations of graphs is a well-studied problem with many applications, for instance as first step for efficiently computing graph decompositions in terms of clique separators. Computing a minimal triangulation can be done in $O(nm)$ time and much effort has been invested to improve this time bound for general and special graphs. We propose a recursive algorithm which works for general graphs and runs in linear time if the input is a claw-free graph and the length of its longest path is bounded by a fixed value k. More precisely, our algorithm runs in $O(f + km)$ time if the input is a claw-free graph, where f is the number of fill edges added, and k is the height of the execution tree; we find all the clique minimal separators of the input graph at the same time. Our algorithm can be modified to a robust algorithm which runs within the same time bound: given a non-claw free input, it either triangulates the graph or reports a claw.

Keywords: claw-free graph, minimal triangulation, clique separator decomposition.

1 Background and Motivation

Chordal graphs are an important class, with properties similar to those of trees, and corresponding efficient algorithms. A graph is *chordal*, or *triangulated*, if it has no induced chordless cycle on 4 or more vertices; any non-chordal graph can be embedded into a chordal graph by adding a set of 'fill' edges, a process called 'triangulation'. Adding a minimum number of fill edges is an NP-complete problem [39], but adding an inclusion-wise minimal set of edges, thus obtaining a 'minimal triangulation', is polynomial.

Minimal triangulations have many applications (see [4], the recent survey [23] and references therein). Originally, the problem stemmed from sparse matrix computation [23,35], where triangulation was needed for Gaussian elimination in sparse symmetric systems, but it is also useful in other fields such as database management [1,38].

* Research partially supported by the French Agency for Research under the DEFIS program TODO, ANR-09-EMER-010.

M.C. Golumbic et al. (Eds.): WG 2012, LNCS 7551, pp. 7–21, 2012.

One of the more surprising applications for minimal triangulation is that it is currently a mandatory first step for efficiently computing a decomposition by clique separators. *Clique separator decomposition* was introduced by Tarjan [37] as hole- and C_4- preserving, and refined to a unique and optimal decomposition using clique *minimal* separators by [28]; this process consists in repeatedly finding a clique minimal separator and copying it into the connected components it defines (see [9] for details). This decomposition has attracted recent attention in the context of characterizing graph classes [12,13], for Bayesian networks [31] and for clustering gene expression data [26].

Regarding minimal triangulation, the seminal paper of Rose, Tarjan and Lueker [36] presented an $O(nm)$ time algorithmic process. Several other $O(nm)$ time algorithms have appeared recently [2,3,6,8]. Efforts have been invested into improving this $O(nm)$ time bound for the general case: [27] offer an $O(n^{2.69})$ time bound, later improved by [24] to $O(n^\alpha log n) = o(n^{2.376})$, where n^α is the time required to do matrix multiplication, currently $n^{2.376}$. For special graph classes, there are surprisingly few results which improve the time bound of the general case: so far, chordal bipartite graphs and hole-and-diamond-free graphs have been shown to have an $O(n^2)$ time algorithm [7]; co-comparability graphs and AT-free claw-free graphs have a linear-time triangulation algorithm [30]; co-bipartite graphs, a subclass of AT-free claw-free graphs, have an even better time, since only a subset of edges needs to be traversed [11].

In this paper, we address the issue of improving the computation of a minimal triangulation for claw-free graphs. For this class, many significant results were obtained during the past twenty years. A particularly active field of research is improving the running times for computing maximum (weight) stable sets in claw-free graphs [20,21,32,33], partly based on decomposing claw-free graphs in an appropriate way [16,17,18].

A further recent result by [15] showed the presence of a hole containing a pair of vertices when there is no clique separator between them. However, not much is known on minimal triangulations and separators for claw-free graphs. Investigating this aspect, we present some interesting properties, which enable us to taylor an algorithm for computing both a minimal triangulation and the clique minimal separators. We show that this algorithm runs in linear time if the input is a claw-free graph where the length of the longest chordless path is bounded. More precisely, the algorithm constructs a tree in the graph and runs in $O(f + km)$, where f is the number of fill edges and k is the length of a longest branch of the tree. In fact, we present a 'robust' algorithm which, given any graph as input, will either triangulate it in $O(f + km)$ or report a claw as negative certificate. Moreover, as the recognition of claw-free graphs in general is in $O(m^{1.69})$ [25], our algorithm may improve finding a claw on some inputs.

The paper is organized as follows: we will give in Section 2 some preliminaries on minimal triangulation, minimal separation, and clique separators. In Section 3, we present and explain our algorithmic process. Section 4 gives the algorithm and proves its correctness and complexity. In Section 5, we discuss the robustness of our algorithm. We conclude with some final remarks.

2 Preliminaries

Basics. All graphs in this work are connected, undirected and finite. A graph is denoted by $G = (V, E)$, with $|V| = n$ and $|E| = m$. We say that a vertex x *sees* a vertex y if $xy \in E$. The *neighborhood* of a vertex x in a graph G is $N_G(x)$, the *closed neighborhood* is $N_G[x] = N_G(x) \cup \{x\}$. The neighborhood of a set X of vertices is $N_G(X) = \cup_{x \in X} N_G(x) - X$. A *clique* is a set of pairwise adjacent vertices; we say that we *saturate* a set X of vertices when we add all the edges necessary to turn X into a clique. A *clique module* is a set X of vertices such that $\forall x, y \in X, N[x] = N[y]$. $G(X)$ denotes the subgraph induced by the set X of vertices, but we will sometimes just denote this by X. A *co-bipartite* graph is a graph whose vertex set can be partitioned into two cliques, K_1 and K_2; we will call an edge *external* if it is neither inside K_1 nor inside K_2. A *claw* is an induced subgraph on 4 vertices $\{x, y_1, y_2, y_3\}$ with 3 edges: xy_1, xy_2, xy_3 where x is called the *center* of the claw. The reader is referred to [22] and [14] for classical graph definitions and results.

Separators. A set S of vertices of a connected graph G is a *separator* if $G(V - S)$ is not connected. A separator S is an *xy-separator* if x and y lie in two different connected components of $G(V - S)$. S is a *minimal xy-separator* if S is an xy-separator and no proper subset S' of S is also an xy-separator. A separator S is said to be *minimal* if there are two vertices x and y such that S is a minimal xy-separator. For any minimal separator S in a connected graph, there are at least two connected components C_1 and C_2 of $G(V - S)$ such that $N(C_1) = N(C_2) = S$ (called *full components*); this means that for any pair (x, y) of vertices from the Cartesian product $C_1 \times C_2$, S is a minimal xy-separator; note also that every vertex of S sees both C_1 and C_2, and that, equivalently, any path from x to y must go through S.

Property 1. [5] In any non-complete graph, there is a clique module whose neighborhood is a minimal separator, called a moplex; the vertex numbered 1 by LexBFS belongs to a moplex.

A *clique minimal separator* is a minimal separator which is a clique.

   The minimal separators included in the neighborhood of a vertex x are called the *substars* of x [6,29]. Computing these substars can be done as follows: let $\{C_1 \ldots C_p\}$ be the connected components of $G(V - N[x])$; then S is a substar of x if (and only if) $S = N_G(C_i)$ for some $i \in [1, p]$. Thus the substars are exactly the neighborhoods of the components defined when x and its neighborhood are removed from the graph.

Chordal Graphs and Minimal Triangulations. A graph is *chordal* (or triangulated) if it contains no chordless induced cycle of length 4 or more. Chordal graphs can be recognized in linear time using Algorithm LexBFS [36]. A graph is chordal if and only if all its minimal separators are cliques [19]. Given a non-chordal graph $G = (V, E)$, the supergraph $H = (V, E + F)$ is a *triangulation* of G if H is chordal. F is the set of *fill edges*, $|F|$ is denoted by f. The triangulation is *minimal* if for any proper subset of edges $F' \subset F$, the graph $(V, E + F')$ fails to be chordal.

There is a strong relationship between minimal triangulations and minimal separators [34]. Minimal separators S and S' are said to be *crossing separators* in a connected graph if S' has at least one vertex in every connected component of $G(V - S)$ (the crossing relation is symmetric). A minimal triangulation can be computed by saturating a maximal set of pairwise non-crossing minimal separators (this set is of size less than n, whereas there may be an exponential number of minimal separators in a non-chordal graph). Saturating a minimal separator S causes all the minimal separators which were crossing with S to disappear. Thus a minimal separator is a clique if and only if it crosses no other minimal separator [6,34]. The process of repeatedly choosing a (non-clique) minimal separator of a graph and saturating it, until there is no non-clique minimal separator left, results in a minimal triangulation. The substars of a given vertex x are pairwise non-crossing [6].

Given a graph $G = (V, E)$ and a minimal triangulation $H = (V, E + F)$ of G, any minimal separator S of H is a minimal separator of G, and $G(V - S)$ has the same connected components as $H(V - S)$. Any clique minimal separator of G is a minimal separator of H [9,34].

Clique Separator Decomposition. The decomposition by clique minimal separators decomposes a graph (in a unique fashion) into *atoms* (also called *MP-subgraphs* [28]), which are characterized as maximal connected subgraphs containing no clique separator. In a chordal graph, the atoms are the maximal cliques.

In [9] it was proved that the decomposition into atoms can be obtained by repeatedly applying the following decomposition step, which we will call *block decomposition step*, until none of the subgraphs obtained contains a clique separator: let $G = (V, E)$ be a graph, S a clique minimal separator of G, $\{C_1 \ldots C_p\}$ be the connected components of $G(V - S)$; decompose the graph into the following subgraphs: $G((C_1) \cup N(C_1)) \ldots G((C_p) \cup N(C_p))$, which we will call *blocks*. Note that $N(C_i)$ is included in S and is a clique minimal separator in its own right.

Property 2. [9] After an application of the block decomposition step on a connected graph G using a clique minimal separator S, all the other minimal separators of G are partitioned into the blocks obtained, *i.e.* each minimal separator is included in exactly one block.

3   Algorithmic Process

In a previous work, we used substars to compute a minimal triangulation [2,6]. The basic step, which is then applied to each vertex of the graph successively, is very simple: choose an unprocessed vertex x; compute the substars of x and saturate them. In this work, we will use a variant of this algorithm: we will combine this basic step with a decomposition step by clique minimal separators. Since after saturation, the substars become clique minimal separators of the resulting graph, we can apply the block decomposition step to all the substars, in order to obtain the corresponding blocks. This results in the following decomposition step for a vertex x:

Decomposition Step 1
- *compute the set $\{C_1 \ldots C_p\}$ of connected components of $G(V - N[x])$ and the corresponding substars $\{S_1 = N(C_1) \ldots S_p = N(C_p)\}$;*
- *saturate all the substars, thus obtaining a graph G';*
- *decompose the resulting graph into the neighborhood piece $G'(N[x])$ and the blocks of the form $G'(C_i \cup S_i)$.*

After applying Decomposition Step 1, we will independently triangulate the neighborhood piece; we will also recursively apply the algorithm to each of the resulting blocks.

In any new block obtained during the recursive decomposition process, we will carefully choose the next vertex x which will be the center of the next substar, such that x is in the new separator but not in any of the other previously defined ones:

Property 3. Let $G = (V, E)$ be a graph, let x_1 be a vertex of G, let C_1 be a connected component of $G(V - N[x_1])$, let $S_1 = N_G(C_1)$ be the corresponding substar of x_1, let B_1 be the block $S_1 \cup C_1$, let x_2 be a vertex of S_1 which is not universal in B_1, let C_2 be a connected component of $V - (N[x_2] \cup S_1)$ in the graph induced by B_1, let $S_2 = N_G(C_2)$ be the corresponding substar of x_2; then $S_2 - S_1$ contains at least one vertex.

Proof. Suppose by contradiction that $S_2 - S_1$ is empty; since S_1 is a minimal separator of G, x_2, which is in S_1, must see some vertex z in C_1; thus z is in $C_1 - C_2$, so $C_1 \neq C_2$, so $S_1 \neq S_2$. Let v be any vertex of C_2; since C_2 is connected, there must be a chordless path P in C_2 from v to z; let w be the first vertex of P to be in $C_1 - C_2$; w by definition is in S_2, as it is in $N(C_2)$, but it is not in S_1, a contradiction.

The benefit is that in the new block $B_2 = S_2 \cup C_2$, all the already computed fill edges are incident to x_2, and thus will not be traversed when searching for the substars of x_2 in block B_2. Moreover, we will be moving along a chordless path during the recursive descent, as the newly chosen vertex x_2 will be adjacent to its father vertex x_1, but not to the father of x_1. The algorithm stops when there are no new substars defined for the chosen vertex x in the new subgraph, which means that x is universal in the block.

By Property 2, after an application of Decomposition Step 1, the remaining minimal separators are partitioned into the various subgraphs obtained. Therefore, when processing these subgraphs separately, we obtain the fill edges defining a minimal triangulation H of G, the minimal separators of H and the clique minimal separators of G.

We can initiate the algorithm with a vertex which belongs to a moplex, which is easy to find according to Property 1; after this minimal separator is saturated, the neighborhood piece is a clique, so no additional effort is required to triangulate it.

This procedure works on general graphs. However, the time complexity may not be interesting, because the successive neighborhood pieces are not easy to triangulate except for the first one, and the substars defined at a given step may overlap. In fact, checking whether the substars are clique separators and if not, computing the fill edges necessary to saturate them, can be costly or require a non-trivial data structure as is the case of the algorithm from [6]. Finally, when the blocks overlap, a given vertex may be processed many times, with possibly huge edge overlaps caused by copying the saturated substars.

For claw-free graphs, however, we will establish the following properties, which will make this an efficient process.

- The first property is that in a claw-free graph, when we initiate the algorithm by choosing a vertex x in a moplex, x defines a single substar (Corollary 1) and thus a single block to start with.
- The second property is that if we initiate our algorithm as described above, at each step of the process, the neighborhood piece obtained is a co-bipartite graph with a universal vertex added to it (Theorem 1 c)). A co-bipartite graph can be triangulated in time proportional to the number of external edges [11]. Moreover, the external edges of the different neighborhood pieces encountered do not overlap, so traversing the fill edges will globally cost at most f, where f is the number of fill edges.
- The third property is an important invariant of Decomposition Step 1: each block obtained remains claw-free, even if the resulting global triangulation is not claw-free (Theorem 1 b)).
- Finally, at each step processing vertex x, the substars are pairwise disjoint, and the blocks are also pairwise disjoint (Theorem 1 a)). As a result, a given vertex will not be processed more than once; moreover, not all vertices will be processed, as only one vertex per substar is processed, which results in a possibly linear complexity.

The algorithm moves along what we will refer to as its *execution tree* T: it will define a succession of centers of substars, forming a partial subgraph which is a tree. We will now prove the properties above, which are special for claw-free graphs.

Lemma 1. *Let G be a connected claw-free graph, let S be a minimal separator of G; then $G(V - S)$ has exactly two components (which are thus both full components).*

Proof. Consider two full components C_1 and C_2 of $G(V - S)$. By definition, every vertex of S sees both C_1 and C_2. If there is a third component C_3 of $G(V - S)$, then at least one vertex $x \in S$ has at least one neighbor $c_3 \in C_3$, forming a claw with two of its neighbors $c_1 \in C_1$ and $c_2 \in C_2$.

Corollary 1. *In a claw-free graph $G = (V, E)$, any vertex x belonging to a moplex defines only one substar S. Moreover, $C = V - N[x]$ induces a connected graph, and $S = N(C)$.*

Lemma 2. *Let $G = (V, E)$ be a claw-free graph, S a minimal separator of G, C_1 and C_2 the full components of $G(V - S)$. For any vertex x of S, the sets $N(x) \cap C_1$ and $N(x) \cap C_2$ are cliques.*

Proof. As C_1 and C_2 are full components of $G(V - S)$, any vertex x of S sees both C_1 and C_2. If there were two non-adjacent vertices in $N(x) \cap C_i$, say $c_1, c_1' \in C_1$, then x would form a claw together with c_1, c_1' and any of its neighbors $c_2 \in C_2$ (or vice versa).

Theorem 1. *Consider a claw-free graph $G = (V, E)$ and apply Decomposition Step 1 using a vertex x. Let G' be the graph obtained from G after saturating all the substars of x. Furthermore, for any connected component C of $G(V - N[x])$, let $S = N_G(C)$ be the corresponding substar of x and $B = G'(S \cup C)$ be the corresponding block. Then we have:*

a) *the substars of x as well as the resulting blocks are pairwise disjoint;*
b) *all the blocks are claw-free subgraphs of G';*
c) *in any block $B = G'(S \cup C)$, let us choose a vertex x' in S; the neighborhood piece $G'(N[x'])$ is a co-bipartite graph with x' as additional universal vertex.*

Proof. Consider a claw-free graph G and adopt the above notations:

a) The substars of x are pairwise disjoint: suppose vertex x has 2 non-disjoint substars, $S_i = N(C_i)$ and $S_j = N(C_j)$, and let y be a vertex in $S_i \cap S_j$; y must see some vertex y_i in C_i, and y must see some vertex y_j in C_j; $\{y, y_i, y_j, x\}$ form a claw with center y. The blocks are pairwise disjoint, as each substar results in a unique block: no substar can define more that 2 connected components by Lemma 1; one component contains x, so the other, C, defines a unique block.
b) Every block B is claw-free: suppose by contradiction that B fails to be claw-free; since the only added edges compared to the input graph in B are the fill edges inside S, the claw in B must have its center in S. Let s be the center of a claw; since S is a clique in B, there must be 2 non-adjacent vertices in C which participate in the claw; but by Lemma 2, the neighborhood of s in C must be a clique, a contradiction.
c) The neighborhood piece $G'(N[x'])$ is a co-bipartite graph with the universal vertex x' added to it: one part of this neighborhood is a minimal separator S which has been saturated, as the processed vertex is chosen inside the minimal separator which defined the new block, and the other part is a clique by Lemma 2.

Example 1. Let us illustrate our process with the simple example shown in Figure 1.
a) Vertex x_0 is chosen first.

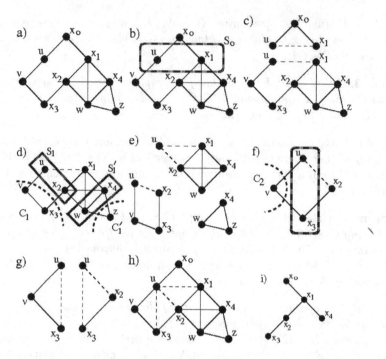

Fig. 1. The decomposition steps applied to a claw-free graph G

b) $N(x_0)$ is a minimal separator, which is $S_0 = \{u, x_1\}$, corresponding to the connected component $V - \{x_0, u, x_1\}$ of $V - N[x_0]$.

c) Fill edge ux_1 is added, and the neighborhood piece $\{x_0, u, x_1\}$ as well as the block $B_1 = V - x_0$ are defined.

d) In block B_1, x_1 is chosen in S_0; $N_{B_1}(x_1) = \{u, x_2, w, x_4\}$; the connected components of $V - N[x_1]$ in B_1 are: $C_1 = \{v, x_3\}$ and $C_1' = \{z\}$; the corresponding minimal separators are $S_1 = \{u, x_2\}$ and $S_1' = \{w, x_4\}$; saturating S_1 adds fill edge ux_2; S_1' is already a clique and thus is a clique minimal separator of the input graph G.

e) Block $B_2 = \{u, x_2, v, x_3\}$ is defined by S_1 and block $B_4 = \{w, x_4, z\}$ is defined from S_1'. The neighborhood piece $P_1 = N[x_1]$ defines a co-bipartite graph $N(x_1) = \{u, x_2, w, x_4\}$, which is already chordal, so no fill edge is added; there is only one minimal separator in this co-bipartite graph, namely the articulation point $\{x_2\}$, so $\{x_2\} \cup \{x_1\} = \{x_1, x_2\}$ is the unique associated (clique) minimal separator of G.

f) In block B_2, vertex x_2 is chosen in $S_1 - S_0$; in B_2, $N(x_2) = \{u, x_3\}$, defining only one connected component: $\{v\}$; the corresponding substar is $\{u, x_3\}$, fill edge ux_3 is added. The vertex x_2 becomes simplicial, so the corresponding neighborhood piece is complete and does not need to be processed.

g) Block $B_3 = \{u, x_3, v\}$ is then recursively defined; vertex x_3 is chosen; x_3 is simplicial and universal, so this branch of the recursive algorithm stops.

Afterwards, it only remains to choose x_4 in block B_4; x_4 is likewise simplicial and universal, so the algorithm terminates.

h) The resulting minimal triangulation is not claw-free, although each block and each piece was claw-free: the triangulation has a claw $\{u, v, x_2, x_0\}$ with center u.

i) The execution tree has a height of 3, requiring 3 linear-time graph searches globally; only 5 vertices are processed, out of a total of 9 vertices. Note that in the input graph, the longest path is of length 6.

4   The Algorithm

We now present our algorithm, consisting in a main algorithm (Claw-Free TRI-DEC), which initiates the process and a recursive algorithm (REC), which is called by the main algorithm. REC in turn uses another algorithm [11] which for a co-bipartite graph computes a minimal triangulation represented by the set F' of fill edges, the set $\mathscr{S}'$ of pairwise non-crossing minimal separators, and the set $\mathscr{K}'$ of clique minimal separators of the co-bipartite graph. We first present the algorithms Claw-Free TRI-DEC and REC in Section 4.1 and, for the sake of completeness, in Section 4.2 also the algorithmic process from [11] to compute a minimal triangulation of a co-bipartite graph.

4.1   Algorithms Claw-Free TRI-DEC and REC

Theorem 2. *Given a claw-free graph $G = (V, E)$, Algorithm Claw-Free TRI-DEC computes a minimal triangulation of G, represented by a set F of fill edges, and the minimal separators of $G' = (V, E + F)$, as well as the set of clique minimal separators of G in $O(f + km)$ time, where $f = |F|$, k is the height of the execution tree and $m = |E|$.*

Proof. To initialize our algorithm, let y be the vertex numbered 1 by an execution of LexBFS on a claw-free graph; note that y belongs to a moplex due to Property 1. By Corollary 1, y has only one substar, which is $S_0 = N(C_0)$, where $C_0 = V - N[y]$; this costs $O(m)$ time.

As the center of each new substar, we select a vertex x_i in $N_G(C_i) - S$, which always exists according to Property 3.

The Decomposition Step 1 is obviously properly applied in REC and produces a neighborhood piece and the necessary blocks with a single graph search. All the resulting blocks are again claw-free by Theorem 1 b) which guarantees a proper recursion. No fill edge is traversed in a given block, as all fill edges are incident to the vertex chosen as center of the next substar, so this search can be done in the input graph in $O(m)$ time.

Because the substars and blocks are pairwise disjoint at each step (Theorem 1 a)), there will be no extra cost incurred by searching the substars to determine which are clique minimal separators and which need to be filled.

The neighborhood piece without x is co-bipartite according to Theorem 1 c); it was shown in [11] that, given a co-bipartite graph with set of external edges

ALGORITHM Claw-Free TRI-DEC
Input : A claw-free graph $G = (V, E)$.
Output: A set $\mathscr{S}$ of minimal separators defining a minimal triangulation of G,
the corresponding set F of fill edges, the set $\mathscr{K}$ of clique minimal
separators of G, and an execution tree T of G.

$y \leftarrow$ vertex 1 by LexBFS ; $C_0 \leftarrow V - N[y]$; $S_0 \leftarrow N_G(C_0)$; $x_0 \leftarrow$ a vertex of S_0;
$F \leftarrow \emptyset$; $T \leftarrow \{y\}$; // T is a tree with only one node;
$\mathscr{S} \leftarrow S_0$; //minimal separators of the triangulation;
$\mathscr{K} \leftarrow \emptyset$; //clique minimal separators of G;
if S_0 *is a clique* then $\mathscr{K} \leftarrow \mathscr{K} + \{S_0\}$;
$\mathrm{REC}(S_0, C_0, x_0)$;

REC
Input : a minimal separator S of G, a set C of vertices which together with S
defined a block in the father graph, and a vertex x of S.
Output: modification of global variables.

$B \leftarrow G(S \cup C)$; // B is the new block;
$K \leftarrow N_G(x) \cap C$; $X \leftarrow K \cup S$; // X is the closed neighborhood of x in B;
$G' \leftarrow G(X - \{x\})$; // G' is the co-bipartite graph;
foreach *connected component C_i of $B(V - X)$* **do**
$\quad\Big|\quad$ $S_i \leftarrow N_G(C_i)$; $\mathscr{S} \leftarrow \mathscr{S} + S$; //$S_i$ is a substar of x;
$\quad\Big|\quad$ if S_i *is not a clique* then $F'_i \leftarrow$ set of fill edges needed to saturate S_i;
$\quad\Big|\quad$ else $\mathscr{K} \leftarrow \mathscr{K} + S$;
$\quad\Big|\quad$ $G' \leftarrow G' + F'_i$; $F \leftarrow F + F'_i$;
$(F', \mathscr{S}', \mathscr{K}') \leftarrow$ set of fill edges, set of minimal separators and set of clique
minimal separators obtained by computing a minimal triangulation of the
co-bipartite graph G';
$F \leftarrow F + F'$;
foreach *minimal separator S in $\mathscr{S}'$* **do** $\mathscr{S} \leftarrow \mathscr{S} + \{S + \{x\}\}$;
foreach *clique minimal separator S in $\mathscr{K}'$* **do** $\mathscr{K} \leftarrow \mathscr{K} + \{S + \{x\}\}$;
foreach *substar S_i of x* **do**
$\quad\Big|\quad$ CHOOSE a vertex x_i in $N_G(C_i) - S$;
$\quad\Big|\quad$ $T \leftarrow$ add node x_i and edge xx_i to T;
$\quad\Big|\quad$ $\mathrm{REC}(S_i, C_i, x_i)$;

E_{ext}, one can compute in $O(|E_{ext}|)$ time: a minimal triangulation $H = (V, E+F)$
of G, the set of minimal separators of H, the set of fill edges F, and the set of
clique minimal separators of G. It is easy to see that if G' is a graph with a
universal vertex x, then S is a minimal separator of G if and only if $S - \{x\}$ is
a minimal separator of $G - \{x\}$, and S is a clique minimal separator of G if and
only if $S - \{x\}$ is a clique minimal separator of $G - \{x\}$. The only fill edges to add
are external edges; no two neighborhood pieces have common external edges: in
each new block $S \cup C$ defined, each external edge has one endpoint in S and the
other in C; the previous neighborhood piece contained S but no vertex of C;
thus we avoid encountering each fill edge more than once. As a result, processing
a neighborhood piece requires $O(m + f')$ time, where f' is the set of fill edges
which must be added to the neighborhood piece to saturate the substars of x,
and globally, processing all the neighborhood pieces costs $O(m + f)$ time.

In the course of Algorithm Claw-Free TRI-DEC on a claw-free graph G, all the minimal separators of G are encountered, and thus all the clique minimal separators of G are encountered: each minimal separator of G belongs either to a neighborhood piece or is a substar at some step; each minimal separator of the co-bipartite graph corresponding to the neighborhood piece is produced [11].

Each time the execution tree T has a branch, the graph is partitioned into several disjoint blocks, by Theorem 1 a). Therefore the set of edges of G is also partitioned, so each of the k layers of T can be processed globally in linear time. Thus if k is the height of T, Algorithm Claw-Free TRI-DEC runs in $O(f + km)$.

Note that in some graphs and with some executions, k may be small compared to the length of a longest chordless path in the graph, as in Example 1.

4.2   Minimal Triangulation of a Co-bipartite Graph

Recently [11], an efficient process for computing a minimal triangulation of a co-bipartite graph was presented. Given a co-bipartite graph $G = (K_1 + K_2, E)$ built on clique sets K_1 and K_2, this algorithm works in the complement $\overline{G} = (K_1 + K_2, \overline{E})$ of G; $\overline{G}$ is thus a bipartite graph.

The process works as follows: first, a maximal chain of the lattice formed by all the maximal bicliques of $\overline{G}$ is computed, thus computing a minimal subgraph of $\overline{G}$ which is a chain graph:

ALGORITHM MAX-CHAIN[11]
Input   : A bipartite graph $G' = (K_1 + K_2, E')$
Output: A maximal chain $\mathscr{C}$
prefix $\leftarrow \emptyset$; $\mathscr{C} \leftarrow \emptyset$;
repeat
| Choose a vertex x of maximum degree in K_1; $X \leftarrow \{x\}$;
| $Y \leftarrow N(x)$;
| $G' \leftarrow$ remove x and $K_2 - Y$ from G';
| $U \leftarrow$ set of universal vertices of G';
| $X \leftarrow X + U$;
| $G' \leftarrow$ remove all vertices of U from G';
| add $(prefix + X, Y)$ to $\mathscr{C}$;
| prefix $\leftarrow prefix + X$;
until G' *is empty*;

Theorem 3. *[11] Algorithm* MAX-CHAIN *computes a maximal chain in* $O(min$ $(|E'|, |\overline{E'}|))$ *time.*

Given a maximal chain of a bipartite graph $((X_1, Y_1), (X_2, Y_2), \ldots, (X_p, Y_p))$, the following inclusions hold: $X_1 \subset X_2 \subset \ldots Y_1$ and $Y_k \subset \ldots Y_2 \subset Y_1$.

It was shown in [10] that the sets $((K_1 - X_1) \cup (K_2 - Y_1)) \ldots ((K_1 - X_p) \cup (K_2 - Y_p))$ form a maximal set of pairwise non-crossing minimal separators of the corresponding co-bipartite graph. A corresponding minimal triangulation of the co-bipartite graph is obtained by adding to the co-bipartite graph any missing edge from the Cartesian products $(K_1 - X_1) \times (K_2 - Y_1), (K_1 - X_2) \times (K_2 - Y_2) \ldots (K_1 - X_k) \times (K_2 - Y_k)$.

Since $X_1 \subset X_2 \subset \ldots Y_1$ and $Y_k \subset \ldots Y_2 \subset Y_1$, we need only to check for the absence of an edge in $(K_1 - X_1) \times (K_2 - Y_1), (K_1 - X_2) \times (Y_1 - Y_2), \ldots, (K_1 - X_k) \times (Y_{k-1} - Y_k)$, which can also be done in $O(min(|E'|, |\overline{E'}|))$ time.

5 Robustness Properties of Algorithm Claw-Free TRI-DEC

Algorithm Claw-Free TRI-DEC can be used with a non-claw-free input, and will sometimes yield a minimal triangulation. If it fails to do so, we can detect a claw.

Let us consider the situation when the input graph G is not claw-free.

On the one hand, a neighborhood piece $G'(N[x])$ may fail to define a co-bipartite graph $G'(N(x))$ with cliques K_1 and K_2; while K_1 is a clique by saturation, K_2 is not necessarily a clique. In a claw-free graph, K_2 is a clique by virtue of Lemma 2; if this is not the case, let v, v' be two non-adjacent vertices in K_2, let y be the father of x in the execution; clearly, $\{x, v, v', y\}$ induce a claw with center x in G. Testing at each step whether these neighborhoods are cliques costs $O(m)$ time, so these claws will be found at no extra cost.

On the other hand, the substars of a vertex x may fail to be disjoint. In this case the graph has a claw: let y be the father of x in the execution; let C_1 and C_2 be two connected components of $G'(V - N[x])$ defining non-disjoint substars and $S_2 = N_{G'}(C_2)$ of x; let v be in $S_1 \cap S_2$, let v_1 be in $C_1 \cap N_{G'}(v)$, let v_2 be in $C_2 \cap N_{G'}(v)$; clearly, $\{v, v_1, v_2, y\}$ induce a claw with center v in G. As above, testing at each step whether these substars are disjoint costs $O(m)$ time, so these claws will be found at no extra cost.

When at each step the neighborhood piece is indeed co-bipartite and the substars are indeed disjoint, the algorithm will run correctly, even if the input graph contains a claw; such a claw could for example be included inside the first minimal separator S_0; the saturation of S_0 will 'hide' that claw, but will not prevent Algorithm Claw-Free TRI-DEC from running correctly.

6 Conclusion

We introduce new structural properties for the much-studied class of claw-free graphs. This leads to a new algorithm, which computes a minimal triangulation of a claw-free graph in $O(f + km)$ time, where f is the number of fill edges, and k is the height of the execution tree. When the graph is P_k-free for some bounded value of k, the algorithm runs in linear time. Even in the case where the graph is not P_k-free, the algorithm runs in linear time if the height of the execution tree is small. In any case, for a claw-free input, the algorithm runs in optimal $O(nm)$ time, but we expect it to run faster than the other minimal triangulation algorithms.

Our algorithm computes, at the same time, the set of all minimal separators of the minimal triangulation, as well as all the clique minimal separators of the input graph; the decomposition by clique minimal separators can be computed at no extra cost. The algorithm also defines a connected dominating set of the graph, since the execution tree obtained is dominating by construction.

It is worth examining whether the algorithm can be streamlined to run in linear time on quasi-line graphs, a subclass of claw-free graphs where the neighborhood of every vertex partitions into two cliques.

The algorithm can be re-written into a robust algorithm, which, in the same time bound, either computes a minimal triangulation of the input graph or reports a claw. It would be interesting to identify some criteria for non-claw-free graphs such that our robust version works without being disturbed by a claw.

A further question is whether the algorithm runs fast with other special classes, such as some sparse graphs.

References

1. Beeri, C., Fagin, R., Maier, D., Yannakakis, M.: On the Desirability of Acyclic Database Schemes. Journal of the ACM (JACM) 30(3) (July 1983)
2. Berry, A.: A wide-range efficient algorithm for minimal triangulation. In: Proceedings of SODA 1999, pp. 860–861 (1999)
3. Berry, A., Blair, J.R.S., Heggernes, P., Peyton, B.W.: Maximum cardinality search for computing minimal triangulations of graphs. Algorithmica 39(4), 287–298 (2004)
4. Berry, A., Blair, J.R.S., Simonet, G.: Preface to Special issue on Minimal Separation and Minimal Triangulation. Discrete Mathematics 306(3), 293 (2006)
5. Berry, A., Bordat, J.P.: Separability generalizes Dirac's theorem. Discrete Applied Mathematics 84(1-3), 43–53 (1998)
6. Berry, A., Bordat, J.-P., Heggernes, P., Simonet, G., Villanger, Y.: A wide-range algorithm for minimal triangulation from an arbitrary ordering. Journal of Algorithms 58(1), 33–66 (2006)
7. Berry, A., Brandstädt, A., Giakoumakis, V., Maffray, F.: The atomic structure of hole- and diamond-free graphs (submitted)
8. Berry, A., Heggernes, P., Villander, Y.: A vertex incremental approach for maintaining chordality. Discrete Mathematics 306(3), 318–336 (2006)
9. Berry, A., Pogorelcnik, R., Simonet, G.: An introduction to clique minimal separator decomposition. Algorithms 3(2), 197–215 (2010)
10. Berry, A., Sigayret, A.: Representing a concept lattice by a graph. Discrete Applied Mathematics 144(1-2), 27–42 (2004)
11. Berry, A., Sigayret, A.: A Peep through the Looking Glass: Articulation Points in Lattices. In: Domenach, F., Ignatov, D.I., Poelmans, J. (eds.) ICFCA 2012. LNCS (LNAI), vol. 7278, pp. 45–60. Springer, Heidelberg (2012)
12. Brandstädt, A., Hoàng, C.T.: On clique separators, nearly chordal graphs, and the Maximum Weight Stable Set Problem. Theoretical Computer Science 389, 295–306 (2007)
13. Brandstädt, A., Le, V.B., Mahfud, S.: New applications of clique separator decomposition for the Maximum Weight Stable Set Problem. Theoretical Computer Science 370, 229–239 (2007)

14. Brandstädt, A., Le, V.B., Spinrad, J.P.: Graph Classes: A Survey. In: SIAM Monographs on Discrete Math. Appl., Philadelphia, vol. 3 (1999)
15. Bruhn, H., Saito, A.: Clique or hole in claw-free graphs. Journal of Combinatorial Theory, Series B 102(1), 1–13 (2012)
16. Chudnovsky, M., Seymour, P.: The Structure of Claw-free Graphs. In: Surveys in Combinatirics 2005. London Math. Soc. Lecture Note Series, vol. 327, pp. 153–171 (2005)
17. Chudnovsky, M., Seymour, P.: Claw-free Graphs IV. Decomposition theorem. Journal of Combinatorial Theory. Ser. B 98, 839–938 (2008)
18. Chudnovsky, M., Seymour, P.: Claw-free Graphs V. Global structure. Journal of Combinatorial Theory. Ser. B 98, 1373–1410 (2008)
19. Dirac, G.A.: On rigid circuit graphs. Abh. Math. Sem. Univ. Hamburg 25, 71–76 (1961)
20. Faenza, Y., Oriolo, G., Stauffer, G.: An algorithmic decomposition of claw-free graphs leading to an $O(n^3)$-algorithm for the weighted stable set problem. In: Proceedings of SODA 2011, pp. 630–646 (2011)
21. Faenza, Y., Oriolo, G., Stauffer, G.: Solving the maximum weighted stable set problem in claw-free graphs via decomposition (submitted)
22. Golumbic, M.C.: Algorithmic Graph Theory and Perfect Graphs. Academic Press (1980)
23. Heggernes, P.: Minimal triangulations of graphs: A survey. Discrete Mathematics 306(3), 297–317 (2006)
24. Heggernes, P., Telle, J.A., Villanger, Y.: Computing Minimal Triangulations in Time $O(n^\alpha log n) = o(n^{2.376})$. SIAM Journal on Discrete Mathematics 19(4), 900–913 (2005)
25. Kloks, T., Kratsch, D., Müller, H.: Finding and Counting Small Induced Subgraphs Efficiently. In: Nagl, M. (ed.) WG 1995. LNCS, vol. 1017, pp. 14–23. Springer, Heidelberg (1995)
26. Kaba, B., Pinet, N., Lelandais, G., Sigayret, A., Berry, A.: Clustering gene expression data using graph separators. In Silico Biology 7(4-5), 433–452 (2007)
27. Kratsch, D., Spinrad, J.: Minimal fill in $O(n^{2.69})$ time. Discrete Mathematics 306(3), 366–371 (2006)
28. Leimer, H.-G.: Optimal decomposition by clique separators. Discrete Mathematics 113, 99–123 (1993)
29. Lekkerkerker, C.G., Boland, J.C.: Representation of a finite graph by a set of intervals on the real line. Fund. Math. 51, 45–64 (1962)
30. Meister, D.: Recognition and computation of minimal triangulations for AT-free claw-free and co-comparability graphs. Discrete Applied Mathematics 146(3), 193–218 (2005)
31. Olesen, K.G., Madsen, A.L.: Maximal prime subgraph decomposition of Bayesian networks. IEEE Transactions on Systems, Man and Cybernetics, Part B: Cybernetics 32(1), 21–31 (2002)
32. Oriolo, G., Pietropaoli, U., Stauffer, G.: A New Algorithm for the Maximum Weighted Stable Set Problem in Claw-Free Graphs. In: Lodi, A., Panconesi, A., Rinaldi, G. (eds.) IPCO 2008. LNCS, vol. 5035, pp. 77–96. Springer, Heidelberg (2008)
33. Oriolo, G., Stauffer, G., Ventura, P.: Stable sets in claw-free graphs: recent achievements and future challenges. Optima 86 (2011)
34. Parra, A., Scheffler, P.: Characterizations and Algorithmic Applications of Chordal Graph Embeddings. Discrete Applied Mathematics 79(1-3), 171–188 (1997)

35. Rose, D.J.: A graph-theoretic study of the numerical solution of sparse positive definite systems of linear equations. In: Graph Theory and Computing, pp. 183–217. Academic Press, NY (1973)
36. Rose, D.J., Tarjan, R.E., Lueker, G.S.: Algorithmic aspects of vertex elimination on graphs. SIAM Journal on Computing 5, 266–283 (1976)
37. Tarjan, R.E.: Decomposition by clique separators. Discrete Mathematics 55, 221–232 (1985)
38. Tarjan, R.E., Yannakakis, M.: Simple linear-time algorithms to test chordality of graphs, test acyclicity of hypergraphs, and selectively reduce acyclic hypergraphs. SIAM Journal on Computing 13, 566–579 (1984)
39. Yannakakis, M.: Computing the minimum fill-in is NP-complete. SIAM J. Alg. Disc. Meth. 2, 77–79 (1981)

Minimum Weighted Clique Cover
on Strip-Composed Perfect Graphs

Flavia Bonomo[1,*], Gianpaolo Oriolo[2], and Claudia Snels[2]

[1] IMAS-CONICET and Departamento de Computación, FCEN,
Universidad de Buenos Aires, Argentina
fbonomo@dc.uba.ar
[2] Dipartimento di Informatica, Sistemi e Produzione,
Università di Roma Tor Vergata, Italia
{oriolo,snels}@disp.uniroma2.it

Abstract. The only available combinatorial algorithm for the minimum weighted clique cover (MWCC) in claw-free perfect graphs is due to Hsu and Nemhauser [10] and dates back to 1984. More recently, Chudnovsky and Seymour [3] introduced a composition operation, strip-composition, in order to define their structural results for claw-free graphs; however, this composition operation is general and applies to non-claw-free graphs as well. In this paper, we show that a MWCC of a perfect strip-composed graph, with the basic graphs belonging to a class $\mathcal{G}$, can be found in polynomial time, provided that the MWCC problem can be solved on $\mathcal{G}$ in polynomial time. We also design a new, more efficient, combinatorial algorithm for the MWCC problem on strip-composed claw-free perfect graphs.

Keywords: claw-free graphs, perfect graphs, minimum weighted clique cover, odd pairs of cliques, strip-composed graphs.

1 Introduction

Given a graph G and a non-negative weight function w defined on the vertices of G, a *weighted clique cover* of G is a collection of cliques, with a non-negative weight y_C assigned to each clique C in the collection, such that, for each vertex v of G, the sum of the weights of the cliques containing v in the collection is at least $w(v)$. A *minimum weighted clique cover* of G (MWCC) is a clique cover such that the sum of the weights of all the cliques in the collection is minimum. When all weights are 1, a (minimum) weighted clique cover is simply called a *(minimum) clique cover*. It is known that for a perfect graph G, the weight $\tau_w(G)$ of a MWCC is the same as $\alpha_w(G)$, the weight of a *maximum weighted stable set* (MWSS) of G, that is, a set of pairwise nonadjacent vertices such that the sum of the weights of the vertices in the set is maximum.

In perfect graphs, the weight of a MWCC can be determined in polynomial time by using Lovász's $\theta_w(G)$ function. If one wants to compute also a MWCC

* Partially supported by ANPCyT PICT-2007-00518, CONICET PIP 112-200901-00178, and UBACyT Grants 20020090300094 and 20020100100980 (Argentina).

M.C. Golumbic et al. (Eds.): WG 2012, LNCS 7551, pp. 22–33, 2012.

of a perfect graph G (and not only the number $\tau_w(G)$), a polynomial time algorithm proposed by Grötschel, Lovász and Schrijver in [7] can be used. This algorithm is not combinatorial and it uses the $\theta_w(G)$ function combined with other techniques; however, for particular classes of perfect graphs, there also exist polynomial time combinatorial algorithms.

This is the case, for instance, for claw-free perfect graphs, where combinatorial algorithms for both the unweighted and the weighted version have been proposed by Hsu and Nemhauser [9,10]. (A graph is *claw-free* if none of its vertices has a stable set of size three in its neighborhood.) The algorithm for the weighted case – in the paper we deal with this case, as it is more general – is essentially a "dual" algorithm as it relies on any algorithm for the MWSS problem in claw-free graphs (we have, nowadays, several algorithms for this, see [11,12,14,5,13,15]), and, in fact, builds a MWCC by a clever use of linear programming complementarity slackness. The computational complexity of the algorithm by Hsu and Nemhauser is $O(|V(G)|^5)$. To the best of our knowledge, this is so far the only available combinatorial algorithm to solve the problem in claw-free perfect graphs.

In the last years a lot of efforts have been devoted to a better understanding of the structure of perfect graphs and of other relevant classes of graphs. Claw-free graphs in particular have been investigated, with an outstanding series of papers by Chudnovsky and Seymour (for a survey see [3]). The results by Chudnovsky and Seymour show that claw-free graphs with stability number greater than three are either *fuzzy circular interval graphs* (a generalization of *proper circular arc graphs*, we do not give the definition, as it is not interesting for this paper) or *strip-composed*, i.e., they are suitable composition of some basic graphs (the formal definition is given in the next section). Understanding this "2-case" structure of claw-free graphs has been the key for several developments for the MWSS problem [4,14,5] and the dominating set problem [8]. In particular, in [14] it is shown that a MWSS of a (non-necessarily claw-free) strip-composed graph, with the basic graphs belonging to a class $\mathcal{G}$, can be found in polynomial time by solving a matching problem, provided that the MWSS problem can be solved on $\mathcal{G}$ in polynomial time. Building upon this result, new algorithms for the MWSS problem in claw-free graphs are given in [14] and [5].

In this paper, we provide an analogous of the result in [14] for the MWCC problem. Namely, we show that a MWCC of a (non-necessarily claw-free) perfect strip-composed graph, with the basic graphs belonging to a class $\mathcal{G}$, can be found in polynomial time, provided that the MWCC problem can be solved on $\mathcal{G}$ in polynomial time. We point out that while the statement of this result goes along the same lines of the result in [14], its proof is by far more challenging. We apply this result to strip-composed claw-free perfect graphs, and provide a $O(|V(G)|^3)$-time algorithm for the MWCC problem that, differently from the $O(|V(G)|^5)$-time dual algorithm by Hsu and Nemhauser, has both a primal (on each basic graph we directly compute a MWCC) and a primal-dual flavour (on the composition of graphs we use a primal-dual algorithm for matching).

We shall consider finite, simple, loopless, undirected graphs. When dealing with multigraphs, we will say so explicitly. Let G be a graph. Denote by $V(G)$ its vertex set and by $E(G)$ its edge set. For a subset $V' \subseteq V(G)$, the j-th *neighborhood* $N_j(V')$ is the set of vertices $u \in V(G)$ at distance j from the set V'. When $V' = \{v\}$ we will write simply $N_j(v)$ and when $j = 1$ we will write just $N(V')$ (resp. $N(v)$). We will denote by $G[V']$ the subgraph of G induced by V', and by $G \setminus V'$ the subgraph of G induced by $V(G) \setminus V'$. Two sets $U, U' \subset V(G)$ are *complete* (*to each other*) if every vertex in U is adjacent to all the vertices in U'. They are *anticomplete* (*to each other*) if no vertex of U is adjacent to a vertex of U'.

A *claw* is a graph formed by a vertex with three neighbors of degree one. An *odd hole* is a chordless cycle of odd length at least 5. If H is a graph, a graph G is called H-*free* if no induced subgraph of G is isomorphic to H.

A graph is *cobipartite* if its vertex set can be covered by two cliques. A clique K of a connected graph G is *distance simplicial* if, for every j, its j-th neighborhood is also a clique. In this case, G is *distance simplicial* w.r.t. K (or simply distance simplicial). Note that a cobipartite graph is distance simplicial w.r.t. each of the two cliques covering its vertex set. Also it is not difficult to see that distance simplicial graphs are perfect.

The *intersection graph* of a family of sets $\mathcal{C}$ is the graph with vertex set $\mathcal{C}$, two sets in $\mathcal{C}$ being adjacent if and only if they intersect. The *line graph* $L(G)$ of a graph or multigraph G is the intersection graph of its edges. A graph H is a line graph if there is a graph or multigraph G such that $H = L(G)$ (G is called a *root graph* of H). A *star* or a *multistar* is the set of edges incident to a vertex v, while a *triangle* or *multitriangle* is a complete graph on three vertices with possibly multiple edges. A *matching* is a set of pairwise nonadjacent edges of a graph (two edges are adjacent if they share a vertex). Note that the multistars and multitriangles of a graph G correspond to the cliques of $L(G)$, while the matchings of G correspond to the stable sets of $L(G)$. Note also that the neighborhood of a vertex in a line graph can be always covered by two cliques. A graph is *quasi-line* if the neighborhood of each vertex is cobipartite. A quasi-line graph is, in particular, claw-free. Moreover, as observed by Hsu and Nemhauser in [10], a claw-free perfect graph is indeed quasi-line.

2   The MWSS Problem on Strip-Composed Graphs

Chudnovsky and Seymour [3] introduced a composition operation in order to define their structural results for claw-free graphs. This composition operation is general and applies to non-claw-free graphs as well.

A *strip* $H = (G, \mathcal{A})$ is a graph G (not necessarily connected) with a multi-family $\mathcal{A}$ of either one or two designated non-empty cliques of G. The cliques in $\mathcal{A}$ are called the *extremities* of H, and H is said a *1-strip* if $|\mathcal{A}| = 1$, and a *2-strip* if $|\mathcal{A}| = 2$. Let $\mathcal{G} = (G^1, \mathcal{A}^1), \ldots, (G^k, \mathcal{A}^k)$ be a family of k vertex disjoint strips, and let $\mathcal{P}$ be a partition of the multi-set of the cliques in $\mathcal{A}^1 \cup \ldots \cup \mathcal{A}^k$. The *composition* of the k strips w.r.t. $\mathcal{P}$ is the graph G that is obtained from the union of $G^1, \ldots, G^k$, by making adjacent vertices of $A \in \mathcal{A}^i$ and $B \in \mathcal{A}^j$ (i, j not necessarily different) if and only if A and B are in the same class of the partition $\mathcal{P}$. In this case we also say that $(\mathcal{G}, \mathcal{P})$, where $\mathcal{G} = \{(G^j, \mathcal{A}^j), j \in 1, \ldots, k\}$, defines a *strip decomposition* of G. Note that we can assume w.l.o.g. that each graph G^i is an induced subgraph of G.

We say that a graph G is *strip-composed* if G is a composition of some set of strips w.r.t. some partition $\mathcal{P}$. Each class of the partition of the extremities $\mathcal{P}$ defines a clique of the composed graph, and is called a *partition-clique*. We denote the extremities of the strip H_i by $\mathcal{A}^i = \{A_1^i, A_2^i\}$ if H_i is a 2-strip and by $\mathcal{A}^i = \{A_1^i\}$ if H_i is a 1-strip. We often abuse notations, and when we refer to a vertex of a strip (or to a stable set of a strip etc.) we indeed consider a vertex (or a stable set etc.) of the graph in the strip.

The composition operation preserves some graph properties. Given a 2-strip $(G, \{A_1, A_2\})$, the graph G_+ is obtained from G by adding two vertices a_1, a_2 such that $N(a_j) = A_j$, for $j = 1, 2$; for a 1-strip $(G, \{A_1\})$ the graph G_+ is obtained from G by adding a vertex a_1 such that $N(a_1) = A_1$. A strip $(G, \mathcal{A})$ is claw-free/quasi-line/line if the graph G_+ is claw-free/quasi-line/line. The composition of claw-free/quasi-line/line strips is a claw-free/quasi-line/line graph (see e.g. [5]).

Suppose we are given a graph G and its strip decomposition $(\mathcal{G}, \mathcal{P})$. In [14] it is shown how to exploit this decomposition in order to solve the MWSS on G.

Theorem 1. *[14] Let G be the composition of strips $H_i = (G^i, \mathcal{A}^i)$ $i = 1, \ldots, k$ w.r.t. a partition $\mathcal{P}$. Suppose that for each $i = 1, \ldots, k$ one can compute a MWSS of H_i in time $O(p_i(|V(G^i)|))$. Then the MWSS problem on G can be solved in time $O(\sum_{i=1}^{k} p_i(|V(G^i)|) + match(|V(G)|))$, where $match(n)$ is the time required to solve the matching problem on a graph with n vertices. If $p_i(|V(G^i)|)$ is polynomial for each i, then the MWSS can be solved on G in polynomial time.*

In order to prove their theorem [14], the authors replace every strip H_i with a suitable simpler *gadget strip* T_i, that is a single vertex for each 1-strip and a triangle for each 2-strip (in this second case the extremities are two different edges of the triangle). Then they define a weight function on the vertices of those simpler strips; for every strip H_i with extremities A_1^i and A_2^i this function depends on the values $\alpha_w(G^i)$, $\alpha_w(G^i \setminus A_1^i)$, $\alpha_w(G^i \setminus A_2^i)$, $\alpha_w(G^i \setminus (A_1^i \cup A_2^i))$ and $\alpha_w(G^i \setminus (A_1^i \triangle A_2^i))$. Thus, if one can compute a MWSS of G^i in polynomial time, then one can compute the weight function of the simpler strips in polynomial time.

They define a suitable partition $\tilde{\mathcal{P}}$ of the extremities of the gadget strips. In this way they obtain a graph $\tilde{G}$ which is the strip-composition of the strips T_i, $i = 1, \ldots, k$, w.r.t. the partition $\tilde{\mathcal{P}}$, and, since the strips are line strips, this graph is line. Moreover, from the construction of the simpler strips and of the weights, it is easy to translate a MWSS of $\tilde{G}$ into a MWSS of G. Finally, as $\tilde{G}$ is a line-graph, they can find a MWSS of $\tilde{G}$ by building the root graph of $\tilde{G}$ and computing a maximum weighted matching in this graph.

3   The MWCC Problem on Strip-Composed Perfect Graphs

Suppose we are given a perfect graph G that is the composition of strips $H_1 = (G^1, \mathcal{A}^1), \ldots, H_k = (G^k, \mathcal{A}^k)$ w.r.t. a partition $\mathcal{P}$, and a non-negative weight function w on $V(G)$. In this section we will show how to exploit the decomposition in order to solve the MWCC on G. We will follow the approach outlined in the previous section for the MWSS; however, as we explain in the following, the task is now much more challenging.

We will compute a MWCC of G in three steps. **Step 1.** We replace each strip H_i by a simple *gadget strip* $\tilde{H}_i = (\tilde{G}^i, \tilde{\mathcal{A}}^i)$ and compose the strips $\tilde{H}_i$ with respect to a suitable partition of the multi-set $\bigcup_{i=1..k} \tilde{\mathcal{A}}^i$ so as to obtain a graph $\tilde{G}$. However, we cannot use the gadget strips defined in the previous section, as the graph $\tilde{G}$ might be imperfect: this will lead us to define four different new gadgets, with different parity properties, that are such that $\tilde{G}$ is odd hole free and line, thus perfect [16]. We also define a suitable weight function $\tilde{w}$ on the vertices of $\tilde{G}$, as well as new weight functions $w^1, \ldots, w^k$ on the vertices of each strip. **Step 2.** Following [16], we may find a MWCC of $\tilde{G}$, w.r.t. the weight $\tilde{w}$, by running a primal-dual algorithm for the maximum weighted matching [6] on the root graph of $\tilde{G}$. **Step 3.** We reconstruct a MWCC of G from a MWCC of $\tilde{G}$ and a MWCC of each of the strips H_i w.r.t. the weight function w^i. Again, this will be more involved than for the MWSS problem, because unfortunately there is not always a direct correspondence between cliques of $\tilde{G}$ and cliques of G. Moreover, for some 2-strips $H_i = (G^i, \mathcal{A}^i)$, besides a MWCC of the strip, we will also need to compute a MWCC of some auxiliary graphs associated to the strip: the graph $G^i_\bullet$ that is obtained from G^i by adding a new vertex x complete to both A^i_1 and A^i_2 and the graph $G^i_=$ that is the graph obtained from G^i by making A^i_1 complete to A^i_2.

In order to give a few more details we need some additional definitions. Let $U, W \subseteq V(G)$. We call a path $P = v_1, \ldots, v_k$ ($k \geq 1$) a *U–W path* if P is chordless, $v_1 \in U$, $v_k \in W$, and $v_i \notin U \cup W$ for $2 \leq i \leq k - 1$. A 2-strip $H_i = (G^i, \mathcal{A}^1 = \{A^i_1, A^i_2\})$ will be called *non-connected* if there is no A^i_1–A^i_2 path, and *connected* otherwise. We say that a connected 2-strip H_i is *even* (resp. *odd*) if every A^i_1–A^i_2 path has even (resp. odd) length. If a connected 2-strip has both even and odd length A^i_1–A^i_2 paths, then we say that H_i is an *even-odd* strip. We call an odd or even-odd strip H_i *odd-short* if every odd A^i_1–A^i_2 path has length one, and we call an even or even-odd strip H_i *even-short* if every even A^i_1–A^i_2 path has length zero (i.e., it consists of a vertex in $A^i_1 \cap A^i_2$). (With the notation of [1], H_i is an odd strip if and only if A^i_1 and A^i_2 are an odd pair of cliques in G^i.)

Theorem 2. *Let G be a perfect graph, composition of strips $H_i = (G^i, \mathcal{A}^i)$ $i = 1, \ldots, k$ w.r.t. a partition $\mathcal{P}$. For each $i = 1, \ldots, k$ let $O(p_i(|V(G^i)|))$ be the time required to compute:*

- *a MWCC of G^i and of $G^i_\bullet$, if H_i is an odd-short strip and $G^i_\bullet$ is an induced subgraph of G (thus perfect);*

- a MWCC of G^i and of $G^i_=$, if $G^i_=$ is an induced subgraph of G (thus perfect), A^i_1 and in A^i_2 belong to the same class of $\mathcal{P}$, and there is an A_1–A_2 path of length two in the strip. In this case, when solving the MWCC on $G^i_=$, one can restrict to the case where the weight function w^i defined on $V(G^i_=)$ is such that $\alpha_{w^i}(G^i_=) = \alpha_{w^i}(G^i_= \setminus (A^i_1 \cup A^i_2))$;
- a MWCC of G^i else.

Then the MWCC problem on G can be solved in time $O(\sum_{i=1}^k p_i(|V(G^i)|) + match(|V(G)|))$, where $match(n)$ is the time required to solve the matching problem on a graph with n vertices. If $p_i(|V(G^i)|)$ is polynomial for each i, then the MWCC can be solved on G in polynomial time.

We devote the rest of this section to provide more details about Theorem 2 and its proof. We first deal with the gadget strips (that in this section we simply call gadgets) that will compose the graph $\tilde{G}$ and establish the relation between $\tau_w(G)$ and $\tau_{\tilde{w}}(\tilde{G})$. We make a heavy use of duality between the MWCC and the MWSS problem: the fact that for every induced subgraph J of G, $\alpha_w(J) = \tau_w(J)$, is due to the perfection of G. We use this relation to easily prove the correctness of the weight function defined on the vertices of each gadget.

To design the gadgets, we delve into three cases: (i) $H_i = (G^i, \mathcal{A}^i)$ is a 1-strip; (ii) $H_i = (G^i, \mathcal{A}^i)$ is a 2-strip with the extremities in the same class of the partition $\mathcal{P}$; (iii) $H_i = (G^i, \mathcal{A}^i)$ is a 2-strip with the extremities in different classes of the partition.

$(i) - (ii)$ For the first two cases the gadget will be a single vertex. In particular we define the trivial 1-strip $\tilde{H}^0_i = (T^i_0, \tilde{\mathcal{A}}^i_0)$, where the graph T^i_0 consists of a single vertex c^i, and $\tilde{\mathcal{A}}^i_0 = \{\{c^i\}\}$. Moreover, for (i) we let $\delta^i_1 = \alpha_w(G^i \setminus A^i_1)$ and define $\tilde{w}(c^i) = \alpha_w(G^i) - \delta^i_1$. For (ii) we let $\delta^i_1 = \alpha_w(G^i \setminus (A^i_1 \cup A^i_2))$ and define $\tilde{w}(c^i) = \max\{\alpha_w(G^i \setminus A^i_1), \alpha_w(G^i \setminus A^i_2), \alpha_w(G^i \setminus (A^i_1 \triangle A^i_2))\} - \delta^i_1$. Finally, if we use $\tilde{H}^0_i$ instead of H_i in the composition, the new partition is $\mathcal{P}' := (\mathcal{P} \setminus \{P\}) \cup \{(P \setminus \mathcal{A}^i) \cup \tilde{\mathcal{A}}^i\}$, where $P \in \mathcal{P}$ was the set containing $\mathcal{A}^i$. We can show that replacing a 1-strip or a 2-strip with both extremities in the same class of $\mathcal{P}$ by its corresponding gadget strip makes the value of the MWSS drop by δ^i_1.

(iii) Let us consider a 2-strip $H_i = (G^i, \mathcal{A}^i)$ with the extremities in different classes of the partition $\mathcal{P}$. We want to introduce a gadget $\tilde{H}_i = (\tilde{G}^i, \tilde{\mathcal{A}}^i)$ and a new weight function $\tilde{w}$ on the vertices of $\tilde{G}^i$ in such a way that, when replacing H_i by $\tilde{H}_i$ in the strip-composition for a suitable partition, the difference between the weights of the MWSS of the original graph and the MWSS of the new graph is δ^i_1, where $\delta^i_1 = \alpha_w(G^i \setminus (A^i_1 \cup A^i_2))$.

This is satisfied by the gadget defined in [14], but it is an even-odd strip, and we need to take into consideration the parity of the strips, otherwise the composition may introduce odd holes. We will introduce three gadgets (an odd strip, an even strip and a non-connected one). None of them will work for all the cases, but depending on the fact that the relation $\alpha_w(G^i \setminus A^i_1) + \alpha_w(G^i \setminus A^i_2) \gtrless \alpha_w(G^i) + \delta^i_1$ is satisfied with $=$, $>$ or $<$. We will see later on that the satisfaction of this relation is strictly related to the parity of the strips. Given a 2-strip $H_i = (G^i, \mathcal{A}^i)$, we define three associated trivial strips as follows:

(a) $\tilde{H}_i^1 = (T_1^i, \tilde{\mathcal{A}}_1^i)$ such that $V(T_1^i) = \{u_1^i, u_2^i\}$, $E(T_1^i) = \emptyset$, $\tilde{\mathcal{A}}_1^i = \{\tilde{A}_1^i, \tilde{A}_2^i\}$ and $\tilde{A}_1^i = \{u_1^i\}$, $\tilde{A}_2^i = \{u_2^i\}$. The new weight function $\tilde{w}$ gives the following weights to the vertices of T_1^i: $\tilde{w}(u_1^i) = \alpha_w(G^i \setminus A_2^i) - \delta_1^i$, $\tilde{w}(u_2^i) = \alpha_w(G^i \setminus A_1^i) - \delta_1^i$.

(b) $\tilde{H}_i^2 = (T_2^i, \tilde{\mathcal{A}}_2^i)$ such that $V(T_2^i) = \{u_1^i, u_2^i, u_3^i\}$, $E(T_2^i) = \{u_1^i u_2^i, u_2^i u_3^i\}$, $\tilde{\mathcal{A}}_2^i = \{\tilde{A}_1^i, \tilde{A}_2^i\}$ and $\tilde{A}_1^i = \{u_1^i, u_2^i\}$, $\tilde{A}_2^i = \{u_3^i\}$. The new weight function $\tilde{w}$ gives the following weights to the vertices of T_2^i: $\tilde{w}(u_1^i) = \alpha_w(G^i) - \alpha_w(G^i \setminus A_1^i)$, $\tilde{w}(u_2^i) = \alpha_w(G^i \setminus A_2^i) - \delta_1^i$, $\tilde{w}(u_3^i) = \alpha_w(G^i \setminus A_1^i) - \delta_1^i$.

(c) $\tilde{H}_i^3 = (T_3^i, \tilde{\mathcal{A}}_3^i)$ such that $V(T_3^i) = \{u_1^i, u_2^i, u_3^i\}$, $E(T_3^i) = \{u_1^i u_2^i, u_2^i u_3^i\}$, $\tilde{\mathcal{A}}_3^i = \{\tilde{A}_1^i, \tilde{A}_2^i\}$ and $\tilde{A}_1^i = \{u_1^i, u_2^i\}$, $\tilde{A}_2^i = \{u_2^i, u_3^i\}$. The new weight function $\tilde{w}$ gives the following weights to the vertices of T_3^i: $\tilde{w}(u_1^i) = \alpha_w(G^i \setminus A_2^i) - \delta_1^i$, $\tilde{w}(u_2^i) = \alpha_w(G^i) - \delta_1^i$, $\tilde{w}(u_3^i) = \alpha_w(G^i \setminus A_1^i) - \delta_1^i$.

If we use either $\tilde{H}_i^1$, $\tilde{H}_i^2$ or $\tilde{H}_i^3$ instead of H_i in the composition, the new partition is $\mathcal{P}' := \mathcal{P} \setminus \{P_1, P_2\} \cup \{(P_1 \setminus \{A_1^i\}) \cup \{\tilde{A}_1^i\}, (P_2 \setminus \{A_2^i\}) \cup \{\tilde{A}_2^i\}\}$, where $P_1, P_2 \in \mathcal{P} : A_1^i \in P_1, A_2^i \in P_2$.

Lemma 1. *Let G be the composition of strips $H_1 = (G^1, \mathcal{A}^1), \ldots, H_k = (G^k, \mathcal{A}^k)$ w.r.t. a partition $\mathcal{P}$, and let w be a non-negative weight function defined on the vertices of G. Suppose that H_1 is a 2-strip with the extremities in different classes of the partition $\mathcal{P}$. For some $j \in \{1, 2, 3\}$, let G' be the composition of strips $\tilde{H}_1^j = (T_j^1, \mathcal{A}_j^1), H_2 = (G^2, \mathcal{A}^2), \ldots, H_k = (G^k, \mathcal{A}^k)$ w.r.t. the partition $\mathcal{P}'$ previously defined. Let w' be the weight function defined on the vertices of G' as $w'(v) = w(v)$ for $v \in \bigcup_{i=2..k} V(G^i)$, and $w'(v) = \tilde{w}(v)$ for $v \in V(T_j^1)$.*

(a) If $j = 1$ and $\alpha_w(G^1 \setminus A_1^1) + \alpha_w(G^1 \setminus A_2^1) = \alpha_w(G^1) + \delta_1^1$, then $\alpha_w(G) = \alpha_{w'}(G') + \delta_1^1$.

(b) If $j = 2$ and $\alpha_w(G^1 \setminus A_1^1) + \alpha_w(G^1 \setminus A_2^1) \geq \alpha_w(G^1) + \delta_1^1$, then $\alpha_w(G) = \alpha_{w'}(G') + \delta_1^1$.

(c) If $j = 3$ and $\alpha_w(G^1 \setminus A_1^1) + \alpha_w(G^1 \setminus A_2^1) \leq \alpha_w(G^1) + \delta_1^1$, then $\alpha_w(G) = \alpha_{w'}(G') + \delta_1^1$.

Lemma 2. *The following relations hold depending of the connection and parity of a 2-strip $H_1 = (G^1, \mathcal{A}^1)$:*

(a) if it is non-connected then $\alpha_w(G^1 \setminus A_1^1) + \alpha_w(G^1 \setminus A_2^1) = \alpha_w(G^1) + \delta_1^1$;

(b) if it is odd and G^1 perfect then $\alpha_w(G^1 \setminus A_1^1) + \alpha_w(G^1 \setminus A_2^1) \geq \alpha_w(G^1) + \delta_1^1$;

(c) if it is even and G^1 perfect then $\alpha_w(G^1 \setminus A_1^1) + \alpha_w(G^1 \setminus A_2^1) \leq \alpha_w(G^1) + \delta_1^1$.

We now give a method to choose one gadget for every 2-strip H_i. If we can calculate the values of the minimum weighted clique covers $\tau_w(G^i)$, $\tau_w(G^i \setminus A_2^i)$, $\tau_w(G^i \setminus A_1^i)$ and $\tau_w(G^i \setminus (A_1^i \cup A_2^i))$ for each strip, we can determine which one of these three relations holds

1. $\tau_w(G^i \setminus A_1^i) + \tau_w(G^i \setminus A_2^i) = \tau_w(G^i) + \tau_w(G^i \setminus (A_1^i \cup A_2^i))$
2. $\tau_w(G^i \setminus A_1^i) + \tau_w(G^i \setminus A_2^i) > \tau_w(G^i) + \tau_w(G^i \setminus (A_1^i \cup A_2^i))$
3. $\tau_w(G^i \setminus A_1^i) + \tau_w(G^i \setminus A_2^i) < \tau_w(G^i) + \tau_w(G^i \setminus (A_1^i \cup A_2^i))$

If 1) holds we can simply use $\tilde{H}_i^1$ as a suitable gadget. If 2) holds we know that the strip is either odd or even-odd and we can use $\tilde{H}_i^2$ as a suitable gadget. If 3) holds we know that the strip is either even or even-odd and we can use $\tilde{H}_i^3$ as a suitable gadget.

Remark 1. Let G be the composition of the strips $H_1, H_2, \ldots, H_k$ with respect to a partition $\mathcal{P}$ and suppose that G is odd hole free. Let G' be the composition of $\tilde{H}_1^j, H_2, \ldots, H_k$ with respect to the partition $\mathcal{P}'$ previously defined. For $j = 0, 1$, G' is odd hole free. If H_1 is odd or even-odd and $j = 2$, then G' is odd hole free. If H_1 is even or even-odd and $j = 3$, then G' is odd hole free.

Strips $\tilde{H}_i^0, \tilde{H}_i^1, \tilde{H}_i^2, \tilde{H}_i^3$ are line strips. So, if we iteratively replace each strip H_i by the suitable gadget $\tilde{H}_i^j$, according to the validity of 1, 2 or 3, the graph $\tilde{G}$ we obtain is odd hole free and a line graph, thus perfect [16]. As a corollary of the previous Lemmas, it follows that $\alpha_w(G) = \alpha_{\tilde{w}}(\tilde{G}) + \sum_{i=1}^k \delta_1^i$. Since both graphs are perfect, by duality the same relation holds for the values of the MWCC of the two graphs.

$\tilde{G}$ is a line, perfect graph. Let H be a multigraph that is a root of $\tilde{G}$. Following [16], we may build a MWCC of $\tilde{G}$, by a primal-dual algorithm for the maximum weighted matching [6]: this is because each maximal clique of $\tilde{G}$ corresponds to either a multistar of H or to a multitriangle of H. We therefore compute a MWCC of $\tilde{G}$, w.r.t. the weight $\tilde{w}$. We now need to "translate" this into a MWCC of G, w.r.t. the weight w. However, there is a catch: unfortunately there are some cliques of $\tilde{G}$ that do not correspond to any clique of G. In order to deal with this problem, we detail the structure of H.

Remark 2. Suppose that $\mathcal{P} = \{P_1, \ldots, P_r\}$. Then H is composed by: a set of vertices $\{x_1, \ldots, x_r\}$, each x_i corresponding to the class P_i of $\mathcal{P}$; an edge $x_j x_\ell$ for each strip H_i such that we use $\tilde{H}_i^3$ in the composition and such that $A_1^i \in P_j$ and $A_2^i \in P_\ell$ (this edge corresponds to the vertex u_2^i of T_3^i); vertices z_j^i and z_ℓ^i and edges $z_j^i x_j$ and $z_\ell^i x_\ell$ for each strip H_i such that we use $\tilde{H}_i^3$ in the composition and such that $A_1^i \in P_j$ and $A_2^i \in P_\ell$ (the edges $z_j^i x_j$ and $z_\ell^i x_\ell$ correspond to the vertices u_1^i and u_3^i of T_3^i, respectively); a vertex $y_{j\ell}^i$ and edges $y_{j\ell}^i x_j$ and $y_{j\ell}^i x_\ell$ for each strip H_i such that we use $\tilde{H}_i^2$ in the composition and such that $A_1^i \in P_j$ and $A_2^i \in P_\ell$ (the edges $y_{j\ell}^i x_j$ and $y_{j\ell}^i x_\ell$ correspond to the vertices u_2^i and u_3^i of T_2^i, respectively); a vertex z_j^i and an edge $z_j^i x_j$ for each strip H_i such that we use $\tilde{H}_i^2$ in the composition and such that $A_1^i \in P_j$ (the edge corresponds to the vertex u_1^i of T_2^i); a vertex z_j^i and an edge $z_j^i x_j$ for each strip H_i such that we use $\tilde{H}_i^0$ in the composition and such that $A_1^i \in P_j$ (the edge corresponds to the vertex c^i of T_0^i); vertices z_j^i and z_ℓ^i and edges $z_j^i x_j$ and $z_\ell^i x_\ell$ for each strip H_i such that we use $\tilde{H}_i^1$ in the composition and such that $A_1^i \in P_j$ and $A_2^i \in P_\ell$ (the edges $z_j^i x_j$ and $z_\ell^i x_\ell$ correspond to the vertices u_1^i and u_2^i of T_1^i, respectively).

The maximal cliques of $\tilde{G}$ correspond to the multistars and multitriangles of H, i.e., the multistars centered at x_j for $j = 1, \ldots, r$, the possible multitriangles $x_i x_j x_\ell$ for i, j, ℓ pairwise distinct elements in $\{1, \ldots, r\}$, and, for each vertex $y_{j\ell}^i$, either the star centered at $y_{j\ell}^i$ or the multitriangle $y_{j\ell}^i x_j x_\ell$ with $j, \ell \in \{1, \ldots, r\}$ and $j \neq \ell$, depending on the existence of edges joining x_j and x_ℓ. To each of these cliques of $\tilde{G}$ we will assign a clique of G, except for the case of cliques involving y-vertices. We have to deal with those cliques in a different way.

To the clique of $\tilde{G}$ corresponding to the multistar centered at x_j in H, we will assign in G the partition-clique $\bigcup_{A_d^i \in P_j} A_d^i$. To the clique of $\tilde{G}$ corresponding to the multitriangle $x_i x_j x_\ell$ in H, we will assign in G the clique induced by $\bigcup_{d \in I_{ij\ell}} (A_1^d \cap A_2^d)$, where $I_{ij\ell}$ is the set of indices d of 2-strips in the decomposition, that have been replaced by $\tilde{H}_d^3$, and having their two extremities belonging to two different sets in $\{P_i, P_j, P_\ell\}$ (we can prove that these intersections are nonempty).

Now we want to show how we deal with the star centered at $y_{j\ell}^i$ and the multitriangles $y_{j\ell}^i x_j x_\ell$. As we have already said, these two structures correspond to cliques in $\tilde{G}$, but the corresponding cliques in $\tilde{G}$ cannot be extended to cliques of G. Thus we have to show that we can rearrange the weight function of the vertices of the strips in order to get a cover with the same value which includes only cliques. First we show that if we have a multitriangle $y_{j\ell}^i x_j x_\ell$ in H, then the 2-strip $(G^i, \mathcal{A}^i)$ is odd-short, and there is a vertex x complete to both extremities of it in G. Then we prove the following lemma. This lemma essentially says that if we have assigned a weight $a > 0$ to the triangle $y_{j\ell}^i x_j x_\ell$ then we can discard this triangle and ask for a MWCC of value $\delta_1^i + a$ in the graph induced by G^i and $A_1^k \cap A_2^k$ for every k such that $A_1^k \cap A_2^k$ is complete to $A_1^i \cup A_2^i$ (this set of vertices form a clique), in such a way that $\bigcup_k (A_1^k \cap A_2^k)$ is covered by a quantity greater or equal to a. W.l.o.g. we may consider that we need to cover just an extra vertex x of weight a complete to $A_1^i \cup A_2^i$.

Lemma 3. Let $H_i = (G^i, \mathcal{A}^i)$ be a 2-strip. Let $G_\bullet^i$ be the graph obtained from G^i by adding a new vertex x complete to both A_1^i and A_2^i. Let w be a non-negative weight function defined on the vertices of G^i. Let $\delta_1^i = \alpha_w(G^i \setminus (A_1^i \cup A_2^i))$. Let a, b_1, b_2 be non-negative numbers such that $b_1 \geq \alpha_w(G^i) - \alpha_w(G^i \setminus A_1^i)$, $a + b_1 \geq \alpha_w(G^i \setminus A_2^i) - \delta_1^i$, $a + b_2 \geq \alpha_w(G^i \setminus A_1^i) - \delta_1^i$, and let w^i be defined as $w^i(v) = w(v)$ for $v \in V(G^i) \setminus (A_1^i \cup A_2^i)$, $w^i(v) = \max\{0, w(v) - b_1\}$ for $v \in A_1^i \setminus A_2^i$, $w^i(v) = \max\{0, w(v) - b_2\}$ for $v \in A_2^i \setminus A_1^i$, and $w^i(v) = \max\{0, w(v) - b_1 - b_2\}$ for $v \in A_1^i \cap A_2^i$. Then $\alpha_{w^i}(G_\bullet^i) = \delta_1^i + a$. In particular, $\alpha_{w^i}(G^i) \leq \delta_1^i + a$.

We underline that the last sentence of Lemma 3 suggests also how to "translate" the weight a possibly assigned to the star centered in $y_{j\ell}^i$ and, in general, how to deal with the strips that have been replaced by $\tilde{H}^2$.

Now we want to show that if we have a weighted clique cover of $\tilde{G}$, we can cover the "residual" weight w^i of each strip $H_i = (G^i, \mathcal{A}^i)$ with a weighted clique cover of value at most δ^i. The following lemmas give the desired result for 1-strips and 2-strips that have been replaced with $\tilde{H}^1$ or $\tilde{H}^3$. In particular, Lemma 6 considers the case of 2-strips with a non empty intersection of the extremities that might cause multitriangles in the root graph of $\tilde{G}$.

Lemma 4. Let $H_i = (G^i, \mathcal{A}^i)$ be a 1-strip and let w be a non-negative weight function defined on the vertices of G^i. Let $\delta_1^i = \alpha_w(G^i \setminus A_1^i)$, let $b \geq \alpha_w(G^i) - \delta_1^i$, and let w^i be defined as $w^i(v) = w(v)$ for $v \in V(G^i) \setminus A_1^i$, $w(v) = \max\{0, w(v) - b\}$ for $v \in A_1^i$. Then $\alpha_{w^i}(G^i) = \delta_1^i$.

Lemma 5. *Let $H_i = (G^i, \mathcal{A}^i)$ be a 2-strip and let w be a non-negative weight function defined on the vertices of G^i. Let $\delta_1^i = \alpha_w(G^i \setminus (A_1^i \cup A_2^i))$. Let b_1, b_2 be numbers such that $b_1 \geq \alpha_w(G^i \setminus A_2^i) - \delta_1^i$, $b_2 \geq \alpha_w(G^i \setminus A_1^i) - \delta_1^i$, and $b_1 + b_2 \geq \alpha_w(G^i) - \delta_1^i$, and let w^i be defined as $w^i(v) = w(v)$ for $v \in V(G^i) \setminus (A_1^i \cup A_2^i)$, $w^i(v) = \max\{0, w(v) - b_1\}$ for $v \in A_1^i \setminus A_2^i$, $w^i(v) = \max\{0, w(v) - b_2\}$ for $v \in A_2^i \setminus A_1^i$, and $w^i(v) = \max\{0, w(v) - b_1 - b_2\}$ for $v \in A_1^i \cap A_2^i$. Then $\alpha_{w^i}(G^i) = \delta_1^i$.*

Lemma 6. *Let $H_i = (G^i, \mathcal{A}^i)$ be an even-short 2-strip such that G^i is perfect, and let w be a non-negative weight function defined on the vertices of G^i. Let $\delta_1^i = \alpha_w(G^i \setminus (A_1^i \cup A_2^i))$. Let b_1, b_2, a be numbers such that $b_1 \geq \alpha_w(G^i \setminus A_2^i) - \delta_1^i$, $b_2 \geq \alpha_w(G^i \setminus A_1^i) - \delta_1^i$, and $a + b_1 + b_2 \geq \alpha_w(G^i) - \delta_1^i$, and let w^i be defined as $w^i(v) = w(v)$ for $v \in V(G^i) \setminus (A_1^i \cup A_2^i)$, $w^i(v) = \max\{0, w(v) - b_1\}$ for $v \in A_1^i \setminus A_2^i$, $w^i(v) = \max\{0, w(v) - b_2\}$ for $v \in A_2^i \setminus A_1^i$, and $w^i(v) = \max\{0, w(v) - b_1 - b_2 - a\}$ for $v \in A_1^i \cap A_2^i$. Then $\alpha_{w^i}(G^i) = \delta_1^i$.*

Finally, we analyze the case of 2-strips with both extremities in the same class of $\mathcal{P}$. Such a strip H_i has been replaced with $\tilde{H}_i^0$, thus every vertex in its extremities is covered by a quantity of at least $\max\{\alpha_w(G^i \setminus A_1^i), \alpha_w(G^i \setminus A_2^i), \alpha_w(G^i \setminus (A_1^i \triangle A_2^i))\} - \delta_1^i$.

Lemma 7. *Let $H_i = (G^i, \mathcal{A}^i)$ be a 2-strip. Let $G_=^i$ be the graph obtained from G^i by adding the edges between A_1^i and A_2^i. Let w be a non-negative weight function defined on the vertices of G^i. Let $\delta_1^i = \alpha_w(G^i \setminus (A_1^i \cup A_2^i))$, let $b \geq \max\{\alpha_w(G^i \setminus A_1^i), \alpha_w(G^i \setminus A_2^i), \alpha_w(G^i \setminus (A_1^i \triangle A_2^i))\} - \delta_1^i$, and let w^i be defined as $w^i(v) \doteq w(v)$ for $v \in V(G^i) \setminus (A_1^i \cup A_2^i)$, $w(v) = \max\{0, w(v) - b\}$ for $v \in A_1^i \cup A_2^i$. Then $\alpha_{w^i}(G_=^i) = \delta_1^i$. Moreover, if $G_=^i$ is perfect, any MWCC of $G_=^i$ w.r.t. w^i does not assign strictly positive weight to the clique $A_1^i \cup A_2^i$.*

Note that the last sentence of the previous lemma implies that, if $G_=^i$ is perfect and there are no two vertices $v_1 \in A_1^i$ and $v_2 \in A_2^i$ having a common neighbor in $V(G^i) \setminus (A_1^i \cup A_2^i)$, then any MWCC of $G_=^i$ w.r.t. w^i is in fact a MWCC of G^i w.r.t. w^i. We also observe that whenever we cannot use Lemma 7 we must be able to compute a MWCC of $G_=^i$ in order to reconstruct a clique cover of G from a clique cover of $\tilde{G}$. This is why we require in Theorem 2 that a MWCC of $G_=^i$ can be computed in time $O(p_i(|V(G^i)|))$ in that case.

4   Application to Strip-Composed Claw-Free Perfect Graphs

As an application of Theorem 2, we give a new algorithm for the MWCC on strip-composed claw-free perfect graphs. Recall that claw-free perfect graphs are in fact quasi-line. In the last decade the structure of quasi-line graphs was deeply investigated, with some results providing a detailed description and characterization of the strips that, through composition, can be part of a quasi-line graph. This is the case of the structure theorem by Chudnovsky and Seymour in [2].

The following algorithmic decomposition theorem from [5] applies to quasi-line graphs. (A *net* is a graph formed by a triangle and three vertices of degree one, each of them adjacent to a distinct vertex of the triangle.)

Theorem 3. *[5] Let G be a connected quasi-line graph. In time $O(|V(G)||E(G)|)$, one can either recognize that G is net-free; or provide a decomposition of G into $k \leq |V(G)|$ quasi-line strips $(G^1, \mathcal{A}^1), \ldots, (G^k, \mathcal{A}^k)$, w.r.t. a partition $\mathcal{P}$, such that each graph G^i is distance simplicial w.r.t. each clique $A \in \mathcal{A}^i$. Moreover, if $\mathcal{A}^i = \{A_1^i, A_2^i\}$, then either $A_1^i = A_2^i = V(G^i)$; or $A_1^i \cap A_2^i = \emptyset$ and there exists j_2 such that $A_2^i \cap N_{j_2}(A_1^i) \neq \emptyset$, $A_2^i \subseteq N_{j_2-1}(A_1^i) \cup N_{j_2}(A_1^i)$ and $N_{j_2+1}(A_1^i) = \emptyset$, where $N_j(A_1^i)$ is the j-th neighborhood of A_1^i in G^i (and, analogously, there exists j_1 such that $A_1^i \cap N_{j_1}(A_2^i) \neq \emptyset$, $A_1^i \subseteq N_{j_1-1}(A_2^i) \cup N_{j_1}(A_2^i)$ and $N_{j_1+1}(A_2^i) = \emptyset$). Besides, each vertex in A has a neighbor in $V(G^i) \setminus A$, for each clique $A \in \mathcal{A}^i$. Finally, if A_1^i and A_2^i are in the same set of $\mathcal{P}$, then A_1^i is anticomplete to A_2^i.*

Now suppose that we are given a strip decomposition obeying to Theorem 3 for a claw-free *perfect* graph G. If we are interested in finding a MWCC of G, following Theorem 2, we must show that for a strip that is distance simplicial we can compute in polynomial time: a MWCC of the strip; a MWCC of $G_\bullet^i$, i.e. G^i plus a vertex complete to both extremities, when the strip $(G^i, \mathcal{A}^i)$ is odd-short; a MWCC of $G_=^i$, i.e. G^i plus the edges joining the extremities A_1^i, A_2^i of the strip, when they are in the same class of the partition and there is an A_1–A_2 path of length two.

We start by briefly describing how to compute a MWCC in distance simplicial graphs (recall that they are indeed perfect). We rely on a property of perfect graphs, namely, there always exists a clique which intersects each MWSS: we will call such a clique *crucial* (crucial cliques are a key ingredient to the algorithm in [10]). Our algorithm relies on the fact that for graphs that are distance simplicial w.r.t. some identifiable clique K, we can inductively compute crucial cliques and decide the value of this clique in a MWCC. The first crucial clique will be $K' := K \cup \{v \notin K : v \text{ is complete to } K\}$: we will suitably update the weight of each vertex, and then find a new crucial clique (w.r.t. the new weights) in an inductive way.

We now show that, for an odd-short distance simplicial strip H_i, we can compute in polynomial time a MWCC of $G_\bullet^i$. Note that, in this case $G_\bullet^i$ is claw-free and, following Theorem 2, perfect. In this case, we prove that $G_\bullet^i$ is cobipartite.

Note that, if $G_\bullet^i$ is cobipartite, then it is distance simplicial w.r.t. each of the two cliques covering its vertex set, so a MWCC can be found as above. We now show that, for a distance simplicial strip H_i, such that the extremities are in the same class of the partition and there is an A_1–A_2 path of length two, we can compute in polynomial time a MWCC for $G_=^i$. Note that, in this case, $G_=^i$ is claw-free and, following Theorem 2, we may assume that it is perfect and that $\alpha_{w^i}(G_=^i) = \alpha_{w^i}(G^i \setminus (A_1^i \cup A_2^i))$ holds, where w^i is the weight function defined on the vertices of G^i (that w.l.o.g. we take strictly positive, i.e., we remove vertices with $w^i(v) = 0$). In this case, we prove that either $G_=^i$ is cobipartite, or every MWCC of G^i is also a MWCC of $G_=^i$. If $G_=^i$ is not cobipartite, then we may simply ignore the edges between the two extremities of the strip and then compute a MWCC in G^i, which is distance simplicial.

We have therefore the following theorem for strip-composed claw-free perfect graphs. We underline that the resulting algorithm never requires the computation of any MWSS on the strips, while it uses a primal-dual algorithm for the maximum weighted matching on the root graph of $\tilde{G}$ (see Section 3).

Theorem 4. *Let G be a claw-free perfect graph with a non-negative weight function w on $V(G)$ and let G be as in case (ii) of Theorem 3. Then we can compute a MWCC of G w.r.t. w in time $O(|V(G)|^3)$.*

References

1. Burlet, M., Maffray, F., Trotignon, N.: Odd pairs of cliques. In: Proc. of a Conference in Memory of Claude Berge, pp. 85–95. Birkhäuser (2007)
2. Chudnovsky, M., Seymour, P.: Claw-free graphs. VII. Quasi-line graphs. J. Combin. Theory, Ser. B (to appear)
3. Chudnovsky, M., Seymour, P.: The structure of claw-free graphs. London Math. Soc. Lecture Note Ser. 327, 153–171 (2005)
4. Eisenbrand, F., Oriolo, G., Stauffer, G., Ventura, P.: Circular one matrices and the stable set polytope of quasi-line graphs. Combinatorica 28(1), 45–67 (2008)
5. Faenza, Y., Oriolo, G., Stauffer, G.: An algorithmic decomposition of claw-free graphs leading to an $O(n^3)$-algorithm for the weighted stable set problem. In: Randall, D. (ed.) Proc. 22nd SODA, San Francisco, CA, pp. 630–646 (2011)
6. Gabow, H.: Data structures for weighted matching and nearest common ancestors with linking. In: Proc. 1st SODA, San Francisco, CA, pp. 434–443 (1990)
7. Grötschel, M., Lovász, L., Schrijver, A.: Geometric Algorithms and Combinatorial Optimization. Springer, Berlin (1988)
8. Hermelin, D., Mnich, M., Van Leeuwen, E., Woeginger, G.: Domination When the Stars Are Out. In: Aceto, L., Henzinger, M., Sgall, J. (eds.) ICALP 2011. LNCS, vol. 6755, pp. 462–473. Springer, Heidelberg (2011)
9. Hsu, W., Nemhauser, G.: Algorithms for minimum covering by cliques and maximum clique in claw-free perfect graphs. Discrete Math. 37, 181–191 (1981)
10. Hsu, W., Nemhauser, G.: Algorithms for maximum weight cliques, minimum weighted clique covers and cardinality colorings of claw-free perfect graphs. Ann. Discrete Math. 21, 317–329 (1984)
11. Minty, G.: On maximal independent sets of vertices in claw-free graphs. J. Combin. Theory, Ser. B 28(3), 284–304 (1980)
12. Nakamura, D., Tamura, A.: A revision of Minty's algorithm for finding a maximum weighted stable set of a claw-free graph. J. Oper. Res. Soc. Japan 44(2), 194–204 (2001)
13. Nobili, P., Sassano, A.: A reduction algorithm for the weighted stable set problem in claw-free graphs. In: Proc. 10th CTW, Frascati, Italy, pp. 223–226 (2011)
14. Oriolo, G., Pietropaoli, U., Stauffer, G.: A New Algorithm for the Maximum Weighted Stable Set Problem in Claw-Free Graphs. In: Lodi, A., Panconesi, A., Rinaldi, G. (eds.) IPCO 2008. LNCS, vol. 5035, pp. 77–96. Springer, Heidelberg (2008)
15. Schrijver, A.: Combinatorial Optimization. In: Polyhedra and Efficiency (3 volumes), Algorithms and Combinatorics, vol. 24. Springer, Berlin (2003)
16. Trotter, L.: Line perfect graphs. Math. Program. 12, 255–259 (1977)

Graph Isomorphism for Graph Classes Characterized by Two Forbidden Induced Subgraphs*

Stefan Kratsch[1] and Pascal Schweitzer[2]

[1] Utrecht University, the Netherlands
s.kratsch@uu.nl
[2] The Australian National University
pascal.schweitzer@anu.edu.au

Abstract. We study the complexity of the Graph Isomorphism problem on graph classes that are characterized by a finite number of forbidden induced subgraphs, focusing mostly on the case of two forbidden subgraphs. We show hardness results and develop techniques for the structural analysis of such graph classes, which applied to the case of two forbidden subgraphs give the following results: A dichotomy into isomorphism complete and polynomial-time solvable graph classes for all but finitely many cases, whenever neither of the forbidden graphs is a clique, a pan, or a complement of these graphs. Further reducing the remaining open cases we show that (with respect to graph isomorphism) forbidding a pan is equivalent to forbidding a clique of size three.

1 Introduction

Given two graphs G_1 and G_2, the Graph Isomorphism problem (GI) asks whether there exists a bijection from the vertices of G_1 to the vertices of G_2 that preserves adjacency. This paper studies the complexity of GI on graph classes that are characterized by a finite number of forbidden induced subgraphs, focusing mostly on the case of two forbidden subgraphs. For a set of graphs $\{H_1, \ldots, H_k\}$ we let $(H_1, \ldots, H_k)$-free denote the class of graphs G that do not contain any H_i as an induced subgraph.

As a first example, consider the class of graphs containing neither a clique K_s on s vertices, nor an independent set I_t on t vertices. Ramsey's Theorem [19] states that the number of vertices in such graphs is bounded by a function $f(s, t)$. Thus the classes (K_s, I_t)-free are finite and Graph Isomorphism is trivial on them. All other combinations of two forbidden subgraphs give graph classes of infinite size, since they contain infinitely many cliques or independent sets.

As a second example, consider the graphs containing no clique K_s on s vertices and no star $K_{1,t}$ (i.e., an independent set of size t with added universal vertex adjacent to every other vertex). On the one hand this class contains all

---

* In this version some proofs are omitted. For these the reader is referred to [12].

M.C. Golumbic et al. (Eds.): WG 2012, LNCS 7551, pp. 34–45, 2012.

graphs of maximum degree less than $\min\{s-1, t\}$, on the other hand, all graphs in $(K_s, K_{1,t})$-free have bounded degree: Indeed, if the degree of a vertex is sufficiently large, its neighborhood must contain a clique of size s or an independent set of size t by Ramsey's Theorem [19], leading to one of the two forbidden subgraphs. Thus, using Luks' algorithm [16] that solves Graph Isomorphism on graphs of bounded degree in polynomial time, isomorphism of $(K_s, K_{1,t})$-free graphs can also be decided in polynomial time.

To systematically study Graph Isomorphism on graph classes characterized by forbidden subgraphs, we ask: *Given a set of graphs* $\{H_1, \ldots, H_k\}$, *what is the complexity of Graph Isomorphism on the class of* $(H_1, \ldots, H_k)$-*free graphs?*

Related Work. The Graph Isomorphism problem is contained in the complexity class NP, since the adjacency preserving bijection (the isomorphism) can be checked in polynomial time. No polynomial-time algorithm is known and it is known that Graph Isomorphism is not NP-complete unless the polynomial hierarchy collapses [5]. More strongly, Graph Isomorphism is in the low hierarchy [21]. This has led to the definition of the complexity class of problems polynomially equivalent to Graph Isomorphism, the so-called GI-complete problems. There is a vast literature on the Graph Isomorphism Problem; for a general overview see [22] or [10], for results on its parameterized complexity see [13].

A question analogous to ours, asking about Graph Isomorphism on any class of $(H_1, \ldots, H_k)$-minor-free graphs, is answered completely by the fact that Graph Isomorphism is polynomially solvable on any non-trivial minor closed class [18]. Recently, the corresponding statement for topological minor free classes was also shown [8]. For the less restrictive family of *hereditary* classes, only closed under vertex deletion (i.e., classes $\mathcal{H}$-free for a possibly infinite set of graphs $\mathcal{H}$), both GI-complete and tractable cases are known: Graph Isomorphism is GI-complete on split graphs, comparability graphs, and strongly chordal graphs [23]. Graph Isomorphism is known to be polynomially solvable for circle graphs and circular-arc graphs [9], interval graphs [2,15], distance hereditary graphs [17], and graphs of bounded degree [16]. For various subclasses of these polynomially solvable cases results with finer complexity analysis are available, but of course the polynomial-time solvability for these subclasses follows already from polynomial-time solvability of the mentioned larger classes. Further results, in particular on GI-completeness, can be found in [4].

Concerning our question, for one forbidden subgraph, the answer, given by Colbourn and Colbourn, can be found in a paper by Booth and Colbourn [4]: If the forbidden induced subgraph H_1 is an induced subgraph of the path P_4 on four vertices, denoted by $H_1 \leq P_4$, then Graph Isomorphism is polynomial on H_1-free graphs, otherwise it is GI-complete.

Apart from the isomorphism problem, other studies aiming at dichotomy results for algorithmic problems on graph classes characterized by two forbidden subgraphs consider the chromatic number [11] and dominating sets [14].

Main Result. Let a graph be *basic* if it is an independent set, a clique, a $P_3 \dot\cup K_1$, or the complement of a $P_3 \dot\cup K_1$ (also called pan). If neither H_1 nor H_2 is basic,

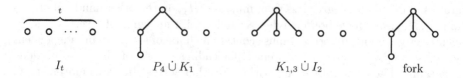

I_t $P_4 \,\dot\cup\, K_1$ $K_{1,3} \,\dot\cup\, I_2$ fork

Fig. 1. The independent sets, the paths of length 3 with independent vertex, the claw with two independent vertices, and the fork, obtained by subdividing an edge in a claw

then we obtain a classification of (H_1, H_2)-free classes into polynomial and GI-complete cases, for all but a small finite number of classes. Theorem 1 justifies the terminology basic by showing that in our context forbidding a basic graph is equivalent to forbidding a complete graph. However, the case of forbidding a clique (alongside a second graph) appears to be structurally different and for a complete classification further new techniques are required.

Technical Contribution. Our main technical contribution lies in establishing tractability of Graph Isomorphism on four types of (H_1, H_2)-free classes (Theorem 4 in Section 4): A structural analysis of the classes enables reductions to the polynomially-solvable case of bounded color valence [1]. This reduction appears necessary since the polynomially-solvable classes of Theorem 4 encompass all classes of graphs of bounded degree, and for these Luks' group-theoretic approach [16] (implicit in [1]) is the only known polynomial-time technique. At the core of the proof of Theorem 4 lie individualization-refinement techniques and recursive structural analysis to allow for a reduction to the bounded color valence case.

However, to put these results in context and obtain the mentioned classification, we have to refine several known results for GI-completeness on bipartite, split, and line graphs (Section 3). In particular, we arrive at a set of four graph properties, which we call *split conditions*, such that Graph Isomorphism remains complete on any class (H_1, H_2)-free unless each property is true for at least one of the two forbidden subgraphs.

Based on this characterization we can state our results in more detail: If on the one hand neither of the two forbidden subgraphs H_1 and H_2 exhibits all four split conditions, then we have a dichotomy of GI on (H_1, H_2)-free classes into polynomial and GI-complete classes; the polynomial cases are due to Theorem 4 (see Section 4) as well as tractability on cographs (i.e., P_4-free graphs) [7], GI-completeness follows by using both known results as well as our strengthened reductions (see Section 3). Suppose on the other hand H_1 and H_2 are both not basic and H_1 simultaneously fulfills all four split conditions, then our hardness and tractability results resolve all but a finite number of cases (i.e., each case is one concrete class (H_1, H_2)-free), as showed in Theorem 6 (see Section 5). For these cases Figure 1 shows the relevant maximal graphs that adhere to all four split conditions.

2   Preliminaries

We write $H \leq G$ if the graph G contains a graph H as an induced subgraph. A graph G is H-free if $H \not\leq G$. It is $(H_1, \ldots, H_k)$-free, if it is H_i-free for all i. A graph class $\mathcal{C}$ is H-free (resp. $(H_1, \ldots, H_k)$-free) if this is true for all $G \in \mathcal{C}$. A graph class $\mathcal{C}$ is *hereditary* if it is closed under taking induced subgraphs. The class $(H_1, \ldots, H_k)$-free is the class of all $(H_1, \ldots, H_k)$-free graphs; each class $(H_1, \ldots, H_k)$-free is hereditary.

By I_t, K_t, P_t, and C_t we denote the independent set, the clique, the path, and the cycle on t vertices; $K_{1,t}$ is the claw with t leaves. By $H \,\dot\cup\, H'$ we denote the disjoint union of H and H'; we use tK_2 for the disjoint union of t graphs K_2. By $\overline{G}$ we denote the (edge) complement of G. The graph $\overline{K_2 \,\dot\cup\, I_2}$, i.e., the same as a K_4 minus one edge, is called diamond.

We recall that GI-completeness is inherited by superclasses while polynomial-solvability of Graph Isomorphism is inherited by subclasses. Also recall that Graph Isomorphism on a class $\mathcal{C}$ is exactly as hard as on $\overline{\mathcal{C}}$, the class of complements of graphs in $\mathcal{C}$. Note that any H-free graph is also H'-free if $H \leq H'$.

Proposition 1. *Let H_1, H_2 be graphs and let $\mathcal{C}$ be any hereditary graph class.*

1. $\overline{(H_1, H_2)\text{-free}} = (\overline{H_1}, \overline{H_2})\text{-free}.$
2. $(H_1, H_2)\text{-free} \subseteq (H'_1, H'_2)\text{-free for any } H'_1, H'_2 \text{ with } H_1 \leq H'_1 \text{ and } H_2 \leq H'_2.$
3. $H_1, H_2 \notin \mathcal{C} \text{ implies } \mathcal{C} \subseteq (H_1, H_2)\text{-free}.$

Definition 1. *The* pan *is the graph $\overline{P_3 \,\dot\cup\, K_1}$, i.e., a vertex and triangle joined by one edge. A graph is* basic, *if it is an independent set, a complete graph, the graph $P_3 \,\dot\cup\, K_1$, or the pan.*

We now show that in the context of the isomorphism problem excluding a basic graph is equivalent to excluding a complete graph or an independent set.

Lemma 1. *Let G be a graph that contains P_4 as an induced subgraph.*

1. *If G is co-connected then it contains I_3 if and only if it contains $P_3 \,\dot\cup\, K_1$.*
2. *If G is connected then it contains K_3 if and only if it contains a pan.*

Proof. By complementarity it suffices to prove Part 2. Fix a P_4 in the graph. Containment of a pan trivially implies containment of a triangle. For the converse, it can be easily verified that there is a pan, if some some triangle contains at least two vertices of the P_4. Else, if a triangle contains one vertex p of the P_4, we can add a vertex of the P_4 adjacent to p to the triangle, obtaining a pan. Else (i.e., if no triangle is incident with the P_4) consider the triangle closest to the P_4. Due to connectivity, there is a vertex that is adjacent to some vertex of the triangle and closer to the P_4. If this vertex is adjacent to exactly one vertex of the triangle, a pan arises. Otherwise we find a closer triangle, which contradicts our initial choice.   □

Theorem 1. *Graph Isomorphism on a class $\mathcal{C}$ of K_3-free graphs is polynomial time equivalent to GI on the subclass of $\mathcal{C}$ that contains all pan-free graphs of $\mathcal{C}$.*

Proof. Since Graph Isomorphism for P_4-free graphs (so-called cographs) is solvable in polynomial time [7], the theorem follows from Lemma 1 and the fact that graph isomorphism can be solved by comparing connected components. □

3   Hardness Results

Our standard method to show GI-completeness for Graph Isomorphism on some graph class $\mathcal{H}$ works by reducing the isomorphism problem of a class $\mathcal{H}'$ for which Graph Isomorphism is known to be GI-complete to a subclass of $\mathcal{H}$. For this we require a mapping $\pi\colon \mathcal{H}' \to \pi(\mathcal{H}') \subseteq \mathcal{H}$ which is computable in polynomial time and for which the images of two graphs are isomorphic if and only if the two original graphs are. We call such a mapping π a *GI-reduction*. To show hardness for a class (H_1, H_2)-free it suffices to provide a GI-reduction π for which no graph $G \in \pi(\mathcal{H}')$ contains H_1 or H_2 as an induced subgraph, implying that $\pi(\mathcal{H}') \subseteq (H_1, H_2)$-free.

Our first reductions prove hardness results for bipartite graphs, split graphs, and line graphs. However, the (previously known) GI-completeness for these particular graph classes is not sufficient. We require hardness for more specific subclasses avoiding specific small graphs. Subsequently, using a more involved reduction, we show that isomorphism of $(P_4 \dot\cup K_1, K_4)$-free graphs is GI-complete.

3.1   Bipartite Graphs

A straightforward GI-reduction consists of subdividing each edge of a graph by a new vertex. Since the obtained graphs are bipartite, this proves that Graph Isomorphism remains GI-complete on bipartite graphs. This also implies that Graph Isomorphism remains GI-complete on (H_1, H_2)-free graphs unless one of the graphs is bipartite, since the class (H_1, H_2)-free contains all bipartite graphs if neither H_1 nor H_2 is bipartite. Let us observe however, that we can draw stronger conclusions namely that Graph Isomorphism remains GI-complete on connected bipartite graphs without induced cycles of length 4, for which the vertices in one of the partition classes have degree two. The following definition allows us to make a first structural observation for the graphs H_1, H_2:

Definition 2. *A path-star is a subdivision of the t-claw* $K_{1,t}$, *for some* $t \in \mathbb{N}$.

Lemma 2. *If neither* H_1 *nor* H_2 *is a disjoint union of path-stars, then Graph Isomorphism on the class* (H_1, H_2)-free *is GI-complete.*

Proof. If a graph is not a disjoint union of path-stars, then it either contains two vertices of degree at least 3 which are in the same connected component, or it contains a cycle. We use that two graphs G_1 and G_2 are isomorphic, if and only if the graphs obtained by subdividing each edge in G_1 and G_2 respectively are isomorphic. For any integer c there is an integer c' such that a graph that has been subdivided c' times neither contains a cycle of length at most c nor two vertices of degree at least three which are at a distance of at most c apart.

Thus, with a finite number of subdivision steps, we can reduce Graph Isomorphism on general graphs to isomorphism on (H_1, H_2)-free graphs. □

Using Part 1 of Proposition 1, we conclude that unless Graph Isomorphism is GI-complete on (H_1, H_2)-free, one of the graphs H_1, H_2 is a forest and one of the graphs is a co-forest.

Lemma 3. *A graph H and its complement $\overline{H}$ are forests, if and only if $H \leq P_4$.*

Proof. For the if part it suffices to observe that P_4 is a self-complementary forest. The only if part is true for graphs of size at most 4. A forest on $n \geq 5$ vertices has at most $n-1$ edges. Since $2(n-1) < \binom{n}{2}$ for $n \geq 5$ the statement follows. □

Graph Isomorphism for P_4-free graphs (cographs) is in P [7]. With the previous two lemmas one can conclude that it remains GI-complete on H-free graphs when H is not an induced subgraph of the P_4; this gives a simple dichotomy for the case of a single forbidden induced subgraph.

Theorem 2 (see [4]). *Let H be a graph. Graph Isomorphism on H-free graphs is in P, if $H \leq P_4$. GI on H-free graphs is GI-complete, if $H \nleq P_4$.*

In the following, we focus on graph classes characterized by two forbidden induced subgraphs. Since isomorphism of P_4-free graphs is in P, we assume from now on that $H_1 \nleq P_4$ and $H_2 \nleq P_4$. Due to Lemmas 2 and 3 and Part 1 of Proposition 1 we may further assume that H_1 is a disjoint union of path-stars and H_2 is the complement of a disjoint union of path-stars.

Being forests, unions of path-stars are bipartite. Since bipartite graphs play a repeated role, we introduce some terminology: For a bipartite graph G, which has been partitioned into two classes , the *bipartite complement* is the graph obtained by replacing all edges that run between vertices from different partition classes by non-edges and vice versa. (Note that the bipartite complement for unpartitioned bipartite graphs is only well defined if the graph is connected.) A *crossing co-cycle* is a set of vertices that form a cycle in the bipartite complement.

Lemma 4. *Isomorphism of graphs that are (H_1, H_2)-free is GI-complete unless H_1 or H_2 can be partitioned as a bipartite graph without crossing co-cycle.*

Proof. Graph Isomorphism is GI-complete on connected graphs. By repeatedly subdividing a connected graph we produce a bipartite graph with an arbitrarily high girth. If at least three subdivisions have been performed on a non-trivial graph, its bipartite complement is connected. Thus taking the bipartite complement of such graphs is a GI-reduction (the bipartite complement of the bipartite complement is the original graph), and we obtain the lemma. □

Lemma 5. *For each $G \in \{3K_2, 2K_2 \mathbin{\dot\cup} I_2, P_4 \mathbin{\dot\cup} I_2\}$, Graph Isomorphism on the class on the class of (H_1, H_2)-free graphs is GI-complete unless one of the graphs H_1 and H_2 is bipartite and does not contain the graph G.*

Proof. Using Lemma 4 this follows, since none of the graphs $3K_2$, $2K_2 \mathbin{\dot\cup} I_2$, and $P_4 \mathbin{\dot\cup} I_2$ can be partitioned as bipartite graph without crossing co-cycle. □

3.2 Split Graphs

We now turn our attention to split graphs. A *split graph* is a graph whose vertices can be partitioned into an independent set and a clique. Recall that the split graphs are exactly the $(2K_2, C_4, C_5)$-free graphs. The reduction that subdivides each edge and connects all newly introduced vertices produces a split graph, and thus proves GI-completeness on that class. As in the previous section we are able to draw further conclusions about the obtained graphs.

Definition 3. *Let G be a split graph that has been partitioned into a clique K and an independent set I. We say that the partition is*

1. *of type 1, if the vertices in class K have at most 2 neighbors in class I.*
2. *of type 2, if the vertices in class K have at most 2 non-neighbors in class I.*
3. *of type 3, if the vertices in class I have at most 2 neighbors in class K.*
4. *of type 4, if the vertices in class I have at most 2 non-neighbors in class K.*

An unpartitioned graph G is said to fulfill split graph condition i (with $i \in \{1, 2, 3, 4\}$), if there is a split partition of the graph that is of type i.

Lemma 6. *For any $i \in \{1, 2, 3, 4\}$, Graph Isomorphism on the class of graphs that fulfill split graph condition i is GI-complete.*

For the proof, we refer the reader to [12]. Since the class of graphs which fulfill condition i is closed under taking induced subgraphs, the lemma implies that Graph Isomorphism on the class (H_1, H_2)-free is GI-complete unless for all $i \in \{1, 2, 3, 4\}$ one of the graphs H_1 or H_2 fulfills split graph condition i.

3.3 Line Graphs

The next graph class we consider is the class of line graphs. The line graph of a graph $G = (V, E)$ is the graph $L(G) = (E, E')$, in which two vertices are adjacent, if they represent two incident edges in the graph G. The line graph of a graph G encodes the isomorphism type of a graph G.

Lemma 7 ([24]). *Let G_1, G_2 be connected graphs such that neither is a triangle. Then G_1 and G_2 are isomorphic if and only if their line graphs are.*

The class of line graphs has a characterization by 9 forbidden subgraphs [3]. However, we reduce to a subclass characterized by three forbidden subgraphs.

Lemma 8. *Line graphs of graphs of girth at least 5 contain no $K_{1,3}$, no C_4, and no diamond.*

Proof. A claw $K_{1,3}$ in a line graph $L(G)$ would correspond to three edges in G that each share an endpoint with a fourth edge; then two of the three edges must share an endpoint (forcing an additional edge in $L(G)$). A C_4 in $L(G)$ corresponds to a C_4 in G. Finally, if three edges of a triangle-free graph G pairwise share an endpoint, then they all three share the same endpoint. A fourth edge can therefore not share an endpoint with exactly two of the edges without forming triangle in G, i.e., there can be no diamond in $L(G)$. □

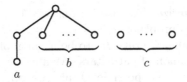

Fig. 2. A claw with a subdivision and added isolated vertices. Every split path-star is a subgraph of such a graph. They are denoted by $H(a, b, c)$ where a is the number of subdivided edges, $a + b$ the degree of the claw, and c the number of isolated vertices.

Since there is essentially a one-to-one correspondence between graphs and their line graphs, and Graph Isomorphism is GI-complete on triangle-free graphs (since $K_3 \nsubseteq P_4$), it is also GI-complete on line graphs of triangle-free graphs.

Lemma 9. *Graph Isomorphism is Graph Isomorphism-complete on the class of* $(diamond, claw, C_4)$-*free graphs.*

Proof. Since Graph Isomorphism is GI-complete on connected graphs of girth at least 5, and since graphs of girth at least 5 do not contain triangles, the lemma follows from Lemmas 7 and 8.   □

3.4   A Reduction to $(P_4 \mathbin{\dot\cup} K_1, K_4)$-Free Graphs

There is a reduction that reduces the class of all graphs to the $(P_4 \mathbin{\dot\cup} K_1, K_4)$-free graphs. For an explicit description and correctness proof of the reduction, we refer the reader to [12]. Our reduction generalizes the reduction to bipartite graphs, making a replacement of the edges while additionally connecting some of the so-created independent sets. In this sense the reduction (as many other established reductions) is part of a larger scheme of GI-reductions, which use finitely many independent sets, cliques, and relationships between them to encode graphs. We obtain the following theorem.

Theorem 3. *Graph Isomorphism is GI-complete on* $(P_4 \mathbin{\dot\cup} K_1, K_4)$-*free graphs.*

4   Structural Results and Polynomially-Solvable Cases

We have previously seen that Graph Isomorphism is GI-complete on graphs that are (H_1, H_2)-free unless each of the four split conditions is fulfilled by one of the two forbidden graphs (Lemma 6). This gives rise to two fundamental cases, namely whether or not one of the two graphs simultaneously fulfills all split conditions. In this section we address the case that neither graph fulfills all split conditions simultaneously. Amongst other conclusions this implies that both graphs must be split graphs or Graph Isomorphism will remain GI-complete. Recall also that one graph must be a path-star while the other must be the complement of a path-star or isomorphism of (H_1, H_2)-free graphs will be GI-complete (w.l.o.g. neither of the graphs is an induced subgraph of P_4 thus only H_i or $\overline{H}_i$ can be a forest or path-star). Using the results of the previous section we are able to fully characterize this case into classes with either polynomial or GI-complete isomorphism problems.

Without loss of generality we take H_1 to be a union of path-stars and H_2 to be the complement of a union of path-stars. We analyze first H_1; the graph H_2 must be a complement of the possible graphs we obtain. Since H_1 is split, i.e., $(2K_2, C_4, C_5)$-free, it is a $2K_2$-free path-star. Therefore it contains at most one non-trivial component and no induced path P_5 on five vertices. Thus if H_1 has a vertex v of degree three or larger, then at most one induced path of length two is emanating from v. Together these observations show that H_1 is an induced subgraph of the type of graph depicted in Figure 2.

We denote by $H(a, b, c)$ the graph that is depicted in Figure 2, with $a \in \{0, 1\}$, $b \in \mathbb{N}$, and $c \in \mathbb{N}$. We require that if $a = 1$ then $b > 0$, otherwise (if $a = 1$ and $b = 0$) we can reinterpret the graph with $b = 2$ and $a = 0$ (thus, $a = 1$ iff the graph contains a P_4). We also require $a + b \geq 1$ since the independent set and the clique fulfill all split conditions. Observe that any induced subgraph of some $H(a, b, c)$ is isomorphic to $H(a', b', c')$ for some values of a', b', c' (it suffices to consider the induced subgraphs of the claw with one subdivision).

We will argue that under these restrictions we may focus on the case that $a = a' = 0$, since Graph Isomorphism remains GI-complete otherwise.

Lemma 10. *Let $H_1 = H(a, b, c)$ and $H_2 = \overline{H(a', b', c')}$ such that neither graph fulfills all split conditions and such that $a + a' \geq 1$. Then GI remains GI-complete on (H_1, H_2)-free graphs.*

For the remaining discussion we may thus assume that $a = 0$ and $a' = 0$.

Theorem 4. *Isomorphism of $(H(0, b, c), \overline{H(0, b', c')})$-free graphs is in P when:*

1. $b = 0$ or $b' = 0$ *(i.e., one of the graphs is a clique or an independent set)*,
2. $c, c' \leq 1$ *and* $b, b' \geq 1$,
3. $c, c' \geq 2$ *and* $b, b' \in \{1, 2\}$,
4. $(c \geq 2, c' \leq 1, b \geq 1, b' \in \{1, 2\})$, *or* $(c' \geq 2, c \leq 1, b' \geq 1, b \in \{1, 2\})$.

In all other cases it is GI-complete.

To prove the theorem we use vertex-colorings of the input graphs. (In the context of Graph Isomorphism the vertex colorings are not assumed to be proper). We say that a vertex-colored graph has *bounded color valences*, if there is a constant D, such that for every color class C every vertex v (possibly in C) has at most D neighbors or at most D non-neighbors in C. In a graph without $H(0, b, c)$ and $\overline{H(0, b', c')}$, bounded color valence within color classes implies bounded color valence overall. For a proof of this, we refer the reader to [12]. Bounding the color valence one can reduce the isomorphism problem to that of graphs of bounded degree.

Theorem 5 (Babai, Luks [1]). *Graph Isomorphism for colored graphs of bounded color valence is solvable in polynomial time.*

To prove Theorem 4 we distinguish cases according to the numbers c and c' in the forbidden subgraphs $H(0, b, c)$ and $\overline{H(0, b', c')}$.

Proof (General proof strategy for Theorem 4.). For the full proof of Theorem 4, we refer the reader to [12] and here, instead, provide a high level description of the general proof-strategy. When proving each of the four cases our strategy is as follows: The starting observation is that a colored $(H(0, b, c), \overline{H(0, b', c')})$-free graph, which has bounded degree or bounded co-degree within each color class, also has bounded color valence between different color classes. This enables the use of Theorem 5. Thus, we now intend to find a canonical (in particular isomorphism-respecting) way of coloring both input graphs, so that the color classes have bounded degree or bounded co-degree. We employ two methods to pick color classes, both of them ensure that the coloring preserves isomorphism. Either we choose the colors of the vertices by properties of the vertices that can be computed in polynomial time. Or we guess an ordered set of vertices of constant size, color the vertices in this set with singleton colors, and then color the remaining vertices according to their adjacencies to the vertices in the ordered set. Guessing a constant number k of vertices increases the running time by a factor of n^k, and can therefore be performed in polynomial time. The second coloring operation is typically referred to as individualization.

In Case 1 we individualize one vertex, and use induction to obtain a canonical coloring with the desired properties. In Case 2 we individualize one vertex, and use a combinatorial argument to show that this gives a canonical coloring with the desired properties. In Case 4, using Lemma 1, we reduce the problem to $(H(0, b, c), K_3)$-free graphs, and then apply induction on c to obtain the canonical coloring. Case 3 is the most interesting one (and rather involved). In this case, by individualizing a finite number of vertices, we can obtain a colored graph, in which each of the color classes is a cluster, or a co-cluster graph. (A cluster is a P_3-free graph or equivalently a disjoint union of cliques.) For our purpose this is not sufficient, as for example a cluster graph can have vertices that simultaneously have large degree and large co-degree. We call a cluster d-diverse if it contains at least d disjoint cliques of size at least d. A d-diverse co-cluster is the complement of a d-diverse cluster. We show that for large d a $(H(0, 2, c), \overline{H(0, 2, c')})$-free graph cannot contain a d-diverse cluster and a d-diverse co-cluster at the same time. With this (possibly taking complements) our situation simplifies to the case where there is one color class A that is a cluster, and all other color classes are of bounded degree or bounded co-degree. After splitting off a bounded number of cliques from A, we can show that for each of the remaining cliques there is only a bounded number of types by which the vertices are connected to the vertices outside the cluster. Using this we replace the cluster by a bounded number of representatives, one for each type, color-encoding the number of vertices of each type. This leaves a graph with bounded color valence and enables us to apply Theorem 5. □

5   The Remaining Cases

In the previous section we investigated the case when neither of the two forbidden induced subgraphs fulfills all split graph conditions. We now consider the case, where one of the two graphs fulfills all split graph conditions. W.l.o.g. we let H_1 be this graph and require that H_1 is a disjoint union of path-stars (otherwise we take complements); there are only few choices for H_1. For the proofs of Lemma 11 and Theorem 6, we refer the reader to [12].

Lemma 11. *If a graph G is a union of path-stars and fulfills all split graph conditions, then it is an induced subgraph of one of the following graphs (depicted in Figure 1): An independent set, $P_4 \cup K_1$, $K_{1,3} \cup I_2$, or the fork.*

Theorem 6. *Suppose H_1 is a nonbasic disjoint union of path-stars and fulfills all split graph conditions. If H_2 has more than 7 vertices, then an application of one of Lemmas 2, 3, 5, or 9, or Theorems 2, 3, or 4 determines that $(H_1 \cup H_2)$-free is GI-complete or polynomial-time solvable. More strongly, this can be concluded unless H_1 is one of the graphs $\{P_4 \cup K_1, K_2 \cup I_2, P_3 \cup I_2\}$ and $\overline{H_2}$ has at most 7 vertices and is a disjoint union of at most 3 paths.*

6 Conclusion

In order to initiate a systematic study of the Graph Isomorphism Problem on hereditary graph classes we considered graph classes characterized by two forbidden induced subgraphs. We presented an almost complete characterization of the case that neither of the two forbidden subgraphs is basic into GI-complete and polynomial cases, leaving only few pairs of forbidden subgraphs. Theorem 4 constitutes the main technical contribution towards this result. Together with the tractability of P_4-free graphs (Theorem 2, [7]) it establishes the polynomially solvable cases. On the other hand suppose H_1 and H_2 are nonbasic and (H_1, H_2)-free is not a polynomial-time solvable case of Theorems 2 or 4. Then, Graph Isomorphism on the class of (H_1, H_2)-free graphs is GI-complete, unless for H_1 and H_2, or for $\overline{H_1}$ and $\overline{H_2}$, one of the graphs is in $\{P_4 \cup K_1, K_2 \cup I_2, P_3 \cup I_2\}$, and the other graph has at most 7 vertices and is the complement of a union of at most 3 paths.

Several further cases, e.g., all cases involving the $\overline{P_6}$ or the $\overline{P_7}$, can be excluded by variants of the reduction used for Theorem 3. Of the remaining cases, in a preprint, Rao [20] resolves positively the case $(P_4 \cup K_1, \overline{P_4 \cup K_1})$-free and its subclasses; similar (modular) decomposition techniques appear to apply to other cases as well. Several of the remaining cases are classes of bounded clique-width [6], which could indicate their tractability.

For the case in which one of the forbidden graphs is basic, our reductions and our polynomial-time algorithms are still applicable and resolve a large portion of the cases. However, as mentioned in the introduction, complete resolution appears to require new techniques. Future steps for studying the hereditary graph classes include the resolution of the remaining cases and analysis of graph classes characterized by more than two forbidden subgraphs.

References

1. Babai, L., Luks, E.M.: Canonical labeling of graphs. In: STOC 1983, pp. 171–183 (1983)

2. Babel, L., Ponomarenko, I.N., Tinhofer, G.: The isomorphism problem for directed path graphs and for rooted directed path graphs. Journal of Algorithms 21(3), 542–564 (1996)
3. Beineke, L.W.: Characterizations of derived graphs. Journal of Combinatorial Theory, Series B 9(2), 129–135 (1970)
4. Booth, K.S., Colbourn, C.J.: Problems polynomially equivalent to graph isomorphism. Technical Report CS-77-04, Comp. Sci. Dep., Univ. Waterloo (1979)
5. Boppana, R.B., Hastad, J., Zachos, S.: Does co-NP have short interactive proofs? Information Processing Letters 25, 127–132 (1987)
6. Brandstädt, A., Dragan, F.F., Le, H., Mosca, R.: New graph classes of bounded clique-width. Theory of Computing Systems 38, 623–645 (2005)
7. Corneil, D.G., Lerchs, H., Stewart Burlingham, L.: Complement reducible graphs. Discrete Applied Mathematics 3(3), 163–174 (1981)
8. Grohe, M., Marx, D.: Structure theorem and isomorphism test for graphs with excluded topological subgraphs. In: STOC, pp. 173–192 (2012)
9. Hsu, W.L.: $O(m \cdot n)$ algorithms for the recognition and isomorphism problems on circular-arc graphs. SIAM Journal on Computing 24(3), 411–439 (1995)
10. Köbler, J., Schöning, U., Torán, J.: The graph isomorphism problem: its structural complexity. Birkhäuser Verlag, Basel (1993)
11. Král, D., Kratochvíl, J., Tuza, Z., Woeginger, G.J.: Complexity of Coloring Graphs without Forbidden Induced Subgraphs. In: Brandstädt, A., Van Le, B. (eds.) WG 2001. LNCS, vol. 2204, pp. 254–262. Springer, Heidelberg (2001)
12. Kratsch, S., Schweitzer, P.: Full version of paper. arXiv:1208.0142 [cs.DS]
13. Kratsch, S., Schweitzer, P.: Isomorphism for Graphs of Bounded Feedback Vertex Set Number. In: Kaplan, H. (ed.) SWAT 2010. LNCS, vol. 6139, pp. 81–92. Springer, Heidelberg (2010)
14. Lozin, V.V.: A decidability result for the dominating set problem. Theoretical Computer Science 411(44–46), 4023–4027 (2010)
15. Lueker, G.S., Booth, K.S.: A linear time algorithm for deciding interval graph isomorphism. Journal of the ACM 26(2), 183–195 (1979)
16. Luks, E.M.: Isomorphism of graphs of bounded valence can be tested in polynomial time. Journal of Computer and System Sciences 25(1), 42–65 (1982)
17. Nakano, S., Uehara, R., Uno, T.: A new approach to graph recognition and applications to distance-hereditary graphs. Journal of Computer Science and Technology 24, 517–533 (2009)
18. Ponomarenko, I.N.: The isomorphism problem for classes of graphs closed under contraction. Journal of Mathematical Sciences 55(2), 1621–1643 (1991)
19. Ramsey, F.P.: On a problem of formal logic. Proceedings of the London Mathematical Society s2-30(1), 264–286 (1930)
20. Rao, M.: Decomposition of (gem,co-gem)-free graphs (unpublished), http://www.labri.fr/perso/rao/publi/decompgemcogem.ps
21. Schöning, U.: Graph isomorphism is in the low hierarchy. Journal of Computer and System Sciences 37(3), 312–323 (1988)
22. Schweitzer, P.: Problems of unknown complexity: Graph isomorphism and Ramsey theoretic numbers. PhD thesis, Universität des Saarlandes, Germany (2009)
23. Uehara, R., Toda, S., Nagoya, T.: Graph isomorphism completeness for chordal bipartite graphs and strongly chordal graphs. Discrete Applied Mathematics 145(3), 479–482 (2005)
24. Whitney, H.: Congruent graphs and the connectivity of graphs. American Journal of Mathematics 54(1), 150–168 (1932)

Optimization Problems
in Dotted Interval Graphs

Danny Hermelin[1], Julián Mestre[2], and Dror Rawitz[3]

[1] Max-Planck-Institut für Informatik, Saarbrücken, Germany
hermelin@mpi-inf.mpg.de
[2] School of Information Technologies, The University of Sydney, Australia
mestre@it.usyd.edu.au
[3] School of Electrical Engineering, Tel Aviv University, Tel Aviv 69978, Israel
rawitz@eng.tau.ac.il

Abstract. The class of *D-dotted interval* (*D*-DI) graphs is the class of intersection graphs of arithmetic progressions with jump (common difference) at most D. We consider various classical graph-theoretic optimization problems in *D*-DI graphs of arbitrarily, but fixed, D.

We show that MAXIMUM INDEPENDENT SET, MINIMUM VERTEX COVER, and MINIMUM DOMINATING SET can be solved in polynomial time in this graph class, answering an open question posed by Jiang *(Inf. Processing Letters, 98(1):29–33, 2006)*. We also show that MINIMUM VERTEX COVER can be approximated within a factor of $(1 + \varepsilon)$ for any $\varepsilon > 0$ in linear time. This algorithm generalizes to a wide class of deletion problems including the classical MINIMUM FEEDBACK VERTEX SET and MINIMUM PLANAR DELETION problems.

Our algorithms are based on classical results in algorithmic graph theory and new structural properties of *D*-DI graphs that may be of independent interest.

1 Introduction

A *dotted interval* $I(s, t, d)$ is an arithmetic progression $\{s, s + d, s + 2d, \ldots, t\}$, where s, t and d are positive integers, and the *jump* d divides $t - s$. When $d = 1$, the dotted interval $I(s, t, d)$ is simply the interval $[s, t]$ over the positive integer line. This paper is mainly concerned with dotted interval graphs. A *dotted interval graph* is an intersection graph of dotted intervals. Each vertex v is associated a dotted interval I_v and we have an edge (u, v) if $I_u \cap I_v \neq \emptyset$. If the jumps of all intervals are at most D, we call the graph *D-dotted-interval* or *D*-DI for short. See Figure 1 for an example.

Dotted interval graphs were introduced by Aumann *et al.* [2] in the context of high throughput genotyping. They used dotted intervals to model microsatellite polymorphisms which are used in a genotyping technique called microsatellite genotyping. The respective genotyping problem translates to MINIMUM COLORING in *D*-DI graphs of small D. Aumann *et al.* [2] showed that MINIMUM COLORING in *D*-DI graphs is NP-hard even for $D = 2$. They also provided

M.C. Golumbic et al. (Eds.): WG 2012, LNCS 7551, pp. 46–56, 2012.
© Springer-Verlag Berlin Heidelberg 2012

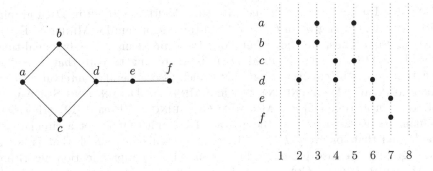

Fig. 1. Example of a 2-DI graph: On the right we have the 2-DI representation of the graph on the left. Notice that graph is clearly not an interval graph since we have hole of length 4.

a $\frac{3}{2}$-approximation algorithm for MINIMUM COLORING in 2-DI graphs, and a $(\frac{7D}{8} + \Theta(1))$-approximation algorithm for general fixed $D \geq 2$. This algorithm was later improved by Jiang [17], and subsequently also by Yanovsky [21]. The current best approximation ratio for MINIMUM COLORING is $\frac{2D+4}{3}$ [21].

Since a dotted interval with jump 1 is a regular interval, dotted interval graphs form a natural generalizations of the well-studied class of interval graphs. Interval graphs have been extensively researched in the graph-theoretic community, in particular from the algorithmic viewpoint, because many real-life problems translate to classical graph-theoretic problems in interval graphs, and because its rich structure allows in many cases designing efficient algorithms for these problems. Substantial research effort has been devoted into generalizing such algorithms to larger classes of graphs. Examples include algorithms proposed for circular arc graphs [13,15], disc graphs [11,16,19,20], rectangle graphs [1,5,9], multiple-interval graphs [4,8], and multiple-subtree graphs [14].

In this paper we study the computational complexity of classical graph-theoretic optimization problems in D-DI graphs. Note that as any graph G is a D-DI graph for large enough D [2], we are interested in studying D-DI graphs for small D; more precisely, we assume $D = O(1)$. Apart from the MINIMUM COLORING problem, Aumann *et al.* [2] also considered the MAXIMUM CLIQUE problem in D-DI graphs, and showed that this problem is fixed parameter tractable with respect to D. Jiang [17] studied the problem of MAXIMUM INDEPENDENT SET in D-DI graphs. He presented a simple $\frac{3}{2}$-approximation algorithm for 2-DI graphs, and a $(\frac{5D}{6} + O(\log d))$-approximation algorithm or D-DI graphs. The question of whether MAXIMUM INDEPENDENT SET in D-DI graphs, for constant D, is NP-hard was left open by Jiang. He also pointed out that the complexity of other classical graph theoretical problems, such as MINIMUM VERTEX COVER and MINIMUM DOMINATING SET, remain open in D-DI graphs.

In this paper we focus mainly on three classical graph-theoretic optimization problems: MAXIMUM INDEPENDENT SET, MINIMUM DOMINATING SET, and MINIMUM VERTEX COVER. We present an $O(Dn^D)$-time algorithm for

MAXIMUM INDEPENDENT SET and MINIMUM VERTEX COVER in D-DI graphs with n vertices, and give an $O(D^2 n^{O(D^2)})$-time algorithm for MINIMUM DOMINATING SET. Thus, we show that both these problems are polynomial-time solvable in D-DI graphs for fixed D. It is interesting to note that a similar situation occurs in circular-arc graphs, which also generalize interval graphs, where MAXIMUM INDEPENDENT SET and MINIMUM DOMINATING SET can be solved in linear time [15] and MINIMUM COLORING is NP-hard [12]. (However, Aumann et al. [2] show that there is a 2-DI graph that is not a circular arc graph, and that for every $D \geq 1$, there is a circular arc graph that is not a D-DI graph.) We also present a linear-time $(1 + \varepsilon)$-approximation algorithm for MINIMUM VERTEX COVER in D-DI graphs. This algorithm can be generalized to a wide range of deletion problems which include among many the classical MINIMUM FEEDBACK VERTEX SET and MINIMUM PLANAR DELETION problems. We assume that the D-DI representation of the input graph is given to us.

2   Preliminaries

2.1   Definitions and Notation

For $i, j \in \mathbb{Z}$ such that $i < j$, we define $[i, j] := \{i, i + 1, \ldots, j - 1, j\}$.

Given a dotted interval $I = \{s, s + d, s + 2d, \ldots, t\}$, we denote its starting and finishing points by $s(I)$ and $t(I)$, respectively. The jump of I is denoted by $d(I)$, and the *offset* of I is defined as $o(I) := s(I) \bmod d(I)$.

Given a set of dotted intervals $\mathcal{I} = \{I_1, \ldots, I_n\}$, we assume that the intervals are ordered by starting point, namely that $s(I_i) \leq s(I_{i+1})$, for every i. Dotted intervals with the same starting point are ordered arbitrarily. Given a dotted interval I_i, we define $\mathcal{I}_{<i} := \{I_j : j < i\}$. Given a point p, and a set of dotted intervals $\mathcal{S} \subseteq \mathcal{I}$, let $\mathcal{S}_p \subseteq \mathcal{S}$ contain the dotted intervals that start at or before p and end at or to the right of p, namely $\mathcal{S}_p := \{I \in \mathcal{S} : p \in [s(I), t(I)]\}$. (Note that it is possible $I \in \mathcal{S}_p$ and $p \notin I$.)

Let $G = (V, E)$ be an undirected graph; for any subset $A \subseteq V$, we use $G[A] = (A, \{(u, v) \in E : u \in A, v \in A\})$ to denote the graph induced by A. Let $w : V \to \mathbb{R}^+$ be a vertex weight function; for any $A \subseteq V$, we use the shorthand notation $w(A) = \sum_{u \in A} w(u)$. A subset $A \subseteq V$ is said to be *independent* if no two vertices in A are connected by an edge in E; the MAXIMUM INDEPENDENT SET problem is to find an independent set of maximum weight. A subset $A \subseteq V$ is said to be *dominating* if every vertex $v \subseteq V$ has at least one neighbor in A; the MINIMUM DOMINATING SET problem is to find a dominating set of minimum weight. A subset $A \subseteq V$ is said to be a *vertex cover* if every edge in E has at least one endpoint in A; the MINIMUM VERTEX COVER is to find a vertex cover of minimum weight.

2.2   Simple Observations

Let $\mathcal{I}$ be a representation of a D-DI graph G, and denote $\ell(D) = \mathrm{lcm}\{2, \ldots, D\}$, the least common multiple of the numbers $2, \ldots, D$.

Observation 1. *Let $I, J \in \mathcal{I}$ be two dotted intervals, and let $i \in I, J$. If $t(I), t(J) \geq i + \ell(D)$, then $i + \ell(D) \in I, J$.*

Given a dotted interval I and an integer i, let

$$I(i) = \{j : j \in I \text{ and } j < i\} \cup \{j - \ell(i) : j \in I \text{ and } j \geq i + \ell(D)\} \ .$$

Namely $I(i)$ is obtained from I by removing the points in $I \cap [i, i + \ell(D) - 1]$ and gluing the two parts of I back together by moving the points in $I \cap [i + \ell(D), t(I)]$ to the left. Let $\mathcal{I}(i) = \{I(i) : I \in \mathcal{I}\}$.

Observation 2. *Let i be an integer such that $[i, i + 2\ell(D) - 1]$ does not contain any starting or finishing point of a dotted interval from $\mathcal{I}$. Then $\mathcal{I}(i)$ is also a representation of G.*

Given an arbitrary D-DI representation, we could apply the above observation repeatedly to obtain an equivalent representation where all intervals with length more than $2\ell(D)$ contain at least one end-point of some dotted interval.

Observation 3. *Any D-DI graph G has a representation $\mathcal{I}$ such that*

$$\max_{I \in \mathcal{I}} t(I) - \min_{I \in \mathcal{I}} s(I) \leq 4 n \ell(D) \ . \tag{1}$$

Hence we may assume that the endpoints of dotted intervals in $\mathcal{I}$ are in $\{1, \ldots, N\}$, where $N \leq 4n \cdot \ell(D)$. By our assumption that $D = O(1)$ this means that $N = O(n)$. We also note that given a representation $\mathcal{I}$ of a D-DI graph G, a representation of G satisfying (1) can be computed in polynomial time.

3 Maximum Independent Set

In this section we present a dynamic programming algorithm for MAXIMUM INDEPENDENT SET on D-DI graphs that runs in $O(D n^D)$ time, for any D. The algorithm can be thought of as a generalization of the well known algorithm for maximum independent set on interval graphs.

The dynamic programming algorithm for MAXIMUM INDEPENDENT SET in interval graphs is based on the following property. Given an interval I_i and an independent set $S \subseteq \mathcal{I}_{<i}$, let I be the interval with the right-most end point in S. If $S' \subseteq \mathcal{I} \setminus \mathcal{I}_{<i}$ is a maximum weight subset such that $\{I\} \cup S'$ is independent, then S' is a maximum weight subset such that $S \cup S'$ is independent. Namely, S can be represented by a single interval for the purpose of finding the best completion of S from $\mathcal{I} \setminus \mathcal{I}_{<i}$. Furthermore, checking whether $S \cup \{I_i\}$ is an independent set can done by checking if I_i intersects I. Our algorithm is based on an extension of this property for D-DI graphs.

First, we show that finding a maximum weight completion of S from $\mathcal{I} \setminus \mathcal{I}_{<i}$ amounts to finding a maximum weight completion of $S_{s(I_i)}$ from $\mathcal{I} \setminus \mathcal{I}_{<i}$.

Lemma 4. *Let $I_i \in \mathcal{I}$ be a dotted interval, and let $\mathcal{S} \subseteq \mathcal{I}_{<i}$ be an independent set. Also, let $\mathcal{S}' \subseteq \mathcal{I} \setminus \mathcal{I}_{<i}$ be an independent set. If $\mathcal{S}'$ is a maximum weight subset such that $\mathcal{S}_{s(I_i)} \cup \mathcal{S}'$ is independent, then $\mathcal{S}'$ is a maximum weight subset of $\mathcal{I} \setminus \mathcal{I}_{<i}$ such that $\mathcal{S} \cup \mathcal{S}'$ is independent.*

Proof. Consider an interval $J \in \mathcal{I} \setminus \mathcal{I}_{<i}$. Any dotted interval $I \in \mathcal{S}$ intersects J then it must satisfy $s(I_i) \in [s(I), t(I)]$, which means that $I \in \mathcal{I}_{s(I_i)}$. It follows that if $\mathcal{S}_{s(I_i)} \cup \{J\}$ is independent, then $\mathcal{S} \cup \{J\}$ is also independent. □

Suppose we are considering the addition of I_i to an independent set $\mathcal{S} \subseteq \mathcal{I}_{<i}$. Clearly, dotted intervals in $\mathcal{S}$ that terminate before $s(I_i)$ may be ignored. We show that, from the view point of I_i, only up to $d-1$ dotted intervals are needed to represent an independent set $\mathcal{S} \subseteq \mathcal{I}_{<i}$.

Lemma 5. *Let $I_i \in \mathcal{I}$ be a dotted interval, and let $\mathcal{S} \subseteq \mathcal{I}_{<i}$ be an independent set. $\mathcal{S} \cup \{I_i\}$ is independent if and only if (i) $\mathcal{S}_{s(I_i)} \cup \{I_i\}$ is independent, and (ii) $|\mathcal{S}_{s(I_i)}| < D$.*

Proof. Any dotted interval $I \in \mathcal{S}$ intersecting I_i must satisfy $s(I_i) \in [s(I), t(I)]$. In addition, observe that any dotted interval $I \in \mathcal{S}_{s(I_i)}$ must contain at least one point in $[s(I_i) - D + 1, s(I_i))$. Hence, $|\mathcal{S}_{s(I_i)}| < D$. □

Our dynamic programming algorithm is based on Lemmas 4 and 5. The dynamic programming table Π is constructed as follows. A state is a pair of a dotted interval I_i and an independent set $\mathcal{P} \subseteq \mathcal{I}_{s(I_i)}$ of size at most $D - 1$. The entry $\Pi(I_i, \mathcal{P})$ stands for the maximum weight of an independent set $\mathcal{S}' \subseteq \mathcal{I} \setminus \mathcal{I}_{<i}$ such that $\mathcal{S}' \cup \mathcal{P}$ is independent. Observe that the optimum is given by $\Pi(I_1, \emptyset)$. The size of the table is $O(n^D)$.

In the base case, we have

$$\Pi(I_n, \mathcal{P}) = \begin{cases} 0 & \mathcal{P} \cup \{I_n\} \text{ is not independent} \\ w(I_n) & \text{otherwise} . \end{cases}$$

For $i < n$, if $\mathcal{P} \cup \{I_i\}$ is not an independent set we have

$$\Pi(I_i, \mathcal{P}) = \Pi(I_{i+1}, \mathcal{P} \cap \mathcal{I}_{s(I_{i+1})}) .$$

On the other hand, if $\mathcal{P} \cup \{I_i\}$ is an independent set, then there are two options. If there exists an index $k > i$ for which the size of $(\mathcal{P} \cup \{I_i\}) \cap \mathcal{I}_{s(I_k)}$ is less than D, then we have

$$\Pi(I_i, \mathcal{P}) = \min \left\{ \Pi(I_{i+1}, \mathcal{P} \cap \mathcal{I}_{s(I_{i+1})}) , \ w(I_i) + \Pi(I_k, (\mathcal{P} \cup \{I_i\}) \cap \mathcal{I}_{s(I_k)}) \right\} ,$$

where $k > i$ is the smallest index for which the size of $(\mathcal{P} \cup \{I_i\}) \cap \mathcal{I}_{s(I_k)}$ is less than D. If such an index does not exist, then

$$\Pi(I_i, \mathcal{P}) = \min \left\{ \Pi(I_{i+1}, \mathcal{P} \cap \mathcal{I}_{s(I_{i+1})}) , \ w(I_i) \right\} .$$

The correctness of the algorithm is implied by Lemmas 4 and 5. Hence it remains to show that the running time of the algorithm is $O(D\, n^D)$. We do so by proving that the running time of computing an entry of Π is $O(D)$. First, checking whether $\mathcal{P} \cup \{I_i\}$ is an independent set takes $O(D)$ time. Filtering out dotted intervals from $\mathcal{P}$ or $\mathcal{P} \cup \{I_i\}$ that do not belong to $\mathcal{I}_{s(I_{i+1})}$ or to $\mathcal{I}_{s(I_k)}$ also requires $O(D)$ time. Also, finding k, if necessary, can be done in $O(D)$ time.

The computation of $\Pi(I_1, \emptyset)$ can be modified to compute a corresponding independent set using standard techniques. The complement of an independent set is a vertex cover, so the complement of the set returned by our algorithm is minimum weight vertex cover. Hence, we get the following theorem.

Theorem 1. *There is an $O(Dn^D)$-time algorithm for* MAXIMUM INDEPENDENT SET *and* MINIMUM VERTEX COVER *in D-DI graphs with n vertices.*

Notice that our algorithm runs in $O(n)$ time when $D = 1$, so Theorem 1 can be viewed as a strict generalization of the classical linear time algorithm for MAXIMUM INDEPENDENT SET in interval graphs.

4   Dominating Set

Using a similar approach to the one used for MAXIMUM INDEPENDENT SET in D-DI graphs, we can solve the MINIMUM DOMINATING SET problem in D-DI graphs in $O(D^2 n^{O(D^2)})$ time, for any D.

Our algorithm for MINIMUM DOMINATING SET is based on the following idea. Let $\mathcal{S}$ be a dominating set of $\mathcal{I}$ and consider the set $\mathcal{S}_{<i} = \mathcal{S} \cap \mathcal{I}_{<i}$. Clearly, $\mathcal{S}_{<i}$ covers some dotted intervals from $\mathcal{I}_{<i}$, but it may be the case that there are dotted intervals in $\mathcal{I}_{<i}$ that do not intersect $\mathcal{S}_{<i}$. Such dotted intervals must end at or after $s(I_i)$. Furthermore, $\mathcal{S}_{<i}$ may cover dotted intervals in $\mathcal{I} \setminus \mathcal{I}_{<i}$.

Given a dotted interval I_i and a subset $\mathcal{S} \subseteq \mathcal{I}_{<i}$, we say that $\mathcal{S}' \subseteq \mathcal{I} \setminus \mathcal{I}_{<i}$ is a *completion* of $\mathcal{S}$ if $\mathcal{S} \cup \mathcal{S}'$ is a dominating set of $\mathcal{I}$. Notice that it may be the case that such a completion for $\mathcal{S}$ does not exist. Also, given a set $\mathcal{T}$, we say that $I \in \mathcal{T}$ is a *left representative of $\mathcal{T}$ with jump $d(I)$ and offset $o(I)$* if

$$t(I) = \min \{t(I') : I' \in \mathcal{T} \text{ and } d(I') = d(I) \text{ and } j(I') = j(I)\} .$$

Similarly, I is a *right representative of $\mathcal{T}$ with jump $d(I)$ and offset $o(I)$* if

$$t(I) = \max \{t(I') : I' \in \mathcal{T} \text{ and } d(I') = d(I) \text{ and } j(I') = j(I)\} .$$

The set of left and right representatives of $\mathcal{T}$ is denoted by $\mathcal{T}^L$ and $\mathcal{T}^R$, and contain one representative from each jump-offset pair realized by intervals in $\mathcal{T}$.

Lemma 6. *Let $I_i \in \mathcal{I}$ be a dotted interval, let $\mathcal{S} \subseteq \mathcal{I}_{<i}$, and let $\mathcal{T} \subseteq \mathcal{I}_{<i}$ be the subset of dotted intervals that are not covered by $\mathcal{S}$. If $\mathcal{S}' \subseteq \mathcal{I} \setminus \mathcal{I}_{<i}$ is a minimum weight subset such that $\mathcal{S}^R_{s(I_i)} \cup \mathcal{S}'$ dominates $\mathcal{T}^L \cup (\mathcal{I} \setminus \mathcal{I}_{<i})$, then $\mathcal{S}'$ is a minimum weight completion of $\mathcal{S}$. Furthermore, $|\mathcal{S}^R_{s(I_i)}| + |\mathcal{T}^L| \leq \frac{1}{2} D(D+1)$.*

Proof. First notice that if $\mathcal{S}$ covers a dotted interval $I \in \mathcal{I} \setminus \mathcal{I}_{<i}$, then $\mathcal{S}^R$ must also cover I. Also, if $\mathcal{S}'$ covers $\mathcal{T}^L$, then it must cover $\mathcal{T}$.

Finally, observe that $\mathcal{T} \subseteq \mathcal{I}_{s(I_i)}$ since otherwise $\mathcal{S}$ cannot be completed. It follows that it cannot be that $I \in \mathcal{S}^R_{s(I_i)}$ and $J \in \mathcal{T}^L$ represent the same jump and offset, since in this case I covers J. Hence, $\mathcal{S}^R_{s(I_i)} \cup \mathcal{T}^L$ contain at most one representative for each pair of jump and offset, and there are $\frac{1}{2}D(D+1)$ such pairs. □

The dynamic programming table Π is constructed as follows. A state is a triple of a dotted interval I_i, and sets $\mathcal{P}$ and $\mathcal{Q}$ such that

- $\mathcal{P}, \mathcal{Q} \subseteq \mathcal{I}_{s(I_i)}$.
- $\mathcal{P} \cap \mathcal{Q} = \emptyset$.
- $\mathcal{P} \cup \mathcal{Q}$ contain at most one dotted interval for every pair of jump and offset.

The entry $\Pi(I_i, \mathcal{P}, \mathcal{Q})$ stands for the minimum weight subset S' such that $\mathcal{P} \cup S'$ dominates $(\mathcal{I} \setminus \mathcal{I}_{<i}) \cup \mathcal{Q}$. Observe that the optimum is given by $\Pi(I_1, \emptyset, \emptyset)$. The size of the table is $n^{O(D^2)}$.

In the base case, we have

$$
\Pi(I_n, \mathcal{P}, \mathcal{Q}) = \begin{cases} 0 & \mathcal{Q} = \emptyset \text{ and } \mathcal{P} \text{ covers } I_n, \\ w(I_n) & \mathcal{Q} \neq \emptyset \text{ and } I_n \text{ covers } \mathcal{Q}, \\ \infty & \text{otherwise.} \end{cases}
$$

For $i < n$, if $\mathcal{Q} \not\subseteq \mathcal{I}_{s(I_i)}$, then

$$
\Pi(I_i, \mathcal{P}, \mathcal{Q}) = \infty .
$$

otherwise,

$$
\Pi(I_i, \mathcal{P}, \mathcal{Q}) = \min \left\{ \ \Pi(I_{i+1}, \mathcal{P}_{s(I_{i+1})}, \mathcal{Q}) \ , \ \Pi(I_{i+1}, (\mathcal{P} \cup \{I_i\})_{s(I_{i+1})}, \mathcal{Q}') \ \right\} \ ,
$$

where $\mathcal{Q}' \subseteq \mathcal{Q}$ is the subset of dotted intervals that are not covered by I_i.

The correctness of our algorithm is implied by Lemma 6. Computing the value $\Pi(I_i, \mathcal{P}, \mathcal{Q})$ can be done in $O(D^2)$. Hence, the running time of the algorithm is $O(D^2 n^{O(D^2)})$. The computation of $\Pi(I_1, \emptyset)$ can be modified to compute a corresponding independent set using standard techniques.

Theorem 2. *There is an $O(D^2 n^{O(D^2)})$-time algorithm for* MINIMUM DOMINATING SET *on D-DI graphs with n vertices.*

5   Deletion Problems

This section presents an EPTAS for a wide class of deletion problems in D-DI graphs. Three classical examples of such problems are MINIMUM VERTEX COVER, MINIMUM FEEDBACK VERTEX SET, and MINIMUM PLANAR DELETION. For ease of presentation, we first describe our algorithm for MINIMUM VERTEX COVER, and then later explain how it generalizes to other deletion problems. We begin by recalling the definition of a path decomposition [18]:

Definition 1. *A* path decomposition *of a given graph G is a path $\mathcal{P}$ whose vertices $V(\mathcal{P}) \subseteq 2^{V(G)}$ are subsets of vertices in G, called* bags, *satisfying the following two properties:*

- $\bigcup_{P \in V(\mathcal{P})} G[P] = G$, *and*
- *for every $v \in V$, the set of bags $\{P \in V(\mathcal{P}) : v \in P\}$ induces a subpath in $\mathcal{P}$.*

The width *of the path decomposition $\mathcal{P}$ is $\max_{P \in \mathcal{P}} |P| - 1$. The* pathwidth *of G is the minimum width of any path decomposition of G.*

It is well known that an interval graph with maximum clique size k has pathwidth $k - 1$. The next lemma shows that this result generalizes quite nicely to D-DI graphs.

Definition 2. *A clique K in a D-DI graph with dotted interval representation $\mathcal{I}$ is a* point clique *if there exists a point $p \in \mathbb{N}$ which is included in every $I_v \in \mathcal{I}$ with $v \in K$.*

Lemma 7. *A D-DI graph with maximum point clique size k has pathwidth at most $Dk - 1$.*

Proof. Let G be a D-DI graph, and let $\mathcal{I}$ denote a set of dotted intervals corresponding to G. Let K_i denote the set of all vertices whose corresponding dotted interval include the integer $i \in \mathbb{N}$. Define a path decomposition $\mathcal{P} := P_1, \ldots, P_N$ for G by $P_i := \bigcup_{j=i}^{i+D-1} K_j$ for all $i \in \{1, \ldots, N\}$, where N is the maximum integer included in any dotted interval of $\mathcal{I}$. Since G has no clique of size $k + 1$, we have $|K_i| \leq k$ for all $i \in \mathbb{N}$. Thus, $|P_i| \leq Dk$ for all $i \in \{1, \ldots, N\}$. We finish the proof by showing that $\mathcal{P}$ is indeed a path decomposition of G.

First observe that any vertex of G is included in some $K_i \subseteq P_i$. Second, since for any edge $\{u, v\} \in E(G)$ we have $i \in I_u \cap I_v$ for some $i \in \{1, \ldots, N\}$, every edge is also completely contained in some K_i, which in turn is contained in P_i; thus, $\bigcup_i G[P_i] = G$. Finally, observe that for any vertex v, if $v \in P_i \cap P_{i+2}$ for any $i \in \{1, \ldots, N - 2\}$, then it must be the case that $v \in P_{i+1}$; otherwise, the jump of I_v must be at least $D + 1$. Thus, the second condition in Definition 1 is also satisfied, and $\mathcal{P}$ is a path decomposition of width at most $Dk - 1$.   □

Aumann *et al.* [2] show that for any $D \in \mathbb{N}$ there exists a finite bipartite graph G which is not a D-DI graph. An interesting corollary of Lemma 7 is that such a statement is true even for trees, a much more restricted class of bipartite graphs.

Corollary 1. *For any $D \in \mathbb{N}$ there is a finite tree T which is not a D-DI graph.*

Proof. Let D be given. Robertson and Seymour [18] argued that for any integer $w \in \mathbb{N}$ there is a finite tree with pathwidth greater than w. By Lemma 7, choosing such a tree for $w := 2D$ gives a tree T which is not a D-DI graph, since the maximum clique size of T is 2.   □

Another interesting corollary of Lemma 7 more relevant to our purposes is that MINIMUM VERTEX COVER can be solved optimally in D-DI graphs of maximum (point) clique size k in time $2^{O(Dk)} \cdot n$. This follows from the well known $2^{O(w)} \cdot n$ algorithm for MINIMUM VERTEX COVER in graph of pathwidth at most w (see e.g. [6]). Recall that by Observation 3, we can assume that $N = O(n)$ so the path decomposition obtain in the proof of Lemma 7 can be computed in linear time.

Corollary 2. *There is a linear-time algorithm for solving* MINIMUM VERTEX COVER *restricted to D-DI graphs given with representations that have maximum point clique size k.*

Theorem 3. *For any fixed $d \in \mathbb{N}$ and $\varepsilon > 0$, there exists a linear time $(1 + \varepsilon)$-approximation algorithm for unweighted* MINIMUM VERTEX COVER *in D-DI graphs.*

Proof. Let G be a given D-DI graph with representation $\mathcal{I}$, and let $k := 1/\varepsilon$. We first greedily compute a maximal set $\mathcal{K} := \{K_1, \ldots, K_t\}$ of pairwise disjoint point cliques of size $k + 1$ in G. (Note that there can be several point cliques of size $k + 1$ at the same point.) Such a set can be computed in linear time. Let $S_1 := \bigcup \mathcal{K}$, and let $G_1 := G[S_1]$ and $G_2 := G[V \setminus S_1]$. Then G_2 has maximum point clique size k, and so by Corollary 2 we can compute an optimal vertex cover S_2 for G in linear time. Our algorithm outputs the set of vertices $S := S_1 \cup S_2$. Clearly, S is a vertex cover of G. We next argue that S is has size at most $(1 + \varepsilon)|\text{OPT}|$, where OPT is a minimum vertex cover of G.

Let OPT_1 and OPT_2 respectively denote minimum vertex covers of the graphs G_1 and G_2. Then $|\text{OPT}_2| = |S_2|$ and $|\text{OPT}_1| + |\text{OPT}_2| \leq |\text{OPT}|$. Observe that for any clique $K \in \mathcal{K}$, we must have $|\text{OPT}_1 \cap K| \geq k$, otherwise OPT_1 would not be a vertex cover of G_1. Since each such K has size $k + 1$, we have

$$|S_1| \leq (1 + 1/k)|\text{OPT}_1| = (1 + \varepsilon)|\text{OPT}_1|.$$

Thus,

$$|S| = |S_1| + |S_2| \leq (1 + \varepsilon)|\text{OPT}_1| + |\text{OPT}_2| \leq (1 + \varepsilon)|\text{OPT}|.$$

$\square$

We next consider other deletion problems. For a graph class (property) $\mathcal{G}$, the MINIMUM $\mathcal{G}$-DELETION problem takes as input a graph G, and the goal is to compute a minimum size subset of vertices S in G such that $G - S \in \mathcal{G}$. We will be interested in this problem for graph classes $\mathcal{G}$ that have finite forbidden subgraph, topological minor, or minor characterizations. We call such a graph class *finitely defined*. For example, if $\mathcal{G}$ is the class of forests (and MINIMUM $\mathcal{G}$-DELETION is MINIMUM FEEDBACK VERTEX SET) then $\mathcal{G}$ has a finite forbidden minor characterization which consists of the single graph K_3; when $\mathcal{G}$ is the set of all planar graphs then it has forbidden minor characterization consisting of $K_{3,3}$ and K_5.

Let $\mathcal{G}$ be a finitely defined graph class. First, notice that for any positive integer w, the MINIMUM $\mathcal{G}$-DELETION problem can be solved in linear time when restricted to graphs of treewidth w; this is due to an extension of Courcelle's

Theorem [10] due to Borie *et al.* [7]. Second, notice that the clique-deletion technique that is applied in the proof of Theorem 3 can be extended to MINIMUM $\mathcal{G}$-DELETION. Specifically, this is done by setting $k := (h-1)/\varepsilon$, where h is the minimum number of vertices in any graph of the forbidden characterization of $\mathcal{G}$. Clearly any solution S for MINIMUM $\mathcal{G}$-DELETION must include at least $k-h+1$ vertices of any clique of size k in the input graph G, since otherwise $G - S$ will contain a graph from the forbidden characterization of $\mathcal{G}$. Using this observation, the argument used in Theorem 3 follows exactly as is.

Theorem 4. *For any fixed $d \in \mathbb{N}$ and $\varepsilon > 0$, and any finitely defined graph class $\mathcal{G}$, there exists a linear time $(1+\varepsilon)$-approximation algorithm for unweighted* MINIMUM $\mathcal{G}$-DELETION *in D-DI graphs.*

6   Concluding Remarks

This paper presents algorithms for a number of classical optimization problems in D-DI graphs. We show an $O(Dn^D)$-time algorithm for MAXIMUM INDEPENDENT SET and MINIMUM VERTEX COVER in D-DI graphs, and give an $O(D^2 n^{O(D^2)})$-time algorithm for MINIMUM DOMINATING SET. We also present a linear-time $(1 + \varepsilon)$-approximation algorithm for unweighted MINIMUM VERTEX COVER in D-DI graphs, that generalizes to a wide range of deletion problems. We note that for MINIMUM VERTEX COVER and many other problems for this class, our algorithm also works for the general weighted case using the local ratio technique [3] for the clique-deletion process in the proof of Theorem 3. However since the Borie *et al.* [7] extension of Courcelle's Theorem does not work for weighted graphs, Theorem 4 in its generality only applies to uniform weights.

Two main open problems stem from our work. The first is to settle the fix parameter tractability of these problems of the problems considered in this paper, when parameterized by D. In particular, is MINIMUM VERTEX COVER parameterized by D in FPT, or is it W[1]-hard? The second question arises from the fact that our algorithms crucially exploit the D-DI representation of the input graph. Thus, the natural question to ask is whether one can in polynomial-time compute a D-DI representation for a given graph G and a fixed D, or to determine that none exists. This can be done efficiently when $D = 1$ since it reduces to finding an interval representation of a given interval graph. We conjecture that finding a D-DI representation is NP-hard for $D \geq 2$.

References

1. Agarwal, P., van Kreveld, M., Suri, S.: Label placement by maximum independent set in rectangles. Computational Geometry: Theory and Applications 11 (1998)
2. Aumann, Y., Lewenstein, M., Melamud, O., Pinter, R.Y., Yakhini, Z.: Dotted interval graphs and high throughput genotyping. In: 16th Annual ACM-SIAM Symposium on Discrete Algorithms, pp. 339–348 (2005)
3. Bar-Yehuda, R., Even, S.: A local-ratio theorem for approximating the weighted vertex cover problem. Annals of Discrete Mathematics 25, 27–46 (1985)

 4. Bar-Yehuda, R., Halldórsson, M.M., Naor, J., Shachnai, H., Shapira, I.: Scheduling split intervals. SIAM J. Comput. 36(1), 1–15 (2006)
 5. Bar-Yehuda, R., Hermelin, D., Rawitz, D.: Minimum vertex cover in rectangle graphs. Comput. Geom. 44(6-7), 356–364 (2011)
 6. Bodlaender, H.L.: A tourist guide through treewidth. Acta Cybern. 11(1-2), 1–22 (1993)
 7. Borie, R., Parker, R., Tovey, C.: Automatic generation of linear-time algorithms from predicate calculus descriptions of problems on recursively constructed graph families. Algorithmica 7, 555–581 (1992)
 8. Butman, A., Hermelin, D., Lewenstein, M., Rawitz, D.: Optimization problems in multiple-interval graphs. ACM Transactions on Algorithms 6(2) (2010)
 9. Chalermsook, P., Chuzhoy, J.: Maximum independent set of rectangles. In: 20th Annual ACM-SIAM SODA, pp. 892–901 (2009)
10. Courcelle, B.: The monadic second-order logic of graphs. I. Recognizable sets of finite graphs. Information and Computation 85(1), 12–75 (1990)
11. Erlebach, T., Jansen, K., Seidel, E.: Polynomial-time approximation schemes for geometric intersection graphs. SIAM Journal on Computing 34 (2005)
12. Garey, M.R., Johnson, D.S., Miller, G.L., Papadimitriou, C.H.: The complexity of coloring circular arcs and chords. SIAM Journal on Algebraic and Discrete Methods 1(2), 216–227 (1980)
13. Golumbic, M.C., Hammer, P.L.: Stability in circular arc graphs. Journal of Algorithms 9(3), 314–320 (1988)
14. Hermelin, D., Rawitz, D.: Optimization problems in multiple subtree graphs. Discrete Applied Mathematics 159(7), 588–594 (2011)
15. Hsu, W.-L., Tsai, K.-H.: Linear time algorithms on circular-arc graphs. Inf. Process. Lett. 40(3), 123–129 (1991)
16. Hunt III, H.B., Marathe, M., Radhakrishnan, V., Ravi, S., Rosenkrantz, D., Stearns, R.: NC-approximation schemes for NP- and PSPACE-hard problems for geometric graphs. Journal of Algorithms 26 (1998)
17. Jiang, M.: Approximating minimum coloring and maximum independent set in dotted interval graphs. Inf. Process. Lett. 98(1), 29–33 (2006)
18. Robertson, N., Seymour, P.: Graph minors. I. Excluding a forest. J. Comb. Theory, Ser. B 35(1), 39–61 (1983)
19. van Leeuwen, E.J.: Approximation Algorithms for Unit Disk Graphs. In: Kratsch, D. (ed.) WG 2005. LNCS, vol. 3787, pp. 351–361. Springer, Heidelberg (2005)
20. van Leeuwen, E.J.: Better Approximation Schemes for Disk Graphs. In: Arge, L., Freivalds, R. (eds.) SWAT 2006. LNCS, vol. 4059, pp. 316–327. Springer, Heidelberg (2006)
21. Yanovski, V.: Approximation algorithm for coloring of dotted interval graphs. Inf. Process. Lett. 108(1), 41–44 (2008)

The Maximum Clique Problem
in Multiple Interval Graphs
(Extended Abstract)*

Mathew C. Francis, Daniel Gonçalves, and Pascal Ochem

LIRMM, CNRS et Université Montpellier 2,
161 rue Ada 34392 Montpellier Cedex 05, France
{francis,goncalves,ochem}@lirmm.fr

Abstract. Multiple interval graphs are variants of interval graphs where instead of a single interval, each vertex is assigned a set of intervals on the real line. We study the complexity of the MAXIMUM CLIQUE problem in several classes of multiple interval graphs. The MAXIMUM CLIQUE problem, or the problem of finding the size of the maximum clique, is known to be NP-complete for t-interval graphs when $t \geq 3$ and polynomial-time solvable when $t = 1$. The problem is also known to be NP-complete in t-track graphs when $t \geq 4$ and polynomial-time solvable when $t \leq 2$. We show that MAXIMUM CLIQUE is already NP-complete for unit 2-interval graphs and unit 3-track graphs. Further, we show that the problem is APX-complete for 2-interval graphs, 3-track graphs, unit 3-interval graphs and unit 4-track graphs. We also introduce two new classes of graphs called t-circular interval graphs and t-circular track graphs and study the complexity of the MAXIMUM CLIQUE problem in them. On the positive side, we present a polynomial time t-approximation algorithm for WEIGHTED MAXIMUM CLIQUE on t-interval graphs, improving earlier work with approximation ratio $4t$.

1 Introduction

Given a family of sets $\mathcal{F}$, a graph G with vertex set $V(G)$ and edge set $E(G)$ is said to be an "intersection graph of sets from $\mathcal{F}$" if $\exists f : V(G) \to \mathcal{F}$ such that for distinct $u, v \in V(G)$, $uv \in E(G) \Leftrightarrow f(u) \cap f(v) \neq \emptyset$. When $\mathcal{F}$ is the set of all closed intervals on the real line, it defines the well-known class of interval graphs. A t-interval is the union of t intervals on the real line. When $\mathcal{F}$ is the set of all t-intervals, it defines the class of graphs called t-interval graphs. This class was first defined and studied by Trotter and Harary [25]. Given t parallel lines (or tracks), if each element of $\mathcal{F}$ is the union of t intervals on different lines, one defines the class of t-track graphs. It is easy to see that this class forms a subclass of t-interval graphs.

These classes of graphs received a lot of attention, for both their theoretical simplicity and their use in various fields like Scheduling [4,13] or Computational

---

* This work was partially supported by the grant ANR-09-JCJC-0041.

M.C. Golumbic et al. (Eds.): WG 2012, LNCS 7551, pp. 57–68, 2012.

Biology [3,9]. West and Shmoys [27] showed that recognizing t-interval graphs for $t \geq 2$ is NP-complete.

Given a circle, the intersection graphs of arcs of this circle forms the class of *circular arc graphs*. We introduce similar generalizations of circular arc graphs. If G has an intersection representation using t arcs on a circle per vertex, then G is called a t-*circular interval* graph. If instead, G has an intersection representation using t circles and exactly one arc on each circle corresponding to each vertex of G, then G is called a t-*circular track* graph. Note that in this case, the class of t-circular track graphs may not be a subclass of the class of t-circular interval graphs. One can see after cutting the circles, that t-circular interval graphs and t-circular track graphs are respectively contained in $(t + 1)$- and $(2t)$-interval graphs.

For all these intersection families of graphs, one can define a subclass where all the intervals or arcs have the same length. We respectively call those subclasses *unit t-interval, unit t-track, unit t-circular interval*, and *unit t-circular track graphs*.

MAXIMUM WEIGHTED CLIQUE is the problem of deciding, given a graph G with weighted vertices and an integer k, whether G has a clique of weight k. The case where all the weights are 1 is MAXIMUM CLIQUE. Zuckerman [28] showed that unless P=NP, there is no polynomial time algorithm that approximates the maximum clique within a factor $O(n^{1-\epsilon})$, for any $\epsilon > 0$. MAXIMUM CLIQUE has been studied for many intersection graphs families. It has been shown to be polynomial for interval filament graphs [12], a graph class including circle graphs, chordal graphs and co-comparability graphs. It has been shown to be NP-complete for B_1-VPG graphs [21] (intersection of strings with one bend and axis-parallel parts [2]), and for segment graphs [7] (answering a conjecture of Kratochvíl and Nešetřil [20]).

MAXIMUM CLIQUE is polynomial for interval graphs (folklore) and for circular arc graphs [11,14]. However, Butman et al. [6] showed that MAXIMUM CLIQUE is NP-complete for t-interval graphs when $t \geq 3$. For t-track graphs, MAXIMUM CLIQUE is polynomial-time solvable when $t \leq 2$ and NP-complete when $t \geq 4$ [19]. Butman et al. also showed a polynomial-time $\frac{t^2-t+1}{2}$ factor approximation algorithm for MAXIMUM CLIQUE in t-interval graphs. Koenig [19] observed that a similar approximation algorithm with a slightly better approximation ratio $\frac{t^2-t}{2}$ exists for MAXIMUM CLIQUE in t-track graphs. Butman et al. asked the following questions:

- Is MAXIMUM CLIQUE NP-hard in 2-interval graphs?
- Is it APX-hard in t-interval graphs for any constant $t \geq 2$?
- Can an algorithm with a better approximation ratio than $\frac{t^2-t+1}{2}$ be achieved for t-interval graphs?

We answer all of these questions in the affirmative. As far as the third question is concerned, Kammer, Tholey and Voepel [18] have already presented an improved polynomial-time approximation algorithm that achieves an approximation ratio of $4t$ for t-interval graphs. In this paper (Section 3), we present a linear time

$2t$-approximation algorithm, and a polynomial time t-approximation algorithm for MAXIMUM WEIGHTED CLIQUE in t-interval graphs (and thus in t-track graphs), t-circular interval graphs, and t-circular track graphs. Then we show in Section 4 that MAXIMUM CLIQUE is APX-complete for many of these families (including 2-interval graphs). Finally in Section 5, we show that for some of the remaining classes (including unit 2-interval graphs) MAXIMUM CLIQUE is NP-complete.

2   Preliminaries

Consider a circle C of length l with a distinguished point O. The coordinate of a point $p \in C$ is the length of the arc going clockwise from O to p. Given two reals p and q, $[p, q]$ is the arc of C going clockwise from the point with coordinate p to the one with coordinate q. In the following, coordinates are understood modulo l.

A *representation* of a t-interval graph G is a set of t functions, $I_1, \ldots, I_t$, assigning each vertex in $V(G)$ to an interval of the real line. For t-track graphs we have t lines $L_1, \ldots, L_t$, and each I_i assigns intervals from L_i. Similarly, for a representation of t-circular interval graphs (resp. t-circular track graphs) we have a circle C (resp. t circles $C_1, \ldots, C_t$) and t functions I_i, assigning each vertex in $V(G)$ to an arc of C (resp. of C_i).

3   Approximation Algorithms

The first approximation algorithms for the MAXIMUM CLIQUE in t-interval graphs and t-track graphs [6,19] are based on the fact that any t-interval representation (resp. t-track representation) of a clique admits a transversal (i.e. a set of points touching at least one interval of each vertex) of size $\tau = t^2 - t + 1$ (resp. $\tau = t^2 - t$) [17]. Scanning the representation of a graph G from left to right (in time $O(tn)$) one passes through the points of the transversal of a maximum clique K of G. At some of those points there are at least $|K|/\tau$ intervals forming a subclique of K. Thus, this gives an $O(tn)$-time τ-approximation. Butman et al. improved this ratio by 2 by considering every pair of points in the representation. The intervals at these points induce a co-bipartite graph, for which computing the maximum clique is polynomial (as computing a maximum independent set of a bipartite graph is polynomial). Then one can see that this gives a polynomial time $(\tau/2)$-approximation algorithm. This actually gives a polynomial exact algorithm for the MAXIMUM CLIQUE in 2-track graphs [19], as $\tau = 2$ in this case. For the other cases, Kammer et al. [18] greatly improved the approximation ratios from roughly $t^2/2$ to $4t$, using the new notion of k-perfect orientability. Actually, earlier observations by Alon [1] and by Bar-Yehuda et al. [4] (about approximating the chromatic number of t-interval graphs) imply that the trivial algorithm (finding the point of the representation belonging to the maximum number of intervals) is a $2t$-approximation algorithm. Using transversal arguments, we can easily improve this ratio for some subclasses. A representation is *balanced* if for each vertex, all its intervals (or arcs) have the same length.

Remark 1. In any balanced t-interval (resp. t-track, t-circular interval, or t-circular track) representation of a clique, the $2t$ interval extremities of the vertex with the smallest intervals form a transversal. Thus, in those classes of graphs MAXIMUM CLIQUE admits a linear time $2t$-approximation algorithm, and a polynomial time t-approximation algorithm.

We shall now show how to achieve the same approximation ratio without restraining to balanced representations.

Theorem 1. *There is a linear time $2t$-approximation algorithm as well as a polynomial time t-approximation algorithm for MAXIMUM WEIGHTED CLIQUE on t-interval graphs, t-track graphs, t-circular interval graphs, and t-circular track graphs.*

Proof. The problem is polynomial when $t = 1$, we thus assume that $t \geq 2$. Let us prove the theorem for t-interval graphs, the proofs for the other classes are exactly the same. Let G be a weighted t-interval graph with weight function $w(u)$ on its vertices, and let K be a maximum weighted clique of G. Let $I_1, \ldots, I_t$ form a t-interval representation of G such that for any vertex $u \in V(G)$, $I_i(u) = [u_i, u_i']$. For any edge uv there exists a i and a $j \in [t]$ such that the point u_i belongs to $I_j(v)$, or such that $v_j \in I_i(u)$. One can thus orient and color the edges of G in such a way that uv goes from u to v in color i if $u_i \in I_j(v)$ for some j. In K there is a vertex u with more weight on its out-neighbors in K than on its in-neighbors in K. Indeed, this comes from the fact that in the oriented graphs obtained from K by replacing each vertex u by $w(u)$ vertices $\overline{u}_i$ and by putting an arc $\overline{u}_i \overline{v}_j$ if and only if there is an arc uv in K, there is a vertex $\overline{u}_i$ with $d^+(\overline{u}_i) \geq d^-(\overline{u}_i)$, which is equivalent to $w(N_K^+(u)) \geq w(N_K^-(u))$. Thus there exists two distinct values i and j such that u has at least weight $(w(K) - w(u))/2t$ on its out-neighbors in color i, and at least weight $(w(K) - w(u))/t$ on its out-neighbors in color i or j. The vertex u and its out-neighbors in a given color clearly induce a clique of G (they intersect at u_i). Thus scanning the representation from left to right looking for the point with the more weights gives a clique of weight at least $w(u) + (w(K) - w(u))/2t > w(K)/2t$, which is a $2t$-approximation.

   Then the graph induced by u and its out-neighbors in color i or j being co-bipartite one can compute its maximum weighted clique in polynomial time (as computing a maximum weighted independent set of a bipartite graph is polynomial). This clique has weight at least $w(u) + (w(K) - w(u))/t > w(K)/t$ (the weight of the subclique of K induced by u and its neighbors in color i or j). Thus, for each vertex u of the graph and any pair u_i and u_j of interval left end, if we compute the maximum weighted clique of the corresponding co-bipartite graph, we obtain a t-approximation.

4   APX-Hardness in Multiple Interval Graphs

The complement of a graph G is denoted by $\overline{G}$. Given a graph G on n vertices with $V(G) = \{x_1, \ldots, x_n\}$ and $E(G) = \{e_1, \ldots, e_m\}$, and a positive integer w, we define $Subd_w(G)$ to be the graph obtained by subdividing each edge of G w times. If $e_k \in E(G)$ and $e_k = x_i x_j$ where $i < j$, we define $l(k) = i$ and $r(k) = j$ (as if x_i and x_j were respectively the left and the right end of e_k). In the following we subdivide edges 2 or 4 times. In $Subd_2(G)$ (resp. $Subd_4(G)$), the vertices subdividing e_k are a_k and b_k (resp. a_k, b_k, c_k, and d_k) and they are such that $(x_{l(k)}, a_k, b_k, x_{r(k)})$ (resp. $(x_{l(k)}, a_k, b_k, c_k, d_k, x_{r(k)})$) is the subpath of $Subd_2(G)$ (resp. $Subd_4(G)$) corresponding to e_k. To prove APX-hardness results we need the following structural theorem, which is of independent interest.

Theorem 2. *Given any graph G,*

- $\overline{Subd_4(G)}$ *is a 2-interval graph,*
- $\overline{Subd_2(G)}$ *is a unit 3-interval graph,*
- $\overline{Subd_2(G)}$ *is a 3-track graph,*
- $\overline{Subd_2(G)}$ *is a unit 4-track graph,*
- $\overline{Subd_2(G)}$ *is a unit 2-circular interval graph (and thus a 2-circular interval graph),*
- $\overline{Subd_2(G)}$ *is a 2-circular track graph, and*
- $\overline{Subd_2(G)}$ *is a unit 4-circular track graph.*

Furthermore, such representations can be constructed in linear time.

Since MAXIMUM INDEPENDENT SET is APX-hard even when restricted to degree bounded graphs [22,5], Chlebík and Chlebíková [8] observed that MAXIMUM INDEPENDENT SET is APX-hard even when restricted to $2k$-subdivisions of 3-regular graphs for any fixed integer $k \geq 0$. Taking the complement graphs, we thus have that MAXIMUM CLIQUE is APX-hard even when restricted to the set $\mathcal{C}_{2k} = \{\overline{Subd_{2k}(G)} \mid$ any graph $G\}$, for any fixed integer $k \geq 0$. Thus, since MAXIMUM CLIQUE is approximable for all the graph classes considered in Theorem 2, we clearly have the next result.

Theorem 3. *MAXIMUM CLIQUE is APX-complete for:*

- *2-interval graph,*
- *unit 3-interval graph,*
- *3-track graph,*
- *unit 4-track graph,*
- *unit 2-circular interval graph (and thus for 2-circular interval graphs),*
- *2-circular track graph, and*
- *unit 4-circular track graph.*

Note that recently, Jiang [15] gave an alternative proof of the fact that MAXIMUM CLIQUE is APX-complete for 3-track graphs by refining the technique used in [6].

Remark 2. To prove that MAXIMUM CLIQUE is NP-hard on B_1-VPG graphs, Middendorf and Pfeiffer [21] proved that for any graph G, $\overline{Subd_2(G)} \in B_1$-VPG. One can thus see that MAXIMUM CLIQUE is actually APX-hard for this class of graphs.

We only prove the first item of Theorem 2 in this extended abstract.

Proof. Recall that each edge $e_k = x_i x_j$ of G where $i < j$, corresponds to the path $(x_i, a_k, b_k, c_k, d_k, x_j)$ in $Subd_4(G)$. We define the representation $\{I_1, I_2\}$ of $\overline{Subd_4(G)}$ as follows (see also Figure 1). For $1 \leq i \leq n$ and $1 \leq k \leq m$:

$$I_1(a_k) = [0, m(l(k) - 1) + k - 1]$$
$$I_1(x_i) = [mi, mn + mi]$$
$$I_2(a_k) = [mn + ml(k) + 1, 4mn + m - ml(k) - k + 1]$$
$$I_1(b_k) = [m(l(k) - 1) + k, mn + m - k]$$
$$I_1(c_k) = [mn + m - k + 1, 3mn + m - mr(k) - k + 1]$$
$$I_1(d_k) = [3mn + m - mr(k) - k + 2, 4mn + mr(k)]$$
$$I_2(b_k) = [4mn + m - ml(k) - k + 2, 5mn + k]$$
$$I_2(x_i) = [4mn + mi + 1, 5mn + mi + 1]$$
$$I_2(d_k) = [5mn + mr(k) + k + 1, 6mn + m + 1]$$
$$I_2(c_k) = [5mn + k + 1, 5mn + mr(k) + k]$$

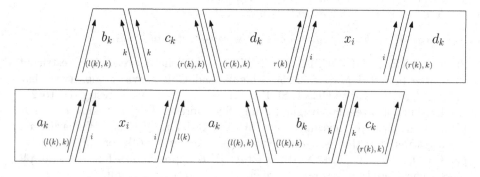

Fig. 1. The 2-interval representation of $\overline{Subd_4(G)}$

Figure 1 (and the other figures of this kind) should be understood in the following way. The leftmost block labeled a_k corresponds to the intervals $I_1(a_k)$, and its shape, together with the label $(l(k), k)$ on the arrow mean that,

- the left end of the intervals $I_1(a_k)$ are the same (coordinate 0), and that
- the right end of the intervals $I_1(a_k)$ are ordered (from left to right) accordingly to $l(k)$, and in case of equality, accordingly to k.

Here we can see that this block is close to the blocks $I_1(b_k)$, and $I_1(x_i)$.

The left end of the interval $I_1(b_k)$ is also ordered (from left to right) accordingly to $(l(k), k)$. Such situation means that $I_1(a_k)$ intersects every $I_1(b_{k'})$ such that $(l(k), k) > (l(k'), k')$, i.e. such that $l(k) > l(k')$ or such that $l(k) = l(k')$ and $k > k'$. Note that since, between $I_2(a_k)$ and $I_2(b_k)$ we have the opposite situation, for any vertex a_k, a_k is adjacent to every $b_{k'}$, except b_k.

The left end of the interval $I_1(x_i)$ is ordered (from left to right) accordingly to i. Such situation means that $I_1(a_k)$ intersects every $I_1(x_i)$ such that $l(k) > i$.

Note that since, between $I_1(x_i)$ and $I_2(a_k)$ we have the opposite situation, for any vertex a_k, a_k is adjacent to every x_i, except $x_{l(k)}$.

We claim that I_1 and I_2 together form a valid 2-interval representation for $\overline{Subd_4(G)}$. We omit the proof in this extended abstract but one can check it with Figure 1. ∎

5 NP-Hardness in Unit 2-Interval and Unit 3-Track Graphs

Valiant [26] has shown that every planar graph of degree at most 4 can be drawn on a grid of linear size such that the vertices are mapped to points of the grid and the edges to piecewise linear curves made up of horizontal and vertical line segments whose endpoints are also points of the grid. It is immediately clear that every planar graph G has a subdivision G' that is an induced subgraph of a grid graph such that each edge of G corresponds to a path of length at most $O(|V(G)|^2)$ (see Figure 2). Note that here, some paths have even length and some have odd length. An *even subdivision* (resp. *odd subdivision*) of G is a graph obtained from G by subdividing each edge e of G an even (resp. odd) number of times, and at most $|V(G)|^{O(1)}$ times.

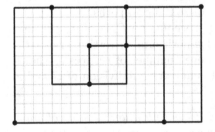 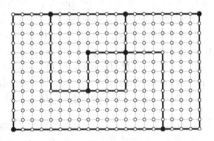

Fig. 2. Embedding a planar graph in a grid

Note that for any integer k, we can embed G in a fine enough grid so that every horizontal and vertical segment in the original drawing of G becomes a path that contains at least k vertices in G'. In Figure 2, we have chosen $k = 5$.

Let $R(w, h)$ be the rectangular grid of height h and width w. A path in $R(w, h)$ that contains only vertices from one row of the grid is called a *horizontal grid-path* and one that contains vertices from only one column is called a *vertical grid-path*. We denote by $R'(w, h)$ the graph obtained by subdividing each edge of $R(w, h)$ once and by adding paths of length 3 between the newly introduced vertices as shown in Figure 3.

Lemma 1. *Any planar graph G, on n vertices and of maximum degree 4, has an even subdivision that is an induced subgraph of $R'(w, h)$ for some values of w and h that are linear in n.*

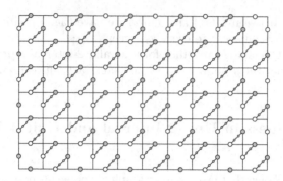

Fig. 3. The graph $R'(11,7)$. The vertices of the grid are not shown.

Proof. Let H be the subdivision of G that is an induced subgraph of the grid $R(w,h)$. Let P_e denote the path in H corresponding to an edge e in G. We assume that P_e is the union of horizontal and vertical grid-paths of length at least 5. We now transform the grid $R(w,h)$ into $R'(w,h)$ by subdividing each edge once and by adding paths of length 3 between the newly introduced vertices as explained before. Clearly, a 1-subdivision of H, which we shall denote by H', is an induced subgraph of $R'(w,h)$. It is also clear that H' is an odd subdivision of G. Let P'_e denote the path in H' corresponding to an edge e of G. Note that P'_e consists of 1-subdivisions of vertical and horizontal grid-paths.

For every edge e of G, we do the following procedure on P'_e in H' to obtain a new graph H'': we replace one of the subdivided horizontal or vertical grid-paths that make up P'_e to obtain P''_e which has an even number of vertices as shown in Figure 4. The new graph H'' so obtained is an even subdivision of G and is also an induced subgraph of $R'(w,h)$.   ∎

Lemma 2. *For any w and h the graph $\overline{R'(w,h)}$ is both a unit 2-interval graph as well as a unit 3-track graph. Thus since those classes are closed under taking induced subgraphs, they also contain the induced subgraphs of $\overline{R'(w,h)}$.*

We omit the proof in this extended abstract.

Theorem 4. *MAXIMUM CLIQUE is NP-complete for unit 2-interval and unit 3-track graphs.*

Proof. It is known that the MAXIMUM INDEPENDENT SET problem is NP-complete even when restricted to planar graphs of degree at most 3 [10]. It is folklore that the instance (G,k) of MAXIMUM INDEPENDENT SET is equivalent to an instance $(H, k+k')$, where H is an even subdivision of G with $|V(G)|+2k'$ vertices. Thus according to Lemma 1, MAXIMUM INDEPENDENT SET is NP-complete on the class of induced subgraphs of $R'(w,h)$. MAXIMUM CLIQUE is thus NP-complete on the class of induced subgraphs of $\overline{R'(w,h)}$. Finally by Lemma 2 this class of graphs is contained in unit 2-interval and unit 3-track graphs. MAXIMUM CLIQUE is thus NP-complete on these classes.   ∎

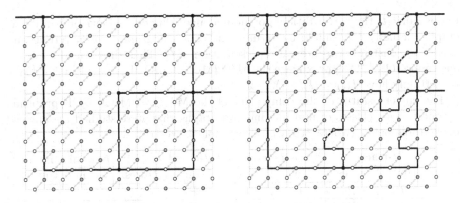

Fig. 4. Modifying the paths in H' to obtain H'': A part of the graph in Figure 2 is shown. The grid vertices are not drawn.

6   Concluding Remarks

The difference between the $4t$-approximation of Kammer et al. [18] and our t-approximation lies in two places. In their paper they proved that t-interval graphs are $2t$-perfectly orientable, but following the lines of Theorem 1 one can see that those graphs are t-perfectly orientable. This improves their approximation for MAXIMUM WEIGHTED INDEPENDENT SET, MINIMUM VERTEX COLORING, and MINIMUM CLIQUE PARTITION in t-interval graphs. For MAXIMUM WEIGHTED INDEPENDENT SET and MINIMUM VERTEX COLORING this reaches the best known ratio of t [4] in a simpler way, and for the other problems it improves the best known approximation ratios [18]. Then Kammer et al. proved that MAXIMUM WEIGHTED CLIQUE can be $2k$-approximated in k-perfectly orientable graphs. Again, following the lines of Theorem 1 one can see that MAXIMUM WEIGHTED CLIQUE can be k-approximated for those graphs. This improves (by 2) their approximation for MAXIMUM WEIGHTED CLIQUE in t-fat objects intersection graphs.

In our approximation algorithm (as in the previous algorithms) we assume that we are given an interval representation. We wonder what we can do if we are not given such representation.

Open Question. Can MAXIMUM (WEIGHTED) CLIQUE be polynomially $c(t)$-approximated in t-interval graphs, for some function c, if we are not given an interval representation?

This would be the case if there is an algorithm that computes, given a t-interval graph G, a $c(t)$-interval representation of G. Actually even when we are given a representation, the approximation ratio might be far from the optimal.

Open Question. Does there exists an approximation algorithm for MAXIMUM (WEIGHTED) CLIQUE in t-interval graphs with a better approximation ratio?

Let us call $f(t)$ the better ratio a polynomial algorithm can achieve on t-interval graphs (actually $f(t)$ should be an infimum). For any graph G on n vertices, it is easy to construct a n-interval representation of G. Thus since for any $\epsilon > 0$, one cannot $O(n^{1-\epsilon})$-approximate the MAXIMUM CLIQUE unless P = NP [28], we certainly have $f(t) = \Omega(t^{1-\epsilon})$.

The current status of the complexity of the MAXIMUM CLIQUE problem for the various classes of multiple interval graphs that were studied are shown in the table below (where "Unres." stands for "Unrestricted").

t	t-track		t-interval		t-circular track		t-circular interval	
	Unit	Unres.	Unit	Unres.	Unit	Unres.	Unit	Unres.
1	P	P	P	P	P	P	P	P
2	P	P	NP-c	APX-c	?	APX-c	APX-c	APX-c
3	NP-c	APX-c	APX-c	APX-c	NP-c	APX-c	APX-c	APX-c
≥ 4	APX-c	APX-c	APX-c	APX-c	APX-c	APX-c	APX-c	APX-c

The entries marked "NP-c" and "?" in this table clearly imply the following questions.

Open Question. Is MAXIMUM CLIQUE for unit 2-interval graphs, unit 3-track graphs or unit 3-circular track graphs APX-hard, or does it admit a PTAS?

Open Question. Is MAXIMUM CLIQUE for unit 2-circular track graphs Polynomial or NP-complete?

Koenig [19] explains that 2-track graphs have a polynomial-time algorithm for MAXIMUM CLIQUE because for any 2-track representation of a clique, there is a transversal of size 2 (i.e. two points such that for every vertex, at least one of its intervals contains one of these points). We note that this is not true for unit 2-circular track graphs as the complete graph on 5 vertices has a unit 2-circular track representation in which each circular track induces a cycle on 5 vertices. This representation clearly does not have a transversal of size 2.

Recently, Jiang and Zhang studied the class of complements of t-interval graphs [16]. In particular they proved that MINIMUM (INDEPENDENT) DOMINATING SET parameterized by the solution size is in W[1] for co-2-interval graphs, and they proved that MINIMUM DOMINATING SET is W[1]-hard for co-3-track graphs.

Following the same line of proof as for Theorem 3 we can prove the following APX-hardness results, for this kind of graph classes.

Theorem 5

(i) *MINIMUM VERTEX COVER is APX-complete in co-2-interval graphs, and the complement classes of all the classes of Theorem 2.*

(ii) *For any graph G, $Subd_3(G)$ is a co-2-interval, a co-unit-3-interval, a co-3-track, a co-unit-4-track, and a co-2-circular track graph, and MINIMUM (INDEPENDENT) DOMINATING SET is APX-hard for these classes of graphs.*

References

1. Alon, N.: Piercing d-intervals. Discrete and Computational Geometry 19, 333–334 (1998)
2. Asinowski, A., Cohen, E., Golumbic, M.C., Limouzy, V., Lipshteyn, M., Stern, M.: Vertex Intersection Graphs of Paths on a Grid. Journal of Graph Algorithms and Applications 16(2), 129–150 (2012)
3. Aumann, Y., Lewenstein, M., Melamud, O., Pinter, R.Y., Yakhini, Z.: Dotted interval graphs and high throughput genotyping. In: Proc. of the 16th Annual Symposium on Discrete Algorithms, SODA 2005, pp. 339–348 (2005)
4. Bar-Yehuda, R., Halldórsson, M.M., Naor, J.S., Shachnai, H., Shapira, I.: Scheduling split intervals. SIAM J. Comput. 36, 1–15 (2006)
5. Berman, P., Fujito, T.: On Approximation Properties of the Independent Set Problem for Degree 3 Graphs. In: Sack, J.-R., Akl, S.G., Dehne, F., Santoro, N. (eds.) WADS 1995. LNCS, vol. 955, pp. 449–460. Springer, Heidelberg (1995)
6. Butman, A., Hermelin, D., Lewenstein, M., Rawitz, D.: Optimization problems in multiple-interval graphs. In: Proceedings of the Eighteenth Annual ACM-SIAM Symposium on Discrete Algorithms, SODA 2007, pp. 268–277 (2007)
7. Cabello, S., Cardinal, J., Langerman, S.: The Clique Problem in Ray Intersection Graph. arXiv (November 2011), http://arxiv.org/pdf/1111.5986.pdf
8. Chlebík, M., Chlebíková, J.: The complexity of combinatorial optimization problems on d-dimensional boxes. SIAM Journal on Discrete Mathematics 21(1), 158–169 (2007)
9. Crochemore, M., Hermelin, D., Landau, G.M., Vialette, S.: Approximating the 2-Interval Pattern Problem. In: Brodal, G.S., Leonardi, S. (eds.) ESA 2005. LNCS, vol. 3669, pp. 426–437. Springer, Heidelberg (2005)
10. Garey, M.R., Johnson, D.S.: Rectilinear steiner tree problem is NP-complete. SIAM J. Appl. Math. 6, 826–834 (1977)
11. Gavril, F.: Algorithms for a maximum clique and a maximum independent set of a circle graph. Networks 3, 261–273 (1973)
12. Gavril, F.: Maximum weight independent sets and cliques in intersection graphs of filaments. Information Processing Letters 73(56), 181–188 (2000)
13. Hochbaum, D.S., Levin, A.: Cyclical scheduling and multi-shift scheduling: Complexity and approximation algorithms. Disc. Optimiz. 3(4), 327–340 (2006)
14. Hsu, W.-L.: Maximum weight clique algorithms for circular-arc graphs and circle graphs. SIAM J. Comput. 14(1), 224–231 (1985)
15. Jiang, M.: Clique in 3-track interval graphs is APX-hard. arXiv (April 2012), http://arxiv.org/pdf/1204.2202v1.pdf
16. Jiang, M., Zhang, Y.: Parameterized Complexity in Multiple-Interval Graphs: Domination, Partition, Separation, Irredundancy. arXiv (October 2011), http://arxiv.org/pdf/1110.0187v1.pdf
17. Kaiser, T.: Transversals of d-Intervals. Discrete Comput. Geom. 18, 195–203 (1997)
18. Kammer, F., Tholey, T., Voepel, H.: Approximation Algorithms for Intersection Graphs. In: Serna, M., Shaltiel, R., Jansen, K., Rolim, J. (eds.) APPROX 2010, LNCS, vol. 6302, pp. 260–273. Springer, Heidelberg (2010)
19. König, F.G.: Sorting with objectives. PhD thesis, Technische Universität Berlin (2009)
20. Kratochvíl, J., Nešetřil, J.: Independent set and clique problems in intersection-defined classes of graphs. Commentationes Mathematicae Universitatis Carolinae 31, 85–93 (1990)

21. Middendorf, M., Pfeiffer, F.: The max clique problem in classes of string-graphs. Discrete Mathematics 108, 365–372 (1992)
22. Papadimitriou, C.H., Yannakakis, M.: Optimization, approximation, and complexity classes. J. Comput. System Sci. 43, 425–440 (1991)
23. Scheinerman, E.R.: The maximum interval number of graphs with given genus. Journal of Graph Theory 11(3), 441–446 (1987)
24. Scheinerman, E.R., West, D.B.: The interval number of a planar graph: Three intervals suffice. Journal of Combinatorial Theory, Series B 35(3), 224–239 (1983)
25. Trotter, W.T., Harary, F.: On double and multiple interval graphs. Journal of Graph Theory 3(3), 205–211 (1979)
26. Valiant, L.G.: Universality considerations in VLSI circuits. IEEE Transactions on Computers 30(2), 135–140 (1981)
27. West, D.B., Shmoys, D.B.: Recognizing graphs with fixed interval number is NP-complete. Discrete Applied Mathematics 8(3), 295–305 (1984)
28. Zuckerman, D.: Linear degree extractors and the inapproximability of max clique and chromatic number. In: Proc. 38th ACM Symp. Theory of Computing, STOC 2006, pp. 681–690 (2006)

Solutions for the Stable Roommates Problem with Payments

Péter Biró[1,*], Matthijs Bomhoff[2], Petr A. Golovach[3,**],
Walter Kern[2], and Daniël Paulusma[3,**]

[1] Institute of Economics, Hungarian Academy of Sciences; H-1112, Budaörsi út 45,
Budapest, Hungary
birop@econ.core.hu
[2] Faculty of Electrical Engineering, Mathematics and Computer Science,
University of Twente, P.O. Box 217, NL-7500 AE Enschede
{m.j.bomhoff,w.kern}@math.utwente.nl
[3] School of Engineering and Computing Sciences, Durham University,
Science Laboratories, South Road, Durham DH1 3LE, UK,
{petr.golovach,daniel.paulusma}@durham.ac.uk

Abstract. The stable roommates problem with payments has as input a graph $G = (V, E)$ with an edge weighting $w : E \to \mathbb{R}_+$ and the problem is to find a stable solution. A solution is a matching M with a vector $p \in \mathbb{R}_+^V$ that satisfies $p_u + p_v = w(uv)$ for all $uv \in M$ and $p_u = 0$ for all u unmatched in M. A solution is stable if it prevents blocking pairs, i.e., pairs of adjacent vertices u and v with $p_u + p_v < w(uv)$. By pinpointing a relationship to the accessibility of the coalition structure core of matching games, we give a simple constructive proof for showing that every yes-instance of the stable roommates problem with payments allows a path of linear length that starts in an arbitrary unstable solution and that ends in a stable solution. This result generalizes a result of Chen, Fujishige and Yang for bipartite instances to general instances. We also show that the problems BLOCKING PAIRS and BLOCKING VALUE, which are to find a solution with a minimum number of blocking pairs or a minimum total blocking value, respectively, are NP-hard. Finally, we prove that the variant of the first problem, in which the number of blocking pairs must be minimized with respect to some fixed matching, is NP-hard, whereas this variant of the second problem is polynomial-time solvable.

1 Introduction

Consider a group of tennis players participating in a doubles tennis tournament. Each two players estimate the expected prize money they could win together by forming a pair in the tournament. Moreover, each player can negotiate his share

* Supported by the Hungarian Academy of Sciences under its Momemtum Programme (LD-004/2010).
** Supported by EPSRC Grant EP/G043434/1.

M.C. Golumbic et al. (Eds.): WG 2012, LNCS 7551, pp. 69–80, 2012.

of the prize money with his chosen partner in order to maximize his own prize money. Can the players be matched together such that no two players have an incentive to leave the matching in order to form a pair together? This example has been given by Eriksson and Karlander [6] to introduce the stable roommates problem with payments. This problem generalizes the stable marriage problem with payments [14] and can be modeled by a weighted graph $G = (V, E)$, i.e., that has an edge weighting $w : E \to \mathbb{R}_+$. A vector $p \in \mathbb{R}^V$ with $p_u \geq 0$ for all $u \in V$ is said to be a *matching payoff* if there exists a matching M in G, such that $p_u + p_v = w(uv)$ for all $uv \in M$, and $p_u = 0$ for each u that is not incident to an edge in M. We then say that p is a payoff *with respect to M*, and we call the pair (M, p) a *matching with payoffs*. A pair of adjacent vertices (u, v) is a *blocking pair* of $p \in \mathbb{R}^V$ if $p_u + p_v < w(uv)$, and their *blocking value* with respect to p is $e_p(u, v)^+ = \max\{0, w(uv) - (p_u + p_v)\}$, which expresses to which extent (u, v) is a blocking pair. We define the *set of blocking pairs* of a vector $p \in \mathbb{R}^V$ as $B(p) = \{(u, v) \mid p_u + p_v < w(uv)\}$, and we define the *total blocking value* of p as $b(p) = \sum_{uv \in E} e_p(u, v)^+$. The problem STABLE ROOMMATES WITH PAYMENTS is to test whether a weighted graph allows a *stable solution*, i.e., a matching with payoffs (M, p) such that $B(p) = \emptyset$, or equivalently, $b(p) = 0$. This problem is well known to be polynomial-time solvable (cf. [6]); recently, an $O(nm + n^2 \log n)$ time algorithm for weighted graphs on n vertices and m edges has been given [3].

We consider two natural questions in our paper:

1. Can we gradually transform an unstable solution into a stable solution assuming that a stable solution exists?
2. Can we find solutions for no-instances that are "as stable as possible"?

Question 1 is of importance, as it will give us insight into the coalition formation process. A sequence of solutions starting from an unstable one and ending in a stable one is called a *path to stability*; we give a precise definition later. Question 2 is relevant when we consider no-instances of STABLE ROOMMATES WITH PAYMENTS. In order to answer it, we generalize this problem in two different ways leading to the following two decision problems. Given a weighted graph G and an integer $k \geq 0$, the BLOCKING PAIRS problem is to test whether G allows a matching payoff p with $|B(p)| \leq k$, and the BLOCKING VALUE problem is to test whether G allows a matching payoff p with $b(p) \leq k$.

Questions 1 and 2 have been studied in two closely related settings that are well known and formed a motivation for our study. The first related setting is similar to ours except that payments are not allowed. Instead, each vertex u in an (unweighted) graph $G(V, E)$ has a linear order on its neighbors expressing a certain preference. Then two adjacent vertices u and v form a *blocking pair* regarding a matching M if either u is not matched in M or else u prefers v to its partner in M, and simultaneously, the same holds for v. This leads to the widely studied problem STABLE ROOMMATES introduced by Gale and Shapley [7]. In this setting, the results are as follows. Answering a question by Knuth [12], Roth and Vande Vate [13] showed the existence of a path to stability for any yes-instance provided that the instance is bipartite. Later, their result was generalized by Diamantoudi et al. [5] to be valid for general instances. Abraham, Biró and Manlove [1] showed that the problem of finding a matching with a minimum number of blocking pairs is NP-complete; note that the problem BLOCKING VALUE cannot be translated to this setting, due to the absence of cardinal utilities.

The second related setting originates from cooperative game theory. A *cooperative game with transferable utilities* (TU-game) is a pair (N, v), where N is a set of n *players* and a *value function* $v : 2^N \to \mathbb{R}$ with $v(\emptyset) = 0$ defined for every *coalition* S, which is a subset of N. In a *matching game* (N, v), the set N of players is the vertex set of weighted graph G, and the value of a coalition S is $v(S) = \sum_{e \in M} w(e)$, where M is a maximum weight matching in the subgraph of G induced by S. The strong relationship between the two settings stems from the fact that finding a *core allocation*, i.e., a vector $x \in \mathbb{R}^N$ with $\sum_{u \in N} x_u = v(N)$ and $\sum_{u \in S} x_u \geq v(S)$ for all $S \subseteq N$ is equivalent to solving the STABLE ROOMMATES WITH PAYMENTS (cf. [6]). The algorithms of Béal et al. [2] and Yang [15] applied to an n-player matching game with a nonempty core find a path to stability with length at most $(n^2 + 4n)/4$ and $2n - 1$, respectively. For matching games, the problems BLOCKING PAIRS and BLOCKING VALUE are formulated as the problems that are to test whether a matching game (N, E) allows an allocation x with $|B(x)| \leq k$, or $b(x) \leq k$, respectively, for some given integer k. Biró, Kern and Paulusma [3] showed that the first problem is NP-complete and that the second is polynomial-time solvable by formulating it as a linear program.

Our Results. In Section 2, we answer Question 1 by showing that any unstable solution for a weighted n-vertex graph G that is a yes-instance of STABLE ROOMMATES WITH PAYMENTS allows a path to stability of length at most $2n$. This generalizes a result of Chen, Fujishige and Yang [4], who show the existence of a path to stability for the aforementioned stable marriage problem with payments, which corresponds to the case when G is bipartite. In Section 3 we answer Question 2 by proving that BLOCKING PAIRS and BLOCKING VALUE are NP-complete. The latter result is somewhat surprising, as the corresponding problem is polynomial-time solvable for matching games; we refer to Table 1 for a survey. In addition, we show that BLOCKING VALUE does become polynomial-time solvable if the desired matching payoff is to be with respect to some specified matching M that is part of the input, whereas this variant of BLOCKING PAIRS turns out to be NP-complete.

Table 1. A comparison of the results for the existence of a path to stability and the problems BLOCKING PAIRS and BLOCKING VALUE in the three different settings of stable roommates (SR), stable roommates with payments (SRwP) and matching games (MG). The three results marked by a * are the new results shown in this paper.

	SR	SRwP	MG
Path to Stability	Yes	Yes*	Yes
BLOCKING PAIRS	NP-complete	NP-complete*	NP-complete
BLOCKING VALUE	n/a	NP-complete*	P

2   Paths to Stability

We first give a useful lemma, which immediately follows from the aforementioned fact that finding a core allocation in a matching game (N, v) defined on a weighted graph $G = (N, E)$ is equivalent to finding a stable solution for G.

Lemma 1 ([6]). *Let G be a weighted graph that forms a yes-instance of* STABLE ROOMMATES WITH PAYMENTS. *Then G allows a stable solution (M^*, p^*) where M^* is a maximum weight matching of G.*

Let $G = (V, E)$ be a graph and M be a matching. If $uv \in M$, then we say that u and v are *partners* in M, denoted $M(u) = v$ and $M(v) = u$. If u is unmatched in M, then we let $M(u) = u$. Let uv be a blocking pair for some payoff p with respect to some matching M; note that $uv \notin M$ by definition. Let p' be a payoff with respect to a matching M'. We say that (M', p') is obtained from (M, p) by *satisfying* blocking pair (u, v) if the following four conditions hold:

(i) $uv \in M'$;

(ii) $p_u \le p'_u$ and $p_v \le p'_v$;

(iii) if $M(u) \ne u$ then $M(u)$ is unmatched in M' (hence $p'_{M(u)} = 0$), and
      if $M(v) \ne v$ then $M(v)$ is unmatched in M' (hence $p'_{M(v)} = 0$);

(iv) $M'(z) = M(z)$ and $p'(z) = p(z)$ for every $z \in V \setminus \{u, v, M(u), M(v)\}$.

That is, the players of a blocking pair become matched to each other by leaving their former partners unmatched (if there were any) and they share the extra utility coming from their cooperation in such a way that neither of them gets worse off. Note that at least one of them strictly improves, i.e., we have $p'_u > p_u$ or $p'_v > p_v$. This is due to the following two arguments. First, by the definition of a blocking pair, $p_u + p_v < w(uv)$. Second, $p'_u + p'_v = w(uv)$, because p' is a payoff with respect to M' and $uv \in M'$ by condition (i).

Let G be a weighted graph that forms a yes-instance of STABLE ROOMMATES WITH PAYMENTS. A *path to stability* for G is a sequence of matchings with payoffs

$$(M^0, p^0), (M^1, p^1), \ldots, (M^k, p^k),$$

where $(M^0, p^0), \ldots, (M^{k-1}, p^{k-1})$ are unstable solutions and (M^k, p^k) is a stable solution, such that (M^{i+1}, p^{i+1}) is obtained from (M^i, p^i) for $i = 0, \ldots, k-1$ by satisfying some blocking pair.

A known proof technique for finding a path to stability is to make use of a so-called reference solution (see e.g. [5,2,11,15]). In our setting, this comes down to the following. We say that (M', p') is obtained from (M, p) by *satisfying* blocking pair (u, v) *with respect to* a payoff p^* of some stable solution (M^*, p^*) that is called a *reference solution*, if in addition to conditions (i)–(iv) also the following condition is satisfied:

(v)   if $p_u \le p^*_u$ then $p'_u \le p^*_u$, and if $p_v \le p^*_v$ then $p'_v \le p^*_v$.

We define $S^*(p) = \{u \in V(G) : p_u > p_u^*\}$ to be the *set of overpaid vertices* in (M, p) with respect to (M^*, p^*). We note that when (M', p') is obtained from (M, p) by satisfying a blocking pair with respect to p^* then $S^*(p') \subseteq S^*(p)$. In order to prove the existence of a path to stability for some graph G that is a yes-instance of STABLE ROOMMATES WITH PAYMENTS, it may be easier to find a path to stability $(M^0, p^0), (M^1, p^1), \ldots, (M^k, p^k)$, where (M^{i+1}, p^{i+1}) is obtained from (M^i, p^i) for $i = 1, \ldots, k$ by satisfying some blocking pair with respect to p^*, in such a way that $S^*(p^{i+1}) \subseteq S^*(p^i)$ for $i = 0, \ldots, k-1$, with strict inclusion occurring after a certain number of steps; the latter property is then to guarantee that an algorithm for solving this problem will eventually terminate in a stable solution.

We will use the approach described above in order to show that any weighted n-vertex graph G that forms a yes-instance of STABLE ROOMMATES WITH PAYMENTS allows a path to stability of length $2n$ that starts in an arbitrary unstable solution. Before we give the proof, we first explain in more detail how our result is connected to results from the literature. Our result is based on the work of Kóczy and Lauwers [11] on the so-called accessibility of the coalition structure core. Their result implies the existence of a path to stability for any TU-game with a nonempty core. In this setting, a path to stability is a sequence of gradual changes that transform a non-core allocation to a core allocation. Recently, Béal et al. [2] and Yang [15] built on the work of Kóczy and Lauwers [11] in order to show the accessibility of the coalition structure core in quadratic time. In particular, Yang [15] obtained a linear upper bound on the length of a path to stability for all TU-games with a nonempty core, which include the matching games with a nonempty core. We can use their proof techniques [2,15] for our setting. Our arguments are slightly different though, because for matching games (N, v) every coalition $S \subseteq N$ may be blocking instead of only pairs $\{u, v\}$ as in our setting. As a consequence, for matching games several blocking pairs may be satisfied in one step by choosing the affected vertices to form a blocking coalition. Moreover, even if the starting solution is a matching with payoffs and the final solution is a stable matching with payoffs, the intermediate solutions in a path to stability for matching games are not necessarily such allocations that can be realized as matchings with payoffs. Therefore, the arguments of Yang [15] for restricting the path length cannot be translated to obtain our linear upper bound. By pinpointing the connection to the setting of cooperative games, we are not only able to generalize the corresponding result of Chen, Fujishige and Yang [4] for the existence of a path to stability for bipartite instances (which are always yes-instances) to general yes-instances, but we could also give a simpler proof of this result with a linear upper bound on the number of blocking pairs that need to be satisfied.

Theorem 1. *Let G be a weighted n-vertex graph that forms a yes-instance of* STABLE ROOMMATES WITH PAYMENTS; *Let (M^0, p^0) be a matching with payoffs. Then there exists a path to stability of length at most $2n$ that starts in (M^0, p^0).*

Proof. Let G be a weighed n-vertex graph that forms a yes-instance of STABLE ROOMMATES WITH PAYMENTS; we also call such a graph G *stable.* Let (M^0, p^0)

be a matching with payoffs. We fix a stable reference solution (M^*, p^*), where we may assume that M^* is a maximum weight matching due to Lemma 1. Note that $|M^*| \leq \frac{n}{2}$ and $|M^0| \leq \frac{n}{2}$. Moreover, $|S^*(p^0)| \leq \frac{n}{2}$, because the vertices u and v of a pair $uv \in M^0$ cannot both belong to $S^*(p^0)$, as otherwise $p_u^0 > p_u^*$, $p_v^0 > p_v^*$ and $w(uv) = p_u^0 + p_v^0$ would imply that uv is blocking for (M^*, p^*).

Input: a matching with payoffs (M^0, p^0) in a weighted stable graph G
Output: a stable solution

Set $i := 0$.

Phase 1: **while** there is a blocking pair uv for (M^i, p^i) such that $uv \in M^*$ **do**
satisfy uv with respect to p^*, $(M^{i+1}, p^{i+1}) \leftarrow (M^i, p^i)$; set $i := i + 1$.

Phase 2: **if** there is a blocking pair uv for (M^i, p^i) **then**
satisfy uv with respect to p^*, $(M^{i+1}, p^{i+1}) \leftarrow (M^i, p^i)$; set $i := i + 1$, and
return to Phase 1.

Return (M^i, p^i).

Fig. 1. The algorithm for finding a path to stability. Contrary to the algorithms of Béal et al. [2] and Yang [15], we do not have to specify the payoff p^{i+1}; any vector p^{i+1} that is a payoff with respect to M^{i+1} and satisfies conditions (ii)-(v) may be chosen.

To obtain a path of stability we run the algorithm displayed in Figure 1. Recall that $S^*(p^{i+1}) \subseteq S^*(p^i)$ for any solution (M^i, p^i) for which the algorithm performs Phase 1 or 2. Now we will prove that whenever we satisfy a blocking pair $uv \notin M^*$ in Phase 2 the above relation is strict. More precisely, let (M^i, p^i) be a solution after a termination of Phase 1, and let (M^{i+1}, p^{i+1}) be the solution obtained after satisfying a blocking pair $u_i v_i \notin M^*$ for (M^i, p^i). Then we will show that $S^*(p^{i+1}) \subset S^*(p^i)$. We first show three claims, where we write $w(M) = \sum_{uv \in M} w(uv)$ for a matching M.

Claim 1. $p_u^* + p_v^* = p_u^i + p_v^i$ for all $uv \in M^*$ and M^i has maximum weight.

We prove Claim 1 as follows. Because no $uv \in M^*$ is blocking for (M^i, p^i) we have $p_u^* + p_v^* = w(uv) \leq p_u^i + p_v^i$ for all $uv \in M^*$. This implies that

$$w(M^*) = \sum_{uv \in M^*} p_u^* + p_v^* \leq \sum_{uv \in M^*} p_u^i + p_v^i \leq w(M^i)$$

However, because M^* is a maximum weight matching, we have equality everywhere, i.e., we have $p_u^* + p_v^* = p_u^i + p_v^i$ for all $uv \in M^*$, and $w(M^i) = w(M^*)$. The latter equality implies that M^i is a maximum weight matching as well.

Claim 2. $p_u^i + p_v^i = p_u^* + p_v^*$ for all $uv \in M^i$.

We prove Claim 2 as follows. The stability of (M^*, p^*) implies that $p_u^i + p_v^i = w(uv) \leq p_u^* + p_v^*$ for all $uv \in M^i$. This leads to

$$w(M^i) = \sum_{uv \in M^i} p_u^i + p_v^i \leq \sum_{uv \in M^i} p_u^* + p_v^* \leq w(M^*).$$

Together with the maximality of M^i that follows from Claim 1, this means that we have equality everywhere again, so $p_u^i + p_v^i = p_u^* + p_v^*$ for all $uv \in M^i$.

Claim 3. If w is unmatched in M^i or M^*, then $p_w^i = p_w^* = 0$.

We prove Claim 3 as follows. Suppose that w is unmatched in M^i. Then $p_w^i = 0$ by definition. We use Claim 2 and the fact that M^* and M^i are maximum weight matchings to obtain $w(M^*) = w(M^i) = \sum_{uv \in M^i} (p_u^i + p_v^i) = \sum_{uv \in M^i} (p_u^* + p_v^*)$. By definition, $w(M^*) = \sum_{u \in V} p_u^*$. Due to these two equalities, $p_w^* = 0$. The case when w is unmatched in M^* can be proven by similar arguments. This completes the proof of Claim 3.

We now consider the pair (u_i, v_i) and write $u = u_i$ and $v = v_i$. Because $uv \notin M^*$ is blocking for (M^i, p^i), and (M^*, p^*) is a stable solution, we deduce that $p_u^i + p_v^i < w(uv) \leq p_u^* + p_v^*$; note that this means that $w(uv) > 0$. If u and v are both unmatched in M^i, then $p_u^* = p_v^* = 0$ by Claim 3. Then $w(uv) \leq 0$, which is not possible. Hence, we are left to analyze two cases.

First suppose that one of u, v, say u, is unmatched in M^i, whereas v is matched by M^i, say $vy \in M^i$. Because u is unmatched, $p_u^i = p_u^* = 0$ by Claim 3. Because we already deduced that $p_u^i + p_v^i < p_u^* + p_v^*$, this means that $p_v^i < p_v^*$. The inequality $p_v^i < p_v^*$ and the equality $p_v^i + p_y^i = p_v^* + p_y^*$ from Claim 2 imply that $p_y^i > p_y^*$, i.e., $y \in S^*(p^i)$. Because y becomes unmatched after satisfying uv by definition, we find that $p^{i+1}(y) = 0$. Hence, $S^*(p^{i+1}) \subset S^*(p^i)$.

Now suppose that both u and v are matched in M^i. Let $xu \in M^i$ and $vy \in M^i$. The equalities $p_x^i + p_u^i = p_x^* + p_u^*$ and $p_v^i + p_y^i = p_v^* + p_y^*$ from Claim 2, together with the aforementioned inequality $p_u^i + p_v^i < p_u^* + p_v^*$, imply that $p_x^i + p_y^i > p_x^* + p_y^*$. Hence, $p_x^i > p_x^*$ or $p_y^i > p_y^*$. This means that x or y is in $S^*(p^i)$. We may assume without loss of generality that $x \in S^*(p^i)$. Because x becomes unmatched after satisfying uv by definition, we find that $p^{i+1}(x) = 0$. Hence, $S^*(p^{i+1}) \subset S^*(p^i)$ also in this case.

Because the number of overpaid vertices decreases after each execution of Phase 2, the algorithm terminates and the returned solution (M^ℓ, p^ℓ) is stable. Consequently, we have shown the existence of a path to stability.

Now we set the linear upper bound for the number of steps ℓ required to reach a stable solution. Each time we satisfy a blocking pair not in M^* in Phase 2, the number of overpaid vertices decreases. Hence, we cannot satisfy more than $|S^*(p^0)| \leq \frac{n}{2}$ of them. Regarding the pairs of M^*, after the first time we satisfy a pair $uv \in M^*$ we may need to satisfy it again only if u or v is involved in a blocking pair xu or uy, respectively, that is not in M^* and that is satisfied in Phase 2. Hence, the satisfaction of a pair xu not in M^* may result that at most two pairs in M^*, involving either x or u, can be subsequently satisfied in Phase 1, but all the other pairs of M^* satisfied in this execution of Phase 1 must be satisfied for the first time. Therefore we have the following upper bounds:

- We satisfy at most $\frac{n}{2}$ pairs not in M^*.
- We satisfy at most $\frac{n}{2}$ pairs of M^* for the first time.
- We satisfy pairs of M^* not for the first time at most $2 \cdot \frac{n}{2} = n$ times.

Thus we satisfy at most $\ell = \frac{n}{2} + \frac{n}{2} + n = 2n$ pairs. This completes our proof. □

Remark. Our proof of Theorem 1 is constructive. The algorithm of Figure 1 constructs a path to stability starting in any unstable solution. Due to the linear upper bound stated in Theorem 1, its running time is $O(n^2)$ time for weighted graphs on n vertices, given a stable reference solution (M^*, p^*) which, if necessary, we can compute in $O(nm + n^2 \log n)$ [3].

3   Blocking Pairs and Blocking Value

We start this section by showing that BLOCKING PAIRS and BLOCKING VALUE are NP-complete. We prove the hardness of these two problems by a reduction from INDEPENDENT SET, in a similar way as was done for the BLOCKING PAIRS problem in the setting of matching games [3]. However, the latter setting and our setting are quite different in nature; in particular, we recall that the BLOCKING VALUE problem is polynomial-time solvable in the setting of matching games [3]. Hence, our hardness proof uses a number of different arguments than the hardness proof for BLOCKING PAIRS in the setting of matching games [3].

Theorem 2. BLOCKING PAIRS *and* BLOCKING VALUE *are* NP-*complete.*

Proof. Clearly, both problems are in NP. In order to prove NP-completeness, we reduce from the INDEPENDENT SET problem. This problem takes as input a graph G with an integer k and is to test whether G contains an *independent set* of size at least k, i.e., a set S with $|S| \geq k$ such that there is no edge in G between any two vertices of S. Garey, Johnson and Stockmeyer [9] show that INDEPENDENT SET is already NP-complete for the class of *3-regular* graphs, i.e., graphs in which all vertices are of degree three. So we may assume that G is 3-regular. We also assume that $k \geq 2$. Let $n = |V|$ and let $V = \{v_1, \dots, v_n\}$.

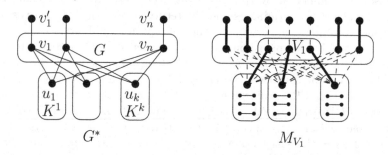

Fig. 2. The graph G^* and an example of a matching M_{V_1}. The edges within the subgraph G of G^* have not been drawn.

From G we construct a weighted graph $G^* = (V^*, E^*)$ on $2n + k(4k + 3)$ vertices. First, we add a set V' of n new vertices $v'_1, \dots, v'_n$, where we add an edge between v_i and v'_i for $i = 1, \dots, n$. So, every v'_i has a unique neighbor in the resulting graph, namely v_i. Now let K be a complete graph on $r = 4k + 3$ vertices; note that r is odd. We add k mutually vertex-disjoint copies $K^1, \ldots, K^k$ of K to the graph constructed so far. In each copy K^i we specify a vertex u_i leading to a set $U = \{u_1, \ldots, u_k\}$. We then finish our construction of G^* by adding an edge $u_h v_i$ for all $1 \le h \le k$ and all $1 \le i \le n$; see Figure 2. It remains to define an edge weighting w on G^*. We let $w(u_h v_i) = \frac{1}{2}$ for all $1 \le h \le k$ and all $1 \le i \le n$, whereas we assign all other edges e of G^* weight $w(e) = 1$.

We make the following observation that is important for the remainder of the proof. By our construction, there exist a matching M_{V_1} for each subset $V_1 \subseteq V$ of size k that can be decomposed as $M_{V_1} = M_1 \cup \cdots \cup M_k \cup M_{U V_1} \cup M_{V_2 V_2'}$, where M_h is a perfect matching of $K^h - u_h$ for $h = 1, \ldots, k$, $M_{U V_1}$ is a perfect matching of $G^*[U \cup V_1]$ and $M_{V_2 V_2'}$ is a perfect matching of $G^*[V_2 \cup V_2']$ for $V_2 = V \setminus V_1$ and its set of neighbors V_2' in V'. We call a matching M_{V_1} as defined above a V_1-*matching*. Note that V_1 has more than one V_1-matching, because we can pick different perfect matchings for the decomposition of M_{V_1} (except for the perfect matching $M_{V_2 V_2'}$ of $G[V_2, V_2']$, which is unique).

For our two NP-hardness reductions, it suffices to show that the following three statements are equivalent.

(i) G has an independent set S of size at most k.
(ii) $|B(p)| \le k$ for some matching payoff p of G^*.
(iii) $b(p) \le k$ for some matching payoff p of G^*.

"(i) $\Rightarrow$ (ii)" Suppose that G has an independent set S of size $|S| \ge k$. Then we may assume without loss of generality that $|S| = k$, as otherwise we could just remove some vertices from S. We pick an arbitrary S-matching M_S and define a payoff p with respect to M_S as follows. We let $p \equiv \frac{1}{2}$ on $K^1 \cup \cdots \cup K^k$, whereas we let $p \equiv 1$ on $V \setminus S$ and $p \equiv 0$ on $S \cup V'$. Because S is an independent set and $p \equiv 1$ on $V \setminus S$, no pair (v_i, v_j) is a blocking pair. This and the definition of p ensures that $B(p) = \{(v_i, v_i') \mid v_i \in S\}$, which has size k.

"(ii) $\Rightarrow$ (iii)" Suppose that $|B(p)| \le k$ for some matching payoff p of G^*. Then $b(p) \le k$, because each blocking pair in $B(p)$ can contribute at most a value of 1 to the total blocking value $b(p)$ as the maximum value of w is 1.

"(iii) $\Rightarrow$ (i)" Suppose that $b(p) \le k$ for some matching payoff p of G^*. Assume that $b(p)$ is minimum over all matching payoffs. Let M be the associated matching. We first show three useful claims.

Claim 1. For all $1 \le h \le k$, every $z \in V_{K^h} \setminus \{u_h\}$ is matched by M.

We prove Claim 1 as follows. First suppose that there exists some complete graph K^h that contains a nonempty subset D of vertices that are not equal to u_h and that are unmatched in M. Let $A = V_{K^h} \setminus \{u_h \cup D\}$. We write $\alpha = |A|$ and $\delta = |D|$. By our construction, the vertices in A can only be matched by M via matching edges in $K^h[A]$. By definition, $p_z + p_{z'} = 1$ for all $zz' \in M$ with $z, z' \in A$. This means that $\sum_{z \in A} p_z = \frac{1}{2}\alpha$. Moreover, $p \equiv 0$ on D by definition, and $\delta \ge 1$ by our assumption. We let E_1 be the set of edges with one end-vertex in A and the other one in D. We let E_2 be the set of edges with both end-vertices in D. By using the properties of A and D, we find that

$$
\begin{aligned}
k \geq b(p) &\geq \sum_{zz' \in E_1} (1 - p_z - p_{z'}) + \sum_{zz' \in E_2} (1 - p_z - p_{z'}) \\
&= \delta \sum_{z \in A} (1 - p_z) + \sum_{zz' \in E_2} 1 \\
&= \alpha\delta - \tfrac{1}{2}\alpha\delta + \frac{1}{2}\delta(\delta - 1) \\
&= \tfrac{1}{2}\alpha\delta + \frac{1}{2}\delta(\delta - 1).
\end{aligned}
$$

Recall that $\delta \geq 1$. We distinguish three cases. If $\delta = 1$, then $\alpha = r - \delta - 1 = r - 2$. Then our deduction implies that $k \geq \tfrac{1}{2}\alpha = \tfrac{1}{2}(r - 2)$, which is equivalent to $r \leq 2k + 2$. If $\delta = 2$, then $\alpha = r - 3$, and we find that $k \geq \alpha + 1 = r - 2$, which is equivalent to $r \leq k + 2$. If $\delta \geq 3$, then we find that $k \geq \tfrac{3}{2}\alpha + \delta \geq \alpha + \delta = r - 1$, which is equivalent to $r \leq k + 1$. Hence, in all three cases, we find that $r \leq 2k + 2$. This is not possible, because $r = 4k + 3 > 2k + 2$. We conclude that $D = \emptyset$. Hence, we have proven Claim 1.

Claim 2. There exists a subset $V_1 \subseteq V$ such that the restriction of M to the edges of $G^*[V_1 \cup U]$ is a perfect matching.

We prove Claim 2 as follows. First suppose that there exists some u_h that is unmatched in M. Then $p_{u_h} = 0$ by definition. Let $A = V_{K^h} \setminus \{u_h\}$. Note that $|A| = r - 1$ is even, because r is odd. Claim 1 tells us that the vertices of A are matched by edges of M. By construction, these matching edges must have both end-vertices in A. Because $p_z + p_{z'} = 1$ for all $zz' \in M$ and $p \geq 0$, this means that there are at least $\tfrac{1}{2}(r - 1)$ vertices in A, whose payoff is at most $\tfrac{1}{2}$. We consider the edges between v and those vertices and deduce that $k \geq b(p) \geq \tfrac{1}{2}(r - 1)(1 - \tfrac{1}{2} - 0)$, which is equivalent to $r \leq 4k + 1$. This is not possible, because $r = 4k + 3$. Hence, every u_h is matched by M.

  Now suppose that u_h forms a matching edge of M together with some other vertex z of K^h. Then M cannot cover all vertices of K^h, because r is odd. This is not possible due to Claim 1. Hence, every u_h forms a matching edge of M with some vertex v_i from V. This gives us the set V_1, and we have proven Claim 2.

Claim 3. $p \equiv \tfrac{1}{2}$ on U.

We prove Claim 3 as follows. Suppose that $p_{u_h} < \tfrac{1}{2}$ for some $1 \leq h \leq k$. By Claim 2, u_h forms a matching edge of M with some vertex v_i. Then $p_{u_h} + p_{v_i} = w(u_h v_i) = \tfrac{1}{2}$. Then $p_{v_i} = \epsilon > 0$. We modify p into a new payoff p' with respect to M by increasing the payoff to u_h with ϵ and decreasing the payoff of v_i to zero. Because G is 3-regular, v_i has 3 neighbors in G. As in the proof of Claim 2, there are at least $\tfrac{1}{2}(r - 1)$ vertices in $K^h - u_h$, whose payoff is at most $\tfrac{1}{2}$. Hence, taking into account the other neighbors of v_i in G^* as well, our modification of p decreases the total blocking value by at most $(k + 4)\epsilon$ but at the same time increases it by at least $\tfrac{1}{2}(r - 1)\epsilon$. Hence, $b(p') \geq b(p) - (k + 4)\epsilon + \tfrac{1}{2}(r - 1)$

$\epsilon = b(p) + (\frac{1}{2}(r - 1) - (k + 4))\epsilon > b(p)$, where the latter inequality follows from the fact that $r \geq 4k+2 \geq 2k+5$, as we assume that $k \geq 2$. However, $b(p') > b(p)$ contradicts the minimality of $b(p)$. Hence, we have proven Claim 3.

We are now ready to argue how to find an independent set of size at least k in G. Let V_1 be the set from Claim 2. By Claim 3 and the fact that the weights $w(e)$ of every edge e between U and V is set to $\frac{1}{2}$, we find that $p \equiv 0$ on V_1. Due to Claim 2, no vertex v_i' with $v_i \in V_1$ can be matched by M. Hence, $p_{v_i'} = 0$ for every $v_i \in V_1$. Because $|U| = k$, we find that $|V_1| = k$. Let E_1' denote the set of edges $v_i v_i'$ with $v_i \in V_1$. Because $|V_1| = k$, we obtain $|E_1'| = k$. Suppose that V_1 contains two adjacent vertices v_i and v_j. Then $b(p) \geq \sum_{zz' \in E_1'}(1 - p_z - p_{z'}) + (1 - p_{v_i} - p_{v_j}) = k + 1$. This is not possible, because $b(p) \leq k$. Hence, no two vertices in V_1 are adjacent. In other words, V_1 is an independent set of size $|V_1| = k$, as desired. This completes the proof of Theorem 2. $\square$

The problems RESTRICTED BLOCKING PAIRS and RESTRICTED BLOCKING VALUE take as input a graph G, an integer k, and a matching M of G, and are to decide whether G has a payoff p with respect to M such that $|B(p)| \leq k$ or $b(p) \leq k$, respectively. We show the following result, the proof of which we omit due to space restrictions.

Theorem 3. *The* RESTRICTED BLOCKING VALUE *problem is polynomial-time solvable, whereas the* RESTRICTED BLOCKING PAIRS *problem is* NP-*complete even for graphs with unit edge weights.*

4   Future Work

We finish our paper by stating the following two open problems. What is the computational complexity of BLOCKING PAIRS and BLOCKING VALUE restricted to input graphs with unit edge weights?

References

1. Abraham, D.J., Biró, P., Manlove, D.F.: "Almost Stable" Matchings in the Roommates Problem. In: Erlebach, T., Persinao, G. (eds.) WAOA 2005. LNCS, vol. 3879, pp. 1–14. Springer, Heidelberg (2006)
2. Béal, S., Rémila, E., Solal, P.: On the number of blocks required to access the coalition structure core. Working Paper, Munich Personal RePEc Archive, MPRA Paper No. 29755 (2011)
3. Biró, P., Kern, W., Paulusma, D.: Computing solutions for matching games. International Journal of Game Theory 41, 75–90 (2012)
4. Chen, B., Fujishige, S., Yang, Z.: Decentralized Market Processes to Stable Job Matchings with Competitive Salaries. Working Paper, Kyoto University, RIMS-1715 (2011)
5. Diamantoudi, E., Miyagawa, E., Xue, L.: Random paths to stability in the roommates problem. Games and Economic Behavior 48, 18–28 (2004)

6. Eriksson, K., Karlander, J.: Stable outcomes of the roommate game with transferable utility. International Journal of Game Theory 29, 555–569 (2001)
7. Gale, D., Shapley, L.S.: College admissions and the stability of marriage. American Mathematical Monthly 69, 9–15 (1962)
8. Garey, M.R., Johnson, D.S.: Computers and Intractability: A Guide to the Theory of NP-Completeness. Freeman, San Francisco (1979)
9. Garey, M.R., Johnson, D.S., Stockmeyer, L.: Some simplified NP-complete graph problems. Theoret. Comput. Sci. 1, 237–267 (1976)
10. Khachiyan, L.G.: A polynomial algorithm in linear programming. Soviet Mathematics Doklady 20, 191–194 (1979)
11. Kóczy, L.Á., Lauwers, L.: The coalition structure core is accessible. Games and Economic Behavior 48, 86–93 (2004)
12. Knuth, D.E.: Mariages stable et leurs relations avec d'autres problèmes combinatoires. Les Presses de l'Université de Montréal, Montréal (1976)
13. Roth, A.E., Vande Vate, J.H.: Random paths to stability in two-sided matching. Econometrica 58, 1475–1480 (1990)
14. Shapley, L.S., Shubik, M.: The assignment game I: the core. International Journal of Game Theory 1, 111–130 (1972)
15. Yang, Y.-Y.: Accessible outcomes versus absorbing outcomes. Mathematical Social Sciences 62, 65–70 (2011)

Which Multi-peg Tower
of Hanoi Problems Are Exponential?

Daniel Berend[1] and Amir Sapir[2,3]

[1] Departments of Mathematics and Computer Science, Ben-Gurion University,
Beer-Sheva, Israel
berend@cs.bgu.ac.il
http://www.cs.bgu.ac.il/~berend
[2] Software Systems Department, Sapir Academic College, Western Negev, Israel*
[3] The Center for Advanced Studies in Mathematics at Ben-Gurion University,
Beer-Sheva, Israel
amirsa@cs.bgu.ac.il
http://www.cs.bgu.ac.il/~amirsa

Abstract. Connectivity properties are very important characteristics of a graph. Whereas it is usually referred to as a measure of a graph's vulnerability, a relatively new approach discusses a graph's *average connectivity* as a measure for the graph's performance in some areas, such as communication. This paper deals with Tower of Hanoi variants played on digraphs, and proves they can be grouped into two categories, based on a certain connectivity attribute to be defined in the sequel.

A major source for Tower of Hanoi variants is achieved by adding pegs and/or restricting direct moves between certain pairs of pegs. It is natural to represent a variant of this kind by a directed graph whose vertices are the pegs, and an arc from one vertex to another indicates that it is allowed to move a disk from the former peg to the latter, provided that the usual rules are not violated. We denote the number of pegs by h. For example, the variant with no restrictions on moves is represented by the Complete K_h graph; the variant in which the pegs constitute a cycle and moves are allowed only in one direction — by the uni-directional graph $Cyclic_h$.

For all 3-peg variants, the number of moves grows exponentially fast with n. However, for $h \geq 4$ peg variants, this is not the case. Whereas for $Cyclic_h$ the number of moves is exponential for any h, for most of the other graphs it is sub-exponential. For example, for a path on 4 vertices it is $O(\sqrt{n}3^{\sqrt{2n}})$, for n disks.

This paper presents a necessary and sufficient condition for a graph to be an H-subexp, i.e., a graph for which the transfer of n disks from a peg to another requires sub-exponentially many moves as a function of n.

To this end we introduce the notion of a shed, as a graph property. A vertex v in a strongly-connected directed graph $G = (V, E)$ is a *shed* if the subgraph of G induced by $V - \{v\}$ contains a strongly connected subgraph on 3 or more vertices. Graphs with sheds will be shown to be much more efficient than those without sheds, for the particular domain of the Tower of Hanoi puzzle. Specifically we show how, given a graph

* Research supported in part by the Sapir Academic College, Israel.

M.C. Golumbic et al. (Eds.): WG 2012, LNCS 7551, pp. 81–90, 2012.

with a shed, we can indeed move a tower of n disks from any peg to any other within $O(2^{\varepsilon n})$ moves, where $\varepsilon > 0$ is arbitrarily small.

Keywords: Tower of Hanoi, directed graphs, connectivity, sub-exponential complexity, shed.

1   Introduction

Given are 3 pegs and a certain number n of disks of distinct sizes. Initially, the disks form a tower: the largest at the bottom of one of the pegs (the source), the second largest on top of it, and so on, until the smallest at the top of that peg. The well-known Tower of Hanoi problem asks: how do we optimally move the tower to another peg (the destination peg), bounded by the *Hanoi rules* (henceforth HR): 1) At each step only one disk is moved. 2) The moved disk must be a topmost one. 3) At any moment, no disk may reside on a smaller one.

The game was composed over a hundred years ago by Lucas [18]. Ever since then, it was studied from numerous points of view. For example, in [1] it is shown that, with a direct approach coding, the string representing the optimal solution is square-free. This line of work was extended in [2]. As computer science education evolved, the Tower of Hanoi problem has been used as a common example, demonstrating the elegance of recursive programming. The reader is referred to [29] for a review of the history of the problem, and to [30] for an extensive bibliography of papers on various lines of research in the field.

Many variants of the original puzzle came up, some of which we will describe here, though not chronologically. Without changing the basic peg configuration (3 pegs, each pair being connected bi-directionally), one direction is solving the problem for any initial and final configurations [12]. Other challenging versions have been proposed and solved in [20],[21],[22],[23],[19]. In another direction a disk may reside on top of a smaller one, with various limitations, [16], [9].

Another version of the original problem, which was discussed in several papers, is where we impose restrictions on the movements between pegs. In [26], [29], [14], the "three-in-a-row" ($\mathtt{Path_3}$) arrangement is studied. The uni-directional cycle ($\mathtt{Cyclic_3}$) has been solved in [3],[11]. As was mentioned, it is natural to represent a variant by a directed graph. Surprising in its simplicity, a necessary and sufficient condition for a variant to be solvable, for any source and destination pegs and any number of disks, is that the corresponding graph is strongly connected [17]. For 3 pegs there are 5 (up to isomorphism) strongly connected variants. A single optimal algorithm for all these variants was obtained in [25], accompanied with an explicit formula for the minimal number of moves for each variant. (Note that individual algorithms and explicit formulas were known beforehand for $\mathtt{K_3}$, $\mathtt{Path_3}$, $\mathtt{Cyclic_3}$, as mentioned above.)

Probably the first multi-peg version is "The Reve's Puzzle" [10, pp. 1–2], in which there are 4 pegs and various specific numbers of disks. It has been generalized to any number of pegs and any number of disks in [27], with solutions in [28] and [13], which were (among several other solutions) proved to be identical in [15]. An analysis of the algorithm reveals, somewhat surprisingly, that the number of moves in the solution grows sub-exponentially as a function of n. In the case of 4 pegs, it grows like $\Theta(\sqrt{n}2^{\sqrt{2n}})$ (cf. [29]). The lower bound issue was considered in [31] and [8], where it has been shown to grow at a rate close to that yielded by the algorithm.

Allowing 4 pegs and above, and imposing movement restrictions, we obtain a huge number of graphs (83 non-isomorphic strongly connected digraphs on 4 vertices already), and no algorithm seems a natural candidate to be optimal. The question whether a variant is sub-exponential had been resolved only for particular ones: Star [29], Cyclic [5] and Path [7]. Whereas the complexity of the majority of the multi-peg variants is sub-exponential, Cyclic is among the few which are exponential.

The fact that, even for the original multi-peg variants, on complete graphs, it is not known whether the proposed algorithms are optimal, indicates that the complexity issue here is non-trivial. Facing the wealth of variants, we would like an easy way to determine, given a variant, whether it is exponential or sub-exponential. This paper presents a simple necessary and sufficient condition for a variant to be sub-exponential.

In Section 2 we describe the problem domain. The main results are introduced in Section 3. The proofs of the theorems are presented in Section 4.

2 Problem Domain and Notations

Any arrangement of pegs and their immediate connections, such that each peg is reachable from each other, constitutes a variant. As mentioned above, it is natural to represent a variant by a digraph G, whose vertices are the pegs, and an arc from one vertex to another designates the ability of moving a disk from the former peg to the latter, provided that the HR are obeyed. In the sequel, when we mention a *variant graph*, we mean a strongly connected simple directed graph on $h \geq 3$ vertices.

A *configuration* is a distribution of the disks among the pegs, satisfying HR.3. A configuration is *perfect* if all disks reside on the same peg. Such a configuration will be denoted by $R_{i,n}$, where n is the number of disks (disk 1 being the smallest and disk n the largest) and i the peg containing the disks.

Given a variant graph G and a positive integer n, the corresponding *configuration graph* $G^{(n)}$ is the graph whose vertices are all configurations of n disks over G, where there is an arc from a vertex to another if one can pass from the former to the latter by a single disk move. More generally, for any pair of configurations there is a corresponding *task*, of passing from the first to the second. Thus, an optimal solution of a task corresponds to a shortest path between the vertices of $G^{(n)}$ representing the task's initial and final configurations. The diameter of $G^{(n)}$ is denoted by $D_n(G)$.

A task is *perfect* if both its initial and final configurations are perfect. We use the notation $R_{i,n} \to R_{j,n}$ both for the task and for a minimal length solution of it. The length of such a minimal solution is expressed by $|R_{i,n} \to R_{j,n}|$. We set $d_{i,j,n}(G) = |R_{i,n} \to R_{j,n}|$. An interesting quantity is $d_n(G) = \max\limits_{i,j} d_{i,j,n}(G)$, which we call the *little diameter* of $G^{(n)}$. When the identity of the graph to which we refer is clear, we may omit its indication. For example, we may write D_n instead of $D_n(G)$.

Formally, a *move* is composed of the disk being moved, the peg on which it resides prior to the move, and the peg to which it is transferred. A *solution* to a task is a sequence of moves accomplishing it. The algorithms constructed in the proofs of our results produce solutions to all perfect tasks.

A variant graph G is *H-exp* if $D_n(G)$ grows exponentially fast as a function of the number of disks, namely there exist $C > 0$ and $\lambda > 1$ such that $D_n(G) \geq C\lambda^n$ for all n. G is *H-subexp* if for every $\varepsilon > 0$ there exists a constant $C = C(\varepsilon)$ such that $D_n(G) \leq C(1 + \varepsilon)^n$. It will follow, in particular, from Theorem 1 below that each graph is either H-exp or H-subexp.

3   Main Results

The main problem we study is how to identify, given a variant graph, whether it is H-subexp or H-exp. We start by proving that the number of moves behaves regularly as a function of the number of disks.

For a variant graph G, denote $\lambda_G = \inf_{n \geq 1} \sqrt[n]{d_{n+1}(G)}$.

Theorem 1. *For any variant graph G*

$$\lim_{n \to \infty} \sqrt[n]{d_n(G)} = \lambda_G \,.$$

Let us recall Corollary 1 of [6]:

Proposition 1. *For any variant graph and any number of disks,*

$$D_n \leq (2n - 1)d_n \,.$$

Combining Theorem 1 and Proposition 1 one can infer

Corollary 1. *For every $\varepsilon > 0$ there exists an n_0 such that*

$$\lambda_G^{n-1} \leq d_n(G) \leq D_n(G) \leq (\lambda_G + \varepsilon)^{n-1}, \qquad n \geq n_0.$$

The main question this paper answers is: what property must G have so that $\lambda_G = 1$? To answer this, we need the following notion.

Definition 1. A *shed* in a strongly connected digraph G (see Fig. 1) is a vertex w with the property that the graph induced by $V(G) - \{w\}$ contains a strongly connected subgraph of size at least 3.

Our main result is

Theorem 2. *A variant graph G with $h \geq 3$ vertices is H-subexp if and only if it contains a shed.*

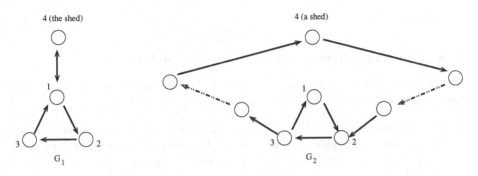

Fig. 1. Two example graphs with sheds

Unlike the proof of sub-exponentiality for K_4 (cf. [29]) or the path P_h [7], the proof for general graphs containing a shed is quite cumbersome. The reason is, intuitively, that the former graphs have (at least) two sheds. This means that there are a lot more options of keeping a block of small disks on some peg for a while, taking care in the meantime of other (large) disks. On the other hand, if the graph has a single shed, then no obvious sub-exponential algorithm comes to mind, and in fact it may seem surprising at first glance that such an algorithm exists at all.

4 Proofs

To prove Theorem 1, we first need

Lemma 1. *For every strongly connected graph G,*

$$d_{m+n-1} \leq d_m d_n \qquad m, n \geq 1.$$

The proof follows, to a certain extent, the idea of the proof of Theorem 2(a) in [5], and is omitted.

Proof of Theorem 1: Putting $b_n = \ln d_{n+1}$, we obtain by Lemma 1:

$$b_{m+n} \leq b_n + b_m, \qquad m, n \geq 0.$$

The sequence $(b_n)_{n=0}^{\infty}$ is thus sub-additive, which implies that the sequence $\frac{b_n}{n}$ converges to its greatest lower bound (cf. [24, p. 198]), and hence so does the sequence

$$e^{b_n/n} = \sqrt[n]{d_{n+1}},$$

thus proving the theorem.

Proof of Corollary 1: Since $\sqrt[n]{d_{n+1}(G)}$ converges to λ_G from above, for any $\varepsilon > 0$ and sufficiently large n

$$\lambda_G \leq \sqrt[n]{d_{n+1}(G)} \leq \lambda_G + \varepsilon,$$

or, equivalently,

$$\lambda_G^{n-1} \le d_n(G) \le (\lambda_G + \varepsilon)^{n-1} .$$

In combination with Proposition 1, this proves the corollary.

Remark 1. Corollary 1 states that d_n and D_n are not too far apart, justifying that it suffices to focus on d_n.

To establish the proof of the 'if' part (which is the main part) of Theorem 2, we will need several definitions. Unless explicitly stated otherwise, G will denote a strongly connected graph with $h \ge 4$ vertices containing a shed w, and G' – a strongly connected subgraph of size at least 3 of the graph left after w is removed. (Note that G may contain more than one shed, and for each choice of a shed there may be several appropriate strongly connected subgraphs of G, but for our purposes any choice of a shed vertex and a strongly connected component in accordance with that shed is suitable.) Taking a shortest path from $w \in V(G) - V(G')$ to G', we denote the *entrance vertex* to G' – the last vertex on this path and the only one that belongs to G' – by e. Similarly, an *exit vertex* $x \in V(G')$ is the first vertex on a shortest path leading from G' to w. These paths will be denoted by $p_{w,e}$ and $p_{x,w}$, respectively. The sequence of moves of disk r along, say, the path $p_{w,e}$ is expressed by $t_{w,e,r}$. $S_{(l)}$ is the l-th move in a sequence of moves S.

Example 1. In each of the graphs G_1 and G_2 in Figure 1, the shed is vertex $w = 4$. (In G_2 one could choose any vertex but 2 and 3 as a shed.) For G_1 we have $e = x = 1$, whereas for G_2 we have $e = 2, x = 3$.

As hinted above, most of the work is required for the 'if' part of Theorem 2. The following lemma shows that some of the tasks are of sub-exponential complexity. Let $\texttt{Inner} = V(G') - \{e, x\}$. We first show that tasks, in which the source is the shed and the destination lies in $\texttt{Inner}$ (or vice versa), are sub-exponential.

Lemma 2. *Let G, G', w be as above and $v \in \texttt{Inner}$. Then $|R_{w,n} \to R_{v,n}|$ and $|R_{v,n} \to R_{w,n}|$ grow sub-exponentially fast as functions of n.*

Sketch of proof: Fix a number $\lambda > \lambda_{G'}$. It suffices to show that for every $\varepsilon > 0$ there exists a $C = C(\varepsilon)$ such that

$$|R_{w,n} \to R_{v,n}| \le C\lambda^{\varepsilon n} , \qquad n = 1, 2, \dots ,$$

and

$$|R_{v,n} \to R_{w,n}| \le C\lambda^{\varepsilon n} , \qquad n = 1, 2, \dots .$$

The proof is based on the procedures $\texttt{ShedToInner}(1 \dots n)$ (Algorithm 1) and $\texttt{InnerToShed}(1 \dots n)$ (of a similar kind), which perform the tasks $R_{w,n} \to R_{v,n}$ and $R_{v,n} \to R_{w,n}$ in $f_{w,v}(n) \le C\lambda^{\varepsilon n}$ and $f_{v,w}(n) \le C\lambda^{\varepsilon n}$ moves, respectively.

The following lemma is required for the proof of the 'only if' part of Theorem 2.

Algorithm 1. `ShedToInner(1...n)`

/* This algorithm moves a tower of n disks from the shed w to vertex v, which is
neither the entrance nor the exit vertex with respect to w. */
if $n \leq \frac{1}{\varepsilon}$ **then**
  $T \leftarrow$ a sequence of moves that performs the task
else
  $m \leftarrow \lceil n(1-\varepsilon) \rceil$
  $T \leftarrow$ `ShedToInner(1...m)`
  for $r \leftarrow m+1$ **to** n **do**
    $T \leftarrow T * t_{w,e,r}$
    $T \leftarrow T *$ `InnerToShed(1...m)`
    $T \leftarrow T *$ `Accumulate`$(e, v, r, m+1...r-1)$
    $T \leftarrow T *$ `ShedToInner(1...m)`
  end for
end if
return T

Lemma 3. *There exists a constant C with the following property. For any graph G, not containing a strongly connected component of size 3 or more, and for any initial configuration, the number of different disks which can participate in any legal sequence of moves is bounded above by $Ch^{\frac{1}{2} \lg h + 2}$, where $h = |V(G)|$, and $\lg h \equiv \log_2 h$.*

Proof: Let T be any legal move sequence and i any peg.
Set $l = \lfloor (h-2)Ch^{\frac{\lg h}{2}} \rfloor + 2$ (where the constant C is sufficiently large; see below). Consider the first move of the disk residing at the l-th place (counting from the top) of peg i before T is started. Right before this move, all smaller disks (residing on peg i prior to T) have to be spread on $h-2$ of the other pegs, making it necessary for at least one peg to accept more than $Ch^{\frac{1}{2} \lg h}$ disks from peg i, contradicting [4, Theorem 1.3] if C is sufficiently large. It follows that the overall number of disks that can participate in any task is bounded above by $h(h-2)Ch^{\frac{1}{2} \lg h} + 2h$, proving the lemma.

Proof of Theorem 2: (a) We prove first the 'if' part of the theorem. Let w and G' be as before. Take $\lambda > \lambda_{G'}$. We will show that, for every pair of vertices $i, j \in V(G)$ and $\varepsilon > 0$, there exists a constant K such that

$$|R_{i,n} \to R_{j,n}| \leq K\lambda^{\varepsilon n}, \qquad n = 1, 2, \dots . \tag{1}$$

Take $v \in V(G') - \{e, x\}$, where e is the entrance vertex and x the exit vertex in G'. Due to Lemma 2, it remains to consider the following cases:

- Case 1: $i = w$, $j \in V(G') - \{v\}$,
- Case 2: $i \in V(G') - \{v\}$, $j = w$,
- Case 3: $i, j \in V(G')$,
- Case 4: $i, j \in V(G)$, where at least one of i, j does not belong to $V(G') \cup \{w\}$.

(In fact, in Case 1 we need only to take care of the case where $j = e$, and in Case 2 only of $i = x$, but this is of no consequence.)

Case 1: We employ ShedToAny$(j, 1 \ldots n)$ (Algorithm 2), which moves a tower of n disks from the shed to any vertex $j \neq v \in G'$. It divides the disks into $p + 1$ sets, where $p = \left\lfloor \frac{n}{\lfloor \varepsilon n \rfloor} \right\rfloor$ of the sets consist of $l = \lfloor \varepsilon n \rfloor$ consecutive disks each, and the remaining (possibly empty) set consists of up to $l - 1$ disks, which are the smallest disks. Then it moves each set to its destination (vertex j) by:

- Moving all the disks from the shed to vertex v.
- Moving all but the largest $\lfloor \varepsilon n \rfloor$ disks in the set from vertex v back to the shed.
- Moving the largest $\lfloor \varepsilon n \rfloor$ disks from vertex v to the destination.

The procedure MoveInG$'(e, v, m + 1 \ldots r - 1)$ performs the task of moving the tower consisting of disks $m + 1, \ldots, r - 1$ from e to v. By Theorem 1, the length of the solution it produces is bounded above by $C' \lambda^{r-m-1}$, where C' is an appropriate constant, determined by G'.

Accumulate$(e, v, r, m + 1 \ldots r - 1)$ adds the 'next' disk to the column already handled. Starting with disks $m + 1 \ldots r - 1$ on peg v and disk r on peg e, utilizing G' only, it unites them so that all disks will reside on peg v. Thus, the number of moves produced by its $(r - m)$-th invocation is bounded above by $2C' \lambda^{r-m}$.

Algorithm 2. ShedToAny$(j, 1 \ldots n)$

/* ShedToAny moves a tower of n disks from the shed to any vertex $j \neq v \in G'$. */
if $n \leq \frac{1}{\varepsilon}$ then
   $T \leftarrow$ a sequence of moves that performs the task
else
   $l \leftarrow \lfloor n\varepsilon \rfloor$; $p \leftarrow \lfloor \frac{n}{l} \rfloor$; $m \leftarrow n - pl$
   $T \leftarrow [\,]$ /* The empty sequence */
   for $r \leftarrow 1$ to p do
      $T \leftarrow T *$ ShedToInner$(1 \ldots n)$
      $T \leftarrow T *$ InnerToShed$(1 \ldots n - l)$
      $T \leftarrow T *$ MoveInG$'(v, j, n - l + 1 \ldots n)$
      $n \leftarrow n - l$
   end for
   $T \leftarrow T *$ ShedToInner$(1 \ldots m) *$ MoveInG$'(v, j, 1 \ldots m)$
end if
return T

ShedToInner$(1 \ldots n)$ and InnerToShed$(1 \ldots n)$ require at most $C\lambda^{\varepsilon n}$ moves each, for an appropriate C. We now bound the number of moves $f_{w,j}(n)$ performed by the procedure ShedToAny$(j, 1 \ldots n)$. For sufficiently large n:

$$f_{w,j}(n) \le (p+1)C'\lambda^l + 2\sum_{r=1}^p C\lambda^{\varepsilon lr} + C\lambda^{\varepsilon n}$$

$$\le ((p+1)C' + 3C)\lambda^{\varepsilon n} + 2C\sum_{r=1}^{p-1}\lambda^{\varepsilon lr}$$

$$\le ((\tfrac{1}{\varepsilon}+2)C' + 3C)\lambda^{\varepsilon n} + 2C\frac{\lambda^{\varepsilon l(\frac{1}{\varepsilon}+1)}}{\frac{1}{2}\lambda^{\varepsilon l}}$$

$$\le ((\tfrac{1}{\varepsilon}+2)C' + 7C)\lambda^{\varepsilon n}.$$

Case 2: We employ `AnyToShed`$(i, 1 \ldots n)$, which moves a tower of n disks from any vertex $i \ne v \in G'$ to the shed. We omit the description of this algorithm, which is similar to `ShedToAny` in the way it works, as well as in its analysis.

Case 3 is a consequence of the first two cases, by first moving all disks from i to w, and then moving them from w to j.

Case 4: Observe that, once G' has been set, any vertex in $V(G) - V(G')$ may serve as a shed. Thus, for any $i \in V(G')$ and $j \in V(G) - V(G')$, both inequalities $|R_{i,n} \to R_{j,n}| \le K\lambda^{\varepsilon n}$ and $|R_{j,n} \to R_{i,n}| \le K\lambda^{\varepsilon n}$ hold. Now let $i, j \in V(G) - V(G')$. Take a vertex $k \in V(G')$. We move a tower of disks from i to j by first moving it from i to k, and then from k to j. This proves the correctness of (1) in this case.

(b) Sketch of proof of the 'only if' part of Theorem 2.

Take any strongly connected graph G without a shed. Start with any configuration C. Denote by v_0 the vertex on which disk 1 resides in C. Without reaching the same configuration more than once, what is the maximal number of moves which may be done without moving disk 1?

Clearly, as long as we do not move disk 1, all other disks may use only the graph induced by $V - \{v_0\}$. This graph has no strongly connected component of size at least 3. By Lemma 3, since in this graph there are $h - 1$ vertices, $m = C(h-1)^{\frac{1}{2}\lg(h-1)+2}$ is an upper bound for the number of disks that may participate in any sequence of moves, yielding exponential growth.

References

1. Allouche, J.-P., Astoorian, D., Randall, J., Shallit, J.: Morphisms, squarefree strings, and the Tower of Hanoi puzzle. Amer. Math. Monthly 101, 651–658 (1994)
2. Allouche, J.-P., Sapir, A.: Restricted Towers of Hanoi and Morphisms. In: De Felice, C., Restivo, A. (eds.) DLT 2005. LNCS, vol. 3572, pp. 1–10. Springer, Heidelberg (2005)
3. Atkinson, M.D.: The cyclic Towers of Hanoi. Inform. Process. Lett. 13, 118–119 (1981)
4. Azriel, D., Berend, D.: On a question of Leiss regarding the Hanoi Tower problem. Theoretical Computer Science 369, 377–383 (2006)
5. Berend, D., Sapir, A.: The Cyclic multi-peg Tower of Hanoi. Trans. on Algorithms 2(3), 297–317 (2006)
6. Berend, D., Sapir, A.: The diameter of Hanoi graphs. Inform. Process. Lett. 98, 79–85 (2006)

7. Berend, D., Sapir, A., Solomon, S.: Subexponential upper bound for the Path multi-peg Tower of Hanoi. Disc. Appl. Math (2012)
8. Chen, X., Shen, J.: On the Frame-Stewart conjecture about the Towers of Hanoi. SIAM J. on Computing 33(3), 584–589 (2004)
9. Dinitz, Y., Solomon, S.: Optimality of an algorithm solving the Bottleneck Tower of Hanoi problem. Trans. on Algorithms 4(3), 1–9 (2008)
10. Dudeney, H.E.: The Canterbury Puzzles (and Other Curious Problems). E. P. Dutton, New York (1908)
11. Er, M.C.: The Cyclic Towers of Hanoi: a representation approach. Comput. J. 27(2), 171–175 (1984)
12. Er, M.C.: A general algorithm for finding a shortest path between two n-configurations. Inform. Sci. 42, 137–141 (1987)
13. Frame, J.S.: Solution to advanced problem 3918. Amer. Math. Monthly 48, 216–217 (1941)
14. Guan, D.-J.: Generalized Gray codes with applications. Proc. Natl. Sci. Counc. ROC(A) 22(6), 841–848 (1998)
15. Klavžar, S., Milutinović, U., Petr, C.: On the Frame-Stewart algorithm for the multi-peg Tower of Hanoi problem. Disc. Appl. Math. 120(1-3), 141–157 (2002)
16. Klein, C.S., Minsker, S.: The super Towers of Hanoi problem: large rings on small rings. Disc. Math. 114, 283–295 (1993)
17. Leiss, E.L.: Solving the "Towers of Hanoi" on graphs. J. Combin. Inform. System Sci. 8(1), 81–89 (1983)
18. Lucas, É.: Récréations Mathématiques, vol. III. Gauthier-Villars, Paris (1893)
19. Lunnon, W.F., Stockmeyer, P.K.: New Variations on the Tower of Hanoi. In: 13th Intern. Conf. on Fibonacci Numbers and Their Applications (2008)
20. Minsker, S.: The Towers of Antwerpen problem. Inform. Process. Lett. 38(2), 107–111 (1991)
21. Minsker, S.: The Linear Twin Towers of Hanoi problem. Bulletin of ACM SIG on Comp. Sci. Education 39(4), 37–40 (2007)
22. Minsker, S.: Another brief recursion excursion to Hanoi. Bulletin of ACM SIG on Comp. Sci. Education 40(4), 35–37 (2008)
23. Minsker, S.: The classical/linear Hanoi hybrid problem: regular configurations. Bulletin of ACM SIG on Comp. Sci. Education 41(4), 57–61 (2009)
24. Pólya, G., Szegő, G.: Problems and Theorems in Analysis, vol. I. Springer (1972)
25. Sapir, A.: The Tower of Hanoi with forbidden moves. Comput. J. 47(1), 20–24 (2004)
26. Scorer, R.S., Grundy, P.M., Smith, C.A.B.: Some binary games. Math. Gazette 280, 96–103 (1944)
27. Stewart, B.M.: Advanced problem 3918. Amer. Math. Monthly 46, 363 (1939)
28. Stewart, B.M.: Solution to advanced problem 3918. Amer. Math. Monthly 48, 217–219 (1941)
29. Stockmeyer, P.K.: Variations on the Four-Post Tower of Hanoi puzzle. Congr. Numer. 102, 3–12 (1994)
30. Stockmeyer, P.K.: Tower of hanoi bibliography (2005), http://www.cs.wm.edu/~pkstoc/biblio2.pdf
31. Szegedy, M.: In How Many Steps the k Peg Version of the Towers of Hanoi Game Can Be Solved? In: Meinel, C., Tison, S. (eds.) STACS 1999. LNCS, vol. 1563, pp. 356–361. Springer, Heidelberg (1999)

h-Quasi Planar Drawings of Bounded Treewidth Graphs in Linear Area⋆

Emilio Di Giacomo, Walter Didimo, Giuseppe Liotta, and Fabrizio Montecchiani

Dip. di Ingegneria Elettronica e dell'Informazione, Università degli Studi di Perugia
{digiacomo,didimo,liotta,montecchiani}@diei.unipg.it

Abstract. We study the problem of computing h-quasi planar drawings in linear area; in an h-quasi planar drawing the number of mutually crossing edges is at most $h - 1$. We prove that every n-vertex partial k-tree admits a straight-line h-quasi planar drawing in $O(n)$ area, where h depends on k but not on n. For specific sub-families of partial k-trees, we present ad-hoc algorithms that compute h-quasi planar drawings in linear area, such that h is significantly reduced with respect to the general result. Finally, we compare the notion of h-quasi planarity with the notion of h-planarity, where each edge is allowed to be crossed at most h times.

1 Introduction

Area requirement of graph layouts is a widely studied topic in Graph Drawing and Geometric Graph Theory. Many asymptotic bounds have been proven for a variety of graph families and drawing styles. One of the most fundamental results in this scenario establishes that every planar graph admits a planar straight-line grid drawing in $O(n^2)$ area and that this bound is worst-case optimal [8]. This has motivated lot of work devoted to discover sub-families of planar graphs that admit planar straight-line drawings in $o(n^2)$ area. Unfortunately, sub-quadratic upper bounds are known only for trees [7] and outerplanar graphs [9], while super-linear lower bounds are known for series-parallel graphs [19]. Bounds for planar poly-line drawings are also known [3,4].

Although planarity is one of the most desirable properties when drawing a graph, many real-world graphs are in fact non-planar. Furthermore, planarity often imposes severe limitations on the optimization of the drawing area, which may sometimes be overcome by allowing either "few" edge crossings or specific types of edge crossings that do not affect too much the drawing readability. So far, only a few papers have focused on computing non-planar layouts in sub-quadratic area. Wood proved that every k-colorable graph admits a non-planar straight-line grid drawing in linear area [22], which implies that planar graphs admit such a drawing. However, the technique by Wood does not provide any guarantee on the type and number of edge crossings. More recently, Angelini *et al.* provided techniques for constructing poly-line *large angle crossing drawings* (*LAC drawings*) of planar graphs in sub-quadratic area [1]. We recall that the study of drawings with large angle crossings started in [13].

⋆ Research supported in part by the MIUR project AlgoDEEP prot. 2008TFBWL4.

M.C. Golumbic et al. (Eds.): WG 2012, LNCS 7551, pp. 91–102, 2012.
© Springer-Verlag Berlin Heidelberg 2012

In this paper we study the problem of computing linear area straight-line drawings of graphs with controlled *crossing complexity*, i.e., drawings where some types of edge crossings are forbidden. We study *h-quasi planar drawings*, i.e., drawings with no h mutually crossing edges; this measure of crossing complexity can be regarded as a sort of planarity relaxation. The combinatorial properties of h-quasi planar drawings have been widely investigated [18,21]. The contributions of the paper are as follows: (i) We prove that every n-vertex partial k-tree (i.e., any graph with bounded treewidth) admits a straight-line h-quasi planar drawing in $O(n)$ area, where h depends on k but not on n (Section 3). (ii) For specific sub-families of partial k-trees (outerplanar graphs, flat series-parallel graphs, and proper simply-nested graphs), we provide ad-hoc algorithms that compute h-quasi planar drawings in $O(n)$ area with values of h significantly smaller than those obtained with the general technique (Section 4). (iii) We compare the notion of h-quasi planarity with that of *h-planarity*, which allows every edge to be crossed at most h times. We prove that h-quasi planarity is, in some cases, less restrictive than h-planarity in terms of area requirement. Namely, while linear area h-quasi planar drawings exist for series-parallel graphs (i.e. partial 2-trees) with $h = 11$, we prove that for any given constant h there exists a family of series-parallel graphs that do not admit a linear area straight-line h-planar drawing (Section 5). For reasons of space, many proofs are omitted in this extended abstract.

2   Preliminaries

A *drawing* Γ of a graph G maps each vertex v of G to a point p_v on the plane, and each edge $e = (u, v)$ to a Jordan arc connecting p_u and p_v not passing through any other vertex; furthermore, any two edges have at most one point in common. If all edges are mapped to straight-line segments, Γ is a *straight-line drawing* of G. If all vertices are mapped to points with integer coordinates, Γ is a *grid drawing* of G. The *bounding box* of a straight-line grid drawing Γ is the minimum axis-aligned box containing the drawing. If the bounding box has side lengths $X - 1$ and $Y - 1$, then we say that Γ is a drawing with *area* $X \times Y$. A drawing Γ is *h-quasi planar* if it has no h mutually crossing edges. A 3-quasi planar drawing is also called a *quasi planar drawing*.

We recall now definitions about track layouts which have been introduced and studied by Dujmović, Pór and Wood [15]. A *vertex coloring* $\{V_i : i \in I\}$ of a graph G is a partition of the vertices of G such that no edge has both endvertices in the same partition set V_i ($i \in I$). The elements of I are *colors* and each set V_i is a *color class*. A *t-track assignment* of G consists of a vertex coloring with t colors and a total ordering $<_i$ of the vertices in each color class V_i. Each pair $(V_i, <_i)$ is a *track* and will be denoted as τ_i. An *X-crossing* in a track assignment consists of two edges (u, v) and (w, z) such that $u, w \in V_i$, $v, z \in V_j$, $u <_i w$ and $z <_j v$, for $i \neq j$. An *edge c-coloring* of G is a partition of the edges of G into c sets, each set called a *color*. A *(c, t)-track layout of* G consists of a t-track assignment of G and an edge c-coloring of G such that no two edges of the same color form an X-crossing. The minimum t such that a graph G admits a (c, t)-track layout is denoted by $tn_c(G)$. A $(1, t)$-track layout is called a *t-track layout*. The *track-number* of G is $tn_1(G)$, simply denoted by $tn(G)$.

A *k-tree*, $k \in \mathbb{N}$, is defined as follows. The clique of size k is a k-tree; the graph obtained from a k-tree by adding a new vertex adjacent to each vertex of a clique of size k is also a k-tree. A *partial k-tree* is a subgraph of a k-tree. A graph has bounded *treewidth* if and only if it is a partial k-tree [5].

3 Compact *h*-Quasi Planar Drawings of Partial *k*-Trees

In this section we first describe a general technique to "transform" a (c, t)-track layout into an h-quasi planar drawing in linear area with $h = c(t - 1) + 1$. We then describe how to compute a $(2, t)$-track layout of a k-tree where t depends on k but not on n. The two results imply that every partial k-tree admits an h-quasi planar drawing in linear area, where h depends on k but not on n.

Lemma 1. *Let G be a graph with n vertices. If G admits a (c, t)-track layout, then G admits an h-quasi planar grid drawing in $O(t^3 n)$ area, where $h = c(t - 1) + 1$.*

Proof. We describe how to use a (c, t)-track layout γ of G to compute an h-quasi planar grid drawing in $O(t^3 n)$ area, where $h = c(t - 1) + 1$. The vertices of each track τ_i $(i = 0, \ldots, t - 1)$ are drawn as points of a horizontal segment s_i whose y-coordinate is $-i$. The idea is to place the t segments s_i on a parabola in such a way that no connection between two segments crosses a third. We place the vertices on s_i from left to right according to $<_i$. As a consequence, no two edges whose endvertices belong to two tracks τ_i and τ_j can cross in the drawing unless they form an X-crossing in γ. We will use this fact to bound the number of mutually crossing edges. More precisely, the vertices are placed on s_i from left to right according to $<_i$, with unit distance between any two consecutive vertices. Each segment has length n^*, where $n^* = \max_i\{|\tau_i|\} - 1$ (thus, the length of s_i is sufficient to host all vertices of τ_i). We denote by p_i and q_i the leftmost and the rightmost point of s_i, respectively. Also, we denote by x_i the x-coordinate of p_i. We place each segment s_i in such a way that $x_i = x_{i-1} + n^* + \Delta_i + 1$, where $\Delta_i = 2(i - 1)n^* + i$ (see Figure 1 for an example).

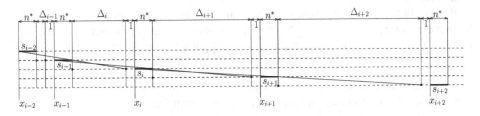

Fig. 1. Illustration of the construction described in Lemma 1

We prove now that the computed drawing Γ is an h-quasi planar grid drawing of G with $h = c(t - 1) + 1$. First of all we prove that no edge in the drawing passes through a vertex in Γ. Let (u, v) be an edge with $u \in \tau_i$ and $v \in \tau_j$, with $i < j$. If (u, v) passed through a vertex w, then w would belong to a track τ_l with $i < l < j$. We prove that segment s_l is in fact completely to the left of the segment $\overline{p_i p_j}$ and therefore w is to the left of (u, v). The proof is by induction on $j - i$. The base case is when $j - i = 2$ (and $l = i + 1$). In this case s_{i+1} is to the left of $\overline{p_i p_{i+2}}$ by construction; namely, the slope of the segment $\overline{p_i q_{i+1}}$ is $-\frac{1}{x_{i+1}+n^*-x_i} = -\frac{1}{2(i+1)n^*+i+2}$, while the slope of the segment $\overline{p_i p_{i+2}}$ is $-\frac{2}{x_{i+2}-x_i} = -\frac{1}{2(i+1)n^*+i+2.5}$, which implies that the whole segment s_{i+1} is to the left of $\overline{p_i p_{i+2}}$. Assume now that $j - i > 2$. All segments s_r with $i < r < j - 1$ are to the left of segment $\overline{p_i p_{j-1}}$ by induction; s_{j-1} is to the left of $\overline{p_{j-2} p_j}$ also by induction. It follows that all segments s_r with $i < r < j - 1$ are to the left of $\overline{p_i p_j}$.

Every edge (u, v) with u drawn on s_i and v drawn on s_j is completely contained in the parallelogram $\Pi_{i,j}$ whose corners are p_i, q_i, p_j, and q_j $(0 \leq i, j \leq t - 1)$. By definition of (c, t)-track layout there are at most c mutually crossing edges inside each parallelogram. We will show that at most $t - 1$ parallelograms $\Pi_{i,j}$ mutually overlap, which implies that there are at most $c(t - 1)$ mutually crossing edges in our drawing. Consider two parallelograms $\Pi_{i,j}$ and $\Pi_{r,l}$ and assume without loss of generality that $i < j$ and $r < l$. It is easy to see that $\Pi_{i,j}$ and $\Pi_{r,l}$ overlap if and only if one of the following three conditions hold: (α) $i < r < j < l$; (β) $i = r$; (γ) $j = l$. The proof that at most $t - 1$ parallelogram mutually overlap in Γ is by induction on t. If $t = 2$, there is a single parallelogram and the statement trivially holds. Assume now that $t > 2$. We denote by Γ_i $(0 \leq i \leq t - 1)$ the subdrawing of Γ induced by the vertices drawn on the segments $s_0, \ldots, s_i$. Suppose, as a contradiction, that there is a set S of at least t mutually overlapping parallelograms in Γ_{t-1}. Partition S into two subsets P and R defined as follows. $P = \{\Pi_{i_1,t-1}, \Pi_{i_2,t-1}, \ldots, \Pi_{i_{|P|},t-1}\}$ is the set of parallelograms having s_{t-1} as rightmost side and $R = S \setminus P$. Since, by induction, there are at most $t-2$ mutually overlapping parallelograms in Γ_{t-2}, P contains at least two parallelograms, i.e., $|P| \geq 2$. Observe that, by conditions α, β, γ, the parallelograms in R have a side s_j with $0 \leq j \leq i_1$ and a side s_l with $i_{|P|}+1 \leq l \leq t-2$. Also, all these parallelograms are present in Γ_{t-2}. By our assumption that S contains at least t parallelograms, it follows that $|R| \geq t - |P|$. Let l be the greatest index among the segments in R $(i_{|P|} + 1 \leq l \leq t - 2)$; we have that each parallelogram in the set $Q = \{\Pi_{i_2,l}, \ldots, \Pi_{i_{|P|},l}\}$ and all the parallelograms in R mutually overlap. Thus, they form a bundle of mutually overlapping parallelograms of size $|R| + |Q| \geq t - |P| + |P| - 1 = t - 1$ in Γ_{t-2}, a contradiction.

We conclude the proof by showing that the area of the computed drawing is $O(t^3 n)$. We have $x_i = x_{i-1} + (2i - 1)n^* + i + 1$. We show by induction that $x_i = x_0 + i^2(n^* + 1) - \frac{i(i-3)}{2}$. This is true for $i = 0$; assume it is true for $i - 1$, we have $x_i = (x_0 + (i-1)^2(n^* + 1) - \frac{(i-1)(4-i)}{2}) + (2i-1)n^* + i + 1 = x_0 + i^2(n^* + 1) - \frac{i(i-3)}{2}$. The width of the drawing is $x_{t-1} + n^* - x_0$ which is $(t - 1)^2(n^* + 1) - \frac{(t-1)(t-4)}{2} + n^* = O(t^2 n^*) = O(t^2 n)$. Since the height is $O(t)$ the statement follows.   □

Lemma 1 implies that every graph with constant track number admits an h-quasi planar grid drawing in linear area with h being a constant. Since it is known that partial k-trees have track number that is constant in n (although depending on k) [14], this implies that every partial k-tree admit an h-quasi planar grid drawing in linear area where the value of h does not depend on n. The current best upper bound on the track number of k-trees is given in [12]. Thus, every k-tree has an h_k-quasi planar drawing in $O(n)$ area with $h_k \in O(1)$. In what follows we will improve this result by presenting a technique that gives better values for h_k.

Now we describe an algorithm, called kTreeLayouter, that computes a $(2, t)$-track layout of a k-tree where t depends on k but not on n. We start by recalling a decomposition technique introduced by Dujmović, Morin, and Wood [14] and by giving some further definitions that will be used to prove our results. Let $G = (V(G), E(G))$ be a graph and let $T = (V(T), E(T))$ be a rooted tree. Let $\{T_\mu \subseteq V(G) \mid \mu \in V(T)\}$ be a set of subsets of $V(G)$ indexed by the nodes of T. The pair $(T, \{T_\mu \mid \mu \in V(T)\})$ is a *tree partition* of G if: (i) $\forall \mu, \nu \in V(T)$, if $\mu \neq \nu$ then $T_\mu \cap T_\nu = \emptyset$; (ii) $\forall (u, v) \in E(G)$, either $\exists$ a node $\mu \in V(T)$ with $u, v \in T_\mu$, or $\exists$ an edge $(\mu, \nu) \in E(T)$ such that $u \in T_\mu$ and $v \in T_\nu$. Let μ be an element of $V(T)$ of a tree partition of G. The *pertinent graph* of μ is the subgraph of G induced by the vertices in T_μ; the pertinent graph of μ is denoted as G_μ. The following result about tree-partitions of k-trees is proved in [14].

Theorem 1. [14] *Let G be a k-tree. There exists a tree-partition $(T, \{T_\mu \mid \mu \in V(T)\})$ of G such that for every node μ of T: (i) The pertinent graph G_μ is a connected partial $(k-1)$-tree. (ii) If μ is a non-root node of T and ν is the parent of μ in T, then the set of vertices in T_ν with a neighbour in T_μ induce a clique of size k in G.*

The clique induced by the vertices in T_ν with a neighbour in T_μ is called the *parent clique* of μ. From now on, we shall only consider tree partitions with the properties of Theorem 1. For reasons of brevity, we shall often use T rather than $(T, \{T_\mu \mid \mu \in V(T)\})$ to denote a tree partition. Let (μ, ν) be an edge of T such that μ is the parent of ν. Let $e = (u, v)$ be an edge of G such that $u \in T_\mu$ and $v \in T_\nu$. Edge e is a *jumping edge*, vertex u is the *parent vertex* of e, and vertex v is the *child vertex* of e. We call *r-prism* a group of r tracks ($r > 1$) and *i-clique* a clique of size i ($i > 0$). Let G be a k-tree, let $\gamma(G)$ be a (c, t)-track layout of G and let Θ be a subset of $k + 1$ tracks of $\gamma(G)$. Let C be an $(k + 1)$-clique of G. C *covers* Θ if C has one vertex in each track of Θ. Let C_0 and C_1 be two $(k + 1)$-cliques of G. C_0 and C_1 are of the same *category* if they cover the same subset of tracks in $\gamma(G)$. The number of distinct categories of $\gamma(G)$ is called the *a-number* of $\gamma(G)$. Let C_0 and C_1 be two $(k + 1)$-cliques of G of the same category. C_0 and C_1 have the same *color* if the vertices of one of them (say C_0) precede (or possibly coincide with) the vertices of the other one (i.e., C_1) on all the tracks covered by the two cliques. This means that no two edges of the two cliques form an *X*-crossing. Notice that, given two cliques of the same color it is possible to order them according to the order of their vertices on the tracks that they cover. The maximum number of colors over all categories of $\gamma(G)$ is called the *b-number* of $\gamma(G)$. Two $(k + 1)$-cliques of G are of the same *type* if they are of the same category and have the same color. The number of distinct types (which is at most $a \cdot b$) is called the *c-number* of $\gamma(G)$. Since the cliques of the same color can be totally ordered, the cliques of the same type can be totally ordered accordingly. We denote such an ordering as $\prec_c$.

Let G be a k-tree, an *equipped tree partition* T of G is a tree partition such that each node μ is equipped with a (g_μ, t_μ)-track layout $\gamma(G_\mu)$ of its pertinent graph G_μ. Let a_μ, b_μ, and c_μ be the a-number, the b-number, and the c-number of $\gamma(G_\mu)$, respectively. We denote by t_T the value $\max_{\mu \in V(T)} t_\mu$. Analogously, we set $a_T = \max_{\mu \in V(T)} a_\mu$, $b_T = \max_{\mu \in V(T)} b_\mu$, and $c_T = \max_{\mu \in V(T)} c_\mu$. In order to compute a $(2, t)$-track layout of a k-tree G, we use a recursive technique based on an equipped tree partition of G. The pertinent graph G_μ of any node μ of the tree-partition T is a partial $(k-1)$-tree. G_μ is augmented to a k-tree and a $(2, t_\mu)$-track layout $\gamma(G_\mu)$ of G_μ with at most

t_T tracks is recursively computed. The maximum number of types of cliques in any $\gamma(G_\mu)$ is c_T. For each type of clique in G_μ the (at most) t_T-tracks of the $(2, t_\mu)$-track layout of each node whose parent clique is of that type are identified with the t_T tracks of a different t_T-prism. We define a total order $\prec_T$ of the nodes of the equipped tree partition T of G. To this aim we first define a total order $\prec_n$ of the children of each node λ of T. The children of λ are first ordered according to the categories of their parent cliques (the categories are ordered arbitrarily), within the same category they are ordered according to their parent clique color (the colors are ordered arbitrarily), within the same type they are ordered according to the order $\prec_c$ of their parent cliques; if they have the same parent clique they are ordered arbitrarily. The total order $\prec_T$ of the nodes of T is the order given by a preorder visit of T where the children of each node are visited according to $\prec_n$.

We are now ready to describe the algorithm kTreeLayouter to compute a $(2, (c_T + 1)t_T)$-track layout of G. We will use $(c_T + 1)$ t_T-prisms denoted as $P_0, \dots,$ P_{c_T}. This results in a number of tracks equal to $(c_T + 1)t_T$. The tracks of prism P_h $(0 \le h \le c_T)$ are denoted as $\tau_{h \cdot t_T + i}$ $(0 \le i \le t_T - 1)$. The nodes of T are processed one per time according to the total ordering $\prec_T$. Let G_μ be the pertinent graph of the current node μ, let λ be the parent of μ and let P_h be the t_T-prism whose tracks contain the vertices of G_λ. Let C be the parent clique of μ and let $\chi_{i,j}$ be its type $(0 \le i \le a_T - 1,$ $0 \le j \le b_T - 1)$, the (at most) t_T-tracks of the $(2, t_\mu)$-track layout $\gamma(G_\mu)$ of G_μ are identified with the t_T tracks of the t_T-prism $P_{h'}$ with $h' = (h + b_T \cdot i + j + 1)$ mod $(c_T + 1)$. Notice that, $h + 1 \le b_T \cdot i + j + 1 \le h + c_T$, which means that the t_T-prism $P_{h'}$ is different from P_h. Consider now a vertex v of G_μ and suppose that v belongs to a track τ_l $(0 \le l \le t_T - 1)$ in $\gamma(G_\mu)$; v is assigned to the track $\tau_{h' \cdot t_T + l}$ of $P_{h'}$. Moreover, the vertices of G_μ are ordered in the tracks of $P_{h'}$ in such a way that: (i) their relative order is the same as the one they have in $\gamma(G_\mu)$; (ii) they follow the vertices on their track that belong to the pertinent graph $G_{\mu'}$ of any node μ' of T that has been processed before μ by the algorithm. It is easy to see that the algorithm kTreeLayouter computes a $((c_T + 1)t_T)$-track assignment $\gamma(G)$. Namely, the edges of each G_μ do not have both endvertices in the same track because $\gamma(G_\mu)$ is a $(2, t_\mu)$-track layout; the jumping edges have endvertices in different tracks because they are in different t_T prisms. To prove that $\gamma(G)$ is a $(2, (c_T + 1)t_T)$-track layout of G, we give a preliminary lemma.

Lemma 2. *Let G be a k-tree and let $\gamma(G)$ be the track assignment computed by algorithm* kTreeLayouter. *Let τ_h and τ_l $(0 \le h, l \le (c_T + 1)t_T - 1)$ be two tracks of $\gamma(G)$. Let $e_0 = (u_0, v_0)$ and $e_1 = (u_1, v_1)$ be two jumping edges of G such that $u_0, u_1 \in \tau_h$, $v_0, v_1 \in \tau_l$, u_0 is the parent vertex of e_0 and u_1 is the parent vertex of e_1. Then e_0 and e_1 do not form an X-crossing.*

Lemma 3. *Let G be a k-tree. The algorithm* kTreeLayouter *correctly computes a $(2, (c_T + 1)t_T)$-track layout $\gamma(G)$ of G.*

Proof. In [15] it has been shown that, given a t-track assignment γ, it is possible to color the edges with c distinct colors so that no two edges of the same color form an X-crossing (i.e., to compute a (c, t)-track layout) if and only if γ has no crossing $(c+1)$-tuple. A set S of $c + 1$ edges in a track assignment γ is called a *crossing $(c + 1)$-tuple* if each pair of edges in S form an X-crossing in γ. Thus, to prove our statement it sufficient to show that there is no crossing 3-tuple in $\gamma(G)$. Consider any three edges $e_0 = (u_0, v_0)$, $e_1 = (u_1, v_1)$, and $e_2 = (u_2, v_2)$ such that u_0, u_1, and u_2 are in the same track τ_h and v_0, v_1, and v_2 are in the same track τ_l $(0 \leq h, l \leq (c_T + 1)t_T - 1)$. Assume first that τ_h and τ_l belong to the same t_T-prism. If e_0, e_1, and e_2 are edges of the same pertinent graph G_μ, then they do not form a crossing 3-tuple because otherwise there would be a crossing 3-tuple in the $(2, t_T)$-track layout $\gamma(G_\mu)$ of G_μ. If e_0, e_1, and e_2 are edges of different pertinent graphs, then at least two of them do not form an X-crossing (give two distinct pertinent graphs on the same t_T-prism, the vertices of one of them follow the vertices of the other one) and therefore they cannot form a crossing 3-tuple.

Assume now that τ_h and τ_l belong to different t_T-prisms (e_0, e_1, and e_2 are jumping edges). At least two among u_0, u_1, and u_2 are either parent vertices or child vertices of their jumping edges. By Lemma 2, at least two among e_0, e_1 and e_2 do not cross.   □

The proof of the upper bound to the value $(c_T + 1)t_T$ is omitted. We can prove that the values of h_k given in Theorem 2 are smaller than those obtained by using the track number upper bound in [12].

Theorem 2. *Every partial k-tree with n vertices admits an h_k-quasi planar grid drawing in $O(t_k^3 n)$ area, where $h_k = 2t_k - 1$ and t_k is given by the following recursive equation:*

$$t_k = (c_{k-1,k} + 1)t_{k-1}$$

$$c_{k,i} = (c_{k-1,k} + 1)(c_{k-1,i} + \frac{c_{k-1,k}}{4} \sum_{j=1}^{i-1} c_{k-1,j} \cdot c_{k-1,i-j}) \quad (i = 1, \ldots, k+1) \quad (1)$$

$$c_{k,k+2} = 0$$

with $t_1 = 2$ and $c_{1,1} = 4$ and $c_{1,2} = 2$.

By Theorem 2, every partial 2-tree admits an 11-quasi planar drawing in $O(n)$ area. Partial 2-trees are SP-graphs [5], which will be further investigated in the next sections.

4   Improved Bounds for Specific Families of Planar Partial k-Trees

According to Theorem 2, every n-vertex partial k-tree admits an h-quasi planar drawing in $O(n)$ area with $h \in O(1)$. In this section we describe some ad-hoc drawing techniques that, still producing drawing in linear area, reduce the value of h for some sub-families of partial k-trees.

Outerplanar graphs. A graph G is *outerplanar* if it admits a planar embedding such that all vertices are on the external face (i.e., an *outerplanar embedding*). It is known that outerplanar graphs are partial 2-trees [5]. Thus, from Theorem 2 they admit an 11-quasi planar drawing in $O(n)$ area. We prove that the value of h can be reduced from 11 to 3, describing an algorithm OuterplanarDrawer, which takes as input an n-vertex outerplanar graph G with a given outerplanar embedding and returns a quasi planar grid drawing of G. The algorithm uses an approach similar to the one described in [17]. It can be divided in two main steps. In the first step it computes a drawing Γ^* of G as follows. Perform a breadth-first-search of G (starting from any vertex) and assign to each vertex v of G two numbers: $level(v)$ which is the depth of v in the BFS tree, and $order(v)$ which is the progressive number of v in the BFS order. For each vertex v of G set $x^*(v) = order(v)$ and $y^*(v) = level(v)$, where $x^*(v)$ and $y^*(v)$ are the x- and y-coordinates of v in Γ^*, respectively. In the second step, it "wraps" the drawing Γ^* of G on two levels, producing the final drawing Γ. For each vertex v of G, it sets $x(v) = x^*(v)$ and $y(v) = y^*(v) \mod 2$, where $x(v)$ and $y(v)$ are the x- and y-coordinates of v in Γ, respectively.

Theorem 3. *Every outerplanar graph with n vertices admits a quasi planar grid drawing in $O(n)$ area.*

Flat series-parallel graphs. A series-parallel graph, or SP-graph, is *flat* if it does not contain two nested parallel components. For an exact definition of flat SP-graphs and decomposition tree see [11]. Flat SP-graphs are a meaningful subfamily of SP-graphs, previously studied in [11]. We lower the value of h for flat SP-graphs from 11 to 5.

Let G be a flat SP-graph and let T be its decomposition tree. We assign to each node ν of T a number, denoted as $level(\nu)$, computed as follows. The root ρ of T has $level(\rho) = 0$. For each non-root node ν, if ν is an S-node then $level(\nu) = level(parent(\nu)) + 1$, else $level(\nu) = level(parent(\nu))$. Using the level numbering of T we assign a number $level(v)$ to each vertex v of G, which is the minimum among the levels of all the nodes of T having v as a pole. We call *jumping edges* those edges whose end-vertices are assigned to different levels. Notice that the level numbering is such that the level number changes in correspondence of the S-nodes. In [11] it has been proved that the leftmost child and the rightmost child of an S-node are both Q-nodes and the edges associated with them are both jumping edges.

We can now describe the drawing algorithm `FlatSPDrawer`, which takes as input a flat SP-graph G and its decomposition tree T and returns a 5-quasi planar grid drawing of G. Also in this case the algorithm has two main steps. In the first step we produce a preliminary drawing Γ^* of G. For each vertex v of G, we set $y^*(v) = level(v)$ and compute x^* as follows. We perform a breadth first search of T, initializing a counter $i = 0$ before starting the visit. For each node ν of T in the BFS order, if ν is a P-node or a Q-node we process its two poles s and t: if the x-coordinate of the source s has not yet been assigned we set $x^*(s) = i$ and increment i by one unit; if the x-coordinate of the sink t has not yet been assigned we set $x^*(t) = i$ and increment i. Notice that if both the poles of ν have not been processed before considering ν (i.e., ν does not share them with its parent), then they receive consecutive x-coordinates. Again, Γ is obtained from Γ^* by setting $x(v) = x^*(v)$ and $y(v) = y^*(v) \mod 2$ for each vertex v of G.

Lemma 4. *Let G be a flat SP-graph and let T be its decomposition tree. Let $l \leq n$ be the number of levels assigned to the nodes of T by the algorithm `LevelNumbering`. The first step of the algorithm `FlatSPDrawer` produces a drawing Γ^* of G on l levels, such that: (i) for every edge e of G either e connects two vertices on the same level, or e is a jumping edge connecting vertices between two consecutive levels; (ii) there are no overlaps among edges; (iii) there are no three mutually crossing edges.*

Proof. By definition a non-jumping edge $e = (u, v)$ has $|level(u) - level(v)| = 0$, and therefore e connects two vertices on the same level. For a jumping edge $e = (u, v)$, we have, by definition, $|level(u) - level(v)| > 0$. Let e be a jumping edge. As already said, the Q node representing e in T is either the leftmost or the rightmost child of a non-root S-node ν. As a consequence, one end vertex of e, say u, is a pole shared by ν and its parent (a P-node); the other end vertex v is a pole shared by two consecutive children of ν (one of which is the leftmost or rightmost child) and not shared with ν. Since the level number changes only in correspondence of the S-nodes, $level(u) = level(parent(\nu))$ and $level(v) = level(\nu)$, i.e., $|level(u) - level(v)| = 1$.

Since every vertex v has a different x-coordinate, there can be only two kinds of crossings: an overlap between two non jumping edges, or a proper crossing between two jumping edges. We prove now that the first case never happen. Let ν be a P-node or a Q-node of T such that $level(\nu) = j$, let s and t be its two poles and let μ be its parent node. We have the following cases: **(1)** ν is a P-node and μ is an S-node. In this case ν and μ do not share a pole (the leftmost and rightmost child of μ are Q-nodes), thus the two poles of ν have consecutive x-coordinates, i.e., $x^*(s) - x^*(t) = 1$ and $y^*(s) = y^*(t)$. **(2)** ν is a Q-node and μ is a P-node. In this case ν represents a transitive edge connecting the two poles of μ which have already been processed when μ was considered; by case 1 we have $x^*(s) - x^*(t) = 1$ and $y^*(s) = y^*(t)$. **(3)** ν is a Q-node and μ is an S-node. If ν is the leftmost/rightmost child of μ, its associated edge is a jumping edge and the two poles have distinct y-coordinates. If ν is not the leftmost/rightmost child of μ, then ν and μ do not share a pole. Also in this case the two poles of ν have consecutive x-coordinates, i.e., $x^*(s) - x^*(t) = 1$ and $y^*(s) = y^*(t)$. If e is a non-jumping edge, then either Case 2 or 3 holds for its corresponding Q-node. In both cases the endvertices of e have consecutive x-coordinates. It follows that there can not be an overlap between two non-jumping edges.

Now we prove that there are no more than 2 mutually crossing jumping edges. We assign to a jumping edge the red color if its corresponding Q-node is the leftmost child of its parent and the blue color if its corresponding Q-node is the rightmost child of its parent. Let $e = (u, v)$ and $e' = (w, z)$ be two jumping edges of the same color. If $level(u) \neq level(w)$ or $level(v) \neq level(z)$ then it is immediate to see that e and e' do not cross. Assume then $level(u) = level(w) = j$ and $level(v) = level(z) = j + 1$. If e and e' share an end vertex they obviously cannot cross. If e and e' do not share an end vertex, u and w are two poles of two different S-nodes ν_u and ν_w. Assume that $x^*(u) < x^*(w)$, which means that ν_u is visited before ν_w in the BFS visit of T. This implies that the Q-node of e is visited before the Q-node of e'. Thus, $x^*(v) < x^*(z)$ and e and e' cannot cross. Hence, there cannot be three mutually crossing edges because red edges can cross only blue edges and vice versa.   □

Theorem 4. *Every flat SP-graph with n vertices admits a 5-quasi planar grid drawing in $O(n)$ area.*

Proper simply-nested graphs. A graph is k-*outerplanar* ($k > 1$) if it admits a planar embedding such that the graph remaining after removing all vertices on the external face is a $(k - 1)$-outerplanar graph. A graph is 1-*outerplanar* if it is outerplanar. In other words a graph is k-outerplanar if it admits a planar embedding such that it can be made empty by removing the vertices on the external face k times. The vertices that are on the external face after i ($0 \leq i \leq k-1$) removals are called vertices of level $i+1$. A *simply-nested graph* is a k-outerplanar graph such that the vertices of levels from 1 to $k-1$ are chordless cycles and level k is either a cycle or a tree. Simply-nested graphs have been widely studied in the literature (see, e.g., [2]). We say that a simply-nested graph is *proper* if level k is a chordless cycle. It is known that k-outerplanar graphs have treewidth at most $3k-1$ [5]. By using the technique of Section 3 we would obtain an h-quasi planar drawing in linear area with h given by Equation 1. Notice that h would be a function of the number of levels k. We show that for simply-nested graphs h can be reduced to 3 (independent of the number of levels). We remark that proper simply-nested graphs may require quadratic area if we want a planar drawing; they include the classical examples used to prove the quadratic area lower bound of planar graphs.

Theorem 5. *Every proper simply-nested graph with n vertices admits a quasi planar grid drawing in $O(n)$ area.*

Sketch of Proof: Let G be a proper simply-nested graph. We describe an algorithm to compute a quasi planar grid drawing of G. Let $C_1, C_2, \ldots, C_k$ be the cycles of levels $1, 2, \ldots, k$, respectively. We assume that all the internal faces of G except possibly the one delimited by C_k are triangles. If this is not the case, we can add edges to guarantee this property. For each cycle C_i we choose a vertex, denoted as v_i, called the *reference vertex* of C_i. The references vertices are chose in such a way that v_i is adjacent to v_{i-1}.

We draw the vertices of each cycle C_i on an isosceles triangle T_i whose basis has length $2(n_i - 2)$ and height 2, where $n_i = |C_i|$. The y-coordinate of the apex of T_i is 3 if i is odd or 0 if i is even. The y-coordinate of the basis of T_i is 1 if i is odd or 2 if i is even. All the vertices of C_i, except v_i, are placed on the basis of T_i on grid points with even x-coordinates so that their left-to-right order coincides with (is opposite to) their counter-clockwise order along C_i if i is odd (if i is even). Vertex v_i is drawn at the apex of T_i. Denote by x_i the x-coordinate of v_i ($i = 1, \ldots, k$) and let $m_i = \max\{n_{i-1}, n_i\}$ ($i = 2, \ldots, k$). The triangles are placed so that $x_1 = n_1 - 2$ and $x_i = x_{i-1} + \lceil \frac{3}{2}(m_i - 1) \rceil$. □

5 Comparing h-Quasi Planarity and h-Planarity

Other definition of crossing complexity are possible, for example h-*planarity* [20]. A drawing of a graph is h-*planar* if no edge has more than h crossings. Straight-line 1-planar drawings are studied in [16]. A natural question deriving from the results of Sections 3 and 4 is whether analogous results also hold for h-planar drawings. Theorem 6 shows that, for every constant h, $\omega(n)$ area is required for SP-graphs, while Theorem 2 implies that every SP-graph admits an 11-quasi planar drawing in $O(n)$ area.

Let G be a graph, we define the h^*-*extension* of G as a graph G^*, constructed by attaching h^* paths of length 2 to each edge of G.

Lemma 5. *Let h be a positive integer, and let G be a planar graph. In any h-planar drawing of the $3h$-extension G^* of G, there are no two edges of G that cross each other.*

Theorem 6. *Let h be a positive integer, for every $n > 0$ there exist a $\Theta(n)$-vertex series-parallel graph such that any h-planar straight-line or poly-line grid drawing requires $\Omega(n2^{\sqrt{\log n}})$ area.*

Proof. Let G be an n-vertex graph of the family defined by Frati in [19], which requires $\Omega(n2^{\sqrt{\log n}})$ area in any planar straight-line or poly-line drawing. By Lemma 5 there exists an n^*-vertex planar graph G^*, with $n^* = \Theta(n)$, such that in any h-planar drawing Γ of G^* the underlying graph G must be drawn planar. Since G^* is still a SP-graph (the $3h$-extension preserves the property of being a SP-graph) the statement follows.   □

With the same argument, we can prove the following theorem for general planar graphs. Notice that it states that quadratic area is necessary if we impose $h(n) \in O(1)$.

Theorem 7. *Let ε be given such that $0 \leq \varepsilon \leq 0.5$ and let $h(n) : \mathbb{N} \to \mathbb{N}$ be a function such that $h(n) \leq n^{0.5-\varepsilon} \; \forall n \in \mathbb{N}$. For every $n > 0$ there exists an $O(n)$-vertex graph G such that any $h(n)$-planar straight-line grid drawing of G requires $\Omega(n^{1+2\varepsilon})$ area.*

6 Concluding Remarks and Open Problems

In this paper we studied the problem of computing compact h-quasi planar drawings of partial k-trees. Indeed, our algorithms can be regarded as drawing techniques that produce drawings with optimal area and with bounded crossing complexity. This point of view is particularly interesting in the case of planar graphs. As recalled in the introduction, planar graphs can be drawn with either optimal crossing complexity (i.e., in a planar way), in which case they may require $\Omega(n^2)$ area [8], or with optimal $\Theta(n)$ area but without any guarantee on the crossing complexity [22]. These two extremal results naturally raise the following question: is it possible to compute a drawing of a planar graph "controlling" both the area and the crossing complexity? In particular, it is possible to compute an h-quasi planar drawing of a planar graph in $o(n^2)$ area and $h \in o(n)$? In Section 4 we showed that $O(n)$ area and $h \in O(1)$ can be simultaneously achieved for some families of planar graphs. In fact our results imply a positive answer to the above question even for general planar graphs.

Theorem 8. *Every planar graph with n vertices admits a $O(\log^{16} n)$-quasi planar grid drawing in $O(n \log^{48} n)$ area.*

Proof. Let G be a n-vertex graph with acyclic chromatic number $\chi_a(G) \leq c$ and queue number $qn(G) \leq q$, then G has track-number $tn(G) \leq c(2q)^{c-1}$ [14]. If G is planar then $qn(G) \in O(\log^4 n)$ [10] and $\chi_a(G) = 5$ [6]. Thus, every planar graph has $tn(G) \in O(\log^{16} n)$ and by Lemma 1 the statement follows.   □

The results in this paper give rise to several interesting open problems. Among them: (1) Reducing the value of h_k given by Equation 1 for other sub-families of partial k-trees. (2) Studying whether planar graphs admits h-quasi planar drawings in $O(n)$ area with $h \in o(n)$, possibly $h \in O(1)$. (3) Studying h-quasi planar drawings in linear area and aspect ratio $o(n)$.

References

1. Angelini, P., Di Battista, G., Didimo, W., Frati, F., Hong, S.-H., Kaufmann, M., Liotta, G., Lubiw, A.: RAC and LAC drawings of planar graphs in subquadratic area. In: ECG 2011, pp. 125–128 (2011)
2. Angelini, P., Di Battista, G., Kaufmann, M., Mchedlidze, T., Roselli, V., Squarcella, C.: Small Point Sets for Simply-Nested Planar Graphs. In: Speckmann, B. (ed.) GD 2011. LNCS, vol. 7034, pp. 75–85. Springer, Heidelberg (2011)
3. Biedl, T.C.: Drawing Outer-Planar Graphs in $O(n \log n)$ Area. In: Goodrich, M.T., Kobourov, S.G. (eds.) GD 2002. LNCS, vol. 2528, pp. 133–148. Springer, Heidelberg (2002)
4. Biedl, T.C.: Small Drawings of Series-Parallel Graphs and Other Subclasses of Planar Graphs. In: Eppstein, D., Gansner, E.R. (eds.) GD 2009. LNCS, vol. 5849, pp. 280–291. Springer, Heidelberg (2010)
5. Bodlaender, H.L.: A partial k-arboretum of graphs with bounded treewidth. TCS 209(1-2), 1–45 (1998)
6. Borodin, O.: On acyclic colorings of planar graphs. Discr. Math. 25(3), 211–236 (1979)
7. Crescenzi, P., Di Battista, G., Piperno, A.: A note on optimal area algorithms for upward drawings of binary trees. CGTA 2, 187–200 (1992)
8. de Fraysseix, H., Pach, J., Pollack, R.: How to draw a planar graph on a grid. Combinatorica 10, 41–51 (1990)
9. Di Battista, G., Frati, F.: Small area drawings of outerplanar graphs. Algorithmica 54, 25–53 (2009)
10. Di Battista, G., Frati, F., Pach, J.: On the queue number of planar graphs. In: Proc. of FOCS 2010, pp. 365–374 (2010)
11. Di Giacomo, E.: Drawing Series-Parallel Graphs on Restricted Integer 3D Grids. In: Liotta, G. (ed.) GD 2003. LNCS, vol. 2912, pp. 238–246. Springer, Heidelberg (2004)
12. Di Giacomo, E., Liotta, G., Meijer, H.: Computing straight-line 3D grid drawings of graphs in linear volume. CGTA 32(1), 26–58 (2005)
13. Didimo, W., Eades, P., Liotta, G.: Drawing graphs with right angle crossings. TCS 412(39), 5156–5166 (2011)
14. Dujmović, V., Morin, P., Wood, D.R.: Layout of graphs with bounded tree-width. SIAM J. on Comp. 34(3), 553–579 (2005)
15. Dujmović, V., Pór, A., Wood, D.R.: Track layouts of graphs. DMTCS 6(2), 497–522 (2004)
16. Hong, S.-H., Eades, P., Liotta, G., Poon, S.: Fáry's Theorem for 1-Planar Graphs. In: Gudmundsson, J., Mestre, J., Viglas, T. (eds.) COCOON 2012. LNCS, vol. 7434, pp. 335–346. Springer, Heidelberg (2012)
17. Felsner, S., Liotta, G., Wismath, S.K.: Straight-Line Drawings on Restricted Integer Grids in Two and Three Dimensions. In: Mutzel, P., Jünger, M., Leipert, S. (eds.) GD 2001. LNCS, vol. 2265, pp. 328–342. Springer, Heidelberg (2002)
18. Fox, J., Pach, J.: Coloring k_k-free intersection graphs of geometric objects in the plane. In: Proc. of SCG 2008, pp. 346–354. ACM (2008)
19. Frati, F.: Lower bounds on the area requirements of series-parallel graphs. DMTCS 12(5), 139–174 (2010)
20. Pach, J., Tóth, G.: Graphs drawn with few crossings per edge. Combinatorica 17, 427–439 (1997)
21. Suk, A.: k-Quasi-Planar Graphs. In: Speckmann, B. (ed.) GD 2011. LNCS, vol. 7034, pp. 266–277. Springer, Heidelberg (2011)
22. Wood, D.R.: Grid drawings of k-colourable graphs. CGTA 30(1), 25–28 (2005)

The Duals of Upward Planar Graphs on Cylinders[*]

Christopher Auer, Christian Bachmaier, Franz J. Brandenburg,
Andreas Gleißner, and Kathrin Hanauer

University of Passau, Germany
{auerc,brandenb,bachmaier,gleissner,hanauer}@fim.uni-passau.de

Abstract. We consider directed planar graphs with an upward planar drawing on the rolling and standing cylinders. These classes extend the upward planar graphs in the plane. Here, we address the dual graphs. Our main result is a combinatorial characterization of these sets of upward planar graphs. It basically shows that the roles of the standing and the rolling cylinders are interchanged for their duals.

1 Introduction

Directed graphs are used as a model for structural relations where the edges express dependencies. Such graphs are often acyclic and are drawn as hierarchies using the framework introduced by Sugiyama et al. [21]. This drawing style transforms the edge direction into a geometric direction: all edges point upward. If only plane drawings are allowed, one obtains *upward planar graphs*, for short **UP**. These graphs can be drawn in the plane such that the edge curves are monotonically increasing in y-direction and do not cross. Hence, **UP** graphs respect the unidirectional flow of information as well as planarity.

There are some fundamental differences between upward planar and undirected planar graphs. For instance, there are several linear time planarity tests [17], whereas the recognition problem for **UP** is $\mathcal{NP}$-complete [13]. The difference between planarity and upward planarity becomes even more apparent when different types of surfaces are studied: For instance, it is known that every graph embeddable on the plane is also embeddable on any surface of genus 0, e.g., the sphere and the cylinder, and vice versa. However, there are graphs with an upward embedding on the sphere with edge curves increasing from the south to the north pole, which are not upward planar [16]. The situation becomes even more challenging if upward embeddability is extended to other surfaces even if these are of genus 0.

Upward planarity on surfaces other than the plane generally considers embeddings of graphs on a fixed surface in $\mathbb{R}^3$ such that the curves of the edges are monotonically increasing in y-direction. Examples for such surfaces are the standing [7, 14, 19, 20, 22] and the rolling cylinder [7], the sphere and the truncated sphere [10,12,15,16], and the lying and standing tori [9,11]. We generalized

---

[*] Supported by the Deutsche Forschungsgemeinschaft (DFG), grant Br835/15-2.

M.C. Golumbic et al. (Eds.): WG 2012, LNCS 7551, pp. 103–114, 2012.

upward planarity to arbitrary two-dimensional manifolds endowed with a vector field which prescribes the direction of the edges [2]. We also studied upward planarity on standing and rolling cylinders, where the former plays an important role for radial drawings [3] and the latter in the context of recurrent hierarchies [4]. We showed that upward planar drawings on the rolling cylinder can be simplified to polyline drawings, where each edge needs only finitely many bends and at most one winding around the cylinder [7]. The same holds for the standing cylinder, where all windings can be eliminated [7]. In accordance to [2], we use the fundamental polygon to define the plane, the standing and the rolling cylinders. The *plane* is identified with $I \times I$, where I is the open interval from -1 to $+1$, i.e., $I \times I$ is the (interior of the) square with side length two. The *rolling (standing) cylinder* is obtained by identifying the bottom and top (left and right) sides. By identifying the boundaries of I, we obtain I_o. Then, the standing and the rolling cylinder are defined by $I_o \times I$ and $I \times I_o$, respectively. Let **RUP** be the set of graphs which can be drawn on the rolling cylinder such that the edge curves do not cross and are monotonically increasing in y-direction. If the edge curves are permitted to be non-decreasing in y-direction, horizontal lines are allowed. Since the top and bottom sides of the fundamental polygon are identified, "upward" means that edge curves wind around the cylinder all in the same direction. Specifically, **RUP** allows for cycles. Accordingly, let **SUP** denote the class of graphs with a planar drawing on the standing cylinder and increasing curves for the edges and let **wSUP** be the corresponding class of graphs with non-decreasing curves. The novelty of **wSUP** graphs are cycles with horizontal curves, whereas **SUP** graphs are acyclic, i.e., $\textbf{SUP} \subsetneq \textbf{wSUP}$. In [2] we established that a graph is in **SUP** if and only if it is upward planar on the sphere. These *spherical* graphs were studied in [10, 12, 15, 16]. Finally, let **UP** be the class of upward planar graphs (in the plane) [8, 18]. Note that for **UP** and **RUP** graphs non-decreasing curves can be replaced by increasing ones and the corresponding classes coincide [2].

Upward planar graphs in the plane and on the sphere or on the standing cylinder were characterized by using *acyclic dipoles*. An acyclic dipole is a directed acyclic graph with a single source s and a single sink t. More specifically, a graph G is **SUP**/spherical if and only if it is a spanning subgraph of a planar acyclic dipole [14, 16, 19]. The idea behind acyclic dipoles is that s corresponds to the south and t to the north pole of the sphere. Moreover, a graph G is in **UP** if and only if the dipole has in addition the (s, t) edge [8, 18].

In contrast, there is no related characterization of **RUP** graphs. Acyclic dipoles cannot be used since **RUP** graphs may have cycles winding around the rolling cylinder. However, the idea behind dipoles can be applied indirectly to **RUP** graphs, namely, to their duals. For this, we generalize acyclic dipoles to *dipoles* which may also contain cycles.

Section 2 provides the necessary definitions. We develop our new characterization of **RUP** and **SUP** graphs in terms of their duals in Sect. 3. In Sect. 4 we obtain related results for **wSUP** graphs. All formal proofs can be found in [1].

2 Preliminaries

The graphs in this work are connected, planar (unless stated otherwise), directed multigraphs $G = (V, E)$ with non-empty sets of vertices V and edges E, where pairs of vertices may be connected by multiple edges. G can be drawn in the plane such that the vertices are mapped to distinct points and the edges to non-intersecting Jordan curves. Then, G has a planar *drawing*. It implies an *embedding* of G, which defines (cyclic) orderings of incident edges at the vertices. In the following, we only deal with embedded graphs and all paths and cycles are simple.

A *face* f of G is defined by a (underlying undirected) circle $C = (v_1, e_1, v_2, e_2, \ldots, v_{k-1}, e_{k-1}, v_k = v_1)$ such that $e_i \in E$ is the direct successor of $e_{i-1} \in E$ according to the cyclic ordering at v_i. The edges/vertices of C are said to be the *boundary* of f and C is a *clockwise traversal* of f. Accordingly, the *counterclockwise traversal* of f is obtained by choosing the predecessor edge at each vertex in the circle. The embedding defines a unique *(directed) dual* graph $G^* = (F, E^*)$, whose vertex set is the set of faces F of G [5]. Let $f \in F$ be a face of G and $e = (u, v) \in E$ be part of its boundary. If the counterclockwise traversal of f passes e in its direction, we say that f *is to the left of* e. If the same holds for e and another face g in clockwise direction, then g *is to the right of* e. For each edge $e \in E$ there is an edge in E^* from the face to the left of e to the face right of e. This definition establishes a bijection between E and E^*. Whenever necessary, we refer to G as the *primal* of G^*. By vertex we mean an element of V, whereas the vertices F of G^* are called faces.

Note that G^* is connected and the dual of G^* is isomorphic to the *converse* G^{-1} of G where all edges are reversed, since G is connected. Hence, an embedding of G implies an embedding of G^*, and vice versa. G and G^{-1} share many properties, see Proposition 1.

An embedding of a graph is an X *embedding* with $X \in \{\mathbf{RUP}, \mathbf{SUP}, \mathbf{wSUP}, \mathbf{UP}\}$ if it is obtained from an X drawing. For every graph in class X, we assume that a corresponding X embedding is given. Given an embedded graph G, a face f *is to the left of a face* g if there is a path $f \rightsquigarrow g$ in G^*. Note that a face can simultaneously lie to the left and to the right of another face, and "to the left" does not directly correspond to the geometric left-to-right relation in a drawing. A cycle in a $\mathbf{RUP}$ embedding winds exactly once around the cylinder [7]. We say that a face $f \in F$ lies *left (right) of a cycle* C if there is another face $g \in F$ such that f is to the left (right) of g and each path $f \rightsquigarrow g$ in the dual contains at least one edge of C. Each edge/face of f's boundary is then also said to *lie to the left (right) of* C.

Next we introduce graphs which represent the high-level structure of a given graph and which inherit its embedding. Let the equivalence class $[v]$ denote the set of vertices of the strongly connected component containing the vertex $v \in V$ and let $\mathbb{V}$ be the set of strongly connected components of G. The *component graph* $\mathbb{G} = (\mathbb{V}, \mathbb{E})$ of G contains an edge $([v], [w]) \in \mathbb{E}$ for each original edge $(v, w) \in E$ with $[v] \neq [w]$. $\mathbb{G}$ is an acyclic multigraph which inherits the embedding of G. A component $\gamma \in \mathbb{V}$ is a *compound*, if it contains more than one vertex or consists of a single vertex with a loop. Its induced subgraph is denoted by $G_\gamma \subseteq G$. For the sake of convenience, we identify G_γ with γ and call both compound. The set of all compounds is denoted by $\mathbb{V}_C$. Each component $[v]$ that is *not a compound* consists of a single vertex v and is called *trivial component*. A trivial component which is a source (sink) in $\mathbb{G}$ is called *source (sink) terminal* and the set of all terminals is denoted by $\mathbb{T} \subseteq \mathbb{V}$. Based on the component graph, we define the *compound graph* $\overline{G} = (\mathbb{V}_C \cup \mathbb{T}, \overline{\mathbb{E}})$, whose vertices are the compounds and terminals. Let $u, v \in \mathbb{V}_C \cup \mathbb{T}$ be two vertices of the compound graph. There is an edge $(u, v) \in \overline{\mathbb{E}}$ if there is a path $u \rightsquigarrow v$ in $\mathbb{G}$ which internally visits only trivial components. Note that $\overline{G}$ is a simple graph. Each edge $\tau \in \overline{\mathbb{E}}$ corresponds to a set of paths in $\mathbb{G}$. Denote by $\mathbb{G}_\tau$ the subgraph of $\mathbb{G}$ which is induced by the set of paths belonging to edge τ. We call τ and its induced graph $\mathbb{G}_\tau$ *transit*. See Fig. 1 for an example, where the fundamental polygon of the rolling cylinder is represented by rectangles with identified bottom and top sides. Based on these definitions, we are now able to define dipoles.

Definition 1. *A graph is a* dipole *if it has exactly one source s and one sink t and its compound graph is a path from s to t.*

Note that similar to the definition of st-graphs [8,18], a dipole is not necessarily planar.

Lemma 1. *Let $G = (V, E)$ be a graph with a source s and a sink t. Then, G is a dipole if and only if every path $s \rightsquigarrow t$ contains at least one vertex of each compound and for every vertex $v \in V$ there are paths $s \rightsquigarrow v$ and $v \rightsquigarrow t$.*

Proposition 1. *A graph G is (i) acyclic, (ii) strongly connected, (iii) upward planar, or (iv) a dipole if and only if the same holds for its converse G^{-1}.*

Thus, in the subsequent statements on the relationship between a graph G and its dual G^*, the roles of G and G^* are interchangeable.

Lemma 2. *A graph G is acyclic if and only if its dual G^* is strongly connected.*

The proof is deduced from the one for polynomial solvability of the feedback arc set problem on planar graphs as given in [5].

3   RUP and SUP Graphs and Their Duals

We consider **RUP** graphs, i.e., upward planar graphs on the rolling cylinder, and characterize them in terms of their duals. Our main result is:

Theorem 1. *A graph G is a **RUP** graph if and only if G is a spanning subgraph of a planar graph H without sources or sinks whose dual H^* is a dipole.*

The theorem is proved by a series of lemmata which are also of interest in their own. For our first observation, consider the **RUP** drawing of graph G in Fig. 1(a), where all vertices within a compound are drawn on a shaded background. The

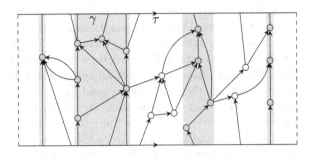

(a) Graph $G \in$ **RUP**

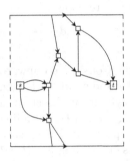

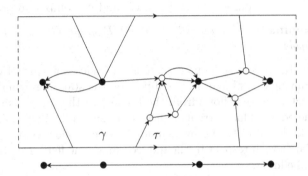

(b) The dual of the second compound γ of G

(c) The component graph $\mathbb{G}$ and the compound graph $\overline{\mathbb{G}}$ of G

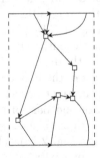

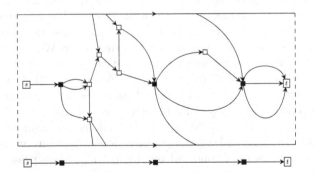

(d) The dual of the second transit τ of G

(e) The component graph $\mathbb{G}^*$ and the compound graph $\overline{\mathbb{G}^*}$ of the dual G^* with $s, t \in \mathbb{T}$

Fig. 1. A **RUP** example

component graph $\mathbb{G}$ of G is displayed in Fig. 1(c) along with its compound graph $\overline{\mathbb{G}}$ below, where the compounds are shaded black. Note that $\overline{\mathbb{G}}$ has the structure of an (undirected) path. Due to Lemma 2, each transit, i. e., edge in $\overline{\mathbb{G}}$, becomes a compound and each compound, i. e., vertex in $\overline{\mathbb{G}}$, becomes a transit in the dual G^* of G. Hence, the path-like structure of $\overline{\mathbb{G}}$ must carry over to the compound graph $\overline{\mathbb{G}}^*$ of G^*. Moreover, since all cycles in the **RUP** drawing have the same orientation, i. e., they all wind around the cylinder in the same direction, the transits in G^* point into the same direction. Also note that G contains neither sources nor sinks, i. e., both the left and right border of the drawing are directed cycles C_l and C_r, respectively. Hence, in the dual G^* of G, the face to the left of C_l is a source s and the face to the right of C_r is a sink t. All these observations together indicate that the compound graph of G^* is a path $s \rightsquigarrow t$, i. e., G^* is a dipole. Indeed, this can be seen for the example in Fig. 1(e), where the component graph of G^* and its compound graph are depicted.

Lemma 3. *The dual G^* of a **RUP** graph G without sources and sinks is a dipole.*

For the following lemma, there is a physical interpretation: Consider an upward drawing of a planar acyclic dipole on the standing cylinder and suppose that an electric current flows from the bottom to the top of the cylinder in direction of the edges. This current induces a magnetic field wrapping around the standing cylinder. Intuitively, by Lemma 4, we can show that a dipole's dual is upward planar with respect to the induced magnetic field, i. e., its embedding is a **RUP** embedding.

Lemma 4. *The embedding of a strongly connected graph G is a **RUP** embedding if and only if its dual G^* is an acyclic dipole.*

The only-if direction follows from Lemmata 2 and 3. For the if direction, we give a sketch of the proof. Let $G^* = (F, E^*)$ be the dual of G and consider a topological ordering $f_1, \ldots, f_k$ of the faces F. We subsequently process the faces according to their topological ordering and construct a drawing of G by placing the edges and vertices of the faces in that order. We start with the only source f_1 in G^*, which corresponds to a directed cycle in G (Fig. 2(a)). In the drawing with the boundaries of all faces $f_1, \ldots, f_i$, we can show that one part of the boundary of face f_{i+1} consists of a directed path $p = (u, \ldots, v)$ along the right border of the current drawing (solid black vertices in Fig. 2(b)). The second part of the boundary, absent in the current drawing, is also a directed path $p' = (u, \ldots, v)$ with the same end points and direction as p (white vertices in Fig. 2(b)). The drawing can be augmented by this path p' while preserving planarity and all edges are monotonically increasing in y-direction (Fig. 2(c)).

Since each **SUP** graph is a subgraph of a planar, acyclic dipole [16], Lemma 4 implies:

Corollary 1. *The dual G^* of a strongly connected **RUP** graph G is in **SUP**.*

Consider again the component graph $\mathbb{G}$ and its compound graph $\overline{\mathbb{G}}$ in Fig. 1(c) of the **RUP** graph G in Fig. 1(a). In the dual G^* of G, compounds and transits of

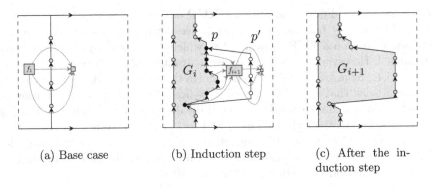

(a) Base case   (b) Induction step   (c) After the induction step

Fig. 2. Inductive construction of a **RUP** drawing from its dual

G swap their roles, i. e., compounds become transits and vice versa, cf. Fig. 1(e). As a compound of G is a strongly connected **RUP** graph, its dual is an acyclic dipole by Lemma 4. For instance, consider the second compound γ in Fig. 1(a), i. e., the vertices on the second shaded area labeled with γ. Its dual is indeed an acyclic dipole as depicted in Fig. 1(b). For the transits, the same holds but with swapped roles, i. e., the dual of a transit is a strongly connected **RUP** graph. As an example, the dual of the second transit τ in Fig. 1(a) is shown in Fig. 1(d) and it is indeed a strongly connected **RUP** graph. The following lemma subsumes these observations.

Lemma 5. *Let G be a **RUP** graph without sources and sinks and $\overline{G} = (\mathbb{V}_C, \overline{\mathbb{E}})$ be its compound graph. Then,*

*(i) the dual of each compound $\gamma \in \mathbb{V}_C$ is a planar, acyclic dipole and, thus, it is in **SUP**.*

*(ii) each transit $\tau \in \overline{\mathbb{E}}$ is a planar, acyclic dipole and, thus, its dual is a strongly connected **RUP** graph.*

Both *(i)* and *(ii)* follow from Lemma 4. For *(ii)* note that the graph induced by a transit is an acyclic dipole.

By Lemma 3 we have seen that the dual of a **RUP** graph that contains neither sources nor sinks is a dipole. Also the converse holds:

Lemma 6. *A graph G without sources and sinks is a **RUP** graph if its dual G^* is a dipole.*

Consider again the example **RUP** graph in Fig. 1(a) and the compound graph $\overline{G^*}$ of its dual G^*. Since G^* is a dipole, $\overline{G^*}$ is a path $p = (s, \tau_1^*, \gamma_1^*, \tau_2^*, \gamma_2^*, \ldots, \tau_4^*, t)$ consisting of compounds γ_i^*, transits τ_j^*, and two terminals s and t. Note that each element on p corresponds to a subgraph in the primal G, i. e., for each γ_i^* there is a transit τ_i in G and for each τ_j^* there is a compound γ_j in G. In the proof of Lemma 6, we construct a **RUP** drawing of G by subsequently processing the elements of p. We start with transit τ_1^*, whose induced subgraph in G^* is an acyclic dipole, and obtain a **RUP** drawing of γ_1 which respects the given embedding by Lemma 4. Then we proceed with γ_1^*, a compound in G^*, for which we obtain a **SUP** drawing of τ_1 which respects the given embedding by Lemma 4. However, this **SUP** drawing is upward only with respect to the x-direction, i.e., from left to right. We transform this drawing, while preserving its embedding, such that it is also upward in y-direction. The so obtained drawing of τ_1 is then attached to the right border of the drawing of γ_1. Then, the drawing of γ_2 is attached to the right side of τ_1 and so forth until we reach t. Note that since all transits τ_j^* point into the same direction in $\overline{\mathbb{G}^*}$, i.e., from s to t, all cycles of the compounds in G have the same orientation in the obtained drawing, i.e., they all wind around the cylinder in the same direction.

Lemmata 3 and 6 both require that the graph at hand contains neither sources nor sinks. At a first glance, this requirement seems to be a strong limitation. However, in the following lemma we show that each **RUP** graph can be augmented by edges such that all sources and sinks vanish while still preserving **RUP** embeddability.

Lemma 7. *A **RUP** graph G is a spanning subgraph of a **RUP** graph H without sources and sinks.*

The proof shows that each source (sink) can be connected to another vertex while preserving the upward planar drawability. We follow the construction of the proof of Theorem 1 in [16], which shows that every graph in **SUP** is a spanning subgraph of a planar dipole. Alternatively, the proof can be obtained using techniques from [7].

The proof of Theorem 1 is now complete. The only-if direction follows from Lemmata 7 and 3 and the if direction is a consequence of Lemma 6 and the fact that every subgraph of a **RUP** graph is a **RUP** graph.

4   wSUP Graphs and Their Duals

We now turn to spherical graphs and upward planar embeddings on the standing cylinder. These graphs were characterized as spanning subgraphs of planar, acyclic dipoles [14,16,19]. We already provided a new characterization for **SUP** in terms of dual graphs in Lemma 4 in combination with Proposition 1. Now we consider graphs which have a weakly upward planar drawing on the standing cylinder. These graphs have not been characterized before.

For a start, consider an upward drawing of a **wSUP** graph. If there are cycles, they must wind around the cylinder horizontally, which leads us to the following observation.

Lemma 8. *Let G be a graph in **wSUP**. Then, all cycles of G are (vertex) disjoint.*

For the characterization of **wSUP** graphs, we use supergraphs which may have an extra source or sink and extend techniques for **SUP** graphs from [16].

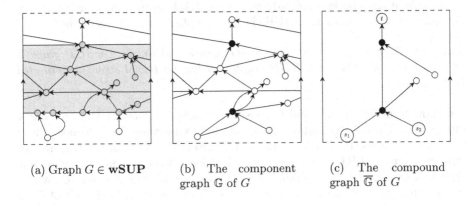

(a) Graph $G \in$ **wSUP**   (b) The component graph $\mathbb{G}$ of G (c) The compound graph $\overline{\mathbb{G}}$ of G

Fig. 3. A **wSUP** example

Lemma 9. *A graph G is a **wSUP** graph if and only if it has a **wSUP** supergraph $H \supseteq G$ with one source and one sink.*

Consider again an upward drawing of a **wSUP** graph G, e. g., the one depicted in Fig. 3. The cycles subdivide the graph into *sections* as in Fig. 3(a), where the intermediate section is shaded gray. In the component graph, cycles are merged into non-trivial strongly connected components (Fig. 3(b)). The corresponding compound graph has a structure as in Fig. 3(c). We proceed section-wise and eliminate sources and sinks as in [16] except for one source in the lowermost section and one sink in the uppermost one, where the lowermost section is not limited by a cycle from below and the uppermost section by a cycle from above. If any of these two sections is empty, a new source or sink is added to the section and connected to the cycle above or below, respectively. This leaves us with a **wSUP** graph with exactly one source and one sink. Conversely, any subgraph of a **wSUP** graph is in **wSUP**.

We are now able to give a first characterization of **wSUP** graphs.

Theorem 2. *A graph G is a **wSUP** graph if and only if it has a supergraph $H \supseteq G$ such that H is a planar dipole whose cycles are (vertex) disjoint.*

The supergraph H of G can be constructed according to Lemma 9 and is a dipole. By Lemma 8, H has only disjoint cycles. We can obtain a **wSUP** drawing for a planar dipole H whose cycles are disjoint by partitioning the dipole into its compounds and transits. Since transits are acyclic dipoles, the induced subgraph can be drawn upward according to its **SUP** embedding. The compounds consist of single cycles only, which we draw horizontally, i. e., such that each winds around the cylinder once. So H has a **wSUP** drawing and the implied embedding is a **wSUP** embedding. Since G is a subgraph of H, also G is in **wSUP**.

Next, we turn to the duals of **wSUP** graphs. Recall from Lemma 4 in combination with Proposition 1 that a graph with one source and one sink is in **SUP** if and only if its dual is a strongly connected **RUP** graph. Introducing vertex disjoint cycles, the characterization via dual graphs now reads as follows.

Theorem 3. *A graph G with exactly one source and sink is a **wSUP** graph if and only if its dual G^* is a **RUP** graph that has no trivial strongly connected components.*

We know from Theorem 1 and Proposition 1 that G^* is in **RUP** since G is a dipole. If G is acyclic, G^* is strongly connected by Lemma 4 and Proposition 1. Otherwise, G consists of compounds and transits. The duals of the transits are strongly connected **RUP** components with at least one edge and, therefore, not trivial. Now consider a compound in G. It consists of a single cycle C, which implies that its dual consists of simple edges from the faces to the left of C to the faces to its right, which themselves are also part of the strongly connected **RUP** components. Hence, all vertices are contained in compounds and G^* has no trivial strongly connected components.

Conversely, a similar argument shows that only single, disjoint cycles can occur in the primal graph of a **RUP** graph without any trivial strongly connected components. Furthermore, by Lemma 3 and Proposition 1, G is a dipole and, therefore, has only one source and one sink. By Theorem 2, G is in **wSUP**.

From Theorem 3 and Lemma 9 we directly obtain the following corollary, which concludes our second characterization of **wSUP** graphs.

Corollary 2. *Every **wSUP** graph G has a **wSUP** supergraph H whose dual H^* is a **RUP** graph without trivial strongly connected components.*

5   Summary

We have shown that a directed graph has a planar upward drawing on the rolling cylinder if and only if it is a spanning subgraph of a planar graph without sources and sinks whose dual is a dipole. This result completes the known characterizations of planar upward drawings in the plane [8, 18] and on the sphere [10, 12, 15, 16]. Every **SUP** graph is a spanning subgraph of a planar, acyclic dipole and every **UP** graph is a spanning subgraph of a planar, acyclic dipole with an st-edge. Moreover, a graph has a weakly upward drawing on the standing cylinder if and only if it is a subgraph of a planar dipole with disjoint cycles.

Concerning dual graphs, the duals of the acyclic components of **RUP** graphs are in **RUP** and the duals of the strongly connected components are in **SUP**. In particular, the dual of a strongly connected **RUP** graph is in **SUP**. Every **wSUP** graph has a planar supergraph whose dual is a **RUP** graph without trivial strongly connected components.

Future work is to investigate whether the characterization by means of dual graphs leads to new insights on the upward embeddability on other surfaces, e. g., the torus. Also, the duals of quasi-upward planar graphs [6] shall be considered.

References

1. Auer, C., Bachmaier, C., Brandenburg, F.J., Gleißner, A., Hanauer, K.: The duals of upward planar graphs on cylinders. Tech. Rep. MIP-1204, Faculty of Informatics and Mathematics, University of Passau (2012), http://www.fim.uni-passau.de/en/research/forschungsberichte/mip-1204.html
2. Auer, C., Bachmaier, C., Brandenburg, F.J., Gleißner, A.: Classification of Planar Upward Embedding. In: Speckmann, B. (ed.) GD 2011. LNCS, vol. 7034, pp. 415–426. Springer, Heidelberg (2011)
3. Bachmaier, C.: A radial adaption of the Sugiyama framework for visualizing hierarchical information. IEEE Trans. Vis. Comput. Graphics 13(3), 583–594 (2007)
4. Bachmaier, C., Brandenburg, F.J., Brunner, W., Fülöp, R.: Drawing recurrent hierarchies. J. Graph Alg. App. 16(2), 151–198 (2012)
5. Bang-Jensen, J., Cutin, G.: Digraphs: Theory, Algorithms and Applications, 1st edn. Springer (2000)
6. Bertolazzi, P., Di Battista, G., Didimo, W.: Quasi-upward planarity. Algorithmica 32(3), 474–506 (2002)
7. Brandenburg, F.J.: On the curve complexity of upward planar drawings. In: Mestre, J. (ed.) Computing: The Australasian Theory Symposium, CATS 2012. Conferences in Research and Practice in Information Technology (CRPIT), vol. 128, pp. 27–36. Australian Computer Society, ACS (2012)
8. Di Battista, G., Tamassia, R.: Algorithms for plane representations of acyclic digraphs. Theor. Comput. Sci. 61(2-3), 175–198 (1988)
9. Dolati, A.: Digraph embedding on t_h. In: Cologne-Twente Workshop on Graphs and Combinatorial Optimization, CTW 2008, pp. 11–14 (2008)
10. Dolati, A., Hashemi, S.M.: On the sphericity testing of single source digraphs. Discrete Math. 308(11), 2175–2181 (2008)
11. Dolati, A., Hashemi, S.M., Kosravani, M.: On the upward embedding on the torus. Rocky Mt. J. Math. 38(1), 107–121 (2008)
12. Foldes, S., Rival, I., Urrutia, J.: Light sources, obstructions and spherical orders. Discrete Math. 102(1), 13–23 (1992)
13. Garg, A., Tamassia, R.: On the computational complexity of upward and rectilinear planarity testing. SIAM J. Comput. 31(2), 601–625 (2001)
14. Hansen, K.A.: Constant width planar computation characterizes ACC^0. Theor. Comput. Sci. 39(1), 79–92 (2006)
15. Hashemi, S.M., Rival, I., Kisielewicz, A.: The complexity of upward drawings on spheres. Order 14, 327–363 (1998)
16. Hashemi, S.M.: Digraph embedding. Discrete Math. 233(1-3), 321–328 (2001)
17. Kaufmann, M., Wagner, D. (eds.): Drawing Graphs. LNCS, vol. 2025. Springer, Heidelberg (2001)
18. Kelly, D.: Fundamentals of planar ordered sets. Discrete Math. 63, 197–216 (1987)
19. Limaye, N., Mahajan, M., Sarma, J.M.N.: Evaluating Monotone Circuits on Cylinders, Planes and Tori. In: Durand, B., Thomas, W. (eds.) STACS 2006. LNCS, vol. 3884, pp. 660–671. Springer, Heidelberg (2006)
20. Limaye, N., Mahajan, M., Sarma, J.M.N.: Upper bounds for monotone planar circuit value and variants. Comput. Complex. 18(3), 377–412 (2009)
21. Sugiyama, K., Tagawa, S., Toda, M.: Methods for visual understanding of hierarchical system structures. IEEE Trans. Syst., Man, Cybern. 11(2), 109–125 (1981)
22. Thomassen, C.: Planar acyclic oriented graphs. Order 5(1), 349–361 (1989)

The (Weighted) Metric Dimension of Graphs: Hard and Easy Cases

Leah Epstein[1], Asaf Levin[2], and Gerhard J. Woeginger[3]

[1] Department of Mathematics, University of Haifa, 31905 Haifa, Israel
lea@math.haifa.ac.il
[2] Faculty of Industrial Engineering and Management,
The Technion, 32000 Haifa, Israel
levinas@ie.technion.ac.il
[3] Department of Mathematics and Computer Science, TU Eindhoven P.O. Box 513,
5600 MB Eindhoven, The Netherlands
gwoegi@win.tue.nl

Abstract. For an undirected graph $G = (V, E)$, we say that for $\ell, u, v \in V$, ℓ separates u from v if the distance between u and ℓ differs from the distance from v to ℓ. A set of vertices $L \subseteq V$ is a feasible solution if for every pair of vertices $u, v \in V$ there is $\ell \in L$ that separates u from v. The metric dimension of a graph is the minimum cardinality of such a feasible solution. Here, we extend this well-studied problem to the case where each vertex v has a non-negative cost, and the goal is to find a feasible solution with a minimum total cost. This weighted version is NP-hard since the unweighted variant is known to be NP-hard. We show polynomial time algorithms for the cases where G is a path, a tree, a cycle, a cograph, a k-edge-augmented tree (that is, a tree with additional k edges) for a constant value of k, and a (not necessarily complete) wheel. The results for paths, trees, cycles, and complete wheels extend known polynomial time algorithms for the unweighted version, whereas the other results are the first known polynomial time algorithms for these classes of graphs even for the unweighted version. Next, we extend the set of graph classes for which computing the unweighted metric dimension of a graph is known to be NPC by showing that for split graphs, bipartite graphs, co-bipartite graphs, and line graphs of bipartite graphs, the problem of computing the unweighted metric dimension of the graph is NPC.

1   Introduction

Let $G = (V, E)$ be a simple, loopless, undirected graph. A vertex $\ell \in V$ is called a *separating landmark* for two vertices $u, v \in V$ with $u \neq v$, if the length of the shortest path from u to ℓ differs from the length of the shortest path from v to ℓ; sometimes we will then also say that vertex ℓ *separates* or *distinguishes* u from v. We denote the number of vertices in G by $n = |V|$ and the number of its edges by $m = |E|$. A subset $L \subseteq V$ is a *landmark set* for the graph G, if for any two vertices $u, v \in V$ with $u \neq v$ there exists a separating landmark $\ell \in L$ that separates u from v. The *metric dimension* $\mathrm{md}(G)$ of the graph G is the cardinality of the smallest landmark set in G. Note that $\mathrm{md}(G)$ is well-defined,

M.C. Golumbic et al. (Eds.): WG 2012, LNCS 7551, pp. 114–125, 2012.
© Springer-Verlag Berlin Heidelberg 2012

as $L = V$ trivially forms a landmark set for G. Additionally, $\mathtt{md}(G) = 0$ iff $|V| = 1$. We consider the problem of computing $\mathtt{md}(G)$ of an input graph G. Applications of this optimization problem arise in diverse areas. See [2] for an application of this problem in network verification, [6] for an application in strategies for the Mastermind game, [9] for an application in metric geometry, [12] for an application in digital geometry, namely in digitizing of images, [11] for an application in robot navigation, and [4] for an application in drug discovery.

This metric dimension problem was introduced by Harary and Melter [9] and by Slater [15], and studied widely in the combinatorics literature. In this line of research, the exact values of the metric dimension or bounds on it for specific graph classes are obtained. We refer to [1,3,4] for results additional to those stated here (see also the survey [5]). It was shown that $\mathtt{md}(G) = 1$ iff G is a path [11]. Tree input graphs (which are not paths) were considered in [9,15,11,4]. It turns out that it is possible to characterize the feasibility of a landmark set for a tree using a notion of *legs*, which are paths of vertices of degree at most 2 connected to a vertex of a higher degree. When the input graph is a cycle then $\mathtt{md}(G) = 2$ [9]. Wheel graphs were mentioned in [9] and further studied in [14]. Melter and Tomescu [12] considered the problem for grid-graphs induced by lattice points in the plane when the distances are measured in the L_1 norm or in the L_∞ norm. Khuller, Raghavachari, and Rosenfeld [11] generalized one result of [12] to lattice points contained inside a d-dimensional rectangle where the distance is according to the L_1 norm (that is, the grid-graph over points in d-dimensional rectangle). The grid-graph with Euclidean metrics was studied in [13] where they relate the problem to the combinatorial coin weighing problem. Recently, Diaz et al. [8] developed a polynomial time algorithm for outerplanar graphs.

As for the complexity of the problem, Khuller, Raghavachari, and Rosenfeld [11] proved that the problem is NP-hard for general graphs and showed that one can apply the greedy algorithm for a set cover instance (where we would like to cover the pairs of vertices in a graph using sets which are defined using a single landmark). Thus, there is an $(2 \ln n + O(1))$-approximation algorithm for general graphs. Beerliova et al. [2] showed that if $P \neq NP$ then there is no $o(\log n)$-approximation algorithm for the problem. These results were strengthened in [10], where it was shown that under another complexity condition there is no $((1 - \varepsilon) \ln n)$-approximation algorithm for any $\varepsilon > 0$. They give an improved $(1 + (1 + o(1)) \ln n)$-approximation algorithm. Diaz et al. [8] showed that the problem is NP-hard even when restricted to planar graphs.

We generalize the problem to a weighted variant. Given a non-negative cost function $c : V \to \mathbf{R}_+$ the goal is to compute a landmark set L such that $\sum_{\ell \in L} c(\ell)$ is minimized. We let $\mathtt{wmd}(G)$ denotes this minimum cost and we say that $\mathtt{wmd}(G)$ is the weighted metric dimension of G. The $\mathtt{wmd}$ problem is to compute a landmark set L of minimum total cost, and to find $\mathtt{wmd}(G)$. Our polynomial time algorithms for special classes of graphs will be for solving $\mathtt{wmd}$ while our NP-hardness proofs will hold even for the unweighted version of computing $\mathtt{md}(G)$ and thus the same holds for the weighted variant as well. We are not aware of any previous work on the weighted version. We say that a feasible landmark set is minimal if it is minimal with respect to inclusion. Note that there is always an optimal solution which is also minimal (since the cost function is non-negative), and thus we sometimes characterize the set of minimal solutions.

The problem is clearly in NP because given a landmark set, we can verify its feasibility in polynomial time. To do this, we find the vectors of the distances of each vertex in V from each of the landmarks. Afterwards, we check that there is no pair of identical vectors. In some of our algorithms we perform an exhaustive enumeration of landmark sets (among a restricted family of vertex subsets), and we can always find a cheapest feasible solution among such a restricted family of subsets (we first ensure that it contains at least one of the optimal solutions).

Our Results. We generalize the known polynomial time algorithms for md to wmd for the cases where G is a path, a tree, a cycle, or a complete wheel. We develop polynomial time algorithms for the weighted problem when G is a cograph, a k-edge-augmented tree (that is, a tree with additional k edges) for a constant value of k, and a (not necessarily complete) wheel. These results are the first polynomial time algorithms even for the unweighted version when G belongs to these classes of graphs. Next, we extend the set of graph classes for which md is known to be NP-complete. We show that md is NP-complete when the input graph for split graphs, bipartite graphs, co-bipartite graphs, and line graphs of bipartite graphs. Omitted proofs and the hardness results will be given in the full version of this paper.

Definitions and Notation. Given a graph $G = (V, E)$, a vertex $v \in V$ is a leaf if its degree is 1, it is an isolated vertex if its degree is 0, if its degree is 2 it is a path vertex, and higher degree vertices are called core vertices. For a pair of vertices u, v we denote by $d_{u,v}$ the length of the shortest path in G from u to v (i.e., the number of edges in it).

2   Paths, Trees, Cycles, and Cographs

In this section we generalize some polynomial solvable cases of $md(G)$ to the weighted case. These simple cases emphasize the differences between the weighted and the unweighted cases. Additionally we consider cographs.

Paths. Assume that G is a path. It suffices to have one landmark vertex at one of the end-vertices of the path [11]. Our algorithm for computing $wmd(G)$ for a path G is as follows: We consider two alternative solutions and pick the better one. The first one has a single vertex as a landmark: a minimum cost end point v of the path (breaking ties arbitrarily). The second solution picks the cheapest pair of distinct vertices v and v'.

Trees. Assume now that the input graph G is a tree which is not a path (and so it has at least one core vertex). We say that a (non-empty) path in the tree between a vertex adjacent to a core vertex u and a leaf v is a *leg of* u if the vertices of the path have degree at most 2 (u is not considered to be a part of the leg). For a vertex u we denote the number of legs of u by leg_u. Assume that $leg_u \geq 2$, then [11] showed that there is a landmark at each of the legs of u except for at most one such leg. In fact, summing $leg_u - 1$ over all core vertices

u with at least two legs, gives a tight bound on $\mathtt{md}(G)$. Here, we generalize their approach. Consider the following algorithm. Compute leg_u for every core vertex u. Each core vertex u with $leg_u \geq 2$ is allocated $leg_u - 1$ landmarks. To place these landmarks, we find a minimum cost set of $leg_u - 1$ vertices in the legs of u, at most one vertex per leg (recall that u does not belong to its legs).

Proposition 1. *The above algorithms solve* $\mathtt{wmd}$ *for any tree G in $O(n)$ time.*

Cycles. We assume that the input graph G is a cycle. We next characterize a minimal (with respect to inclusion) feasible landmark set L. We say that a pair of vertices u, v are opposite if their distance is exactly $\frac{n}{2}$, and otherwise they are non-opposite (in which case the shortest path from u to v is unique). Note that if G is an odd-length cycle, then every pair of vertices are non-opposite.

Lemma 1. *A minimal feasible landmark set is a pair of non-opposite vertices.*

Our algorithm for solving $\mathtt{wmd}$ for a cycle G simply finds the cheapest pair of non-opposite vertices in the cycle. By the above lemma, it finds an optimal solution. Note that the running time of the algorithm is $O(n)$ by first identifying the cheapest set of three vertices (breaking ties arbitrarily), and finding the cheapest pair of non-opposite vertices among them (breaking ties arbitrarily). Our tie breaking rule is justified because even if the set of minimum weight vertices has more than three vertices, it is sufficient to consider only three of them, and similar arguments apply for the other extreme cases.

Dealing with Disconnected Input Graphs. Consider a disconnected graph $G = (V, E)$. A connected component of G is called *non-trivial connected component* if it contains at least two vertices, and otherwise it is an isolated vertex. We denote by $(V_1, E_1), \ldots, (V_p, E_p)$ the non-trivial connected components of G (for $p \geq 0$), and by $v_1, \ldots, v_t$ its isolated vertices (for $t \geq 0$), where if $p = 0$ or $t = 0$ then there is no non-trivial connected component or an isolated vertex, respectively. Without loss of generality, we assume that $c(v_t) = \max_{i:1 \leq i \leq t} c(v_i)$. In this section we show that it is sufficient to solve the weighted metric dimension problem for each non-trivial connected component of G. The time complexity of solving $\mathtt{wmd}$ is $O(n + m)$ plus the total running times of solving $\mathtt{wmd}$ for each of the non-trivial connected components of G.

Proposition 2. *An optimal solution L for* $\mathtt{wmd}$ *is achieved by the union solutions for the non-trivial connected components (V_i, E_i) of G and $v_1, v_2, \ldots, v_{t-1}$.*

Cographs. For two graphs $G_1 = (V_1, E_1)$ and $G_2 = (V_2, E_2)$ with $V_1 \cap V_2 = \emptyset$, the disjoint union $G_1 \cup G_2$ is the graph $(V_1 \cup V_2, E_1 \cup E_2)$. The product $G_1 \times G_2$ of these two graphs is obtained by first taking the disjoint union of G_1 and G_2 and then adding all the edges $\{v_1, v_2\}$ with $v_1 \in V_1$ and $v_2 \in V_2$. A graph G is a *cograph*, if (i) G consists of a single vertex or (ii) G is the disjoint union of two cographs, or (iii) G is the product of two cographs. An equivalent characterization states that G is a cograph iff it does not contain the path P_4 on four vertices as an induced subgraph. Note that the complement graph $G' = (V, V \times V - E)$ of a cograph $G = (V, E)$ is a cograph as well.

The *cotree* of a cograph G is a rooted binary tree whose leaves correspond to single-vertex graphs and whose inner vertices correspond to subgraphs of G. Every inner vertex of the cotree is labeled either by $\cup$ (union) or by $\times$ (product) and has exactly two children: if it is labeled by $\cup$ then it corresponds to the disjoint union of the two cographs that correspond to its two children, and if it is labeled by $\times$ then it corresponds to the product of the two cographs. Corneil, Perl, and Stewart [7] showed how to compute a cotree for a given cograph in linear time $O(m + n)$. Note that the number of inner vertices of the cotree is $n - 1$. By Proposition 2, we conclude that we may restrict ourselves to connected cographs. Since a connected cograph is a product of two non-empty graphs, the distance between a pair of vertices v, v' of a connected cograph G is either 1 or 2. We define a *binary landmark set* L of an arbitrary cograph G (not necessarily a connected one) to be a set of vertices such that for every pair of vertices $v, v' \in V \setminus L$ there is a landmark $\ell \in L$ such that either both $\{v, \ell\} \in E$ and $\{v', \ell\} \notin E$ or both $\{v, \ell\} \notin E$ and $\{v', \ell\} \in E$. In this case we say that ℓ separates v from v'. Given a connected cograph, a set of vertices is a feasible landmark set iff it is a feasible binary landmark set. In the remainder of this section we will present a linear time algorithm for computing a binary landmark set of a minimum total cost. For a cograph G and its complement graph G', a set $L \subseteq V$ is a binary landmark set for G iff L is a binary landmark set for G'.

We adapt our decomposition of the problem for disconnected graphs to the problem of computing a binary landmark set.

Lemma 2. *Assume that G is a disjoint union of G_1 and G_2 where $G_i = (V_i, E_i)$. (i) If L is a feasible binary landmark set for G, then $L_i = L \cap V_i$ is a feasible binary landmark set for G_i, for $i = 1, 2$. (ii) Assume that L_1 and L_2 are feasible binary landmark set for G_1 and G_2, respectively. Then, $L = L_1 \cup L_2$ is a feasible binary landmark set for G iff there exists $i \in \{1, 2\}$ such that in G_i every vertex $v \in V_i \setminus L_i$ is adjacent to a vertex of L_i.*

Our algorithm for solving wmd of a cograph uses the cotree structure. Let a *point* be a vertex which is not a landmark. If a subgraph is a disjoint union of two subgraphs, then we treat recursively each of the two subgraphs but we keep track of the existence of a point which is not adjacent to any landmark in its subgraph. If a subgraph is the product of two subgraphs, then we transform our problem to the complement of the subgraph, and apply the first case. Since by moving from a graph to its complement, the role of a point which is not adjacent to any landmark switches with the role of a point which is adjacent to all landmarks, we will keep track of the number of points (zero or one) of each of these types, and every subgraph in the cotree corresponds to four optimization problems. Thus, wmd can be solved in linear time $O(m + n)$ for cographs.

3 k-Edge-Augmented Trees

In this section we consider the class of connected graphs for which a removal of at most k edges results in a spanning tree. We call this class of graphs k-*edge-augmented trees*. Our polynomial time algorithm for computing wmd first applies a preprocessing step which handles the tree-like part, and then uses an exhaustive enumeration approach for selecting an optimal landmark set in a reduced problem. Clearly our algorithm is polynomial only if k is a constant and it is unlikely that there is an algorithm which is polynomial in n and k for solving this problem (since any connected graph is also a k-edge-augmented tree for a sufficiently large value of k, and the problem is NP-hard for general graphs).

Preprocessing step. Our preprocessing step uses the following procedure which can be applied on a vertex u of degree at least 3 with p legs (where a leg is a path consisting of at least one vertex, starting at a neighbor of u and ending at a leaf, where all vertices except for the leaf have degree 2) for $p \geq 2$. Consider subsets of $p - 1$ vertices which belong to the legs of u, at most one vertex per leg. We choose L_u to be a set of minimum cost among the sets which satisfy these constraints. We will place landmarks at L_u and remove from the graph every vertex belonging to a leg of u which contains a vertex of L_u (that is, the leg which does not contain a vertex of L_u is not removed). In the remaining graph we change the cost of u to zero, and thus allow the solution to place a landmark at u without increasing its cost. We apply this preprocessing on one vertex at a time until there is no vertex which has at least two legs, and in the remaining graph every vertex has at most one leg. Since we already dealt with trees in Proposition 1, we assume that the input graph is not a tree. The following lemma holds in fact for trees as well (but in order to prove it for trees as well, the special case of a spider graph should be treated separately).

Lemma 3. *Denote by $G = (V, E)$ the input graph which is not a tree, and by $G' = (V', E')$ the graph obtained after applying the procedure at a vertex u (that is, the set of vertices after removing all the legs of u but one). Let L_u denote the set of landmarks which we placed on the removed legs of u, and let L' denote the set of landmarks in an optimal solution of the remaining graph. Then, $L = L_u \cup L' \setminus \{u\}$ is an optimal landmark set in G.*

The case of $k = 1$. The unweighted version of computing $\mathrm{md}(G)$ for the case of 1-edge-augmented tree is discussed in [4] (where such graphs are called unicycles). Here we consider the weighted case. If $k = 1$, then at the end of the preprocessing phase we are left with a cycle C where some of its vertices have legs (at most one leg for each vertex of C). Recall that some of the vertices of G may have zero cost resulting from the preprocessing step, and if we choose to place a landmark at such a vertex u, then the solution returned by the algorithm skips it (but has at least one landmark in the tree-like part connected to u which was removed in the preprocessing step). In what follows we only consider the graph G resulting from the preprocessing. We next characterize a minimal landmark set for this type of graphs. This characterization shows that any minimal landmark set for the resulting graph has at most three vertices. By enumerating all subsets consisting of two or three vertices we can choose the cheapest feasible landmark set and solve wmd in polynomial time.

We denote by n_C the number of vertices in C. For a vertex v, we let v' be its *cycle vertex* which is defined as follows. v' is the closest vertex in C to v, that is if $v \in C$ then $v' = v$ and otherwise v belongs to some leg of a vertex in

C, and we let v' denote this vertex of C. Consider a path P from u to v, then its C-length is defined as the number of edges of C which belong to P. We say that a path P from u to v is a clockwise path if it traverses the edges in $P \cap C$ in a clockwise order, and otherwise it is a counterclockwise path. We say that two vertices are non-opposite in G if their cycle vertices are distinct, and the C-length of the shortest path from u to v is not equal to $\frac{n_C}{2}$.

Lemma 4. *Let $L \subseteq V$ be a feasible minimal landmark set. There is no cycle vertex u such that L contains a pair of vertices v_1, v_2 whose cycle vertex is u.*

In what follows, we focus on a minimal landmark set, and thus assume that it does not contain two vertices with a common cycle vertex.

Lemma 5. *(i) If $L \subseteq V$ contains a pair of non-opposite vertices u_1, u_2, then for every pair of vertices $x_1, x_2 \in C$, there exists $w \in \{u_1, u_2\}$ that separates x_1 from x_2. (ii) If L consists of at least three vertices, no pair of which have the same cycle vertex, then for every pair of vertices $x, y \in C$ there is $\ell \in L$ that separates x from y.*

We next define a *covering of the legs by landmarks*. We say that a landmark ℓ clockwise-covers (counterclockwise-covers) the leg of a vertex u if one of the following conditions hold: Either ℓ is one of the vertices of the leg of u, or the clockwise path (counterclockwise path) from u to ℓ has C-length of at least 1 and at most $\frac{n_C+1}{2}$ (u cannot cover its own leg). We say that a leg is *covered* by $L \subseteq V$ if there is a landmark in L which clockwise-covers the leg and (perhaps another) landmark in L which counterclockwise-covers the leg.

Lemma 6. *Let L be a set of vertices, where $|L| \geq 3$ and no two of which have the same cycle vertex. L is a feasible landmark set iff every leg is covered by L.*

Lemma 7. *Let L be a minimal landmark set. Then $|L| \leq 3$.*

To summarize, our algorithm for computing $\mathtt{wmd}(G)$ where G is a 1-edge-augmented tree is to apply the preprocessing step, and afterwards try all possibilities of sets L such that $|L| \leq 3$, and for each of them test its feasibility in polynomial time, and among the feasible solution we pick a cheapest one. Clearly, the algorithm runs in polynomial time and computes a cheapest minimal feasible landmark set. Therefore, we have established the following.

Proposition 3. *There is a polynomial time algorithm for solving $\mathtt{wmd}$ for 1-edge-augmented trees.*

The General Case. Assume that $G = (V, E)$ is the graph resulting from applying the preprocessing step, so in G every vertex has at most one leg. The case of $k = 1$ is already solved, and here we assume that $k \geq 2$ is a fixed constant. We next define a subgraph of G called the base graph $G_b = (V_b, E_b)$ resulting from G by removing the vertices of all legs. We next characterize the structure of this base graph. That is, we will show that it consists of $O(k)$ edge disjoint paths connecting core vertices where all internal vertices are path vertices.

The Structure of the Base Graph. For a vertex $u \in V_b$, we denote by $deg_b(u)$ its degree in G_b. Note that G_b contains no leaves, and thus the degree of every vertex is at least 2. Moreover, since G_b is a connected subgraph of G and every cycle of G belongs to G_b as well, G_b results from a tree T by adding exactly k edges. Thus the tree T has $|V_b| - 1$ edges, and thus the number of edges in G_b is $|V_b| + k - 1$. Therefore, $\sum_{u \in V_b} deg_b(u) = 2|V_b| + 2k - 2$ and thus $\sum_{u \in V_b}(deg_b(u) - 2) = 2k - 2$.

Lemma 8. *The base graph G_b is decomposed into $q \leq 3k - 3$ edge disjoint paths, where in G_b every internal vertex of a path has degree 2, and the end-vertices of such a path are core vertices.*

Given a vertex $v \in V$, we define its base vertex as the vertex $v' \in V_b$ which is the closest to v. Given the path decomposition defined above, we associate each vertex v in V with one of the paths in the following way. If the base vertex of v belongs to exactly one of the paths then we associate v with this path. Otherwise, the base vertex v' of v is a core vertex in G_b, and we associate all the vertices in G whose base vertex is v' with one of the paths incident to v'.

Bounding the Number of Landmarks Associated with One Path. Next, we consider a minimal feasible landmark set L, and one specific path P in the path decomposition of G_b. Our goal is to bound the number of landmarks in L which are associated with P. The following lemma generalizes Lemma 4 to the case of $k \geq 2$.

Lemma 9. *(i) Let L be a minimal feasible landmark set, and let P be a path in the path decomposition of G_b. Then, the number of vertices in L which are associated with P is at most six. (ii) Let L be a minimal landmark set of a graph G which results from a k-edge-augmented tree by the preprocessing step (where $k \geq 2$). Then $|L| \leq 18k - 18$.*

To summarize, our algorithm for computing $\mathsf{wmd}(G)$ where G is a k-edge-augmented tree (for $k \geq 2$) is to apply the preprocessing step, afterwards, try all possibilities of sets L such that $|L| \leq 18k - 18$, for each of them test its feasibility in polynomial time, and among the feasible solution we pick a cheapest one. Clearly, the algorithm runs in polynomial time (for a constant value of k) and computes a cheapest minimal feasible landmark set.

4   Wheels

Complete Wheels. In this section we consider complete wheels. A (complete) wheel on n vertices $\{1, 2, \ldots, n\}$ is defined as follows. There is a cycle C over the vertices $1, 2, \ldots, n-1$ (the clockwise order of the vertices along C is $1, 2, \ldots, n-1, 1$), and vertex n is adjacent to all other vertices. Vertex n is called the *hub* of the wheel, whereas the other vertices are called *cycle-vertices*. The distances in G are either 1 or 2, clearly the distance between every cycle-vertex and the hub is 1, the distance of every cycle-vertex and its two neighbors along the cycle is 1, and between every pair of (non-adjacent) cycle-vertices are reachable via a two-edge path through the hub. Consider a feasible landmark set L. A gap between consecutive landmarks is defined as follows: If $\ell, \ell' \in L$ are cycle vertices such that along the clockwise path from ℓ to ℓ' there is no other landmark, then the set of internal vertices of the clockwise path from ℓ to ℓ' is the *gap* (between ℓ and ℓ'), and we say that the gap is adjacent to ℓ and ℓ'. The length of the gap is the number of vertices in the gap. Here, we characterize the lengths of the gaps in a feasible solution L.

Lemma 10. *Assume that G is a wheel over at least 8 vertices. Let $L \subseteq V$. L is a feasible landmark set iff the following three conditions hold with respect to L: 1. There is no gap of length at least 4. 2. There is at most one gap of length 3. 3. Two gaps of length at least two are not adjacent to a common vertex of $L \cap C$.*

Theorem 1. *Given a complete wheel G, wmd can be solved in linear time.*

A minimal landmark set L does not contain the hub, and based this characterization, our algorithm is a simple dynamic program. By [14], for a complete wheel G on n vertices, $\mathtt{md}(G) = \Theta(n)$, and since it is a $(n-1)$-edge-augmented tree, a k-edge-augmented tree needs $\Omega(k)$ landmarks in any feasible solution.

The general case A (non-complete) wheel on n vertices $\{1, 2, \ldots, n\}$ is defined as follows. There is a cycle C over the vertices $1, 2, \ldots, n-1$ (the clockwise order of the vertices along C is $1, 2, \ldots, n-1, 1$), and vertex n is adjacent to some of the other vertices. Vertex n is called the *hub* of the wheel, whereas the other vertices are called *cycle-vertices*. The neighbors of n are called *connectors*, and the edges incident at n are called *chords*. We let *layer j* be the set of vertices of distance j from n, and denote it by V_j (i.e., $V_j = \{u \in V : d_{u,n} = j\}$), and thus $V_0 = \{n\}$ and V_1 is the set of connectors). Let $L \subseteq V$, we say that a cycle vertex u is close to $\ell \in L$ if $d_{\ell,u} < d_{\ell,n} + d_{n,u}$. We say that u is close to L if there is $\ell \in L$ such that u is close to ℓ. In this section we consider wheels with at least 22 connectors, and present a polynomial time algorithm for solving wmd for such a graph. Since wheels with at most 21 connectors are k-edge-augmented trees for a value of k such that $k \leq 21$, we conclude that we will obtain a polynomial time algorithm for solving wmd on (arbitrary) wheels. Thus let $G = (V, E)$ be a wheel with at least 22 connectors. We first characterize a minimal landmark set.

Lemma 11. *(i) Let L be a feasible landmark set. Then for every j there is at most one vertex of V_j which is not close to L. Moreover, the vertices in V which are not close to L form a shortest path from some vertex v to the hub. (ii) Let $\ell \in V$. The set of vertices which are close to ℓ is a subpath P of C containing ℓ. Moreover, consider P as a clockwise path from u to v, then the subpath of P from u to ℓ (including u, ℓ) has at most two connectors, and the subpath of P from ℓ to v has at most two connectors. Thus, P contains at most four connectors. (iii) L is a feasible landmark set iff the following two conditions hold: For every layer V_j, there is at most one vertex of V_j which is not close to L. For every $\ell \in L$ and every j, if $u_1, u_2 \in V_j$ are close to ℓ and $d_{\ell,u_1} = d_{\ell,u_2}$, then there is*

$\ell' \in L \setminus \{\ell\}$ which is close to at least one of the vertices u_1 or u_2. (iv) A minimal landmark set L does not contain the hub, and $|L| \geq 6$.

Let L^* be a fixed optimal solution. We say that two landmarks ℓ_1 and ℓ_2 are consecutive if the clockwise path from ℓ_1 to ℓ_2 does not contain an additional landmark, such a path is called the *natural path between the landmarks*. Given a landmark $\ell \in L^*$, we say that a pair of cycle-vertices x, y is a bad pair with respect to ℓ (or of ℓ), if x, y are close to ℓ, belong to a common layer, $d_{x,\ell} = d_{y,\ell}$, and the clockwise path from x to y traverses ℓ. Recall that in this case there must be a landmark $\ell' \neq \ell$ which is close to at least one of x and y, and we say that ℓ' covers the bad pair x, y of ℓ. A *minimal bad pair of a landmark* ℓ is a bad pair x, y such that $d_{x,\ell}$ is minimal among all bad pairs of ℓ.

Lemma 12. (i) Let x, y be a bad pair with respect to a landmark ℓ. Let $\ell' \neq \ell$ be a landmark which is close to at least one of x and y. Let ℓ_1 and ℓ_2 be landmarks such that ℓ_1 and ℓ are consecutive landmarks and ℓ and ℓ_2 are consecutive landmarks. Then the landmark ℓ_1 is close to x, or the landmark ℓ_2 is close to y (or both). (ii) Let $\ell' \neq \ell$ be a landmark that covers a minimal bad pair x, y of ℓ such that either ℓ', ℓ or ℓ, ℓ' are consecutive, then ℓ' covers every bad pair of ℓ.

We say that a minimal bad pair x, y of ℓ which is covered by ℓ' is *covered from the left by* ℓ' if ℓ' and ℓ are consecutive, and otherwise if ℓ and ℓ' are consecutive we say that x, y are covered from the right by ℓ'. By Lemma 12, a bad pair which is covered is either covered from the right or from the left, by some ℓ'.

Corollary 1. A landmark set $L \subseteq V$ is a minimal feasible landmark set iff it satisfies the following three conditions. 1. $n \notin L$. 2. The set of vertices which are not close to L forms a shortest path from n to some vertex v (possibly $v = n$), in particular, if $v \neq n$ then the path contains exactly one connector and no landmark. 3. For every $\ell \in L$, there is $\ell' \in L \setminus \{\ell\}$ which covers the minimal bad pair of ℓ either from the left or from the right (if there exists a bad pair of ℓ).

If there is a cycle-vertex which is not close to L^*, then all such cycle-vertices form a path not containing a landmark. We guess a pair of consecutive landmarks in L^* such that if there exists a cycle-vertex not close to L^*, then all such vertices appear along the natural path between the two guessed landmarks. Without loss of generality assume that the two guessed landmarks are 1 and $k \leq n - 1$ and the natural path between them is the clockwise path from k to 1. The number of possibilities for the selection of $1, k$ is $O(n^2)$. We verify that the set of vertices along the natural path from k to 1 which are not close to 1 or to k (if such a cycle-vertex exists) form a shortest path from some vertex to n (and it contains at most one connector since all connectors are in V_1). If this condition does not hold, then this possibility is impossible, and we stop considering it. In what follows, we consider a possibility which passed this test. Since the number of connectors along the natural path from k to 1 is at most 5 connectors, the clockwise path from 1 to k contains at least 17 connectors, and it is not a shortest path.

Let $1 < v < k$. We define $low(v)$ as the minimum index i such that $1 \leq i \leq v$ and the clockwise path from i to v is a unique shortest path. We let $high(v)$

be the maximum index i such that $v \leq i \leq k$ and the clockwise path from v to i is the unique shortest path. $\delta(v)$ is the minimum number $i \geq 1$ such that $v - i \geq low(v)$ and $v + i \leq high(v)$ belong to a common layer. If such i does not exist, we let $\delta(v) = \infty$. The motivation for this definition of $\delta(v)$ is to identify the minimal bad pair of v (if it exists).

Claim. Let $1 < v < k$ be a landmark such that ℓ_1, v are consecutive landmarks and v, ℓ_2 are consecutive landmarks ($1 \leq \ell_1 < v < \ell_2 \leq k$ since $1, k$ are landmarks), assume that v has a bad pair, and let x, y be the minimal bad pair of v. If $\{x, y\}$ is not contained in the clockwise path from 1 to k, then either 1 or k covers x, y. Otherwise, x, y is covered iff at least one of the following conditions holds: (i) $x \leq high(\ell_1)$, (ii) $y \geq low(\ell_2)$.

For the vertex 1, we define the parameters $Low(1), High(1), \Delta(1)$ as follows. $Low(1)$ is the minimum index i such that the clockwise path from i to 1 is the unique shortest path (by definition $n - 1$ satisfies this condition, hence $Low(1)$ is well-defined, and thus the vertices $i, i+1, \ldots, n-1$ are close to 1). $High(1)$ is the maximum index i such that the clockwise path from 1 to i is the unique shortest path (similarly to the existence of $Low(1)$, the value of $High(1)$ is well-defined, vertices $2, 3, \ldots, High(1)$ are close to 1, and $High(1) < k$). $\Delta(1)$ is the minimum number $i \geq 1$ such that $i + 1$ and $n - i \geq k + 1$ are both close to 1 (and have the same distance from 1) and belong to a common layer. If such i does not exist, we let $\Delta(1) = \infty$. It can be the case that there exists a bad pair of 1 even if $\Delta(1) = \infty$ but in this case we show later that k covers all such bad pairs.

For the vertex k, we define similar parameters $Low(k), High(k), \Delta(k)$ as follows. $Low(k)$ is the minimum index $i > 1$ such that the clockwise path from i to k is the unique shortest path (by definition $k - 1$ satisfies this condition, hence $Low(k)$ is well-defined). $High(k)$ is the maximum index $i \leq n - 1$ such that the clockwise path from k to i is the unique shortest path (note that it is possible that vertex 1 is also close to k, using the clockwise path from k to 1, but we are not interested whether this holds or not). $Delta(k)$ is the minimum number $i \geq 1$ such that $k + i \leq n - 1$ and $k - i \geq 1$ are both close to k (and have the same distance from k) and belong to a common layer. If such i does not exist, we let $\Delta(k) = \infty$. We also let $low(1) = 1$, $high(1) = High(1)$, $low(k) = Low(k)$ and $high(k) = k$. We define $\delta(1)$ and $\delta(k)$ in the following way. We let $\delta(1) = \infty$ if $High(k) \geq n - \Delta(1)$. Otherwise, $\delta(1) = \Delta(1)$. The motivation is to define $\delta(1)$ to be infinite if there is no bad pair of 1, or if k covers the minimal bad pair of 1. Similarly, we define $\delta(k) = \infty$ if $Low(1) \leq k + \Delta(k)$. Otherwise, we let $\delta(k) = \Delta(k)$. The differences between Δ and δ, and the property that $\Delta(1), \Delta(k)$ can be infinite even if there is a bad pair of 1, and k, respectively, follow since $High(k) \geq n - \Delta(1)$ iff the minimal bad pair of 1 is covered by k, and $Low(1) \leq k + \Delta(k)$ iff the minimal bad pair of k is covered by 1.

We define a function $F : \{1, 2, \ldots, k\} \times \{left, right\} \to \mathbf{R}^+$ as follows. $F(v, right)$ ($F(v, left)$) is the minimum cost of a set L such that $1, v \in L$, every $1 < i < v$ is close to at least one of the vertices in L, and for every $\ell \in L \setminus \{v\}$ ($\ell \in L$, respectively) such that $\delta(\ell)$ is finite the minimal bad pair of ℓ is covered by a vertex in $L \setminus \{\ell\}$. We compute the values of F using the following dynamic program. If $\delta(1) = \infty$, then $F(1, right) = F(1, left) = c(1)$, and if $\delta(1)$ is finite, then $F(1, left) = \infty$ and $F(1, right) = c(1)$. For $\ell > 1$ we define sets of feasible values of $\ell' < \ell$ (to be consecutive landmarks) $\alpha(\ell) = \{\ell' \in \{1, 2, \ldots, \ell - 1\} : high(\ell') \geq low(\ell) - 1\}$, $\beta(\ell) = \{\ell' \in \alpha(\ell) : high(\ell') \geq \ell - \delta(\ell)\}$, $\gamma(\ell) = \{\ell' \in \alpha(\ell) : low(\ell) \leq \ell' + \delta(\ell')\}$. The recursive formula is as follows.

1. $F(\ell, left) = \min\{\min_{\ell' \in \beta(\ell)} F(\ell', left), \min_{\ell' \in \beta(\ell) \cap \gamma(\ell)} F(\ell', right)\} + c(\ell)$,
2. $F(\ell, right) = \min\{\min_{\ell' \in \alpha(\ell)} F(\ell', left), \min_{\ell' \in \gamma(\ell)} F(\ell', right)\} + c(\ell)$.

If $\delta(k) = \infty$, then we are looking for $F(k, right)$, and otherwise we are looking for $F(k, left)$. Then, by backtracking we compute the optimal landmark set. We conclude that the algorithm computes an optimal solution in polynomial time. Thus, we proved that for a wheel G, wmd can be solved in polynomial time.

References

1. Babai, L.: On the order of uniprimitive permutation groups. Annals of Mathematics 113(3), 553–568
2. Beerliova, Z., Eberhard, F., Erlebach, T., Hall, A., Hoffmann, M., Mihalák, M., Ram, L.S.: Network discovery and verification. IEEE Journal on Selected Areas in Communications 24(12), 2168–2181 (2006)
3. Cáceres, J., Hernando, M.C., Mora, M., Pelayo, I.M., Puertas, M.L., Seara, C., Wood, D.R.: On the metric dimension of cartesian products of graphs. SIAM Journal on Discrete Mathematics 21(2), 423–441 (2007)
4. Chartrand, G., Eroh, L., Johnson, M.A., Oellermann, O.R.: Resolvability in graphs and the metric dimension of a graph. Discrete Applied Mathematics 105(1-3), 99–113 (2000)
5. Chartrand, G., Zhang, P.: The theory and applications of resolvability in graphs: A survey. Congressus Numerantium 160, 47–68 (2003)
6. Chvátal, V.: Mastermind. Combinatorica 3(3), 325–329 (1983)
7. Corneil, D.G., Perl, Y., Stewart, L.K.: A linear recognition algorithm for cographs. SIAM Journal on Computing 14(4), 926–934 (1985)
8. Diaz, J., Pottonen, O., Serna, M., van Leeuwen, E.J.: On the complexity of metric dimension. CoRR, abs/1107.2256. Proc. of ESA 2012 (to appear, 2012)
9. Harary, F., Melter, R.A.: The metric dimension of a graph. Ars Combinatoria 2, 191–195 (1976)
10. Hauptmann, M., Schmied, R., Viehmann, C.: Approximation complexity of metric dimension problem. Journal of Discrete Algorithms (2011) (to appear)
11. Khuller, S., Raghavachari, B., Rosenfeld, A.: Landmarks in graphs. Discrete Applied Mathematics 70(3), 217–229 (1996)
12. Melter, R.A., Tomescu, I.: Metric bases in digital geometry. Computer Vision, Graphics, and Image Processing 25, 113–121 (1984)
13. Sebö, A., Tannier, E.: On metric generators of graphs. Mathematics of Operations Research 29(2), 383–393 (2004)
14. Shanmukha, B., Sooryanarayana, B., Harinath, K.S.: Metric dimension of wheels. Far East Journal of Applied Mathematics 8(3), 217–229 (2002)
15. Slater, P.J.: Leaves of trees. Congressus Numerantium 14, 549–559 (1975)

Determining the $L(2,1)$-Span
in Polynomial Space

Konstanty Junosza-Szaniawski[1], Jan Kratochvíl[2],
Mathieu Liedloff[3], and Paweł Rzążewski[1]

[1] Warsaw University of Technology, Faculty of Mathematics and Information Science,
Koszykowa 75, 00-662 Warszawa, Poland
{k.szaniawski,p.rzazewski}@mini.pw.edu.pl
[2] Department of Applied Mathematics, and Institute for Theoretical Computer
Science, Charles University, Malostranské nám. 25, 118 00 Praha 1, Czech Republic
honza@kam.ms.mff.cuni.cz
[3] Laboratoire d'Informatique Fondamentale d'Orléans, Université d'Orléans,
45067 Orléans Cedex 2, France
mathieu.liedloff@univ-orleans.fr

Abstract. An $L(2,1)$-labeling of a graph is a mapping from its vertex
set into a set of integers $\{0,..,k\}$ such that adjacent vertices get labels
that differ by at least 2 and vertices in distance 2 get different labels.
The main result of the paper is an algorithm finding an optimal $L(2,1)$-
labeling of a graph (i.e. an $L(2,1)$-labeling in which the largest label is
the least possible) in time $O^*(7.4922^n)$ and polynomial space. Moreover,
a new interesting extremal graph theoretic problem is defined and solved.

1   Introduction

The Frequency Assignment Problem is the problem of assigning channels (represented by nonnegative integers) to each radio transmitter in a network so that
no transmitters interfere with each other. Hale [11] formulated this problem in
terms of so-called T-coloring of graphs. According to [10], Roberts was the first
one who proposed a modification of this problem, which is called an $L(2,1)$-
labeling problem. It asks for a labeling with nonnegative integer labels in which
no vertices in distance 2 in a graph have the same label and labels of adjacent
vertices differ by at least 2.

The span of an $L(2,1)$-labeling is the difference between the largest and smallest labels used. By $\lambda(G)$ we denote the $L(2,1)$-span of a graph G, which is the
smallest span over all $L(2,1)$-labelings of G.

The problem of $L(2,1)$-labeling has been extensively studied (see [4] for a
survey on the problem and its generalizations). A considerable attention has
been given to bounding the value of $\lambda(G)$ by some function of G.

Griggs and Yeh [10] proved that $\lambda(G) \leq \Delta^2 + 2\Delta$ (where Δ denotes the largest
vertex degree in G) and conjectured, that $\lambda(G) \leq \Delta^2$ for every graph G. There
are several results supporting this conjecture, for example Gonçalves [9] proved
that $\lambda(G) \leq \Delta^2 + \Delta - 2$ for graphs with $\Delta \geq 3$. Havet *et al.* [12] have settled

M.C. Golumbic et al. (Eds.): WG 2012, LNCS 7551, pp. 126–137, 2012.

the conjecture in affirmative for graphs with $\Delta \geq 10^{69}$. For graphs with smaller Δ, the conjecture still remains open.

The second main direction of research in $L(2, 1)$-labeling was to analyze its computational complexity. Griggs and Yeh [10] proved that computing $\lambda(G)$ is an NP-hard problem. Fiala et al. [8] later proved that deciding $\lambda(G) \leq k$ remains NP-complete for every fixed $k \geq 4$ (for $k \leq 3$ the problem is polynomial). NP-completeness for planar inputs was proved by Bodlaender et al. [2] for $k = 8$, by Janczewski et al. [15] for $k = 4$ and finally by Eggeman et al. [6] for all $k \geq 4$. When k is part of the input, the problem remains NP-complete even for series-parallel graphs (see Fiala *et al.* [7]).

Král' [20] presented an exact algorithm for a more general problem called Channel Assignment, which solves $L(2, 1)$-labeling problem in time $O^*(4^n)$ [‡]. Havet *et al.* [13] presented an algorithm for computing $\lambda(G)$, which works in time $O(3.8730^n)$. This algorithm has been improved [18,19], achieving a complexity bound $O(3.2361^n)$. Cygan and Kowalik published [5] an algorithm solving Channel Assignment Problem, which restricted to $L(2, 1)$-labeling works in time $O^*(3^n)$. The currently fastest algorithm for $L(2, 1)$-labeling with a complexity bound $O(2.6488^n)$ has been presented in [16]. All the algorithms mentioned above use exponential amount of memory. Havet *et al.* [13] presented a branching algorithm which determines if $\lambda(G) \leq k$ in time $O^*((k-2.5)^n)$ and polynomial space.

Until now, no algorithm for $L(2, 1)$-labeling with time complexity $O(c^n)$ for some constant c and polynomial space complexity has been published. However, there are such algorithms for a related problem of classical graph coloring. The first one, with time complexity $O(5.2830^n)$, was shown by Bodlaender and Kratsch [3]. The best currently known algorithm for graph coloring with polynomial space complexity is by Björklund *et al.* [1], using the inclusion-exclusion principle. Its time complexity is $O(2.2461^n)$.

The main idea of the algorithm presented in this paper was invented (but not published) by Havet *et al.* [14] and independently by Junosza-Szaniawski and Rzążewski [17]. Both groups obtained the algorithm with time complexity $O((9+\epsilon)^n)$ for arbitrarily small positive constant ϵ. In this joint paper we develop the idea to obtain the the the following theorem.

Theorem 1. *The $L(2, 1)$-span of a graph on n vertices can be computed in time $O(7.4922^n)$ and polynomial space.*

To prove the complexity bound of the algorithm, in Section 4 we consider an extremal combinatorial problem, which is highly related to famous results concerning the maximum number of maximal independent sets, obtained independently by Miller and Muller [21], and Moon and Moser [22].

2   Preliminaries

A graph is a pair $G = (V, E)$, where V is a finite set, called the set of vertices, and E is a family of 2-element subsets of V. A k-$L(2, 1)$-labeling of a graph G is any function $\psi : V \to \{0, \ldots k\}$ such that

---

[‡] In the O^* notation we omit polynomially bounded terms.

1. $|\psi(v) - \psi(w)| \geq 1$ for all $v, w \in V(G)$ such that $\exists u \in V$ $vu, uw \in E(G)$,
2. $|\psi(v) - \psi(w)| \geq 2$ for all $v, w \in V(G)$ such that $vw \in E(G)$.

By $\lambda(G)$ we denote the smallest k such that there exists a k-$L(2,1)$-labeling of G. We will define $L(2,1)$-labeling in a more general way and for that we will need graphs with two kinds of edges. A triple $G = (V, R, B)$ is called a *red-black graph* if V is finite, R, B are disjoint families of 2-element subsets of V fulfilling the condition: if $vw \in B$ and $vu \in B$, then $uw \in R \cup B$ for any vertices v, u, w. For a red-black graph $G = (V, R, B)$ we refer to V, R, B, $R \cup B$ by $V(G)$, $R(G)$, $B(G)$, $E(G)$, respectively. We call $R(G)$ the set of red edges and $B(G)$ the set of black edges, while $E(G)$ is the set of edges.

An *R-closure* of a graph G is a red-black graph H, such that $V(H) = V(G)$, $B(H) = E(G)$ and $R(H) = E(G^2) \setminus E(G) = \{uw \in V(G): \exists v \in V(G) \ uv \in E(G) \text{ and } vw \in E(G) \text{ and } uw \notin E(G)\}$.

For a red-black graph H we define a k-L-labeling as a function $\varphi : V(H) \to \{1, \ldots, k\}$ (we do not use 0 as a label for convenience) fulfilling

1. $|\varphi(v) - \varphi(w)| \geq 1$ for all $v, w \in V(G)$ such that $vw \in R(G)$
2. $|\varphi(v) - \varphi(w)| \geq 2$ for all $v, w \in V(G)$ such that $vw \in B(G)$.

Notice that a $(k+1)$-L-labeling of the R-closure of a graph G corresponds to a k-$L(2,1)$-labeling of G.

Remark 1. If ψ is a k-$L(2,1)$-labeling of a graph G and H is an R-closure of G, then a labeling φ defined by $\varphi(v) = \psi(v) + 1$ is a $(k+1)$-L-labeling of H and vice-versa.

For technical reasons we need to consider two more restrictions. Suppose that the given instance is a red-black graph G and sets P, Q. A labeling $\varphi: V(G) \to \{1, \ldots, k\}$ is a k-L_Q^P-labeling of G if it is a k-L-labeling of G and $Q \cap \varphi^{-1}(1) = \emptyset$ and $P \cap \varphi^{-1}(k) = \emptyset$. A function φ is an L_Q^P-labeling of G if it is a k-L_Q^P-labeling of G for some k. Note that every L_Q^P labeling of G is in fact a $L_{Q \cap V(G)}^{P \cap V(G)}$-labeling of G. However, we will not restrict the definition to sets $P, Q \subseteq V(G)$, as it makes the description of the algorithm simpler.

Let $\Lambda_Q^P(G)$ denote the smallest possible $k \geq 0$ admitting the existence of a k-L_Q^P-labeling of G. In particular, for a graph with no vertices we have $\Lambda_Q^P((\emptyset, \emptyset, \emptyset)) \overset{def.}{=} 0$.

The considered generalized problem asks to compute $\Lambda_Q^P(G)$. Any k-L_Q^P-labeling of G with $k = \Lambda_Q^P(G)$ is called *optimal*. We observe that even if φ is an optimal L_Q^P-labeling of G, then any of the sets $\varphi^{-1}(1)$ and $\varphi^{-1}(\Lambda_Q^P(G))$ may be empty. In the extremal case, if $P = Q = V(G)$, then $\varphi^{-1}(1) = \varphi^{-1}(k) = \emptyset$ for all k and feasible k-L_Q^P-labelings φ of G.

Notice that if H is an R-closure of a graph G then $\Lambda_\emptyset^\emptyset(H) = \lambda(G) + 1$, by Remark 1. In this way we shall use our algorithm to compute an $L(2,1)$-span of a given input graph.

Let $N(v) = \{u \in V(G): (u, v) \in E(G)\}$ denote the set of neighbors (the *neighborhood*) of a vertex v. A *red neighborhood* (*black neighborhood*, respectively) of a vertex v, denoted by $N_R(v)$ ($N_B(v)$, respectively), is the set of vertices w such that $vw \in R(G)$ ($vw \in B(G)$, respectively). The neighborhood, red neighborhood, black neighborhood of a set Y of vertices in G are denoted by $N(Y) = \bigcup_{v \in Y} N(v)$, $N_R(Y) = \bigcup_{v \in Y} N_R(v)$, $N_B(Y) = \bigcup_{v \in Y} N_B(v)$, respectively.

For a red-black graph $G = (V, R, B)$ and for a set of vertices $V' \subseteq V(G)$ let $G[V'] = (V', \{e \in R(G) : e \subseteq V'\}, \{e \in B(G) : e \subseteq V'\})$ denote the *subgraph of G induced by the set of vertices V'*.

A set of vertices Y is *independent* if no two vertices in this set are adjacent (the color of edges is not important here). An independent set Y is *R-maximal* if every vertex v such that $N_R(v) = N(v)$ is either in Y or has a neighbor in Y. A pair (X, Y) of disjoint subsets of $V(G)$ is *proper* if Y is independent. A proper pair is *R-maximal* if Y is R-maximal.

A triple of sets (X, Y, Z) is a *balanced partition* of G if

1. The sets X, Y, Z form a partition of $V(G)$.
2. The set Y is independent.
3. All sets X, Y, Z are non-empty.
4. $|X| \le \frac{|V(G)|}{2}$ and $|Z| \le \frac{|V(G)|}{2}$.

A triple of sets (X, Y, Z) is a *correct partition* of G if it is a balanced partition of G and Y is an R-maximal independent set. Note that if (X, Y, Z) is a correct partition of G, then (X, Y) is an R-maximal proper pair. Hence we obtain the following Remark.

Remark 2. The number of R-maximal proper pairs in G is an upper bound for the number of correct partitions of G.

3   Algorithm

The algorithm presented in this section is based on the classical *Divide-and-Conquer* paradigm. The main idea is based on two key observations, described in Lemmas 1 and 2.

Lemma 1. *Let φ be a k-L_Q^P-labeling of G. For $h \in \{1, \ldots, k\}$ let $X_h = \bigcup_{i=1}^{h-1} \varphi^{-1}(i)$, $Y_h = \varphi^{-1}(h)$ and $Z_h = \bigcup_{i=h+1}^{k} \varphi^{-1}(i)$. There exists $h \in \{1, \ldots, k\}$ such that one of the following cases occurs:*

1. *$X_h = \emptyset$ and $|Y_h| \ge |V(G)|/2$,*
2. *$Z_h = \emptyset$ and $|Y_h| \ge |V(G)|/2$,*
3. *the triple (X_h, Y_h, Z_h) is a balanced partition of G.*

Proof. We shall prove that if neither the case 1 nor the case 2 occurs, then we can choose h in such a way, that the case 3 occurs. Notice that for every $h \in \{1, \ldots, k\}$ the sets X_h, Y_h and Z_h clearly form a partition of $V(G)$ and Y_h is independent. Let h be the smallest number, such that $|X_h \cup Y_h| \ge |V(G)|/2$. Clearly $Y_h \ne \emptyset$, because otherwise we would choose $h-1$. By the choice of h, since the cases 1 and 2 did not occur, $X_h \ne \emptyset$ and $Z_h \ne \emptyset$. Hence $|X_h| \le |V(G)|/2$. Moreover, the fact that $|X_h \cup Y_h| \ge |V(G)|/2$ implies that $|Z_h| \le |V(G)|/2$. $\square$

Lemma 2. *Let* $k = \Lambda_Q^P(G)$ *and* $h \in \{2, \ldots, k-1\}$. *There exists an optimal* L_Q^P-*labeling* φ *of* G, *such that the set* $\varphi^{-1}(h)$ *is* R-*maximal.*

Proof. Let φ be an optimal L_Q^P-labeling of G. If $\varphi^{-1}(h)$ is R-maximal, then φ is the optimal labeling we are looking for. Otherwise, there exists at least one vertex v such that:

1. for all $w \in N(v)$ holds $vw \in R(G)$ and $\varphi^{-1}(w) \neq h$
2. $\varphi^{-1}(v) \neq h$

Note that we can change the label of v to h obtaining an optimal L_Q^P-labeling φ' of G. Applying such a relabeling recursively, we finally obtain a labeling satisfying our criteria.   □

Moreover, we observe that if the graph is disconnected, we can label each of its connected components separately. Hence for all graphs G and sets P, Q

$$\Lambda_Q^P(G) = \max\{\Lambda_Q^P(C) \colon C \text{ is a connected component of } G\}. \tag{1}$$

Therefore we may assume that the input graph G is connected.

The algorithm partitions the vertex set $V(G)$ into all possible triples of sets X, Y, Z, which form a correct partition of G. The graphs $G[X]$ and $G[Z]$ are then labeled recursively. Due to restrictions related to the sets P and Q, the cases when $X = \emptyset$ or $Z = \emptyset$ have to be considered separately.

The labeling of the whole G is constructed from the labelings found in the recursive calls. The sets of labels used on the sets X and Z are separated from each other by the label used for the R-maximal independent set Y. This allows to solve the subproblems for $G[X]$ and $G[Z]$ independently from each other. Iterating over all such partitions of $V(G)$, the algorithm computes the minimum k admitting the existence of a k-L_Q^P-labeling of G, which is by definition $\Lambda_Q^P(G)$. The pseudo-code of our algorithm is given by Algorithm 1.

Lemma 3. *For a red-black graph* $G = (V, \emptyset, \emptyset)$ *and any sets* P, Q, *it holds that* $\Lambda_Q^P(G) \leq 3$.

Proof. The labeling $\varphi \colon V \to \{1, 2, 3\}$ such that $\varphi(v) = 2$ for every $v \in V$ is a feasible 3-L_Q^P-labeling of G.   □

Lemma 4. *For any graph* G *and sets* P, Q, *the algorithm call* **Find-Lambda(**G, P, Q**)** *returns* $\Lambda_Q^P(G)$.

Proof. The proof proceeds by the induction on $|V(G)|$. If $V(G) = \emptyset$, the correct result is given in line 1 (by the definition of $\Lambda_Q^P(\emptyset, \emptyset, \emptyset)$). If $\Lambda_Q^P(G) \leq 3$, the result is found by the exhaustive search in line 3. Notice that if $|V(G)| \leq 1$, then $\Lambda_Q^P(G) \leq 3$ by Lemma 3.

Assume that the statement is true for all graphs G' and all sets P', Q', such that $|V(G')| < n$, where $n > 1$. Let G be a graph on n vertices and P, Q be sets. We may also assume that $\Lambda_Q^P(G) > 3$, because otherwise it would be labeled by the exhaustive search in line 3. Let k be the value returned by the algorithm call

---

Algorithm 1. Find-Lambda(G, P, Q)

Input: Red-black graph G, Sets P, Q

1   **if** $V(G) = \emptyset$ **then return** 0
2   **for** $k \leftarrow 1$ **to** 3 **do**
3   if *there exists a k-L_Q^P-labeling of G* **then return** k

4   **foreach** *connected component C of G* **do**
5      $k[C] \leftarrow \infty$
6   **foreach** *independent set $Y \subseteq V(C)$ such that $|Y| \geq |V(C)|/2$* **do**
7        $k_1 \leftarrow$ **Find-Lambda**$(C - Y, N_B(Y), Q)$
8        $k_2 \leftarrow$ **Find-Lambda**$(C - Y, P, N_B(Y))$
9   **if** $Y \cap P \neq \emptyset$ **then** $k_1 \leftarrow k_1 + 1$
10   **if** $Y \cap Q \neq \emptyset$ **then** $k_2 \leftarrow k_2 + 1$
11       $k[C] \leftarrow \min(k[C], k_1 + 1, k_2 + 1)$

12   **foreach** *correct partition (X, Y, Z) of C* **do**
13        $k_X \leftarrow$ **Find-Lambda**$(C[X], N_B(Y), Q)$
14        $k_Z \leftarrow$ **Find-Lambda**$(C[Z], P, N_B(Y))$
15        $k[C] \leftarrow \min(k[C], k_X + 1 + k_Z)$

16 **return** $\Lambda_Q^P(G) = \max\{k[C] : C$ is a connected component of $G\}$

Find-Lambda(G, P, Q). To show that $k = \Lambda_Q^P(G)$ it is enough to show that that $k[C] = \Lambda_Q^P(C)$ for every connected component C (for $k[C]$ defined as in the algorithm). Having proven this, we have $k = \max\{k[C] : C$ is a component of $G\} = \max\{\Lambda_Q^P[C] : C$ is a component of $G\} = \Lambda_Q^P(G)$ (by (1)). Let C be a connected component of G.

First we will show that $k[C] \geq \Lambda_Q^P(C)$, i.e. there exists a k-L_Q^P-labeling of C. Assume that $k[C]$ was set in the line 11. Consider the independent set Y and the iteration of the loop in lines 6–11 for which k was set. Let $k_1' =$ **Find-Lambda**$(C - Y, N_B(Y), Q)$ and $k_2' =$ **Find-Lambda**$(C - Y, P, N_B(Y))$. By the inductive assumption there exist a k_1'-$L_Q^{N_B(Y)}$-labeling φ' of $C - Y$ and a k_2'-$L_{N_B(Y)}^P$-labeling φ'' of $C - Y$. Notice that $k[C] \in \{k_1' + 1, k_1' + 2, k_2' + 1, k_2' + 2\}$ and at least one of the following cases occurs.

Case 1: $k[C] = k_1' + 1$ and $Y \cap P = \emptyset$. In this case we can extend φ' in the following way

$$\varphi(v) = \begin{cases} \varphi'(v) & \text{if } v \in V(C) \setminus Y \\ k_1' + 1 & \text{if } v \in Y \end{cases}$$

obtaining a $k[C]$-L_Q^P-labeling φ of C.

Case 2: $k[C] = k_1' + 2$ and $Y \cap P \neq \emptyset$. In this case we can extend φ' in the following way

$$\varphi(v) = \begin{cases} \varphi'(v) & \text{if } v \in V(C) \setminus Y \\ k_1' + 1 & \text{if } v \in Y \end{cases}$$

obtaining a $k[C]$-L_Q^P-labeling φ of C (note that due to restriction on P the label $k_1' + 2$ is counted as used, despite the fact that no vertex has this label).

Case 3: $k[C] = k_2' + 1$ and $Y \cap Q = \emptyset$. In this case we can extend φ' in the following way

$$\varphi(v) = \begin{cases} 1 & \text{if } v \in Y \\ \varphi'(v) + 1 & \text{if } v \in V(C) \setminus Y \end{cases}$$

obtaining a $k[C]$-L_Q^P-labeling φ of C.

Case 4: $k[C] = k_2' + 2$ and $Y \cap Q \neq \emptyset$. In this case we can extend φ' in the following way

$$\varphi(v) = \begin{cases} 2 & \text{if } v \in Y \\ \varphi'(v) + 2 & \text{if } v \in V(C) \setminus Y \end{cases}$$

obtaining a $k[C]$-L_Q^P-labeling φ of C.

Now assume that that $k[C]$ was set in line 15. Consider the correct partition (X, Y, Z) and the iteration of the loop in lines 12–15 for which k was set. Let k_X and k_Z be defined as in lines 13 and 14 for this iteration. Hence $k[C] = k_X + 1 + k_Z$. By the inductive assumption there exists k_X-$L_Q^{N_B(Y)}$-labeling φ_X of $C[X]$ and k_Z-$L_{N_B(Y)}^P$-labeling φ_Z of $C[Z]$. We can define a $k[C]$-L_Q^P-labeling of C in the following way:

$$\varphi(v) = \begin{cases} \varphi_X(v) & \text{if } v \in X \\ k_X + 1 & \text{if } v \in Y \\ k_X + 1 + \varphi_Z(v) & \text{if } v \in Z \end{cases}$$

Notice that sets X, Y, Z are non-empty since (X, Y, Z) is a correct partition of C. Hence $k_X, k_Z \geq 1$ and $\varphi^{-1}(1) \cap Q = \varphi^{-1}(k_X + 1 + k_Z) \cap P = \emptyset$.

Now we will show that $k[C] \leq \Lambda_Q^P(C)$. Let φ be an optimal L_Q^P-labeling of C. Recall that $\Lambda_Q^P(C) > 3$. One of the following cases occurs.

Case 1: $\varphi^{-1}(1) \geq |V(C)|/2$. Consider the iteration of the loop in lines 6–11 for $Y = \varphi^{-1}(1)$. By the inductive assumption the algorithm **Find-Lambda** sets $k_2 = \Lambda_{N_B(Y)}^P(C - Y)$. Notice that $\Lambda_{N_B(Y)}^P(C - Y) = \Lambda_Q^P(C) - 1$ and $\varphi^{-1}(1) \cap Q = \emptyset$. Hence the condition in line 10 is not fulfilled and k_2 is equal to $\Lambda_Q^P(C) - 1$. By the condition in line 11 we have $k[C] \leq k_2 + 1 \leq \Lambda_Q^P(C)$.

Case 2: $\varphi^{-1}(1) = \emptyset$ and $\varphi^{-1}(2) \geq |V(C)|/2$. Consider the iteration of the loop in lines 6–11 for $Y = \varphi^{-1}(2)$. By the inductive assumption the algorithm **Find-Lambda** sets $k_2 = \Lambda_{N_B(Y)}^P(C - Y)$. Notice that $\varphi^{-1}(1) \cap Q \neq \emptyset$ since otherwise we could decrease the label of every vertex by one, obtaining $(\Lambda_Q^P(C) - 1)$-L_Q^P-labeling of C. Hence $\Lambda_{N_B(Y)}^P(C - Y) = \Lambda_Q^P(C) - 2$ and the algorithm **Find-Lambda** in line 10 sets k_2 to $\Lambda_Q^P(C) - 2 + 1$. By the condition in line 11 $k[C] \leq k_2 + 1 \leq \Lambda_Q^P(C) - 2 + 1 + 1 = \Lambda_Q^P(C)$.

Case 3: $\varphi^{-1}(\Lambda_Q^P(C)) \geq |V(C)|/2$ is symmetric to the Case 1.

Case 4: $\varphi^{-1}(\Lambda^P_Q(C)) = \emptyset$ and $\varphi^{-1}(\Lambda^P_Q(C) - 1) \geq |V(C)|/2$ is symmetric to the Case 2.

Notice that that neither the case $\varphi^{-1}(1) = \varphi^{-1}(2) = \emptyset$ nor $\varphi^{-1}(\Lambda^P_Q(C)) = \varphi^{-1}(\Lambda^P_Q(C) - 1) = \emptyset$ may occur, since the labeling φ is optimal. Hence the only case remaining is:

Case 5: $0 < |\varphi^{-1}(1)| \leq |V(C)|/2$ or $0 < |\varphi^{-1}(1) \cup \varphi^{-1}(2)| \leq |V(C)|/2$ and $0 < |\varphi^{-1}(\Lambda^P_Q(C))| \leq |V(C)|/2$ or $0 < |\varphi^{-1}(\Lambda^P_Q(C)) \cup \varphi^{-1}(\Lambda^P_Q(C) - 1)| \leq |V(C)|/2$.

Let h be defined as in Lemma 1, i.e. $h = \min\{j \in \{1, \ldots, \Lambda^P_Q(C)\}$ such that $|\bigcup_{i=1}^j \varphi^{-1}(i)| \geq |V(C)|/2\}$. Note that $1 < h < \Lambda^P_Q(C)$, because of the conditions of the Case 5. The triple $(\bigcup_{i=1}^{h-1} \varphi^{-1}(i), \varphi^{-1}(h), \bigcup_{i=h+1}^{\Lambda^P_Q(C)} \varphi^{-1}(h))$ is a balanced partition of G.

Let φ' be an optimal L^P_Q-labeling of C, constructed from φ as in the proof of the Lemma 2. Let $X = \bigcup_{i=1}^{h-1} \varphi'^{-1}(i)$, $Y = \varphi'^{-1}(h)$ and $Z = \bigcup_{i=h+1}^{\Lambda^P_Q(C)} \varphi'^{-1}(h)$. Notice that

– Y is R-maximal
– $\emptyset \neq \varphi^{-1}(h) \subseteq Y$ and therefore $Y \neq \emptyset$
– $X \subseteq \bigcup_{i=1}^{h-1} \varphi^{-1}(i)$ and therefore $|X| \leq |V(C)|/2$
– $Z \subseteq \bigcup_{i=h+1}^{\Lambda^P_Q(C)} \varphi^{-1}(h))$ and therefore $|Z| \leq |V(C)|/2$.

Moreover $X \neq \emptyset$ and $Z \neq \emptyset$, because otherwise Case 2 or Case 4 would occur. Hence (X, Y, Z) is a correct partition of C and it is considered in the iteration of the loop in lines 12–15. In the iteration for (X, Y, Z) the algorithm sets $k_X = \Lambda^{N_B(Y)}_Q(C(X)) = h - 1$ in line 13 and $k_X = \Lambda^P_{N_B(Y)}(C(Z)) = \Lambda^P_Q(C) - h$ in line 14 (by the inductive assumption). Hence $k[C] \leq k_X + 1 + k_Z = h - 1 + 1 + \Lambda^P_Q(C) - h = \Lambda^P_Q(C)$. This finishes the proof. $\qquad\square$

4   The Number of R-Maximal Proper Pairs

This section is purely combinatorial. We consider the maximum number of R-maximal proper pairs, which will be used later to bound the complexity of the algorithm Find-Lambda. Let $\rho(G)$ denote the number of R-maximal proper pairs in a red-black graph G. Let $\rho(n)$ denote the maximum value of $\rho(G)$ over all connected red-black graphs on n vertices.

In this section we prove the following Theorem.

Theorem 2. *The maximum number of R-maximal proper pairs in a connected red-black graph on n vertices is $\Theta(\sqrt{8}^n)$.*

First let us prove the bound for *red graphs*, i.e. red-black graphs with no black edges. Let $\rho_R(n)$ denote the maximum value of $\rho(G)$ over all red graphs on n vertices. Note that R-maximal independent sets in a red graph are just maximal independent sets. Therefore R-maximal proper pairs in a red graph are the pairs (X, Y) of disjoint set, where Y is a maximal independent set. The proof of Theorem 3 is inspired by an elegant proof by Wood [23].

Theorem 3. *If a graph G is red, then $\rho(G) = O\left(\sqrt[5]{80}^n\right) = O(2.4023^n)$, where n is the number of vertices in G.*

Proof. We shall prove the statement by induction on the number of vertices n. If $n \leq 2$, the statement is obviously true. Assume that $n \geq 3$ and the statement is true for all red graphs with less than n vertices.

Let G be a red graph on n vertices, such that $\rho(G) = \rho_R(n)$. Let v be the vertex of G having the smallest degree (denoted by δ). Notice that for every R-maximal proper pair (X, Y), at least one of the vertices in $N[v]$ must be in Y (since Y is a maximal independent set). Let $w \in N[v] \cap Y$. Since the set Y is independent, none of the vertices from $N(w)$ belongs to it. However, each of them can be in X or not. Hence we obtain the following recursion: $\rho_R(n) \leq \sum_{w \in N[v]} 2^{\deg w} \rho_R(n - \deg w - 1)$. Let d be the element from $\{\delta, \ldots, n-1\}$ maximizing the expression $2^d \rho_R(n - d - 1)$. Then $\rho_R(n) \leq \sum_{w \in N[v]} 2^d \rho_R(n - d - 1) = (\delta + 1)2^d \rho_R(n - d - 1) \leq (d + 1)2^d \rho_R(n - d - 1)$.

From this we obtain that $\rho_R(n) = O\left(\left(\sqrt[d+1]{(d+1)2^d}\right)^n\right)$. One can easily verify that this value is maximized for $d = 4$. Hence $\rho_R(n) = O\left(\sqrt[5]{80}^n\right) = O(2.4023^n)$. $\qquad\square$

The proof for the lower bound $\rho_R(n)$ is also analogical to the case of maximal independent sets (see [21,22]). Consider a red graph H_k consisting of k disjoint copies of a complete graph K_5. A direct computation shows that $\rho(H_k) = \Theta\left((5 \cdot 2^4)^k\right) = \Theta\left(\sqrt[5]{80}^n\right)$, which proves that $\rho_R(n) = \Theta\left(\sqrt[5]{80}^n\right)$.

It is interesting to mention that the same bound applies if we restrict ourselves to connected red graphs. Let H'_m be a graph H_m with one additional vertex adjacent to exactly one vertex from each copy of K_5. It is easy to check that $\rho(H'_m) = \Theta\left(\sqrt[5]{80}^n\right)$.

Having the bound on $\rho_R(n)$, we can now proceed to bounding the number of R-maximal proper pairs in all connected red-black graphs. In the analysis of the number of R-maximal proper pairs we shall use the concept of partitioning graphs to stars, used by Havet *et al.* [13] to bound the number of 2-packings in a connected graph.

For a red-black graph G let G_B denote the graph induced by the set of vertices belonging to black edges, i.e. $G[\bigcup_{e \in B(G)} e]$. By G_R we denote the subgraph induced by the set of vertices $V(G) \setminus V(G_B)$. We say that a graph G is *black* if $G = G_B$.

Lemma 5. *The maximum number of R-maximal proper pairs in a black graph G without isolated vertices is upper-bounded by $O(\sqrt{8}^n) = O(2.8285^n)$.*

Proof. Let us consider a partition of the vertex set of G with disjoint black stars $S_1, S_2, \ldots, S_d$, each containing at least 2 vertices. To construct such a partition, let us consider a graph $G' = (\bigcup_{e \in B(G)} e, B(G))$ and consider each connected component C of G' separately. Let T be a spanning tree of C and let v and

u be, respectively, the end-vertex and its neighbor on a longest path in T. All neighbors of u in T except at most one are leaves in T. We add the star S induced by u and all its neighbors which are leaves in T to our partition and proceed recursively with the tree $T \setminus S$.

Let $u_k(n)$ denote the maximum number of k-element independent sets Y in G. Note that at most one vertex from each black star S_i can be in Y, since the graph G is red-black. Therefore we observe that $u_k(n) \leq \sum_{1 \leq i_1 < ... < i_k \leq d} \prod_{\ell=1}^{k} |S_{i_\ell}|$. It can be easily proven (see Havet *et al.* [13]) that this expression has the largest value for $|S_1| = |S_2| = \ldots = |S_d|$. Thus $u_k(n) \leq \sum_{1 \leq i_1 < ... < i_k \leq d} 2^k = \binom{n/2}{k} 2^k$.

Notice that every vertex that is not in Y can be included in X or not. Finally, we obtain the following formula: $\rho(G) = \sum_{k=0}^{n} u_k(n) 2^{n-k} = O(\sqrt{8}^n) = O(2.8285^n)$. □

Proof of Theorem 2. We shall construct proper pairs (X, Y) in two steps:

1. From G_B select an independent set Y_B and a set X_B disjoint with Y_B.
2. From G_R select a maximal independent set Y_R such that $Y_R \cap N(Y_B) = \emptyset$, and a set X_R disjoint with Y_R.
3. Return $X = X_B \cup X_R$ and $Y = Y_B \cup Y_R$.

Notice that such constructed pairs (X, Y) are exactly R-maximal proper pairs in G. Since the graph G_B has no isolated vertices, for $|V(G_B)| = n'$ we obtain the following formula. $\rho(n) = \rho(G) = O(\sqrt{8}^{n'} \cdot \rho_R(n - n')) = O(\sqrt{8}^{n'} \cdot 2.4023^{n-n'}) = O(\sqrt{8}^n) = O(2.8285^n)$.

To show that this bound is best possible, let us consider the graph M_k consisting of k disjoint black edges and a vertex v connected with a red edge to one vertex from each black edge.

A direct calculation shows that $\rho(M_k) = \Theta(8^k) = \Theta(8^{n/2})$. Thus $\rho(n) = \Theta(\sqrt{8}^n)$. This finishes the proof of Theorem 2. □

It is easy to observe that R-maximal proper pairs in connected black graphs are exactly the proper pairs, considered in [16].

Theorem 4 (J.-S.,L.,K.,Rossmanith, Rz. [16]). *Let $\rho_B(n)$ denote the maximum value of $\rho(G)$ over all graphs G, which are an R-closure of some connected graph H on n vertices. Then $\rho_B(n) = O(2.6488^n)$ and $\rho_B(n) = \Omega(2.6117^n)$.*

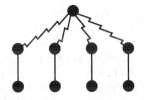

Fig. 1. Graph M_4, straight edges are black and zig-zag edges are red

5   Complexity of the Algorithm

A direct estimation of the computation complexity of our algorithm gives a running-time $O((9 + \epsilon)^n)$. By using the bounds from Section 4, we can improve this running-time upper bound, as claimed in the next Lemma :

Lemma 6. *The algorithm* **Find-Lambda** *computes* $\Lambda_Q^P(G)$ *of a red-black graph in time* $O((8 + \epsilon)^n)$ *and polynomial space, where n is the number of vertices in G and ϵ is an arbitrarily small positive constant.*

Proof. Verifying if a given set Y is independent can be performed in polynomial time. We can check if a given function $\varphi \colon V(G) \to \mathbb{N}$ is an L_Q^P-labeling of G in polynomial time as well. The algorithm **Find-Lambda** first checks in constant time if $V(G) = \emptyset$. Then it exhaustively checks if there exists a k-L_Q^P-labeling of G for $k \in \{1, 2, 3\}$. There are 3^n functions $\varphi \colon V(G) \to \{1, 2, 3\}$, so this step is performed in time $n^{O(1)} \cdot 3^n$.

Then for every connected component C of G the algorithm checks all independent sets of size at least $|V(C)|/2$ (there are no more than 2^n such sets) and all correct partitions of C (by Theorem 2 and Remark 2 there are at most $\sqrt{8}^n$ considered partitions). The algorithm is called recursively for at most two sets of size at most $n/2$ per component. Hence we obtain the following inequality for the complexity:

$$T(n) \leq (2^n + \sqrt{8}^n) n^{O(1)} T(n/2) \tag{2}$$

The solution of this recursion is bounded by $O(8^n n^{O(1) \log n}) = O(8^n 2^{O(1) \log^2 n})$, which is bounded by $O((8 + \epsilon)^n)$, for all $\epsilon > 0$. The space complexity of the algorithm is clearly polynomial.   □

Proof of Theorem 1. Notice that if we are looking for the $L(2, 1)$-labeling of G, we can assume that G is connected (otherwise we would label each of its components separately) and the initial graph given to the algorithm is an R-closure of G (R-closure of a graph can be found in polynomial time).

Hence the complexity of the algorithm is bounded by

$$T'(n) \leq (2^n + 2.6488^n) n^{O(1)} T(n/2), \tag{3}$$

where T is given by inequality (2). From (3) this we obtain $T'(n) = O(7.4922^n)$.   □

References

1. Björklund, A., Husfeldt, T., Koivisto, M.: Set Partitioning via Inclusion-Exclusion. SIAM J. Comput. 39, 546–563 (2009)
2. Bodlaender, H.L., Kloks, T., Tan, R.B., van Leeuwen, J.: Approximations for lambda-Colorings of Graphs. Computer Journal 47, 193–204 (2004)
3. Bodlaender, H.L., Kratsch, D.: An exact algorithm for graph coloring with polynomial memory. UU-CS 2006-015 (2006)

4. Calamoneri, T.: The $L(h,k)$-Labelling Problem: An Updated Survey and Annotated Bibliography. Computer Journal 54(8), 1344–1371 (2011)
5. Cygan, M., Kowalik, L.: Channel assignment via fast zeta transform. Inf. Proc. Letters 111, 727–730 (2011)
6. Eggeman, N., Havet, F., Noble, S.: k-$L(2,1)$-Labelling for Planar Graphs is NP-Complete for $k \geq 4$. Disc. Appl. Math. 158, 1777–1788 (2010)
7. Fiala, J., Golovach, P., Kratochvíl, J.: Distance Constrained Labelings of Graphs of Bounded Treewidth. In: Caires, L., Italiano, G.F., Monteiro, L., Palamidessi, C., Yung, M. (eds.) ICALP 2005. LNCS, vol. 3580, pp. 360–372. Springer, Heidelberg (2005)
8. Fiala, J., Kloks, T., Kratochvíl, J.: Fixed-parameter complexity of λ-labelings. Disc. Appl. Math. 113, 59–72 (2001)
9. Gonçalves, D.: On the L(p; 1)-labelling of graphs. Disc. Math. 308, 1405–1414 (2008)
10. Griggs, J.R., Yeh, R.K.: Labelling graphs with a condition at distance 2. SIAM J. Disc. Math. 5, 586–595 (1992)
11. Hale, W.K.: Frequency assignemnt: Theory and applications. Proc. IEEE 68, 1497–1514 (1980)
12. Havet, F., Reed, B., Sereni, J.-S.: $L(2,1)$-labellings of graphs. In: Proc. of SODA 2008, pp. 621–630 (2008)
13. Havet, F., Klazar, M., Kratochvíl, J., Kratsch, D., Liedloff, M.: Exact algorithms for $L(2,1)$-labeling of graphs. Algorithmica 59, 169–194 (2011)
14. Havet, F., Klazar, M., Kratochvíl, J., Kratsch, D., Liedloff, M.: Exact Algorithms for $L(p,q)$-labelings of graphs (manuscript)
15. Janczewski, R., Kosowski, A., Małafiejski, M.: The complexity of the $L(p,q)$-labeling problem for bipartite planar graphs of small degree. Discrete Mathematics 309, 3270–3279 (2009)
16. Junosza-Szaniawski, K., Kratochvíl, J., Liedloff, M., Rossmanith, P., Rzążewski, P.: Fast Exact Algorithm for $L(2,1)$-Labeling of Graphs. In: Ogihara, M., Tarui, J. (eds.) TAMC 2011. LNCS, vol. 6648, pp. 82–93. Springer, Heidelberg (2011)
17. Junosza-Szaniawski, K., Rzążewski, P.: Determining $L(2,1)$-span in Polynomial Space. arXiv:1104.4506v1 [cs.DM]
18. Junosza-Szaniawski, K., Rzążewski, P.: On Improved Exact Algorithms for $L(2,1)$-Labeling of Graphs. In: Iliopoulos, C.S., Smyth, W.F. (eds.) IWOCA 2010. LNCS, vol. 6460, pp. 34–37. Springer, Heidelberg (2011)
19. Junosza-Szaniawski, K., Rzążewski, P.: On the Complexity of Exact Algorithm for $L(2,1)$-labeling of Graphs. Inf. Proc. Letters 111, 697–701 (2011)
20. Král', D.: An exact algorithm for channel assignment problem. Discrete Applied Mathematics 14, 326–331 (2005)
21. Miller, R.E., Muller, D.E.: A problem of maximum consistent subsets. IBM Research Report RC-240. Thomas J. Watson Research Center, New York (1960)
22. Moon, J.W., Moser, L.: On cliques in graphs. Israel J. Math. 3, 23–28 (1965)
23. Wood, D.R.: On the number of maximal independent sets in a graph. Disc. Math. and Theoretical Computer Science 13, 17–20 (2011)

On the Minimum Degree Up to Local Complementation: Bounds and Complexity

Jérôme Javelle[2], Mehdi Mhalla[1,2], and Simon Perdrix[1,2]

[1] CNRS
[2] Laboratoire d'Informatique de Grenoble, Grenoble University

Abstract. The local minimum degree of a graph is the minimum degree reached by means of a series of local complementations. In this paper, we investigate on this quantity which plays an important role in quantum computation and quantum error correcting codes.

First, we show that the local minimum degree of the Paley graph of order p is greater than $\sqrt{p} - \frac{3}{2}$, which is, up to our knowledge, the highest known bound on an explicit family of graphs. Probabilistic methods allows us to derive the existence of an infinite number of graphs whose local minimum degree is linear in their order with constant 0.189 for graphs in general and 0.110 for bipartite graphs. As regards the computational complexity of the decision problem associated with the local minimum degree, we show that it is NP-complete and that there exists no l-approximation algorithm for this problem for any constant l unless $P = NP$.

1   Introduction

For any undirected graph G, the local complementation is an operation which consists in complementing the neighborhood of a given vertex of a graph. It that has been introduced by Kotzig [Kot68] and the study of this quantity is motivated by several applications: Bouchet [Bou90, Bou94] and de Fraysseix [dF81] used local complementation to give a characterization of circle graphs, and Oum [Oum08] links the notion of "vertex minor of a graph" to the equivalence classes up to local complementation. One of the most important results is established by Bouchet in [Bou87]: deciding whether two graphs are equivalent up to local complementations can be done in polynomial time.

In the field of quantum information theory, the rate of some quantum codes obtained by graph concatenation can be bounded by the minimum degree up to local complementation (called "local minimum degree" and denoted δ_{loc}) of the constructed graphs [BCG+11]. Another application of δ_{loc} is the preparation of graph states (quantum states represented by a graph), which are a very powerful tool used for measurement-based quantum computing [RB01] and blind quantum computing [BFK09], for example. In [HMP06], it has been proven that the complexity of preparation of a graph state is bounded by its local minimal degree. Threshold quantum secret sharing protocols from graph states (first introduced in [MS08]) can be built from graph states with the methods described

M.C. Golumbic et al. (Eds.): WG 2012, LNCS 7551, pp. 138–147, 2012.
© Springer-Verlag Berlin Heidelberg 2012

in [JMP11], and the local minimum degree of the corresponding graphs gives, under additional parity conditions, a value for the threshold that can be reached with these graph states. Moreover, we also focus on bipartite graphs which are of high interest for entanglement purification [ADB05] and the study of Schmidt measure [Sev06], for example.

In this paper, several techniques from different backgrounds are used. We consider a family of graphs defined from quadratic residues, the Paley graphs Pal_p, and the bound that we give on $\delta_{loc}(Pal_p)$ is closely related to a fundamental result in algebraic geometry (see Lemma 2). Probabilistic methods are also used to prove the existence of graphs with large local minimum degree. In particular, we use the asymmetric version of the Lovász Local Lemma [Lov75] (see Lemma 4) to prove the existence of an infinite family of graphs with linear δ_{loc}. We also use this family to derive a polynomial reduction to a problem from coding theory in order to find the computational complexity of finding the local minimum degree of a graph in the general case.

In section 2, we recall the definition of the local minimum degree, main notion of this paper, and we give an explicit family of graphs Pal_p of order p such that $\delta_{loc}(Pal_p) \geq \sqrt{p} - \frac{3}{2}$, which is, up to our knowledge, the best known lower bound for any family of graphs. The next section is dedicated to the proof of the existence of graphs with linear δ_{loc}. In the last section, we show that the decision problem associated with δ_{loc} is NP-complete even on the family of bipartite graphs, and we show that there exists no approximation algorithm up to a constant factor for this problem unless $P = NP$.

2   Definitions

Local complementation is defined as follows:

Definition 1. *The local complementation of a graph G with respect to one of its vertices u results in a graph $G * u = G \Delta K_{\mathcal{N}(u)}$ where Δ stands for the symmetric difference between edges and $K_{\mathcal{N}(u)}$ is the complete graph on the neighbors of u.*

The transitive closure of a graph with respect to the local complementation forms an equivalence class. In [Bou87], Bouchet gives a polynomial algorithm that tells whether any two graphs are in the same equivalence class with respect to local complementation. For a given graph G, the quantity we will focus on is the minimum degree of the graphs in its equivalence class. This value is called the local minimum degree and is written $\delta_{loc}(G)$. Its formal definition follows:

Definition 2. *Given a graph G, $\delta_{loc}(G) = min\left\{ \delta(G') \mid G \equiv_{LC} G' \right\}$ where $\delta(G')$ is the minimal degree of G' and the equivalence relation $G_1 \equiv_{LC} G_2$ is verified when G_1 can be changed into G_2 by a series of local complementations.*

In [HMP06], a characterization of the quantity δ_{loc} has been established by means of the odd and even neighborhoods of subsets of vertices of a graph defined as follows:

Definition 3. *Let G be an undirected graph and D a subset of its vertices.*

$$Odd(D) = \{ \, v \in V(G) \mid |\mathcal{N}(v) \cap D| = 1 \bmod 2 \, \} \tag{1}$$
$$Even(D) = \{ \, v \in V(G) \mid |\mathcal{N}(v) \cap D| = 0 \bmod 2 \, \} \tag{2}$$

The local minimum degree is related to the size of the smallest set of the form $D \cup Odd(D)$:

Property 1 ([HMP06]). Let G be an undirected graph.

$$\delta_{loc}(G) = \min \{ \, |D \cup Odd(D)| \mid D \neq \varnothing, D \subseteq V(G) \, \} - 1 \tag{3}$$

3   Local Minimum Degree of Paley Graphs

It is challenging to find a family of graphs with "high" local minimum degree. The family of hypercubes, for example, has a logarithmic local minimal degree [HMP06].

In the following, we prove that a Paley graph of order n has a δ_{loc} greater than $\sqrt{n}$. This value is only a lower bound, and we do not know whether it is reached. This family is defined with quadratic residues over a finite field. Up to our knowledge, there is no known family of graphs whose local minimum degree is greater than the square root of their order.

For any prime p such that $p = 1 \bmod 4$, the Paley graph Pal_p is a graph on p vertices where each vertex is an element of $\mathbb{F}_p$. There is an edge between two vertices i and j if and only if $i - j$ is a square in $\mathbb{F}_p$.

Theorem 1. *For any prime $p = 1 \bmod 4$,*

$$\delta_{loc}(Pal_p) \geq \sqrt{p} - \frac{3}{2} \tag{4}$$

where Pal_p is the Paley graph of order p.

The rest of this section is dedicated to the proof of Theorem 1. To this end, we give a bound on the size of the sets of the form $D \cup Odd(D)$ in Paley graphs. The size of such sets is characterized as follows:

Lemma 1. *For any non-empty set $S \subseteq V(Pal_p)$ and any $i \in V(Pal_p)$,*

$$\left| \sum_{i=0}^{p-1} \chi_L \left(f_S(i) \right) \right| = \Big| \, |S \cup Odd(S)| - |S \cup Even(S)| \, \Big| \tag{5}$$

where $f_S(i) = \prod_{j \in S}(i - j)$ and χ_L is the Legendre character ($\chi_L(x) = x^{\frac{p-1}{2}}$ mod p).

Proof. First, note that $\chi_L(0) = 0$, $\chi_L(x) = 1$ if x is a quadratic residue in $\mathbb{F}_p$ and $\chi_L(x) = -1$ otherwise. Since the Legendre character is multiplicative, $\left|\sum_{i=0}^{p-1} \chi_L(f_S(i))\right| = \left|\sum_{i=0}^{p-1} \prod_{j \in S} \chi_L(i-j)\right|$. If $i \in S$ the quantity $\prod_{j \in S} \chi_L(i-j)$ equals 0. Otherwise, the product equals $(-1)^{|S|-|\mathcal{N}(i) \cap S|}$, which is $(-1)^{|S|}$ if $i \in Even(S) \setminus S$ and $-(-1)^{|S|}$ if $i \in Odd(S) \setminus S$. Then, the sum over all vertices i is the difference between the exclusive odd and even neighborhood of the set S: $\left|\sum_{i=0}^{p-1} \prod_{j \in S} \chi_L(i-j)\right| = \big||Odd(S) \setminus S| - |Even(S) \setminus S|\big|$. The last expression can be written $\big||S \cup Odd(S)| - |S \cup Even(S)|\big|$. $\qquad\square$

A well-known result from algebraic geometry related to the hyperelliptic curve of equation $y^2 = \prod_{j \in S}(x - j)$ can be found in [Wei48] or [Sch04], for example, and is reformulated by Joyner in [Joy06]:

Lemma 2 ([Joy06], Proposition 1). *For any non-empty set $S \subseteq \mathbb{F}_p$, let $f_S(x) = \prod_{j \in S}(x - j)$. Then*

$$\left|\sum_{i \in \mathbb{F}_p} \chi_L(f_S(i))\right| \le (|S| - 1)\sqrt{p} + 1 \tag{6}$$

This allows us to derive a bound on the sets of type $S \cup Odd(S)$ and $S \cup Even(S)$ in Paley graphs.

Lemma 3. *Let Pal_p be the Paley graph of order p. For all $S \subseteq V(P_p)$, $S \ne \varnothing$, we have $\sqrt{p} - \frac{1}{2} \le |S \cup Odd(S)|$ and $\sqrt{p} - \frac{1}{2} \le |S \cup Even(S)|$.*

Proof. We consider the case $|S \cup Odd(S)| \le |S \cup Even(S)|$, the other case can be treated a similar way. Lemma 1 states that $|S \cup Odd(S)| - |S \cup Even(S)| = -\left|\sum_{i \in \mathbb{F}_p} \chi_L(f_S(i))\right|$. On the other hand, the equality $|S \cup Odd(S)| + |S \cup Even(S)| = p + |S|$ is always true. Thus adding both equalities, $p + |S| - \left|\sum_{i \in \mathbb{F}_p} \chi_L(f_S(i))\right| = 2|S \cup Odd(S)|$. Thanks to Lemma 2, we derive $p + |S| - (|S| - 1)\sqrt{p} - 1 \le 2|S \cup Odd(S)|$.

If $|S| \le \sqrt{p}$ then the left-hand side of the previous inequality can be bounded: $p + |S| - (|S| - 1)\sqrt{p} - 1 = p + |S|(1 - \sqrt{p}) + \sqrt{p} - 1 \ge 2\sqrt{p} - 1$. Thus, $\sqrt{p} - \frac{1}{2} \le |S \cup Odd(S)|$, otherwise $|S| > \sqrt{p}$ and the previous inequality is obviously true. $\qquad\square$

Proof of Theorem 1: The characterization given by Property 1 and the bounds on the size of sets of the form $D \cup Odd(D)$ obtained in Lemma 3 imply that the local minimum degree for Paley graphs is greater than the square root of the order of the graph. This ends the proof of Theorem 1.

It is significant and interesting to notice that the conjecture of the existence of an infinite family of Paley graphs with linear δ_{loc} is equivalent to the Bazzi-Mitter conjecture [BM06]. However, it is already known that not all Paley graphs have a linear δ_{loc}: there exists no $p_0 \in \mathbb{N}$ such that for all $p > p_0$, $\delta_{loc}(Pal_p)$ is linear in p thanks to Theorem 7 of [Joy06].

4   Existence of Graphs with Linear Local Minimum Degree

In this section, we give a proof of the existence of bipartite graphs for which the local minimum degree is linear in the order of the graph. The proof uses the asymmetric version of Lovász Local Lemma [Lov75]:

Lemma 4 (Asymmetric Lovász Local Lemma). *Let $\mathcal{A} = \{A_1, \cdots, A_n\}$ be a set of bad events in an arbitrary probability space and let $\Gamma(A)$ denote a subset of $\mathcal{A}$ such that A is independent from all the events outside A and $\Gamma(A)$. If for all A_i there exists $\sigma(A_i) \in [0,1)$ such that $Pr(A_i) \leq \sigma(A_i) \prod_{B_j \in \Gamma(A_i)} (1 - \sigma(B_j))$ then we have $Pr(\overline{A_1}, \cdots, \overline{A_n}) \geq \prod_{A_j \in \mathcal{A}} (1 - \sigma(A_j))$.*

We apply the Local Lovász Lemma (Lemma 4) on random bipartite graphs to show the existence of bipartite graphs with linear local minimum degree.

Theorem 2. *There exists $\nu_0 \in \mathbb{N}$ such that for all $\nu > \nu_0$ there exists a bipartite graph of order $n = 2\nu$ whose local minimum degree is greater than $0.110n$.*

Proof. Let G_B be a bipartite graph of order $n = 2\nu$ with two independent sets of size ν and where any possible edge exists with probability $\frac{1}{2}$. An event which implies that a graph G has a linear δ_{loc} is: "$\forall D \subseteq V(G), |D \cup Odd(D)| > cn$" for some $c \in]0, 1]$. In the case of G_B, it is sufficient to verify the previous event for sets D such that $D \subseteq V_1$ or $D \subseteq V_2$. Indeed, G_B is bipartite, therefore $|D \cup Odd(D)| \geq |(D \cap V_1) \cup Odd(D \cap V_1)|$. Therefore we consider the "bad" events A_D^1 and A_D^2 defined as follows: if $D \subseteq V_1$ (resp. V_2), A_D^1 (resp. A_D^2) = "$|D \cup Odd(D)| \leq cn$".

We want to compute $Pr(A_D^1)$ with $D \subseteq V_1$. Let $|D| = d\nu$ for some $d \in]0, 1]$. For any $u \in V_2$, $Pr(\text{"}u \in Odd(D)\text{"}) = \frac{1}{2}$. Thus, $Pr(|Odd(D)| \leq x) = (\frac{1}{2})^\nu \sum_{k=0}^x \binom{\nu}{k} \leq (\frac{1}{2})^\nu 2^{\nu H(\frac{x}{\nu})}$ where $H : t \mapsto -t \log_2(t) - (1 - t) \log_2(1 - t)$ is the binary entropy function. Then, $Pr(A_D^1) = Pr(\text{"}|D \cup Odd(D)| \leq cn\text{"}) = Pr(\text{"}|D| + |Odd(D)| \leq cn\text{"}) = Pr(\text{"}|Odd(D)| \leq cn - |D|\text{"}) \leq 2^{\nu(H(2c-d)-1)}$.

Let $\sigma(A_D^1) = \frac{1}{r\binom{\nu}{d\nu}}$ for some $r \in \mathbb{R}$ that will be chosen later. First, we verify that $Pr(A_D^1) \leq \sigma(A_D^1) \prod_{D' \in V_1, D'' \in V_2} (1 - \sigma(A_{D'}^1))(1 - \sigma(A_{D''}^2))$. The product of the right-hand side of the previous equation can be written $p = \prod_{|D'|=1}^\nu$

$$\left(1 - \frac{1}{r\binom{\nu}{|D'|}}\right)^{2\binom{\nu}{|D'|}} = \left[\prod_{|D'|=1}^\nu \left(1 - \frac{1}{r\binom{\nu}{|D'|}}\right)^{r\binom{\nu}{|D'|}}\right]^{\frac{2}{r}}.$$ The function $f : x \mapsto$

$\left(1 - \frac{1}{x}\right)^x$ verifies $f(x) \geq \frac{1}{4}$ when $x \geq 2$, therefore $p \geq \left(\frac{1}{4}\right)^{\nu * \frac{2}{r}} = 2^{-\frac{4\nu}{r}}$ for any $r \geq 2$. Thus, it is sufficient to have $2^{\nu(H(2c-d)-1)} \leq \frac{1}{r\binom{\nu}{d\nu}} 2^{-\frac{4\nu}{r}}$. Rewriting this inequality gives $r\binom{\nu}{d\nu} 2^{(2c-1)\nu - d\nu + \frac{4\nu}{r}} \leq 1$. Thanks to the bound $\binom{\nu}{d\nu} \leq 2^{\nu H(\frac{d\nu}{\nu})}$ and after applying the logarithm function and dividing by ν, it is sufficient that $\frac{\log_2 r}{\nu} + H(d) + H(2c - d) - 1 + \frac{4}{r} \leq 0$. Therefore, if we take $r = \nu$ and $\nu \to +\infty$, the asymptotic condition on the value of c is $H(d) + H(2c - d) - 1 \leq 0$. Since this bound must be verified for all $d \in (0, 1]$, it must be true for the value of d for which the function $d \mapsto H(d) + H(2c - d) - 1$ is minimum. Usual techniques show that the minimum is reached for $d = c$, and a numerical analysis shows that $c = 0.110$ satisfies the condition $Pr(A_D^1) \leq \sigma(A_D^1)p$ for some $r \in \mathbb{R}$ and $\nu > \nu_0$. A similar reasoning is used to prove $Pr(A_D^2) \leq \sigma(A_D^2)p$ for all $D \in V_2$.

The conditions and the choice of the weights $\sigma(A_D^1)$ and $\sigma(A_D^1)$ allow us to use the Lovász Local Lemma (Lemma 4), and we derive $Pr\left(\{\overline{A_D^1} \mid D \in V_1\}, \{\overline{A_D^2} \mid D \in V_2\}\right) \geq p > 0$, which proves that $Pr\left(\delta_{loc}(G_B) \geq cn\right) > 0$ for any $c \leq 0.110$ and for $\nu > \nu_0$. Then there exists at least one bipartite graph G_B of order n such that $\delta_{loc}(G_B) \geq 0.110n$. □

The general case of a random graph without the bipartite constraint leads to a slightly better constant:

Theorem 3. *There exists $n_0 \in \mathbb{N}$ such that for all $n > n_0$ there exists a graph of order n whose local minimum degree is greater than $0.189n$.*

Due to its similarity to the above proof, the proof of this theorem is given in Appendix.

5   NP-Completeness of the Local Minimum Degree Problem

In this section, we show that given a graph G and an integer d, deciding whether $\delta_{loc}(G) \leq d$ is NP-complete even for the family of bipartite graphs. This result is established through a reduction to the problem of the shortest word of a linear code [Var97] and uses the families of graphs whose existence has been proven in the previous section.

Lemma 5. *Let $G = (V, E)$ be a bipartite graph. Let $V = V_1 \cup V_2$ where V_1 and V_2 are the two parties of the graph G. There exists $D_0 \subseteq V$ such that $\delta_{loc}(G) + 1 = |D_0 \cup Odd(D_0)|$ and $D_0 \subseteq V_1$ or $D_0 \subseteq V_2$.*

Proof. Let $D \subseteq V$ such that $|D \cup Odd(D)| = \delta_{loc}(G) + 1$. We write $D = D_1 \cup D_2$ with $D_1 \subseteq V_1$ and $D_2 \subseteq V_2$. $D \neq \varnothing$, then without loss of generality, we assume that $D_1 \neq \varnothing$. G is bipartite, then $Odd(D_1) \subseteq V_2$ and $Odd(D_2) \subseteq V_1$. Thus $Odd(D_1 \cup D_2) = Odd(D_1) \cup Odd(D_2)$, and $\delta_{loc}(G) + 1 = |D \cup Odd(D)| = |D_1 \cup Odd(D_1) \cup D_2 \cup Odd(D_2)| \geq |D_1 \cup Odd(D_1)| \geq \delta_{loc}(G) + 1$. The bounds are tight, therefore $|D_1 \cup Odd(D_1)| + 1 = \delta_{loc}(G)$. □

Theorem 4. *Given a graph G and an integer d, deciding whether $\delta_{loc}(G) \leq d$ is NP-complete for the family of bipartite graphs.*

Proof. The problem is in NP since a set of the form $D \cup Odd(D)$ with $D \neq \varnothing$ and $|D \cup Odd(D)| = \delta_{loc}$ is a YES certificate. We do a reduction to the problem of the shortest codeword. Let $A \in \mathcal{M}_{n+k,k}(\mathbb{F}_2)$ be the generating matrix of a binary code. Using oracle for the problem related to the quantity δ_{loc} on bipartite graphs, we answer the problem of finding the shortest word of A.

If $dim(Ker(A)) \neq 0$ then $\min_{X \in \mathbb{F}_2^k, X \neq 0}\{w(AX)\} = 0$, where w is the Hamming weight function. Otherwise, $\min_{X \in \mathbb{F}_2^k, X \neq 0}\{w(AX)\} = \min_{X \in \mathbb{F}_2^k, X \neq 0}$ $\{w(X) + w(A'X)\}$ where A is written in the form $\begin{pmatrix} I_k \\ A' \end{pmatrix}$. Thus, A' is of size $n \times k$.

We want to construct a bipartite graph G (Figure 1) on which the oracle call is performed. To this purpose, we build two auxiliary graphs $G_{A'}$ and G_B in a first time. Let $G_{A'} = (V_{A'_1} \cup V_{A'_2}, E_{A'})$ be the bipartite graph defined as follows: the sets $V_{A'_1}$ of size k and $V_{A'_2}$ of size n denote both sides of the bipartition of $G_{A'}$, and for all $x \in V_{A'_1}$ and $x' \in V_{A'_2}$, $(x, x') \in E_{A'}$ if and only if $A'_{x',x} = 1$. After that, thanks to Theorem 2, there exists $n_0 \in \mathbb{N}$ such that for all $n > n_0$ there exists a bipartite graph $G_B = (V_{B_1} \cup V_{B_2}, E_B)$ of order $10(n+1)$ such that $\delta_{loc}(G_B) > n+1$. The sets V_{B_1} and V_{B_2} denote both sides of the bipartition of G_B. Let u be any vertex of V_{B_1}. Consider the bipartite graph $G = (V_1 \cup V_2, E)$ (Figure 1) defined as follows: $V_1 = V_{1L} \cup V_{1R}$ with $V_{1L} = V_{A'_1} \times \{u\}$ and $V_{1R} = V_{A'_2} \times V_{B_2}$, and $V_2 = V_{A'_2} \times V_{B_1}$. For all $(x, y) \in V_1$ and $(x', y') \in V_2$, $\big((x,y),(x',y')\big) \in E$ if and only if $\big((x, x') \in E_{A'} \wedge y = y'\big) \vee \big((y, y') \in E_B \wedge x = x'\big)$.

Both independent sets V_1 and V_2 form a partition of the vertices of the graph. Thanks to Lemma 5, there exists a non-empty set $D_0 \subseteq V(G)$ such that $\delta_{loc}(G) + 1 = |D_0 \cup Odd(D_0)|$ and $D_0 \subseteq V_1$ or $D_0 \subseteq V_2$.

Suppose that $D_0 \subseteq V_2$. Therefore $\delta_{loc}(G) = |D_0 \cup Odd(D_0)| - 1 \geq \delta_{loc}(G_B) > n+1 \geq \delta(G) + 1 \geq \delta_{loc}(G)$. This leads to a contradiction, therefore $D_0 \subseteq V_1$.

Suppose that $D_0 \cap V_{1R} \neq \varnothing$. Let $v \in D_0 \cap V_{1R}$. Then $\delta_{loc}(G) = |D_0 \cup Odd(D_0)| - 1 \geq |\{v\} \cup Odd(\{v\})| - 1 \geq \delta_{loc}(G_B) > n+1 \geq \delta(G) + 1 \geq \delta_{loc}(G)$. This also leads to a contradiction, therefore $D_0 \subseteq V_{1L}$.

The reader will notice that since $D_0 \subseteq V_{1L}$, $|Odd(D_0)|$ in the graph G can be written $w(A'X_{D_0})$ where X_{D_0} is the vector representation of the set D_0. Moreover, since V_{1L} is an independent set, $|D_0 \cup Odd(D_0)| = |D_0| + |Odd(D_0)| = w(X_{D_0}) + w(A'X_{D_0})$. By definition of D_0, we have $\delta_{loc}(G) + 1 = \min_{X \in \mathbb{F}_2^k, X \neq 0}\{w(AX)\}$, which ends the reduction to the shortest codeword problem which is NP-complete [Var97]. □

Notice that a constructive version of NP-completeness on non-necessarily bipartite graphs can be done by replacing the graph G_B by a Paley graph in the above reduction.

Since finding the local minimum degree is hard, one can wonder whether there exists a l-approximation algorithm for this problem for some constant l. The previous reduction also shows that such an algorithm does not exist unless $P = NP$, even for the family of bipartite graphs.

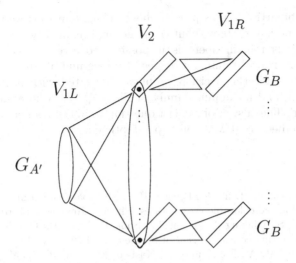

Fig. 1. Construction of the graph G from the bipartite graph $G_{A'}$ (ellipses) and several copies of the bipartite graph G_B (rectangles). $V_1 = V_{1L} \cup V_{1R}$

Theorem 5. *There exists no approximation algorithm with a constant factor for the problem of finding the local minimum degree of bipartite graphs, unless* $P = NP$.

Proof. In the proof of Theorem 4, the value of $\delta_{loc}(G)$ where G is constructed as described in Figure 1 is the same as the shortest word of the linear code described by its generating matrix A. This is true for any A, therefore for any constant l, any l-approximation of $\delta_{loc}(G)$ is a l-approximation of the Hamming weight of the shortest word of A. Under the hypothesis $P \neq NP$, since finding the shortest codeword of a linear code is known to have no approximation algorithm with a constant factor [DMS03, CW09], there exists no polynomial approximation algorithm with a constant factor for the problem of finding the local minimum degree of bipartite graphs. □

6   Conclusion

After having shown that the local minimum degree of the family of Paley graphs is greater than the square root of their order, we proved that there exist an infinite family of graphs whose local minimum degree is linear in their order (with constant at least 0.189 in general and 0.110 for bipartite graphs). Then, a study of the computational complexity of the decision problem associated with δ_{loc} with a polynomial reduction to the problem of the shortest word of a linear code shows its NP-completeness, even on bipartite graphs. It is also impossible to find an approximation algorithm with any constant factor for this problem, unless $P = NP$. The specificity of the reduction performed lies in the fact that the construction of an instance for the problem associated with δ_{loc} uses the existence of a family of bipartite graphs proven above. Thus, in a way, we proved that a polynomial reduction exists without constructing it explicitly.

Some questions remain open: is it possible to give an explicit family of graphs with linear local minimum degree? Can we find a constructive proof of NP-completeness for the decision problem associated with δ_{loc} on bipartite graphs? Can we find an infinite family of Paley graphs whose local minimum degree is linear? The answer of the last question would provide an answer to the Bazzi-Mitter conjecture [BM06] on hyperelliptic curves.

References

[ADB05]   Aschauer, H., Dur, W., Briegel, H.J.: Multiparticle entanglement purification for two-colorable graph states. Physical Review A 71, 012319 (2005)

[BCG+11] Beigi, S., Chuang, I., Grassl, M., Shor, P., Zeng, B.: Graph concatenation for quantum codes. Journal of Mathematical Physics 52 (2011)

[BFK09]   Broadbent, A., Fitzsimons, J., Kashefi, E.: Universal blind quantum computation. In: Proceedings of FOCS, pp. 517–526 (2009)

[BM06]   Bazzi, L.M.J., Mitter, S.K.: Some randomized code constructions from group actions. IEEE Transactions on Information Theory 52(7), 3210–3219 (2006)

[Bou87]   Bouchet, A.: Digraph decompositions and eulerian systems. SIAM J. Algebraic Discrete Methods 8, 323–337 (1987)

[Bou90]   Bouchet, A.: κ-transformations, local complementations and switching. Cycles and Rays (1990)

[Bou94]   Bouchet, A.: Circle graph obstructions. J. Comb. Theory Ser. B 60, 107–144 (1994)

[CW09]   Cheng, Q., Wan, D.: A deterministic reduction for the gap minimum distance problem. In: Proceedings of the 41st Annual ACM Symposium on Theory of Computing, STOC 2009, pp. 33–38 (2009)

[dF81]   de Fraysseix, H.: Local complementation and interlacement graphs. Discrete Mathematics 33(1), 29–35 (1981)

[DMS03]   Dumer, I., Micciancio, D., Sudan, M.: Hardness of approximating the minimum distance of a linear code. IEEE Transactions on Information Theory 49(1), 22–37 (1999); Preliminary version in FOCS 1999

[HMP06]   Høyer, P., Mhalla, M., Perdrix, S.: Resources Required for Preparing Graph States. In: Asano, T. (ed.) ISAAC 2006. LNCS, vol. 4288, pp. 638–649. Springer, Heidelberg (2006)

[JMP11]   Javelle, J., Mhalla, M., Perdrix, S.: New protocols and lower bound for quantum secret sharing with graph states. arXiv:1109.1487 (September 2011)

[Joy06]   Joyner, D.: On quadratic residue codes and hyperelliptic curves. ArXiv Mathematics e-prints (September 2006)

[Kot68]   Kotzig, A.: Eulerian lines in finite 4-valent graphs and their transformations. In: Colloqium on Graph Theory, pp. 219–230. Academic Press (1968)

[Lov75]   Lovász, L.: Problems and results on 3-chromatic hypergraphs and some related questions. In: Colloquia Mathematica Societatis Janos Bolyai, pp. 609–627 (1975)

[MS08]   Markham, D., Sanders, B.C.: Graph states for quantum secret sharing. Physical Review A 78, 042309 (2008)

[Oum08]   Oum, S.-I.: Approximating rank-width and clique-width quickly. ACM Trans. Algorithms 5, 10:1–10:20 (2008)

[RB01]   Raussendorf, R., Briegel, H.: A one-way quantum computer. Physical Review Letters 86(22), 5188–5191 (2001)

[Sch04]   Schmidt, W.M.: Equations over finite fields: an elementary approach, 2nd edn. Kendrick Press (2004)

[Sev06]   Severini, S.: Two-colorable graph states with maximal schmidt measure. Physics Letters A 356, 99 (2006)

[Var97]   Vardy, A.: Algorithmic complexity in coding theory and the minimum distance problem. In: STOC, pp. 92–109 (1997)

[Wei48]   Weil, A.: On some exponential sums. Proceedings of the National Academy of Sciences 34, 204–207 (1948)

A Proof of Theorem 3

Theorem 3 *There exists $n_0 \in \mathbb{N}$ such that for all $n > n_0$ there exists a graph of order n whose local minimum degree is greater than $0.189n$.*

Proof. Let G be a graph of order n where any possible edge exists with probability $\frac{1}{2}$. We are looking for the greatest value of c such that $Pr\left(\delta_{loc}(G) \geq cn\right) > 0$. Thus, we want that "$\forall D \subseteq V(G), |D \cup Odd(D)| > cn$". Consequently, the events to avoid are A_D: "$|D \cup Odd(D)| \leq cn$". Obviously, it is sufficient to consider only the events A_D with $D \leq cn$.

For all D sucht that $|D| \leq cn$, we want to get an upper bound on $Pr(A_D)$. Let $|D| = dn$ for some $d \in (0, c]$. For all $u \in V \setminus D$, $Pr("u \in Odd(D)") = \frac{1}{2}$. If D is fixed, the events "$u \in Odd(D)$" when u is outside D are independent. Therefore, if the event A_D is true, any but at most $(c - d)n$ vertices outside D are contained in $Odd(D)$. There are $(1 - d)n$ vertices outside D, then $Pr(A_D) = \left(\frac{1}{2}\right)^{(1-d)n} \sum_{k=0}^{(c-d)n} \binom{(1-d)n}{k} \leq \left(\frac{1}{2}\right)^{(1-d)n} 2^{(1-d)nH\left(\frac{c-d}{1-d}\right)} = 2^{(1-d)n\left[H\left(\frac{c-d}{1-d}\right)-1\right]}$ where $H : t \mapsto -t\log(t) - (1 - t)\log(1 - t)$ is the binary entropy function.

Let $\sigma(A_D) = \frac{1}{r\binom{n}{|D|}}$. Let $p = \prod_{|D'| \leq cn}(1 - \sigma(A_{D'}))$. In order to apply the Lóvasz Local Lemma (Lemma 4), we want to have $Pr(A_D) \leq \sigma(A_D)p$. The product p verifies $p = \prod_{|D'|=1}^{cn}\left(1 - \frac{1}{r\binom{n}{|D'|}}\right)^{\binom{n}{|D'|}} = \left[\prod_{|D'|=1}^{cn}\left(1 - \frac{1}{r\binom{n}{|D'|}}\right)^{r\binom{n}{|D'|}}\right]^{\frac{1}{r}}$.

The function $f : x \mapsto \left(1 - \frac{1}{x}\right)^x$ verifies $f(x) \geq \frac{1}{4}$ when $x \geq 2$, therefore $p \geq \left(\frac{1}{4}\right)^{\frac{cn}{r}} = 2^{-\frac{2cn}{r}}$ for any $r \geq 2$. Thus, it is sufficient that $2^{(1-d)n\left[H\left(\frac{c-d}{1-d}\right)-1\right]} \leq \frac{1}{r\binom{n}{dn}}2^{-\frac{2cn}{r}}$. Rewriting this inequality with the bound $\binom{n}{dn} \leq 2^{nH\left(\frac{dn}{n}\right)}$ and applying the logarithm function and dividing by n gives the following sufficient condition: $(1 - d)\left[H\left(\frac{c-d}{1-d}\right) - 1\right] + H(d) + \frac{2c}{r} + \frac{\log_2 r}{n} \leq 0$. Taking $r = n$, the condition becomes asymptotically $(1 - d)\left[H\left(\frac{c-d}{1-d}\right) - 1\right] + H(d) \leq 0$.

Numerical analysis shows that this condition is true for any $c \leq 0.189$ and for all d such that $0 < d \leq cn$. Therefore, Lemma 4 ensures that $Pr\left(\left\{\overline{A_D} \mid |D| \leq cn\right\}\right) \geq p > 0$, which proves the existence of at least one graph G of order n such that $\delta_{loc}(G) \geq 0.189n$. This ends the proof of Theorem 3. □

On the Stable Degree of Graphs

Haiko Müller

University of Leeds, School of Computing
Leeds, LS2 9JT, United Kingdom
h.muller@leeds.ac.uk

Abstract. We define the stable degree $s(G)$ of a graph G by $s(G) = \min_U \max_{v \in U} d_G(v)$, where the minimum is taken over all maximal independent sets U of G. For this new parameter we prove the following. Deciding whether a graph has stable degree at most k is NP-complete for every fixed $k \geq 3$; and the stable degree is hard to approximate. For asteroidal triple-free graphs and graphs of bounded asteroidal number the stable degree can be computed in polynomial time. For graphs in these classes the treewidth is bounded from below and above in terms of the stable degree.

1 Introduction

An *asteroidal triple*, or AT for short, is a set of three pairwise non-adjacent vertices in a graph such that any two of them are connected by a path that avoids the neighbourhood of the third. Graphs without asteroidal triples are *AT-free* [7]. This applies for instance to interval graphs, the intersection graphs of intervals on the real line. More precisely, interval graphs are exactly the chordal AT-free graphs [13]. Unlike the subclass of interval graphs, the whole class of AT-free graphs is not contained in the class of perfect graphs. For instance, C_5, the chordless cycle on five vertices, is AT-free, but not perfect. AT-free graphs form an interesting class of graphs due to their structural properties and also when studying the complexity on AT-free graphs for problems being NP-complete in general [5,12].

An independent set of vertices in a graph is *asteroidal* if each three-element subset forms an AT [11]. The maximal size of an asteroidal set in a graph is its *asteroidal number*. Lots of the polynomial time algorithms for AT-free graphs, *i.e.* graphs of asteroidal number at most two, generalise to graphs of bounded asteroidal number [6,5].

The *treewidth* is a parameter that measures the tree-likeness of a graph. Definitions are given in 2.2. Lots of the polynomial time algorithms for trees generalise to graphs of bounded treewidth. This applies to all problems that can be defined in monadic second order logic [8]. The *pathwidth* is a parameter similar to treewidth. For AT-free graphs, both parameters coincide [14], but are still hard to compute [1].

We introduce a new parameter, the stable degree of a graph. In Sections 3 and 4 we bound the treewidth in terms of the asteroidal number and stable degree.

M.C. Golumbic et al. (Eds.): WG 2012, LNCS 7551, pp. 148–159, 2012.

In Section 5 we show that the stable degree is hard to compute in general, but if we restrict the input to AT-free graphs, or even graphs of bounded asteroidal number, then the stable degree can be computed by a polynomial time algorithm. As an immediate consequence, this enables new constant-factor approximations for the treewidth of AT-free graphs and graphs of bounded asteroidal number. In both cases these approximation algorithms are not better than the best known algorithms [3,4], see Section 5.5.

2   Preliminaries

For a vertex v of a graph $G = (V, E)$ let $N_G(v) = \{u \mid \{u, v\} \in E\}$ denote its *open neighbourhood*. The *closed neighbourhood* of v is $N_G[v] = \{v\} \cup N_G(v)$. Both concepts generalise to sets $U \subseteq V$ as follows: $N_G[U] = \bigcup_{v \in U} N_G[v]$ and $N_G(U) = N_G[U] \setminus U$. The *degree* of a vertex is the cardinality of its open neighbourhood, $d_G(v) = |N_G(v)|$. We omit the subscript G for neighbourhoods and degrees if there is no ambiguity about the graph G.

The set U is *independent* in G if $U \cap N(u) = \varnothing$ for all $u \in U$, and U is *dominating* in G if $N[U] = V$. An *independent dominating set* is both independent and dominating. An independent set is maximal (with respect to set inclusion) if and only if it is dominating.

2.1   Degrees

We introduce a new graph parameter based on the notion of degree. The *stable degree* of a graph G is defined by

$$s(G) = \min_{U} \max_{v \in U} d_G(v)$$

where the minimum is taken over all maximal independent sets U of G.

We recall some parameters of a graph $G = (V, E)$ with complement $(V, \overline{E})$:

minimum degree $\delta(G) = \min\{d_G(v) \mid v \in V\}$
2nd smallest degree $\delta_2(G) = 0$ if $|V| \le 1$ and
    $\delta_2(G) = \min\{d_G(v) \mid v \in V \wedge \exists u \in V \setminus \{v\} \, (d_G(u) \le d_G(v))\}$ otherwise
degeneracy $d(G) = \max\{\delta(G[U]) \mid U \subseteq V\}$
Ramachandramurthi-bound $\gamma_R(G) = |V| - 1$ if G is a complete graph and
    $\gamma_R(G) = \min\{\max\{d_G(u), d_G(w)\} \mid \{u, w\} \in \overline{E}\}$ otherwise
maximum degree $\Delta(G) = \max\{d_G(v) \mid v \in V\}$

For all graphs G the following inequalities hold: $\delta(G) \le \delta_2(G) \le d(G) \le \Delta(G)$, $\delta_2(G) \le \gamma_R(G) \le \Delta(G)$ and $\delta_2(G) \le s(G) \le \Delta(G)$ [15]. For more information on these parameters and their use in lower bounding the treewidth of graphs we refer to [2].

2.2   Tree Decomposition

A pair (X, T) is a *tree decomposition* of a graph $G = (V, E)$ if $T = (I, F)$ is a tree and $X : I \to 2^V$ maps the *nodes* of T to *bags*, i.e. subsets of V, such that

- for all $v \in V$ there is an $i \in I$ such that $v \in X(i)$,
- for all $e \in E$ there is an $i \in I$ such that $e \subseteq X(i)$
- for all $v \in V$, $T(v)$ is connected, where $T(v)$ is the subgraph of T induced by the $i \in I$ with $v \in X(i)$.

The *width* of (X, T) is $\max\{|X(i)| \mid i \in I\} - 1$, and the *treewidth* $\mathrm{tw}(G)$ of G is the minimal width of a tree decomposition of G.

The *pathwidth* $\mathrm{pw}(G)$ of G is the minimal width of a tree decomposition (X, T) of G where T is a path. For all AT-free graphs G we have $\mathrm{tw}(G) = \mathrm{pw}(G)$ by a result from [14].

In [15] Ramachandramurthi showed that γ_R is a lower bound on the treewidth. In Section 4 we use his idea to prove a lower bound in terms of the stable degree.

2.3   Asteroidal Sets

A set $A \subseteq V$ is *asteroidal* in $G = (V, E)$ if for every vertex $u \in A$ there is a connected component $G[C]$ of $G - \mathrm{N}[u]$ containing $A \setminus \{u\}$. Consequently every asteroidal set is independent, and every independent set of size at most two is asteroidal. By $\mathrm{an}(G)$ we denote the *asteroidal number* of G that is the maximum cardinality of an asteroidal set in the graph G.

For different and non-adjacent vertices u and v of $G = (V, E)$ let $C(u, v)$ induce the connected component of $G - \mathrm{N}[u]$ containing v. We can use this notation to characterise asteroidal sets: an independent set A is asteroidal if and only if $C(u, v) = C(u, w)$ holds for every triple of different vertices $u, v, w \in A$.

The *interior* of an asteroidal set A in (V, E) is the subset of $V \setminus \mathrm{N}[A]$ of vertices that belong to the same connected component of $G - \mathrm{N}[u]$ as $A \setminus \{u\}$ for all $u \in A$. For $|A| > 1$ let $C(u, A)$ denote the set of vertices in this connected component, i.e. $C(u, A) = C(u, v)$ for all $v \in A \setminus \{u\}$. This enables us to define interior $I(A)$ formally by $I(A) = \bigcap_{u \in A} C(u, A)$. Furthermore we set $I(\varnothing) = V$ and $I(\{u\}) = V \setminus \mathrm{N}[u]$ for each vertex $u \in V$. A subset A of an asteroidal set B is asteroidal too, and we have $I(B) \cup (B \setminus A) \subseteq I(A)$ since $C(u, A) = C(u, B)$ for all $u \in A$.

A subset $A \subseteq D$ is a *cell* of the independent set D if A is asteroidal and $I(A) \cap D = \varnothing$. For two cells A and B of D, $A \subseteq B$ implies $A = B$ because $B \setminus A \subseteq I(A) \cap D$.

3   Upper Bound on Treewidth

Theorem 1. *For all non-empty graphs G we have* $\mathrm{tw}(G) < \mathrm{an}(G) \cdot s(G)$.

For $G = (V, \varnothing)$ we have $\mathrm{tw}(G) \leq 0$, $\mathrm{an}(G) = \min\{2, |V|\}$ and $s(G) = 0$.

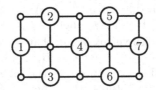

Fig. 1. $A = \{1, 2, 6, 7\}$ is an asteroidal set of this graph with interior $I(A) = \{3, 4, 5\}$. $A \cup \{4\}$ is independent but not asteroidal. Its cells are the sets $\{1, 2, 4\}$ and $\{4, 6, 7\}$.

Proof. Let D be a maximal independent set of $G = (V, E)$, and let $\mathcal{C}$ be the collection of cells of D. We construct a tree-decomposition (X, T) of G with $T = (\mathcal{C} \cup D, F)$ and X defined by

$$X(A) = \mathrm{N}(A) \quad \text{for all } A \in \mathcal{C} \qquad\qquad X(v) = \mathrm{N}[v] \quad \text{for all } v \in D.$$

If D was chosen such that $s(G) = \max\{\mathrm{d}_G(v) \mid v \in D\}$, we have $|X(v)| \leq \mathrm{d}_G(v) + 1$ for all $v \in D$, which implies $|X(v)| \leq s(G) + 1$. For all $A \in \mathcal{C}$ we have $|X(A)| \leq |A| \cdot s(G)$, and hence $|X(A)| \leq \mathrm{an}(G) \cdot s(G)$ since A is asteroidal. So the width of (X, T) will be less than $\mathrm{an}(G) \cdot s(G)$, since $E \neq \varnothing$ implies $s(G) \geq 1$.

It remains to show that for each D there is an F such that (X, T) is indeed a tree-decomposition of G. Since D is a maximal independent set of G we have $V = \bigcup_{v \in D} \mathrm{N}[v]$ and therefore $V = \bigcup_{i \in \mathcal{C} \cup D} X(i)$. We prove that (X, T) has the remaining properties of a tree-decomposition by induction on $|\mathcal{C}|$.

In the base case D is an asteroidal set of G. So we have $\mathcal{C} = \{D\}$. We make T a star with centre D and a leaf u for each vertex $u \in D$. Let $\{u, v\}$ be an edge of G. If there is a vertex $w \in \{u, v\} \cap D$ then we have $\{u, v\} \subseteq X(w)$. Otherwise there are vertices c and d in D that are adjacent to u and v because D is a dominating set of G. In this case we have $\{u, v\} \subseteq X(D)$. Next we prove that, for every vertex $v \in V$, the bags containing v induce a subtree $T(v)$ of T. This is obvious for $v \in D$ because $X(v)$ is the only bag containing v. Each vertex $v \in V \setminus D$ belongs to the central bag $X(D)$ and since T is a star, the subgraph $T(v)$ is connected.

In the inductive step there is a vertex $v \in D$ such that different connected components of $G - \mathrm{N}[v]$ contain vertices in D. That is, D is not asteroidal. Let $B_1, B_2, \ldots, B_k$ induce the connected components of $G - \mathrm{N}[v]$. For $j = 1, 2, \ldots, k$ we define $G_j = G[\mathrm{N}[v] \cup B_j]$, $D_j = \{v\} \cup (B_j \cap D)$, and $\mathcal{C}_j$ to be the set of cells of D_j in G_j. We have $D = \bigcup_{j=1}^{k} D_j$ and $\mathcal{C} \supseteq \bigcup_{j=1}^{k} \mathcal{C}_j$. Consider an asteroidal set $A \subseteq D$ that is not asteroidal in any G_j. Then A contains vertices in different connected components of $G - \mathrm{N}[v]$, which implies $v \in I(A)$. That is, $A \notin \mathcal{C}$ and therefore $\mathcal{C} = \bigcup_{j=1}^{k} \mathcal{C}_j$.

By induction hypothesis there is, for each $j = 1, 2, \ldots, k$, a set F_j of edges such that $T_j = (\mathcal{C}_j \cup D_j, F_j)$ is a tree, and the pair (X_j, T_j) is a tree-decomposition of G_j. Let $T = (\mathcal{C} \cup D, F)$ be the tree defined by $F = \bigcup_{j=1}^{k} F_j$.

We show that (X, T) is a tree-decomposition of G. For each edge $\{u, w\}$ of G there is an index j such that $\{u, w\}$ is an edge of G_j. By induction hypothesis there is an $i \in C_j \cup D_j$ such that $\{u,w\} \subseteq X(i)$. Finally we show that $T(w)$ is a tree for every vertex $w \in V$. This is obvious for $w = v$ because $X(v)$ is the only bag containing v. For each vertex $w \neq v$ that is not adjacent to v there is a unique index $j \in \{1,2,\ldots,k\}$ such that $w \in B_j$. All the bags containing w are contained in B_j, and by induction hypothesis the indices $\{i \mid w \in X(i)\}$ induce a subtree $T_j(w)$ of T_j. Clearly $T_j(w)$ is the subtree $T(w)$ of T. If w is adjacent to v then the elements of $I_j(w) = \{i \in C_j \cup D_j \mid w \in X(i)\}$ induce a subtree $T_j(w)$ of T_j for every $j \in \{1,2,\ldots,k\}$. Since $v \in I_j(w)$ for each j, the union of all the $T_j(w)$ is the tree $T(w)$, which is a subtree of T. □

Corollary 1. *For all non-empty AT-free graphs G we have* $\mathrm{pw}(G) < 2 \cdot s(G)$.

Proof. For all AT-free graphs G we have $\mathrm{pw}(G) = \mathrm{tw}(G)$ [14]. □

4   Lower Bound on Treewidth

In Lemma 1 we give the treewidth of chain graphs, which form a subclass of AT-free graphs. This result is used in the proof of Theorem 2, which provides a lower bound on the treewidth of a graph in terms of its stable degree and its asteroidal number.

4.1   Chain Graphs

A connected bipartite graph $G = (A, B, E)$ is a *chain graph* if the vertices in A can be numbered $a_1, a_2, \ldots, a_p$ such that $\mathrm{N}(a_{i-1}) \supseteq \mathrm{N}(a_i)$ holds for all indices i with $1 < i \leq p$.

Let $G = (A, B, E)$ be a chain graph with $A = \{a_i \mid 1 \leq i \leq p\}$ and $B = \{b_j \mid 1 \leq j \leq q\}$ as above. We define $\Pi(G)$ to be the set of all pairs (s,t) with $1 < s \leq p$ and $1 \leq t < q$ such that $(a_s, b_{t+1}, a_{s-1}, b_t)$ is a P_4 of G, but not a C_4.

Lemma 1. *For every chain graph G with $\Pi(G) \neq \varnothing$ we have* $\mathrm{tw}(G) = \min\{\mathrm{d}(a_s) + \mathrm{d}(b_t) - 1 \mid (s,t) \in \Pi(G)\}$.

We omit the proof due to space restrictions. A chain graph $G = (A, B, E)$ with $\Pi(G) = \varnothing$ is complete bipartite. In this case we have $\mathrm{tw}(G) = \min\{|A|, |B|\}$.

4.2   Construction

A tree decomposition is *small* if no bag is contained in another bag. If (X, T) is not small then T has an edge $\{i, j\}$ such that $X(i) \subseteq X(j)$ or $X(i) \supseteq X(j)$. We can *contract* the edge $\{i, j\}$ to obtain tree decomposition of the same graph and the same width, but with smaller index set I. To do so we choose a new index $l \notin I$, define $X(l) = X(i) \cup X(j)$, replace I by $\{l\} \cup I \setminus \{i, j\}$, and modify T such that $\mathrm{N}(l) = \mathrm{N}(\{i, j\})$. Iteration leads to a small tree decomposition.

Lemma 2. *Let (X, T) be a small tree decomposition of a graph G. Then G has a vertex that is contained in exactly one bag.*

input : A tree decomposition (X,T) of width t of a graph $G = (V,E)$ with
          $\text{an}(G) \le a$

output : A maximal independent set D of G with $\text{d}_G(v) \le at^2$ for all $v \in D$

```
1  begin
2  |   D ← ∅;
3  |   while V ≠ ∅ do
4  |   |   while there is a contractible edge of (X,T) do contract it;
5  |   |   choose a vertex v ∈ V that appears in exactly one bag of (X,T);
6  |   |   D ← D ∪ {v}; V ← V \ N_G[v];
7  |   |   for i ∈ I do X(i) ← X(i) \ N_G[v]
```

Theorem 2. *For all non-empty graphs G we have $s(G) \le \text{an}(G) \cdot \text{tw}(G)^2$.*

Proof. Let $G = (V,E)$ be a graph and let (X,T) be its tree decomposition of width w. We consider the set $D \subseteq V$ constructed by the algorithm above. Throughout the algorithm (X,T) is a tree decomposition of the shrinking graph G, and the width of (X,T) does not increase.

The set D is independent in G because we remove in Line 6 the closed neighbourhood of v from G for every vertex v added to D. The algorithm terminates when $V = \varnothing$ holds. Therefore D is a maximal independent set of G.

To bound the degree of a vertex $v \in D$ we consider the sets $U = \text{N}_G(v)$ and $W = \text{N}_G(U)$, and define a partial order $\sqsubseteq$ on W such that $\text{N}_G(w_1) \cap U \subset \text{N}_G(w_2) \cap U$ implies $w_1 \sqsubseteq w_2$ for all vertices $w_1, w_2 \in W$. For different vertices $w_1, w_2 \in W$ with $\text{N}_G(w_1) \cap U = \text{N}_G(w_2) \cap U$ we ensure that $\sqsubseteq$ becomes antisymmetric by fixing $w_1 \sqsubset w_2$ or $w_2 \sqsubset w_1$ accordingly, for instance based on a given linear order on V.

The set U splits into new and old neighbours of v. The *new neighbours* are in the unique bag of (X,T) containing v when v is chosen. There are at most t new neighbours. The *old neighbours* are adjacent to v and a vertex w that was added to D before v. These old neighbours of v were new neighbours of w and removed from G together with w (Line 6).

Let $C \subseteq W$ be a maximal chain of $(W, \sqsubseteq)$, *i.e.* C is a set of $\sqsubseteq$-comparable vertices, and $\subseteq$-maximal with this property. Let B be the new neighbours in U of vertices in C. We define a bipartite graph $H = (B, C, F)$ with $F = E \cap \{\{b,c\} \mid b \in B, c \in C\}$. By maximality we have $v \in C$. Therefore H is a chain graph. We define subsets $B_1 \subseteq B$ and $C_1 \subseteq C$ as follows:

- If $\Pi(H) = \varnothing$ and $|B| \le |C|$ then $B_1 = B$ and $C_1 = \varnothing$.
- If $\Pi(H) = \varnothing$ and $|B| > |C|$ then $B_1 = \varnothing$ and $C_1 = C$.
- If $(r,s) \in \Pi(H)$ and $\text{tw}(H) = \text{d}_H(b_r) + \text{d}_H(c_s) - 1$ then $B_1 = \text{N}_H(c_s)$ and $C_1 = \text{N}_H(b_r)$.

In all three cases let $B_2 = B \setminus B_1$ and $C_2 = C \setminus C_1$. We have $|B_1| + |C_1| \le t$ because H is a subgraph of G, which implies $\text{tw}(H) \le \text{tw}(G)$. Moreover we have $|B_2| \le t \cdot |C_1|$ since there is no edge of H with endpoints in B_2 and C_2, that is, all vertices in B_2 are new neighbours of vertices in C_1. This implies $|B| \le t^2$ since $t \ge 1$ because of $E \ne \varnothing$.

Next let $A \subseteq W$ be an antichain of $(W, \sqsubseteq)$, *i.e.* A is a set of $\sqsubseteq$-incomparable vertices. For different vertices $w_1, w_2 \in A$ there is a vertex $u_2 \in N(w_2) \cap U \backslash N(w_1)$. It establishes a path (w_2, u_2, v) in $G - N[w_1]$. Since such a path exists for all $w_2 \in A \backslash \{w_1\}$ we have $A \backslash \{w_1\} \subseteq C(w_1, v)$. Since this holds for all $w_1 \in A$ the set A is asteroidal in G. Consequently we have $|A| \le a$ for every antichain A of $(W, \sqsubseteq)$.

By Dilworth's theorem W can be covered by k chains of of $(W, \sqsubseteq)$, where k is the maximum size of an antichain. U is the union of the B-sets for the chains in the cover. With $k \le a$ this implies $d_G(v) \le at^2$ for all $v \in D$. □

There might be a better lower bound on the stable degree:

Conjecture 1. For every graph G we conjecture $s(G) \le \mathrm{an}(G) \cdot \mathrm{tw}(G)$.

For AT-free graphs we can prove this conjecture:

Theorem 3. *For every AT-free graph G we have $s(G) \le 2\,\mathrm{tw}(G)$.*

Proof. We assume a tree decomposition (X, T) of G where $T = (I, F)$ is a path with $I = \{1, 2, \ldots, \ell\}$ and $F = \{\{i-1, i\} \mid 1 < i \le \ell\}$. We construct D by the algorithm as before and choose v always from the bag indexed by the maximum leaf of T. Let $l(v) = \min\{i \in I \mid v \in X(i)\}$ and $r(v) = \max\{i \in I \mid v \in X(i)\}$ for each vertex v. To prove $s(G) \le 2\,\mathrm{tw}(G)$ it suffices to show $N[v] \subseteq X(l(v)) \cup X(r(v))$ for all $v \in D$. Assume a neighbour $v' \in N(v) \backslash (X(l(v)) \cup X(r(v)))$. Then v' and v belong to a bag $X(i)$ with $l(v) < i < r(v)$, contradicting the fact that v appears in exactly one bag when chosen. □

5   Computing the Stable Degree

5.1   Polynomial Cases: $k \le 2$

We define the decision problems SD and k-SD for every $k \in \mathbb{N}$ by

$$\mathrm{SD} = \{(G, k) \mid s(G) \le k\} \qquad k\text{-SD} = \{G \mid s(G) \le k\}.$$

Lemma 3. *The problem k-SD can be solved in polynomial time for $k \in \{0, 1, 2\}$.*

Proof. If C induces a component of G then $s(G) \le \max\{\Delta(G[C]), s(G - C)\}$. On this observation we base the following reduction rule for k-SD:

Low-degree component: If C induces a component in G with $\Delta(G[C]) \le k$ then we replace G by $G - C$ because $G \in k\text{-SD} \iff G - C \in k\text{-SD}$.

In a graph $G = (V, E)$ with $s(G) \le k$ the set $D_k = \{v \in V \mid d(v) \le k\}$ is dominating. For $k = 0$ and $k = 1$ this necessary condition for $G \in k\text{-SD}$ is also sufficient. For $k = 2$, more reduction rules are required:

Pendant vertex: Let $(x, y_1, y_2, y_3, \ldots, y_l, z)$ be a path in G with $d(x) = 1$, $d(y_i) = 2$ for $i = 1, 2, 3, \ldots, l$ and $d(z) > 2$. Then we replace G by the graph $G' = G - \{x, y_1, y_2, \ldots, y_l, z\}$ because $G \in$ 2-SD $\iff G' \in$ 2-SD.

Long path: Let $(x, y_1, y_2, \ldots, y_l, z)$ be a path in G with $d(x) > 2$, $d(y_i) = 2$ for $i = 1, \ldots, l$, $d(z) > 2$ and $l \notin \{0, 2\}$. Then $G \in$ 2-SD if and only if $G - \{x, y_1, y_2, \ldots, y_k, z\} \in$ 2-SD.

Unique neighbour: Let (x, y_1, y_2, z) be a path in G, with $d(x) > 2$, $d(y_1) = 2$, $d(y_2) = 2$ and $d(z) > 2$. If $N(z) \cap D_2 = \{y_2\}$ then $G \in$ 2-SD if and only if $G - \{y_1, y_2, z\} \in$ 2-SD.

If none of these reduction rules apply then the minimum degree $\delta(G)$ of $G = (V, E)$ is at least two. We will show that our necessary condition for $G \in k$-SD is also sufficient. Let $X = \{v \in V \mid d(v) > 2\}$ and $Y = \{v \in V \mid d(v) = 2\}$. Clearly, if there is a vertex in X without neighbour in Y then $G \notin$ 2-SD. Otherwise we will show that $G \in$ 2-SD.

We construct an auxiliary bipartite graph $H = (X \cup Z, F)$ where Z is the set of edges of $G[Y]$ and $F = \{\{x, z\} \mid x \in X, z \in Z, N_G(x) \cap z \neq \varnothing\}$. We have $d_H(x) \geq 2$ for all $x \in X$ and $d_H(z) = 2$ for all $z \in Z$. This implies $|N_H(A)| \geq |A|$ for all $A \subseteq X$ and therefore H has an X-saturating matching. The X-saturating matching of H corresponds to an X-saturating matching M of G. The M-saturated vertices in Y form an independent set that dominates X. Therefore this subset extends to a maximal independent subset of Y, and we have $G \in$ 2-SD. □

5.2 Hardness for $k \geq 3$

Lemma 4. *For every $k \geq 3$, the problem k-SD is NP-complete.*

Proof. Clearly the problem k-SD is in NP. To show the NP-hardness we reduce from a restricted version of SAT, where every boolean variable x appears in at most two clauses positively, that is as x, and in at most two clauses negatively as $\overline{x}$ [16].

Let φ be a formula in CNF with this property. For each variable x_i appearing in φ, $1 \leq i \leq n$, we create a truth assignment component which is a $K_{2,2}$ with partite sets $\{x_i^1, x_i^2\}$ and $\{\overline{x}_i^1, \overline{x}_i^2\}$. For the clause c_j of φ, $1 \leq j \leq m$, we create a satisfaction test component which consists of a single vertex c_j. We add the edge $\{x_i^l, c_j\}$ if the clause c_j contains the lth appearance of the positive literal x_i, and we create the edge $\{\overline{x}_i^l, c_j\}$ if the clause c_j contains the lth appearance of the negative literal $\overline{x}_i$. We complete the construction of the reduction graph G by adding all edges $\{c_j, c_l\}$ for $j \neq l$.

Every vertex v in a truth assignment component of G has degree at most three, and every vertex in a satisfaction test component has degree at least m. We may assume $k < n < m$.

Let $a : \{x_1, x_2, \ldots, x_n\} \to \{\text{true}, \text{false}\}$ be a satisfying truth assignment of φ. Then $D = \{x_i^l \mid a(x_i) = \text{true}, l \in \{1, 2\}\} \cup \{\overline{x}_i^l \mid a(x_i) = \text{false}, l \in \{1, 2\}\}$ is a maximal independent set of G, and therefore $s(G) \leq k$.

On the other hand, let D be a maximal independent set of G with $d_G(v) \leq k$ for all $v \in D$. By this degree condition D contains only vertices from truth assignment components of G. Since D is independent, these are either x_i^1 and x_i^2 or $\bar{x}_i^1$ and $\bar{x}_i^2$. We define $a(x_i) = \text{true}$ if $x_i^1 \in D$ and $a(x_i) = \text{false}$ if $\bar{x}_i^1 \in D$. Assume that a clause c_j is not satisfied. Then the vertex c_j has no neighbour in D, contradicting the fact that D is maximal.   □

In fact the reduction shows that the stable degree is hard to approximate: $\varphi \in$ SAT implies $s(G) \leq 3$ and $\varphi \notin$ SAT implies $s(G) \geq m$. Since SAT remains NP-complete when restricted to formulae with more than m clauses (for fixed value of m), we have the following lemma.

Lemma 5. *There is no polynomial time algorithm approximating the stable degree by a constant factor, unless* $\text{P} = \text{NP}$.

5.3 Bounded Cliquewidth

For fixed values of k the problem k-SD can be formulated in MSOL. Therefore its restriction to graphs of bounded tree- or cliquewidth can be solved in linear time [8,9].

5.4 Bounded Asteroidal Number

In this subsection we develop an algorithm computing the stable degree of graphs of bounded asteroidal number, such as (unit) interval graphs or AT-free graphs, which have unbounded tree- and cliquewidth. We start with technical lemmas on connected components and cells. Remember that $C(u,v)$ induces the connected component of $G - N[u]$ containing v.

Lemma 6. *For all independent triples* $\{u, v, w\}$ *of a graph,* $C(v,u) \neq C(v,w)$ *implies* $C(v,u) \subseteq C(w,u)$ *and* $C(v,w) \subseteq C(u,w)$.

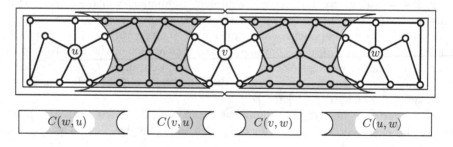

Fig. 2. An example illustrating Lemma 6

Proof. Let x be a vertex in $C(v, u)$, which means there is a path $(x, \ldots, u)$ in $G[C(v, u)]$. In G this path avoids $N[w]$ since $C(v, u) \neq C(v, w)$. Therefore it exists in $G[C(w, u)]$ as well, and hence $C(v, u) \subseteq C(w, u)$ holds. By a symmetric argument we have $C(v, w) \subseteq C(u, w)$. $\qquad\square$

Remember that an asteroidal subset A of an independent set D is a cell of D if the interior of A does not contain a vertex in D.

Lemma 7. *Let D be an independent set of $G = (V, E)$, $v \in V \setminus N[D]$, and let B be the union of all subsets $A \subseteq D$ such that $A \cup \{v\}$ is a cell of $D \cup \{v\}$. Then B is a cell of D and $v \in I(B)$.*

Proof. Let C be the set of all subsets $A \subseteq D$ such that $A \cup \{v\}$ is a cell of $D \cup \{v\}$. For every $A \in C$ and every $u \in A$ we have $A \setminus \{u\} \subseteq C(u, v)$ because $A \cup \{v\}$ is asteroidal. For vertices u and w in different sets in C we have $C(u, w) = C(u, v)$ and $C(w, u) = C(w, v)$ by Lemma 6. Hence B is an asteroidal set of G and $v \in I(B)$.

To show that B is a cell of D we assume a vertex $x \in I(B) \cap D$. Because $x \notin B$, the set $\{v, x\}$ is not asteroidal, and therefore we have $x \in N[v]$, which contradicts $x \in D$ and $v \in V \setminus N[D]$. $\qquad\square$

Lemma 8. *Let B be a cell of an independent set D of $G = (V, E)$. Each vertex $v \in I(B)$ defines a partition C of B such that $A \cup \{v\}$ is a cell of $D \cup \{v\}$ for each $A \in C$.*

Proof. Let $C_1, C_2, \ldots, C_k$ induce the connected components of $G - N[v]$. Then $C = \{B \cap C_i \mid 1 \leqslant i \leqslant k\}$ is a partition of B, and for every set $A \in C$, $A \cup \{v\}$ is asteroidal. By Lemma 7, $I(A \cup \{v\}) \subseteq I(B)$, which implies $I(A \cup \{v\}) \cap D = \varnothing$ because B is a cell of D. Since $v \notin I(A \cup \{v\})$ we conclude $I(A \cup \{v\}) \cap (D \cup \{v\}) = \varnothing$, and hence $A \cup \{v\}$ is a cell of $D \cup \{v\}$. $\qquad\square$

Corollary 2. *1. Every vertex in an independent set D belongs to a cell of D.*
   2. For every independent set D of (V, E), $V \setminus N[D]$ is the disjoint union of the interiors of the cells of D.
   3. An independent set D is maximal independent if and only if $I(A) = \varnothing$ holds for every cell A of D.

Let $G = (V, E)$ be a graph, A an asteroidal set of G, and let $B \subseteq I(A)$ induce some connected components of $G[I(A)]$. We define

$$s(A, B) = \min_{U} \max_{u \in U} d_G(u)$$

where the minimum is taken over all maximal independent sets U of $G[B]$. Then $s(G) = s(\varnothing, V)$. The following recurrence allows us to compute the values of $s(A, B)$:

$$s(A, \varnothing) = 0$$
$$s(A, B) = \max_{D \in \mathcal{D}(B)} s(A, D) \qquad\qquad \text{if } G[B] \text{ is disconnected}$$
$$s(A, B) = \min_{v \in B} \max \left(d_G(v), \max_{C \in \mathcal{C}(A \cup \{v\})} s(C, I(C)) \right) \quad \text{if } G[B] \text{ is connected}$$

where $\mathcal{C}(S)$ is the set of cells of the independent set S, and $\mathcal{D}(B)$ is the set of vertex sets that induce a connected component of $G[B]$. The recurrent equation above directly translates into an algorithm. The correctness follows from Lemmas 7 and 8.

A bottom-up dynamic programming algorithm first sorts all objects (A, B) by the cardinality of B and processes them in increasing order of $|B|$. It store the values $s(A, B)$ computed in a table that is an array indexed by B.

We bound the running time of the algorithm in terms of $k = \text{an}(G)$, $n = |V|$ and $m = |E|$. The algorithm considers at most $\sum_{i=0}^{k} \binom{n}{i} = \mathcal{O}(n^k)$ asteroidal sets A of G. For each A it considers at most $n+1$ subsets $B \subseteq I(A)$, namely $B = I(A)$ and all set B that induce a connected component of $G[I(A)]$. For a fixed pair (A, B), it needs time $\mathcal{O}(n+m)$ to organise the look up of values computed before if $G[B]$ is disconnected. This is mainly for computing the connected components of $G[B]$. If $G[B]$ is connected, the algorithm minimises over $\mathcal{O}(n)$ vertices $v \in B$, and spends $\mathcal{O}(n + m)$ time per vertex v to organise the table look-up. That is, the algorithm runs in time $\mathcal{O}(n^{k+1}m)$.

Theorem 4. *For graphs G with* $\text{an}(G) \leq k$, *$s(G)$ can be computed in time* $\mathcal{O}(n^{k+1}m)$.

5.5 Approximating Treewidth

By Theorems 1 and 3 we have

$$\tfrac{1}{2} \cdot s(G) \leq \text{tw}(G) < 2 \cdot s(G)$$

for AT-free graphs G, and in general Theorems 1 and 2 imply

$$\sqrt{s(G)/\text{an}(G)} < \text{tw}(G) < \text{an}(G) \cdot s(G).$$

These lower and upper bounds enable us to extend the algorithm from the previous subsection such that it approximates the treewidth of AT-free graphs by a factor of 4 in the worst case. In contrast, the algorithm developed in [3] guarantees an approximation factor of 2 for AT-free graphs. Theorem 1 and Conjecture 1 would imply

$$\frac{1}{\text{an}(G)} \cdot s(G) \leq \text{tw}(G) < \text{an}(G) \cdot s(G),$$

and the constant factor approximation would generalise to graphs G of bounded asteroidal number. Its ratio would be $\text{an}(G)^2$ in the worst case, which would beat the $8 \, \text{an}(G)$ factor from [4] if the bound on the asteroidal number is less than eight. With Theorem 2 instead of Conjecture 1 we obtain an approximation ratio of $s(G)^{1/2} \, \text{an}(G)^{3/2}$ via the stable degree.

6   Conclusions

Graph problems definable in MSOL can be solved in linear time when restricted to graphs of bounded treewidth [8]. We showed that inside AT-free graphs, bounded treewidth can be replaced by bounded stable degree. This allows us to concentrate on the hard cases when we consider problems on AT-free graphs for which the complexity status is still unknown, such as vertex colouring or Hamiltonicity.

References

1. Arnborg, S., Corneil, D.G., Proskurowski, A.: Complexity of finding embeddings in a k-tree. SIAM Journal on Algebraic and Discrete Methods 8, 277–284 (1987)
2. Bodlaender, H.L., Koster, A.M.C.A.: Treewidth computations II. Lower bounds. Information & Computation 209, 1103–1119 (2011)
3. Bouchitté, V., Todinca, I.: Approximating the treewidth of AT-free graphs. Discrete Applied Mathematics 131, 11–37 (2003)
4. Bouchitté, V., Kratsch, D., Müller, H., Todinca, I.: On treewidth approximations. Discrete Applied Mathematics 136, 183–196 (2004)
5. Broersma, H.-J., Kloks, T., Kratsch, D., Müller, H.: Independent sets in asteroidal triple-free graphs. SIAM Journal on Discrete Mathematics 12, 276–287 (1999)
6. Broersma, H.-J., Kloks, T., Kratsch, D., Müller, H.: A generalization of AT-free graphs and a generic algorithm for solving triangulation problems. Algorithmica 32, 594–610 (2002)
7. Corneil, D.G., Olariu, S., Stewart, L.: Asteroidal triple-free graphs. SIAM Journal on Discrete Mathematics 10, 399–430 (1997)
8. Courcelle, B.: The monadic second-order logic of graphs. III: Tree-decompositions, minors and complexity issues. RAIRO. Informatique Théorique et Applications 26, 257–286 (1992)
9. Courcelle, B., Makowsky, J.A., Rotics, U.: Linear time solvable optimization problems on graphs of bounded clique-width. Theory of Computing Systems 33, 125–150 (2000)
10. Kloks, T., Kratsch, D., Spinrad, J.: On treewidth and minimum fill-in of asteroidal triple-free graphs. Theoretical Computer Science 175, 309–335 (1997)
11. Kloks, T., Kratsch, D., Müller, H.: On the structure of graphs with bounded asteroidal number. Graphs and Combinatorics 17, 295–306 (2001)
12. Kratsch, D.: Domination and total domination on asteroidal triple-free graphs. Discrete Applied Mathematics 99, 111–123 (2000)
13. Lekkerkerker, C.G., Boland, J.C.: Representation of a finite graph by a set of intervals on the real line. Fundamenta Mathematicae 51, 45–64 (1962)
14. Möhring, R.H.: Triangulating graphs without asteroidal triples. Discrete Applied Mathematics 64, 281–287 (1996)
15. Ramachandramurthi, S.: A Lower Bound for Treewidth and its Consequences. In: Mayr, E.W., Schmidt, G., Tinhofer, G. (eds.) WG 1994. LNCS, vol. 903, pp. 14–25. Springer, Heidelberg (1995)
16. Tovey, C.A.: A simplified NP-complete satisfiability problem. Discrete Applied Mathematics 8, 85–89 (1984)

A $9k$ Kernel for Nonseparating Independent Set in Planar Graphs*

Łukasz Kowalik and Marcin Mucha

Institute of Informatics, University of Warsaw, Poland
{kowalik,mucha}@mimuw.edu.pl

Abstract. We study kernelization (a kind of efficient preprocessing) for NP-hard problems on planar graphs. Our main result is a kernel of size at most $9k$ vertices for the PLANAR MAXIMUM NONSEPARATING INDEPENDENT SET problem. A direct consequence of this result is that PLANAR CONNECTED VERTEX COVER has no kernel with at most $9/8k$ vertices, assuming $\mathbf{P} \neq \mathbf{NP}$. We also show a very simple $5k$-vertices kernel for PLANAR MAX LEAF, which results in a lower bound of $5/4k$ vertices for the kernel of PLANAR CONNECTED DOMINATING SET (also under $\mathbf{P} \neq \mathbf{NP}$).

1 Introduction

Many NP-complete problems, while most likely not solvable efficiently, admit kernelization algorithms, i.e. efficient algorithms which replace input instances with an equivalent, but often much smaller one. More precisely, a *kernelization algorithm* takes an instance I of size n and a parameter $k \in \mathbb{N}$, and after time polynomial in n it outputs an instance I' (called a *kernel*) with a parameter k' such that I is a yes-instance iff I' is a yes instance, $k' \leq k$, and $|I'| \leq f(k)$ for some function f depending only on k. The most desired case is when the function f is polynomial, or even linear (then we say that the problem admits a polynomial or linear kernel). In such a case, when the parameter k is relatively small, the input instance, possibly very large, is "reduced" to a small one. In this paper by the size of the instance $|I|$ we always mean the number of vertices.

In the area of kernelization of graph problems the class of planar graphs (and more generally H-minor-free graphs) is given special attention. This is not only because planar graphs are models of many real-life networks but also because many problems do not admit a (polynomial) kernel for general graphs, while restricted to planar graphs they have a polynomial (usually even linear) kernel. A classic example is the $335k$-vertex kernel for the PLANAR DOMINATING SET due to Alber et al. [1]. In search for optimal results, and motivated by practical applications, recently researchers try to optimize the constants in the linear function bounding the kernel size, e.g. the current best bound for the size of the kernel for the DOMINATING SET is $67k$ [2]. Such improvements often require nontrivial auxiliary combinatorial results which might be of independent interest.

* Work supported by the National Science Centre (grant N N206 567140).

M.C. Golumbic et al. (Eds.): WG 2012, LNCS 7551, pp. 160–171, 2012.

Our paper fits into this framework. We focus on kernelization of the following problem:

MAXIMUM NONSEPARATING INDEPENDENT SET (NSIS)   **Parameter:** k
Input: a graph $G = (V, E)$ and an integer $k \in \mathbb{N}$
Question: Is there an independent set I of size at least k such that $G[V - I]$ is connected?

In what follows, $|V|$ is denoted by n. This problem is closely related with CON-NECTED VERTEX COVER (CVC in short), where given a graph $G = (V, E)$ and an integer k we ask whether there is a set $S \subseteq V$ of size at most k such that S is a vertex cover (i.e. S touches every edge of G) and S induces a connected subgraph of G. The CVC problem has been intensively studied, in particular there is a series of results on kernels for planar graphs [6,10] culminating in the recent $\frac{11}{3}k$ kernel [8]. It is easy to see that C is a connected vertex cover iff $V - C$ is a nonseparating independent set. In other words, (G, k) is a yes-instance of CVC iff $(G, n - k)$ is a yes-instance of NSIS. In such a case we say that NSIS is a *parametric dual* of CVC. An important property of a parametric dual, discovered by Chen et al [2], is that if the dual problem admits a kernel of size at most αk, then the original problem has no kernel of size smaller than $\alpha/(\alpha - 1)k$, unless P=NP.

As we will see, the NSIS problem in planar graphs is strongly related to the MAX LEAF problem: given a graph G and an integer k, find a spanning tree with at least k leaves.

Our Kernelization Results. We study PLANAR MAXIMUM NONSEPARAT-ING INDEPENDENT SET (PLANAR NSIS in short), which is the NSIS problem restricted to planar graphs. We show a kernel of size at most $9k$ for PLANAR NSIS. This implies that PLANAR CONNECTED VERTEX COVER has no kernel of size smaller than $9/8k$, unless P=NP. This is the first non-trivial lower bound for the kernel size of the PLANAR CVC problem. Our kernelization algorithm is very efficient: it can be implemented to run in $O(n)$ time. As a by-product of our considerations we also show a $5k$ kernel for both MAX LEAF and PLANAR MAX LEAF, which in turn implies a lower bound of $5/4k$ for its parametric dual, i.e. PLANAR CONNECTED DOMINATING SET.

Our Combinatorial Results. Some of our auxiliary combinatorial results might be of independent interest. We mention two of them here. Kleitman and West [7] showed that an n-vertex graph of minimum degree three contains a spanning tree with at least $n/4$ leaves. We generalize their result to graphs that contain no separator consisting of only degree two vertices. We also show that every n-vertex outerplanar graph contains an independent set I and a collection of vertex-disjoint cycles $\mathcal{C}$ such that $9|I| \geq 4n - 3|\mathcal{C}|$.

Previous Results. As the CVC is NP-complete even in planar graphs [5], so is NSIS. To the best of our knowledge there is no prior work on the parameter-ized complexity of MAXIMUM NONSEPARATING INDEPENDENT SET. The reason for that is simple: a trivial reduction from INDEPENDENT SET (add a vertex connected to all the vertices of the original graph) shows that NSIS is W[1]-hard, i.e. existence of an algorithm of complexity $O(f(k) \cdot |V|^{O(1)})$ is very unlikely (and so is the existence of a polynomial kernel). However, by general results on kernelization for sparse graphs [4], one can see that NSIS admits a $O(k)$ kernel for apex-minor-free graphs, so in particular for planar graphs. However, the general approach does not provide a good bound on the constant hidden in the asymptotic notation. Observe that this constant is crucial: since we deal with an NP-complete problem, in order to find an exact solution in the reduced instance, most likely we need exponential time (or at least superpolynomial, because for planar graphs $2^{O(\sqrt{k})}$-time algorithms are often possible), and the constant appears in the exponent.

MAX LEAF has been intensively studied. Although there is a $3.75k$ kernel even for general graphs due to Estivill-Castro et al. [3], some of their reductions do not preserve planarity. Moreover, the algorithm and its analysis are extremely complicated, while our method is rather straightforward.

Yet Another Equivalent Formulation of NSIS. Consider the NSIS problem again. It is easy to see that if graph G has two nontrivial (i.e. with at least two vertices) connected components, then the answer is NO. Furthermore, an instance (G, k) consisting of a connected component C and an independent set I is equivalent to the instance $(G[C], k - |I|)$. Hence, w.l.o.g. we may assume that the input graph G is connected. It is easy to see that the MAXIMUM NONSEPARATING INDEPENDENT SET problem for connected graphs is equivalent to the following problem, which we name MAXIMUM INDEPENDENT LEAF SPANNING TREE:

MAXIMUM INDEPENDENT LEAF SPANNING TREE   **Parameter:** k
Input: a graph $G = (V, E)$ and an integer $k \in \mathbb{N}$
Question: Is there a spanning tree T such that the set of leaves of T contains a subset of size k that is independent in G?

In what follows, we will use the above formulation, since it directly corresponds to our approach.

Terminology and Notation. By $N_G(v)$ we denote the set of neighbors of vertex v, $G[S]$ denotes the subgraph of graph G induced by a set of vertices S. If $G = (V, E)$ is a connected graph and $S \subset V$, then we say that S is a separator if $G[V - S]$ is disconnected. By a d-vertex we mean a vertex of degree d.

2   A Simple $12k$ Kernel for Planar NSIS

In this section we describe a relatively simple algorithm that finds a $12k$ kernel for PLANAR NSIS. This is achieved by the following three steps. First, in Section 2.1 we show a reduction rule and a linear-time algorithm which, given an instance (G, k), returns an equivalent instance (G', k') such that $|V(G')| \leq |V(G)|$, $k' \leq k$, and moreover G' has no separator consisting of only 2-vertices. Second, in Section 2.2 we show that G' has a spanning tree T with at least $|V(G')|/4$ leaves (and it can be found in linear time). Denote the set of leaves of T by L. Third, in Section 2.3 we show that the graph $G'[L]$ is outerplanar. It follows that $G'[L]$ has an independent set of size at least $|L|/3$ (which can be easily found in linear time) and, consequently, T has at least $|V(G')|/12$ leaves that form an independent set. Hence, if $k' \leq |V(G')|/12$ our algorithm returns the answer YES (and the relevant feasible solution if needed). Otherwise $|V(G')| < 12k' \leq 12k$ so (G', k') is indeed the desired kernel.

2.1   The Separator Rule

Now we describe our main reduction rule, which we call *separator rule*. It is easier for us to prove the correctness of the rule for the CVC problem and then convert it to a rule for the NSIS problem.

> **Separator Rule.** Assume there is a separator S consisting of only 2-vertices. As long as S contains two adjacent vertices, remove one of them from S (note that S is still a separator). Next, choose any $v \in S$ such that the two neighbors a, b of v belong to distinct connected components of $G[V - S]$. If $\deg(a) = \deg(b) = 1$, remove a from G. If $\deg(a) = 1$ and $\deg(b) \geq 2$, remove a from G and decrease the parameter k by 1. Proceed analogously when $\deg(b) = 1$ and $\deg(a) \geq 2$. Finally, when $\deg(a), \deg(b) \geq 2$, contract the path avb into a single vertex v' and decrease k by 2.

We say that a reduction rule for a parameterized problem P is *correct* when for every instance (G, k) of P it returns an instance (G', k') such that:

a) (G', k') is an instance of P,
b) (G, k) is a yes-instance of P iff (G', k') is a yes-instance of P,
c) $k' \leq k$.

Lemma 1. *The separator rule is correct for* PLANAR *CVC.*

Proof. Since the separator rule modifies the graph by removing a vertex or contracting a path it is planarity preserving, so a) holds. The condition c) is easy to check so we focus on b), i.e. the equivalence of the instances.

The case when $\deg(a) = \deg(b) = 1$ are trivial so we skip the argument.

Now assume $\deg(a) = 1$, $\deg(b) \geq 2$ (the case $\deg(b) = 1$, $\deg(a) \geq 2$ is symmetric). If C is a minimum connected vertex cover of G, $|C| \leq k$, then $v \in C$ and $a \notin C$. Since $G[C]$ is connected, $\deg(v) = 2$ and $\deg(b) \geq 2$, also $b \in C$. It follows that $C \setminus \{v\}$ is a connected vertex cover of G' of size at most $k' = k - 1$. In the other direction, if C' is a connected vertex cover of G', $|C'| \leq k' = k - 1$, then $b \in C'$ and clearly $C' \cup \{v\}$ is a connected vertex cover of G.

Finally, assume $\deg(a), \deg(b) \geq 2$. Let A and B be the connected components of $G[V - S]$ that contain a and b, respectively. Let a_0 (resp. b_0) be any neighbor of a (resp. b) distinct from v (a_0 and b_0 exist since $\deg(a), \deg(b) \geq 2$). Note that $a_0 \in A \cup S$, $b_0 \in B \cup S$ and $a_0, b_0 \in G'$.

Let us first assume that G' has a connected vertex cover C', $|C'| \leq k'$. We show that G' has a connected vertex cover D', $|D'| \leq k'$ such that $v' \in D'$. Then it is easy to check that $C = (D' \setminus \{v'\}) \cup \{a, v, b\}$ is the required connected vertex cover of G, and $|C| = |D'| + 2 \leq k$.

If $v' \in C'$ we just put $D' = C'$ so suppose that $v' \notin C'$. Then $a_0, b_0 \in C'$. Since $G'[C']$ is connected, there is a path P from a_0 to b_0 in $G'[C']$, possibly of length 0. Since $a_0 \in A \cup S$ and $b_0 \in B \cup S$ we infer that P contains a vertex $w \in S \cap C'$. It follows that $D' = C' \setminus \{w\} \cup \{v'\}$ is a connected vertex cover of G' of size at most k'.

Let us now assume that (G, k) has a connected vertex cover C, $|C| \leq k$. If $\{a, v, b\} \subseteq C$, then clearly $C' = (C \setminus \{a, v, b\}) \cup \{v'\}$ is a connected vertex cover of G' with $|C'| \leq k - 2 = k'$. On the other hand, if $\{a, v, b\} \not\subseteq C$, then $G[C \cup \{a, v, b\}]$ contains a cycle (because C is connected). Since S is a separator, this cycle has to contain some $w_1 \in S$ other than v. In this case we claim that $C' = (C \setminus \{w_1\}) \cup \{a, v, b\}$ is a connected vertex cover of G' and $|C'| \leq k'$. It is clear that C' is a connected vertex cover. The size bound follows from the fact, that C has to contain two out of the three vertices $\{a, v, b\}$. $\square$

Now, we convert the separator rule to the **dual separator rule** as follows. Let (G, ℓ) be an instance of (PLANAR) NSIS. Put $k = |V(G)| - \ell$, apply the separator rule to (G, k) and get (G', k'). Put $\ell' = |V(G')| - k'$ and return (G', ℓ').

Corollary 1. *The dual separator rule is correct for* PLANAR NSIS.

Proof. The condition c) is easy to check while Lemma 1 implies a) and b). $\square$

It is clear that the dual separator rule can be implemented in linear time. However, we would like to stress a stronger claim: there is a linear-time algorithm that given a graph G applies the separator rule as long as it is applicable. This algorithm can be sketched as follows. First we remove all 1-vertices that are adjacent to 2-vertices, and we modify k as described in the separator rule. Second, we find all maximal paths that contain 2-vertices only. For every such path, if it contains at least 3 vertices (and hence two of them form a separator), we replace it by a path of two vertices. It is easy to implement these two steps in linear time. Now, every 2-vertex has at most one neighboring 2-vertex. We remove all 2-vertices that do not have neighbors of degree 2 and for each pair of adjacent 2-vertices we remove exactly one of them. Next, we pick a connected component A of the resulting graph and we mark all its vertices. Then we consider all degree 2 neighbors of this component (that has been removed). If such a neighbor v has an unmarked neighbor then it connects A with another component B. We apply the separator rule to the vertex v (in constant time) and we mark v as *processed*. Then we mark all the vertices of B. As a result the components A and B are joined into a new component A. In any case, the vertex v is not considered any more. We continue the procedure as long as the graph gets connected. All the removed 2-vertices that are not marked as processed are put back in the graph. Since every vertex of G is marked at most once, the whole algorithm works in linear time.

2.2 Finding a Spanning Tree with Many Leaves

Kleitman and West [7] showed how to find a spanning tree with at least $n/4$ leaves in a graph of minimum degree 3. In this section we generalize their result by proving the following theorem.

Theorem 1. *Let G be a connected n-vertex graph that does not contain a separator consisting of only 2-vertices. Then G has a spanning tree with at least $n/4$ leaves. Moreover, such a tree can be found in linear time.*

We will slightly modify the approach of Kleitman and West so that vertices of smaller degree are allowed.

First note that it suffices to show a simplified case where G has no edge uv such that both u and v are 2-vertices (and we still assume the nonexistence of a separator consisting of only 2-vertices). Indeed, if the theorem holds for the simplified case, we just remove from G all the edges uv such that both u and v are 2-vertices, call the new graph $\tilde{G}$. Note that $\tilde{G}$ is connected and $\tilde{G}$ does not contain a separator consisting of only 2-vertices (otherwise the old G contains such a separator). Hence we get a spanning tree T of $\tilde{G}$ with at least $|V(\tilde{G})|/4$ leaves by applying the simplified case. However, since $|V(\tilde{G})| = |V(G)|$ and T is also a spanning tree of G, so T is also the required tree for the general claim. Hence in what follows we assume that G has no edge with both endpoints of degree 2.

In order to build a spanning tree T our algorithm begins with a tree consisting of an arbitrarily chosen vertex (called a *root*), and then the spanning tree is built by a sequence of *expansions*. To expand a leaf $v \in T$ means to add the vertices of $N_G(v) \setminus V(T)$ to T and connect them to v in T. Note that in a tree T built from a root by a sequence of expansions, if a vertex in $V(G) - V(T)$ is adjacent with $v \in V(T)$, then v is a leaf.

The order in which the leaves are expanded is important. To describe this order, we introduce three operations (operations O1 and O3 are the same as in [7], but O2 is modified):

(O1) Applies when there is a leaf $v \in V(T)$ such that $|N_G(v) \setminus V(T)| \geq 2$. Then v is expanded.

(O2) Applies when there is a leaf $v \in V(T)$ such that $|N_G(v) \setminus V(T)| = 1$ (let $N_G(v)\setminus V(T) = \{x\}$), and moreover $|N_G(x)\setminus V(T)| = 0$ or $|N_G(x)\cap V(T)| \geq 2$. Then v is expanded.

(O3) Applies when there is a leaf $v \in V(T)$ such that $|N_G(v) \setminus V(T)| = 1$ (let $N_G(v) \setminus V(T) = \{x\}$), and moreover $|N_G(x) \setminus V(T)| \geq 2$. Then v is expanded and afterwards x is expanded.

Now we can describe the algorithm for Theorem 1, which we call GENERIC:

1. choose an arbitrary vertex $r \in V$ and let $T = \{r\}$,
2. apply O1-O3 as long as possible, giving precedence to O1.

We claim that GENERIC returns a spanning tree of G. Assume for a contradiction that at some point the algorithm is able to apply none of O1-O3 but $V(G) \neq V(T)$. Consider any leaf $v \in T$ such that $N_G(v) \not\subseteq V(T)$. Such a leaf exists because $V(G) \neq V(T)$ and T is built by a sequence of expansions . Since O1 does not apply, $|N_G(v) \setminus V(T)| = 1$. Let $N_G(v) \setminus V(T) = \{x\}$. Since O2 does not apply, $N_G(x) \cap V(T) = \{v\}$. Since neither O2 nor O3 apply, $|N_G(x) \setminus V(T)| = 1$. It follows that $\deg_G(x) = 2$. Moreover, since there are no edges between 2-vertices, the neighbor of x outside of T is not a 2-vertex. It follows that $\bigcup_{v \in L(T)} N_G(v) \setminus V(T)$ is a separator consisting of 2-vertices, which is the desired contradiction.

It remains to show that if a spanning tree T was constructed, then it has at least $n/4$ leaves. It can be done exactly as in the work of Kleitman and West [7]. However, we do it in a different way, in order to introduce and get used to some notation that will be used in later sections, where we describe an improved kernel.

We say that a leaf u of T is *dead* if $N_G(u) \setminus V(T) = \emptyset$. Note that after performing O2 there is at least one new dead leaf: if $|N_G(x) \setminus V(T)| = 0$ then x is a dead leaf, and if $|N_G(x) \cap V(T)| \geq 2$ then all of $(N_G(x) \cap V(T)) \setminus \{v\}$ are dead leaves, because of O1 precedence. For any tree $\hat{T}$, by $L(\hat{T})$ we denote the set of leaves of $\hat{T}$.

Let X_i be the set of the inner vertices of T that were expanded by an operation of type Oi. Let X be the set of the inner vertices of T; note that $X = X_1 \cup X_2 \cup X_3$. Since T is rooted, the standard notions of parent and children apply. For a positive integer i, let P_i denote the set of vertices of T with exactly i children.

Since every vertex besides r is a child of some vertex, we have $\sum_{d \geq 1} d|P_d| = n - 1$. Since the set of vertices with one child is equal to $X_2 \cup (X_3 \cap P_1)$ and $|X_3 \cap P_1| = |X_3 \cap P_{\geq 2}|$ it follows that

$$|X_2| + |X_3 \cap P_{\geq 2}| + \sum_{d \geq 2} d|P_d| = n - 1. \tag{1}$$

Since during O2 at least one leaf dies, $|X_2| \leq |L(T)|$. Similarly, since after expanding a vertex from $X_3 \cap P_{\geq 2}$ the cardinality of $L(T)$ increases, $|X_3 \cap P_{\geq 2}| \leq |L(T)| - 1$. Finally, $\sum_{d \geq 2} d|P_d| \leq \sum_{d \geq 2} 2(d-1)|P_d| = 2(|L(T)| - 1)$. After plugging these three bounds to (1) we get $|L(T)| > n/4$, as required. This finishes the proof of Theorem 1.

2.3   Outerplanarity

Lemma 2. *If G is a planar graph and T is a spanning tree of G, then the graph $G[L(T)]$ is outerplanar.*

Proof. Fix a plane embedding of G and consider the induced plane subgraph $G' = G[L(T)]$. Since T is connected, all vertices of $L(T)$ lie on the same face of G'. Therefore G' is outerplanar.   □

Corollary 2. *If G is a planar graph and T is a spanning tree of G then there is a subset of leaves of T of size at least $|L(T)|/3$ that is independent in G (and it can be found in linear time).*

Proof. It is well-known that outerplanar graphs are 3-colorable and the 3-coloring can be found in linear time. So, by Lemma 2 we can 3-color $G[L(T)]$ and we choose the largest color class.   □

3   A 9k Kernel

In this section we present an improved kernel for the MAXIMUM INDEPENDENT LEAF SPANNING TREE problem. Although the analysis is considerably more involved than that of the $12k$ kernel, the algorithm is almost the same. We need only to force a certain order of the operations O1-O3 in step 2. As before, the algorithm always performs the O1 operation if possible (we will refer to this as *the O1 rule*). Moreover, if more than one O1 operation applies then we choose the one which maximizes the number of vertices added to T (we will refer to this as *the largest branching rule*). If there is still more than one such operation applicable then among them we choose the one which expands a vertex that was added to T later than the vertices which would be expanded by other operation (we will refer to this as *the DFS rule*). Similarly, if there are no O1 operations applicable but more than one O2/O3 operations apply, we also use the DFS rule. The algorithm GENERIC with the order of operations described above will be called BRANCHING.

Note that the algorithm BRANCHING is just a special case of GENERIC, so all the claims we proved in Section 2 apply. Let us think where the bottleneck in this analysis is. There are two sources of trouble: first, if there are many O2/O3 operations we get a spanning tree with few leaves: in particular there might be only O2 operations and O3 operations that add just two leaves (consider a cubic graph which can be built by joining a number of diamonds by edges to form a cycle) and we get roughly $n/4$ leaves. Second, if the outerplanar graph $G[L(T)]$ is far from being bipartite (i.e. has many short odd cycles) then we get a small independent set: in particular, when $G[L(T)]$ is a collection of disjoint triangles, the maximum independent set in $G[L(T)]$ is of size exactly $|L(T)|/3$. However, we will show that these two extremes cannot happen simultaneously. More precisely, we prove the following two theorems.

Theorem 2. *Let G be a connected n-vertex graph that does not contain a separator consisting of only 2-vertices. Then G has a spanning tree T such that if $\mathcal{C}$ is a collection of vertex-disjoint cycles in $G[L(T)]$, then*

$$|L(T)| \geq \frac{n + 3|\mathcal{C}|}{4}.$$

Moreover, T can be found in linear time.

Theorem 3. *Every n-vertex outerplanar graph contains*

- *an independent set* I, *and*
- *a collection of vertex-disjoint cycles* $\mathcal{C}$

such that $9|I| \geq 4n - 3|\mathcal{C}|$.

Note that Theorem 3 is tight, which is easy to see by considering an outerplanar graph consisting of disjoint triangles. From the above two theorems and Lemma 2 we easily get the following corollary.

Corollary 3. *Let G be a connected n-vertex graph that does not contain a separator consisting of only 2-vertices. Then G has a spanning tree T such that $L(T)$ has a subset of size at least $n/9$ which is independent in G.*

By a similar reasoning as in the beginning of Section 2 we get a $9k$-kernel for the MAXIMUM INDEPENDENT LEAF SPANNING TREE problem. In what follows, we prove Theorem 2. Because of the space constraints the proof of Theorem 3 is deferred to the journal version.

Proof of Theorem 2

Note that similarly as in Theorem 1 it suffices to show a simplified case when G has no edge uv such that both u and v are 2-vertices. Indeed, if the theorem holds for the simplified case, as before we create a new graph $\tilde{G}$ by removing from G all the edges with both endpoints of degree 2 and as before $\tilde{G}$ does not contain a separator consisting of only 2-vertices. Then we apply the simplified case and we get a spanning tree T such that for any collection $\mathcal{C}$ of vertex-disjoint cycles in $\tilde{G}[L(T)]$, we have $|L(T)| \geq (|V(\tilde{G})| + 3|\mathcal{C}|)/4$ and T is a spanning tree of G as well. Moreover, no edge of $E(\tilde{G}) \setminus E(G)$ belongs to a cycle in $G[L(T)]$ for otherwise both of its endpoints have degree at least 3 in G. Hence if $\mathcal{C}$ is a collection of vertex-disjoint cycles in $G[L(T)]$ then it is also a collection of vertex-disjoint cycles in $\tilde{G}[L(T)]$, so the desired inequality holds. Hence in what follows we assume that G has no edge with both endpoints of degree 2. Since we proved that in this case the algorithm GENERIC returns a spanning tree, and each execution of BRANCHING is just a special case of an execution of GENERIC we infer that BRANCHING returns a spanning tree of G, which will be denoted by T.

Let $\mathcal{C}$ be an arbitrary collection of vertex-disjoint cycles in $G[L(T)]$. Our general plan for proving the claim of Theorem 2 is to show that if $|\mathcal{C}|$ is large then we have few O2/O3 operations — by (1) this will improve our bound on $|L(T)|$. To be more precise, let us introduce several definitions.

Recall the O2 operation: it adds a single vertex x to T and at least one leaf of T dies. We choose exactly one of these dead leaves and we *assign* it to x. However, if the vertex x dies we always assign x to itself (so if some other leaves die during this operation, they are unassigned). Let L_u be the set of unassigned leaves of T. Clearly, $|X_2| = |L(T)| - |L_u|$. In order to show that there are few O2 operations, we will show that $|L_u|$ is big.

Let $x_1, x_2, \ldots x_{|X|}$ be the inner vertices of T in the order of expanding them (in particular $x_1 = r$). A *run* is a maximal subsequence $x_b, x_{b+1}, \ldots, x_e$ of vertices from $P_{\geq 2}$, i.e. the nodes in T that have at least two children.

Lemma 3. *Vertices of any run* $R = x_b, \ldots, x_e$ *form a subtree of* T *rooted at* x_b.

Proof. Assume that a vertex $x \in \{x_{b+1}, \ldots, x_e\}$ has the parent x_p outside the run. Then $p < b$ and x was a leaf in T while x_b was being expanded. Hence, by the definition of a run, $x_{b-1} \in P_1$, and in particular x_{b-1} was expanded by O2 or O3. However, when this operation was performed, it was possible to expand x by O1, a contradiction with the O1 rule. Hence every vertex $x \in \{x_{b+1}, \ldots, x_e\}$ has the parent in the run, which is equivalent to the claim of the lemma.   □

In what follows, the subtree from Lemma 3 is denoted by T_R. Moreover, let $\text{ch}(T_R)$ denote the set of children of the leaves of T_R, i.e.

$$\text{ch}(T_R) = \{v \in V(T) \setminus V(T_R) \; : \; T_R \text{ contains the parent of } v\}.$$

We say that a run R *opens* a cycle C in $\mathcal{C}$ if the first vertex of C that was added to T belongs to $\text{ch}(T_R)$. The following lemma shows a relation between cycles in $\mathcal{C}$ and runs.

Lemma 4. *Every cycle in* $\mathcal{C}$ *is opened by some run.*

Proof. Consider any cycle $C \in \mathcal{C}$ and let v be the first vertex of C that is added to T. Note that v is not added by O2, for otherwise just after adding v to T, v has at least two neighbors and by the O1 rule v would be the next vertex expanded and hence not a leaf of T, a contradiction. It follows that v is added by O1 or O3 and consequently $v \in \text{ch}(T_R)$ for some run R.   □

Now we can sketch our idea for bounding the number of O3 operations (#O3). Both after O1 and O3 the cardinality of $L(T)$ increases. Hence, if we fix the number of leaves in the final tree, then if $|X_1|$ is large then #O3 should be small. Since a run contains at most one vertex of $|X_3|$ (e.g. by Lemma 3), it means that a tree T_R with a large number of children contains plenty of vertices from $|X_1|$. We will show that if a run opens many cycles, then indeed $|\text{ch}(T_R)|$ is large. Let $\mathcal{C}_R$ denote the set of cycles in $\mathcal{C}$ opened by R.

Lemma 5. *Let* R *be a run. For any cycle* $C \in \mathcal{C}_R$ *one of the following conditions holds:*

    (i) $|\text{ch}(T_R) \cap V(C)| + |L_u \cap V(C)| \geq 4$, *or*
    (ii) $|\text{ch}(T_R) \cap V(C)| + |L_u \cap V(C)| = 3$ *and* $|R \cap P_{\geq 3}| \geq 1$.

Proof. Let v_1 be the vertex of C that is added first to the tree T. By the definition of $\mathcal{C}_R$, $v_1 \in \text{ch}(T_R)$. We see that at least one neighbor of v_1, call it v_2, is in $\text{ch}(T_R)$, for otherwise just after expanding the last vertex of R the vertex v_1 can be expanded by O1, while the algorithm chooses O2/O3, a contradiction with the O1 rule.

Let w be the neighbor of v_2 on C that is distinct from v_1. Assume $w \notin \text{ch}(T_R)$. Then just after expanding the last vertex of R we have $N(v_2) \setminus T = \{w\}$, since if $|N(v_2) \setminus T| \geq 2$ then it is possible to expand v_2 by O1. Hence if v_2 is assigned then it is assigned to w. However, then w is added to T by O2 so w dies during this operation (otherwise w is expanded because of the DFS rule so $w \notin L(T)$), and hence w is assigned to w and $v_2 \in L_u$. To conclude, $w \in \mathrm{ch}(T_R)$ or $v_2 \in L_u$.

If we denote by u the neighbor of v_1 on C that is distinct from v_2, by the same argument we get $u \in \mathrm{ch}(T_R)$ or $v_1 \in L_u$.

It follows that (i) holds, unless $u = w$ (i.e. C is a triangle), $v_1, v_2 \notin L_u$ and $w \in \mathrm{ch}(T_R)$. Let us investigate this last case. We see that $|\mathrm{ch}(T_R) \cap V(C)| = 3$. We will show that $|R \cap P_{\geq 3}| \geq 1$. Since $v_2, w \in \mathrm{ch}(T_R)$, they could not be added by O2 and hence v_1 is assigned to a vertex $x \notin V(C)$. Note that x is added to T after v_2 and w. Assume w.l.o.g. that v_2 was added to T before w. We consider two cases. If v_2 was not added to T by expanding the parent of v_1, then the parent p of v_2 has at least three children (otherwise instead of expanding p the algorithm can expands v_1 and add at least three children, a contradiction with the largest branching rule), so $|R \cap P_{\geq 3}| \geq 1$ as required. Finally, if v_2 was added to T by expanding the parent p of v_1, then $p \in P_{\geq 3}$ for otherwise just after expanding p O1 is applicable to v_1 so either v_1 or v_2 is expanded by the DFS rule. This concludes the proof. □

By applying Lemma 5 to all cycles of a single run R we get the following corollary.

Corollary 4. *For any run R that opens at least one cycle,*

$$|\mathrm{ch}(T_R) \cap V(\mathcal{C}_R)| + |L_u \cap V(\mathcal{C}_R)| + |R \cap P_{\geq 3}| \geq 3|\mathcal{C}_R| + 1.$$

Lemma 6. *For any run R,*

$$|L_u \cap V(\mathcal{C}_R)| + \sum_{d \geq 2}(2d - 3)|R \cap P_d| - |R \cap X_3| \geq 3|\mathcal{C}_R|. \tag{2}$$

Because of the space limitations the proof of Lemma 6 is deferred to the journal version. Now we are ready to prove the claim of Theorem 2, i.e. that $|L(T)| \geq (n + 3|\mathcal{C}|)/4$.

Let us add $\sum_{d \geq 2}(2d - 3)|P_d|$ to both sides of (1):

$$|X_2| + |X_3 \cap P_{\geq 2}| + 3\sum_{d \geq 2}(d - 1)|P_d| = n - 1 + \sum_{d \geq 2}(2d - 3)|P_d|. \tag{3}$$

Since $|X_2| = |L(T)| - |L_u|$ and $\sum_{d \geq 2}(d - 1)|P_d| = |L(T)| - 1$ we get

$$4|L(T)| = n + 2 + |L_u| + \sum_{d \geq 2}(2d - 3)|P_d| - |X_3 \cap P_{\geq 2}|. \tag{4}$$

By Lemma 4 after summing (2) over all runs we get

$$|L_u| + \sum_{d \geq 2}(2d - 3)|P_d| - |X_3 \cap P_{\geq 2}| \geq 3|\mathcal{C}|. \tag{5}$$

The claim follows immediately after plugging (5) to (4).

4   A Simple 5k Kernel for (PLANAR) MAX LEAF

In this section we show a simple kernelization algorithm for MAX LEAF and PLANAR MAX LEAF. Below we describe three simple rules, which preserve planarity.

- **(1, 2)-rule** If there is a 1-vertex u adjacent with a 2-vertex v then remove v.
- **Adjacent 2-vertices Rule** Assume that there are two adjacent 2-vertices u and v. If uv is a bridge, contract uv, otherwise remove uv.
- **Trivial Rule** If G consists of a single edge, return YES if $k \leq 2$.

It is quite clear that the above rules are correct for (PLANAR) MAX LEAF (see e.g. [9], Rules 1-3 for a proof). Note that if none of our rules applies to a connected graph G, then every edge of G has an endpoint of degree at least 3.

Theorem 4. *Let G be a connected graph in which every edge has an endpoint of degree at least 3. Then G has a spanning tree with at least $n/5$ leaves.*

The proof of Theorem 4 is deferred to the journal version.

Let G' be the graph obtained from G by applying our three rules as long as one of them applies. By Theorem 4, if $k \leq n/5$ we can return the answer YES. Hence $n < 5k$ and G' is a 5k-kernel for PLANAR MAX LEAF and MAX LEAF.

Acknowledgments. We are very grateful for the reviewers for numerous comments. We also thank Michal Debski for helpful discussions.

References

1. Alber, J., Fellows, M.R., Niedermeier, R.: Polynomial-time data reduction for dominating set. J. ACM 51(3), 363–384 (2004)
2. Chen, J., Fernau, H., Kanj, I.A., Xia, G.: Parametric duality and kernelization: Lower bounds and upper bounds on kernel size. SIAM J. Comput. 37(4), 1077–1106 (2007)
3. Estivill-Castro, V., Fellows, M.R., Langston, M.A., Rosamond, F.A.: FPT is P-Time Extremal Structure I. In: ACiD 2005, pp. 1–41 (2005)
4. Fomin, F.V., Lokshtanov, D., Saurabh, S., Thilikos, D.M.: Bidimensionality and kernels. In: Charikar, M. (ed.) SODA, pp. 503–510. SIAM (2010)
5. Garey, M.R., Johnson, D.S.: The rectilinear steiner tree problem in NP complete. SIAM Journal of Applied Mathematics 32, 826–834 (1977)
6. Guo, J., Niedermeier, R.: Linear Problem Kernels for NP-Hard Problems on Planar Graphs. In: Arge, L., Cachin, C., Jurdziński, T., Tarlecki, A. (eds.) ICALP 2007. LNCS, vol. 4596, pp. 375–386. Springer, Heidelberg (2007)
7. Kleitman, D.J., West, D.B.: Spanning trees with many leaves. SIAM J. Discrete Math. 4(1), 99–106 (1991)
8. Kowalik, L., Pilipczuk, M., Suchan, K.: Towards optimal kernel for connected vertex cover in planar graphs. CoRR abs/1110.1964 (2011)
9. Prieto-Rodriguez, E.: Systematic kernelization in FPT algorithm design. Ph.D. thesis, University of Newcastle (2005)
10. Wang, J., Yang, Y., Guo, J., Chen, J.: Linear Problem Kernels for Planar Graph Problems with Small Distance Property. In: Murlak, F., Sankowski, P. (eds.) MFCS 2011. LNCS, vol. 6907, pp. 592–603. Springer, Heidelberg (2011)

Parameterized Algorithms
for Even Cycle Transversal

Pranabendu Misra[2], Venkatesh Raman[1], M.S. Ramanujan[1],
and Saket Saurabh[1]

[1] The Institute of Mathematical Sciences, Chennai, India
{vraman,msramanujan,saket}@imsc.res.in
[2] Chennai Mathematical Institute
pranabendu@cmi.ac.in

Abstract. We consider a decision version of the problem of finding the minimum number of vertices whose deletion results in a graph without even cycles. While this problem is a natural analogue of the Odd Cycle Transversal problem (which asks for a subset of vertices to delete to make the resulting graph bipartite), surprisingly this problem is not well studied. We first observe that this problem is NP-complete and give a constant factor approximation algorithm. Then we address the problem in parameterized complexity framework with the solution size k as a parameter. We give an algorithm running in time $O^*(2^{O(k)})$ for the problem and give an $O(k^2)$ vertex kernel. (We write $O^*(f(k))$ for a time complexity of the form $O(f(k)n^{O(1)})$, where $f(k)$ grows exponentially with k.)

1 Introduction

Cycle hitting set problems like Feedback Vertex Set and Odd Cycle Transversal (OCT) are very well studied in graph theory, algorithms and complexity. In the Feedback Vertex Set problem and the Odd Cycle Transversal problem, we are given a graph $G = (V, E)$ and a positive integer k as inputs. The objective in these two problems is to check if there exists a subset $S \subseteq V$ of size at most k such that $G - S$ does not have a cycle (i.e. it is a forest) and does not have an odd cycle (it is bipartite), respectively. Both these problems are well studied in the realm of parameterized and approximation algorithms. In this paper, we consider a natural analogue of the Odd Cycle Transversal, namely the Even Cycle Transversal (EvenCT) problem and study this problem in the realm of parameterized complexity.

Even Cycle Transversal (EvenCT)

Instance: An undirected graph $G = (V, E)$ and a positive integer k.
Parameter: k.
Problem: Does G have a set S of size at most k such that $G - S$ does not contain an even cycle?

M.C. Golumbic et al. (Eds.): WG 2012, LNCS 7551, pp. 172–183, 2012.

We start with some basic definitions in parameterized complexity. For a decision problem with an input of size n and a parameter k, the goal in parameterized complexity is to design an algorithm with a runtime $f(k)n^{O(1)}$ where f is a function of k alone. Problems which admit such algorithms are said to be *fixed parameter tractable* (FPT). A decision problem is said to have a *kernelization* algorithm, if there is a polynomial time algorithm that takes an instance (I, k) and returns an equivalent instance (I', k') such that, $k' \le k$ and $|I'| \le g(k)$, where g is a function of k alone. The theory of parameterized complexity was developed by Downey and Fellows [2]. For recent developments, see the book by Flum and Grohe [4].

The parameterized complexity of OCT was a long standing open problem, which was resolved in 2003 by Reed et al. [11], who developed an algorithm for the problem running in time $O^*(3^k)$. However, there has been no further improvement over this algorithm in the last 9 years though several reinterpretations of the algorithm have been published [7,9]. Only recently, an algorithm with a running time $O^*(2.32^k)$ has been obtained [10]. Thus, it is rather surprising that while OCT has attracted so much attention, EVENCT has so far been largely been ignored. Recently, Kakimura et al. [8] studied a generalization of EVENCT, called SUBSET EVEN CYCLE TRANSVERSAL. In this problem, apart from a graph $G = (V, E)$ and a positive integer k, we are also given a $T \subseteq V$ as the input. The objective is to determine whether there exists a vertex set $S \subseteq V$ of size at most k such that $G - S$ does not contain any even cycle whose intersection with T is non-empty. Observe that for $T = V$ this is precisely EVENCT. Their paper [8], shows that SUBSET EVEN CYCLE TRANSVERSAL is FPT, and thus EVENCT is also FPT. However this algorithm utilizes graph minor machinery, and the dependence of this algorithm on the parameter k is at least triply exponential. Hence, a second motivation is to find out if the special case EVENCT has a better FPT algorithm.

In this paper, we do a systematic study of EVENCT and obtain the following results.

- We show that EVENCT can be solved in time $2^{O(k)}n^{O(1)}$. To this end, we study the class of graphs $\mathcal{G}_e$, which is all graphs which do not contain even cycles as subgraphs. These graphs have been studied by Thomassen [13], who obtained a more general result that if a graph G does not contain cycles of length 0 modulo p for some fixed integer p then the treewidth of G is upper bounded by a function of p. This implies that the treewidth of a graph which does not contain even cycles is bounded by some constant. We show that, in fact $\mathcal{G}_e$ is even more structured. Any pair of two odd cycles in a graph from this class may intersect in at most one vertex. These kind of graphs are known as cactus graphs and their treewidth at most 2.

- We also show that EVENCT admits a quadratic kernel. That is, we obtain a polynomial time algorithm that, given an instance (G, k) of EVENCT returns an equivalent instance $(G' = (V', E'), k')$ such that $k' \le k$ and $|V(G')| + |E(G')| = O(k^2)$. The kernelization algorithm is along similar lines as earlier kernelization algorithms for FEEDBACK VERTEX SET [12] and the problem of Θ_c-DELETION [5].

This problem adds to the growing list of parity problems that are studied from the point of view of parameterized complexity. We also mention in passing that EVENCT is NP-complete and show that it has an approximation algorithm with factor 10.

2   Preliminaries

Let $G = (V, E)$ be an undirected graph. A *cut vertex* is a vertex v such that $G - \{v\}$ contains two or more connected components than G does. A *block* in a graph is a maximal connected subgraph without a cut vertex. Thus blocks in a graph G are either an isolated vertex, an edge or a maximally-2-connected component. A *pendant block* is a block such that it has only one cut vertex. Removing a pendant block from a graph means, deleting all the vertices in that block except the cut vertex.

A *cactus graph* is a graph where each edge is part of at most one cycle. Equivalently a graph is a cactus graph if and only if each of its block is isomorphic to either K_1, K_2 or a cycle. An *odd cactus graph* is a cactus graph without any even cycles.

Lemma 1. [*][3] *If there are two cycles C_1 and C_2 in G such that C_1 and C_2 intersect in at least 2 vertices, then there is an even cycle in G.*

A set of vertices $S \subseteq V(G)$ is called an *ect* if $G - S$ does not contain an even cycle.

Lemma 2. *Let G be a graph and S be an ect of G. Then $(G - S)$ is an odd cactus graph.*

Proof. Lemma 1 implies that any pair of cycles in $(G - S)$ intersect in at most 1 vertex. Hence, $(G - S)$ is a cactus graph. Additionally, since $(G - S)$ excludes even cycles, it is an odd cactus graph.

3   NP-Completeness and Constant Factor Approximation

3.1   NP-Completeness

By a simple reduction from the NP-Complete FEEDBACK VERTEX SET problem, the NP-hardness of the problem is clear.

Theorem 1. [*] *The EVENCT problem is NP-Complete.*

---

[3] The proofs of results marked [*] will appear in the full version of the paper.

3.2   An Approximation Algorithm

We will show that EVENCT has a 10-approximation algorithm. First we prove the following easy lemma.

Lemma 3. [*] EVENCT *on cactus graphs is solvable in polynomial time.*

The Diamond Hitting Problem. The *diamond hitting problem* is defined as follows. Given a graph G, is there a subset S of at most k vertices of G such that, $(G - S)$ is a cactus graph. We need the following result by Fiorini et al.

Lemma 4 ([3], Theorem 6.6). *There is a factor 9 approximation algorithm for the diamond hitting problem.*

Using Lemma 3 and Lemma 4, we prove the following theorem.

Theorem 2. *There is a factor 10 approximation algorithm for* EVENCT.

Proof. Let G be the input graph. Let S be an optimum ect of G. Since $(G - S)$ is a cactus graph, S is also a solution to the diamond hitting problem on G. Hence, if T were an optimum solution to the diamond hitting problem on G, then $|T| \leq |S|$. We first apply Lemma 4 to obtain a factor 9 approximation to the diamond hitting problem on G. Let S_1 be the approximate solution obtained. Observe that $|S_1| \leq 9|S|$.

Since $H = G - S_1$, is a cactus graph, we can apply Lemma 3 to obtain an optimum ect S_2 for H in polynomial time. Since H is a subgraph of G, $|S_2| \leq |S|$. Therefore, the set $S_1 \cup S_2$ which is clearly an ect for G, has size at most $10|S|$.

This algorithm forms a critical part of the kernelization procedure in Section 5.

4   FPT Algorithm

In this section we give an algorithm for EVENCT which runs in time $O^*(2^{O(k)})$. Towards this, we first apply the technique of iterated compression. The idea is to start with a solution of size $k + 1$ and try to 'compress' it to a solution of size k if possible.

We start with the compression version of the problem.

4.1   Compression Version

COMPRESSION EVENCT
Input: Graph G, a positive integer k and an ect S of G such that $|S| = k + 1$
Parameter: k
Question: Does G have an ect of size at most k ?

We fix an ect of G of size at most k (assuming one exists) and denote it by S^*. Let $Y = S \cap S^*$ and $N = S - Y$. We attempt to find an ect of the graph $(G - S) \cup N$ such that it has size at most $k - |Y|$ and is contained in $(G - S)$. Since the ect has to be disjoint from the set N, the graph N cannot contain even cycles and hence must be a cactus graph.

We give a branching algorithm to find the ect S^*. This algorithm works in two phases. In the first phase, we use an appropriate branching, at the end of which, every connected component C in $G - S$ is such that $C \cup N$ has no even cycles and all the edges from C to N are incident on the same connected component of N. In the second phase, we use the structure provided by the first phase to deal with the remaining even cycles in $(G - S) \cup N$.

R ecall that $G - S$ is an odd cactus graph. Let C be a connected component of $G - S$ such that either $C \cup N$ contains an even cycle, or there are at least 2 edges which have one end point in C, and the other end point in distinct components of N. In both cases, our aim is to find a 'minimal' path in C that satisfies the property (of forming an even cycle with N or having adjacency in more than two components of N). This will then suggest ways of branching (by picking vertices from this 'minimal' path).

Let P be any path in C and let r be any vertex of C. Let $v \in P$ be a vertex such that there is a path from r to v, which intersects P only at v, and let $R_P(r)$ be the set of all such vertices in P.

Lemma 5. *For any vertex r and any path P in C, $|R_P(r)| \leq 2$.*

Proof. Suppose that there are 3 vertices $\{u, v, w\}$ in $R_P(r)$. By definition of $R_P(r)$, there is a path from r to u which doesn't intersect $\{v, w\}$; similarly for v and w. Let T_1 be a minimal subtree of $(G - P) \cup R_P(r)$, that contains $\{u, v, w, r\}$, such that $\{u, v, w\}$ is the set of leaves of T_1. Let T_2 be a minimal sub-path of P which connects $\{u, v, w\}$. Then in $T_1 \cup T_2$, there are two cycles that have two or more common vertices. But then by Lemma 1, there is an even cycle in $(G - S)$, a contradiction.

Observation 1. *Given a vertex r and a path P in a component C of $G - S$, r is not reachable from any vertex in $P - R_P(r)$.*

Now, to identify a path with the required property, we define a partial order on all the paths in the component. Towards this end, we fix a root r_C for the component and measure distances of the paths from r_C. Let $dist(u, v)$ be the length of a shortest path between the two vertices u and v. Suppose for P $|R_P(r_C)| = 2$, then the *depth* of P, denoted by $d(P)$, is defined as the ordered pair $(dist(u, r_C), dist(v, r_C))$, where $R_P(r_C) = \{u, v\}$ and $dist(u, r_C) \geq dist(v, r_C)$. If $|R_P(r_C)| = 1$, then $d(P) = (dist(u, r_C), dist(u, r_C))$. As it will be clear from the context, we drop the symbol r_C from $R_P(r_C)$ hereafter. Given two paths P_1 and P_2, we say that P_1 is *smaller* than P_2 if either $d(P_1) > d(P_2)$ in the lexicographic order or, $d(P_1) = d(P_2)$ and P_1 is a sub-path of P_2. This relation induces a partial order on the set of all paths of a component C.

In the rest of the paper, by $P \cup N$ we refer to the graph obtained by considering the path P, the graph induced on N and all the edges between the vertices in P and the vertices in N. We call a path P a *branching path* if either $P \cup N$ contains an even cycle, or vertices in P have adjacencies to more than one connected component of N. Note that in the latter case, $P \cup N$ has fewer connected components than N has. By a *minimal branching path*, we mean a minimal element in the partial order defined above.

The following lemma shows that we can compute a branching path in polynomial time.

Lemma 6. [*] *Let H be a connected odd cactus graph such that $H \cup N$ either contains an even cycle or $H \cup N$ has fewer connected components than N has. Then we can find a path P of H in polynomial time such that P is a branching path.*

The next lemma shows that we can compute a minimal branching path in polynomial time. Furthermore, such a path Q has the crucial property that there is a subset of vertices A_Q in Q such that $|A_Q| \leq 6$ and there exists an optimum ect that either doesn't intersect Q or intersects only in a subset of A_Q. We call A_Q the set of important vertices in Q.

Lemma 7. [*] *Suppose P is a branching path in a connected component C of $G-S$. Then there is a branching path Q and a subset of its vertices A_Q containing at most 6 vertices such that, if any vertex v in $(Q - A_Q)$ is in some solution T, then there is a solution T' of size at most $|T|$ which doesn't contain v. Given P, we can find Q and A_Q in polynomial time.*

Using the above two lemmas we show the following,

Lemma 8. COMPRESSION EVENCT *is fixed-parameter tractable.*

Proof. We are given as input G and a solution S of size at most $k + 1$. We wish to construct a new solution S^* of size at most k. We consider partitions of S into two sets Y and N, where $Y = S^* \cap S$ and $N = S - Y$.

For each such partition we delete Y from G. We then attempt to find an ect R of $(G - S) \cup N$ such that $R \subseteq V(G - S)$ and $|R| \leq k - |Y|$. For such a solution to exist, we require N to be an odd cactus graph. This can be easily checked by taking a block decomposition of N and checking for blocks not isomorphic to K_1, K_2 or an odd cycle.

To find an ect of $(G - S) \cup N$ we use a branching algorithm which we will analyze with the measure

$$\mu = (k - |Y| + \textit{number of components in } N.)$$

It is easy to see that $\mu \leq 2(k - |Y|) + 1$. During the branching we update $(G - S)$ and N by deleting vertices or moving vertices from $(G - S)$ to N.

We first apply the following preprocessing rule. Let C be a connected component of $G - S$ such that there is at most one edge with one endpoint in C and the other in N. Then, no vertex of C is part of an even cycle in $(G - S) \cup N$, and hence we delete C from $G - S$.

Let C be a connected component of $G - S$ such that $C \cup N$ has an even cycle or has fewer connected components than N has. We apply Lemma 6 to find a branching path Q. We then apply Lemma 7 on Q to find a minimal branching path P and A_P and branch on the vertices in A_P. Since $|A_P| \leq 6$, there are at most 7 branches, the first six branches each correspond to one of the vertices in A_p being added to the solution R and in the final branch, no vertex is chosen and we move the entire path P to N. Note that picking no vertex from P may not be feasible if $P \cup N$ contains an even cycle. When we add a vertex into the solution set R and delete it from the graph, μ decreases by 1. Also, since P was a branching path, moving P to N, reduces the number of connected components in N by at least 1. Hence, in every branch the measure μ drops by at least 1.

We do the above branching for every component in $(G - S)$. Eventually we'll reach a state where no component of $G - S$ satisfies the conditions of Lemma 6. Hence all the edges from each component are to the same connected component of N. Furthermore, if more than two edges from a component C are adjacent to a component of N, then it can be easily argued that there is an even cycle in $C \cup N$. Hence, every component of $G - S$ has edges to exactly one component of N and exactly two such edges. Let B_i be the set of vertices in a component C_i of $G - S$ such that they have a neighbour in N. Note that $|B_i| \leq 2$. We will show the following,

Claim. There exists an optimum solution to the compression EVENCT instance that, for any component C_i of $G - S$, is either disjoint from C_i or intersects C_i in a subset of B_i.

Proof. The claim follows from the observation that every even cycle intersecting C_i must also intersect b_1 and b_2, where b_1 and b_2 are the vertices in B_i.

For a pair of connected components C_i, C_j of $(G - S)$, let $H_{ij} = C_i \cup C_j \cup N$. We can check if H_{ij} contains an even cycle. If we find one, by the above claim we have a four-way branching on the vertices in $B_i \cup B_j$ and in each branch we add a vertex to R and delete it from the graph. In every branch μ drops by 1. We do this for every pair of connected components in $(G - S)$. Finally, let $(G - S) \cup N$ be the remaining graph. We will show that there are no even cycles left in this graph.

Claim. [∗] If there is a cycle in $(G - S) \cup N$ passing through some $l > 2$ connected components of $(G - S)$, then there is a cycle passing through at most $l - 1$ components of $G - S$.

Claim. [∗] If there is a cycle in $(G - S) \cup N$ passing through two connected components C_1, C_2 of $(G - S)$, then there is an even cycle in $C_1 \cup C_2 \cup N$.

The above two claims imply that $(G - S) \cup N$ now has no even cycles. For, if there was an even cycle, then it must pass through some l connected components of $G - S$. If $l = 1$, then there would be a connected component C of $G - S$ such that $C \cup N$ contains an even cycle, a contradiction. Otherwise if $l \geq 3$, then we apply the first claim repeatedly to conclude that there are two components C_1 and C_2 such that $C_1 \cup C_2 \cup N$ contains a cycle which passes through both C_1 and C_2. Then by the second claim, there is an even cycle in $C_1 \cup C_2 \cup N$, a contradiction. Hence, we conclude that at the end of the above branching procedure we have constructed a solution R and the remaining graph is even cycle free.

For a partition of S into Y and N, the above branching takes time $O^*(7^{2(k-|Y|)})$. Summing over all the 2^{k+1} partitions of S we see that the overall algorithm takes time $\sum_{i=0}^{k} \binom{k+1}{i} O^*(7^{2k-2i}) = O^*(50^k)$.

Finally the following claim proves the correctness of the algorithm.

Claim. [*] Suppose S' were some solution to this EVENCT compression instance of size at most k. Let S^* be the solution constructed by the above algorithm. Then $|S^*| \leq k$.

4.2   FPT Algorithm for EVENCT

Theorem 3. EVENCT *parameterized by solution size has an* **FPT** *algorithm running in time* $O^*(50^k)$.

Proof. Suppose G is the input graph and k is the parameter. We arbitrarily order the vertices of G and define a sequence of graphs $G_{k+2}, G_{k+3}, \ldots, G_n$, where G_i is the induced subgraph on the first i vertices of G.

It is easy to see that if for some i, G_i doesn't have an ect of size at most k, then G doesn't have an ect of size at most k. And if G has an ect of size k, then each G_i has an ect of size at most k. We start with the graph G_{k+2} and let S_{k+2} be the first $k + 1$ vertices. Clearly S_{k+2} is an ect of size $k + 1$ for G_{k+2}. We use Lemma 8 to compute a solution S of size at most k for G_{k+2}. If we fail, then there is no ect of size k for G and so we answer NO. And if we succeed then $S_{k+3} = S \cup \{v_{k+3}\}$ is an ect of size $k + 1$ for G_{k+3}. We then repeat the above process.

Note that there are at most $O(n)$ iterations and each iteration takes time $O^*(50^k)$. So this algorithm runs in time $O^*(50^k)$.

5   Polynomial Kernel

In this section we give a quadratic kernel for EVENCT. We first introduce reduction rules which allow us to establish a bound on the maximum degree of the graph. During this process we may introduce parallel edges in the graph, which we can remove via a simple preprocessing. Following that we use the bounded degree of the graph to design further reduction rules, to obtain the kernel.

5.1   Bounding the Degree of the Graph

Definition 1. *For a vertex v, a set of even cycles is called a* flower *passing through v if the even cycles intersect only at v. If the number of even cycles in the set is t, then this set is called a t-flower passing through v. Each even cycle of a flower is called a* petal.

We will now use an approach similar to that in [5] and [12] to design reduction rules to eliminate vertices of high degree. We begin by proving a few lemmas which will be used crucially in designing a reduction rule that removes high degree vertices.

Lemma 9. [*] *Given a graph G such that $tw(G) = r$ for some constant r, and a vertex v in G, we can find in linear time, the size of the largest flower passing through v in G.*

Lemma 10. [*] *Given a graph $G = (V, E)$ let $v \in V$ be a vertex such that $G - \{v\}$ is an odd cactus graph, and let k be an integer. Then, in polynomial time, we can either compute an ect X of size $O(k)$ in G such that X is disjoint from v or conclude that there is a $k + 1$-flower passing through v.*

Lemma 11. [*] *There is a polynomial time procedure, which, for every vertex v in the graph, either returns a set S_v such that $v \notin S_v$, $|S_v| = O(k)$ and S_v is a solution or conclude that there is a $k + 1$-flower passing through v.*

We now move ahead to the description of the reduction rules.

Reduction 1. *If there are more than 2 parallel edges between a pair of vertices, delete all but 2 of them.*

Reduction 2. *For every vertex v, apply Lemma 11. If some vertex has a $(k + 1)$-flower passing through it, then delete this vertex from the graph and reduce k by 1.*

Correctness. The correctness follows from the fact that such a vertex must be part of every solution of size at most k.

Definition 2. *For a vertex v, let $C_1^v, \ldots, C_r^v$ be the set of connected components of $G - (S_v \cup \{v\})$ which are adjacent to v in G.*

Observation 2. *A vertex v cannot have more than 2 neighbors in any component C_i^v.*

Proof. The proof follows from Lemma 1.

Reduction 3. *If there is a component C_i^v which does not have an edge to S_v, then we delete the vertices in this component.*

Correctness. The correctness follows from the fact the vertices of this component cannot be part of an even cycle in the given graph.

Definition 3. *Define a graph $G_v = (A, B, E)$ to be the bipartite graph where the vertices in A correspond to the components $C_1^v, \ldots, C_r^v$, and the vertices in B correspond to the vertices in S_v. We add an edge between two vertices $a_i \in A$ and $u \in B$ if there is an edge from the component C_i^r to the vertex u.*

We now state without proof (a slightly weaker form of) the well known q-expansion lemma which we will use to define the high degree vertices. This lemma is a generalization of a lemma in [12]. A q-star is defined as a star with q leaves. We say that a vertex v has a q-star in the vertex set T if v is the center of a q-star and the leaves of the star all lie in T.

Lemma 12. *Let q be a positive integer and let G be a bipartite graph with vertex bipartition $A \uplus B$. If there are no isolated vertices in B and $|B| > q|A|$, then there are non-empty vertex sets $S \subseteq A$ and $T \subseteq B$ such that S has $|S|$ many vertex disjoint q-stars in T and S separates the vertices of T from the rest of the graph. Furthermore, the sets S and T can be computed in polynomial time.*

Reduction 4. *If $deg(v) \geq 8|S_v|$, then compute a 3-expansion set $X \subseteq B$ in the graph $G_v = (A, B, E)$. Delete the components corresponding to the neighbors of X, and add parallel edges between v and u for every vertex $u \in X$.*

Note that, in each application of this rule, the degree of v decreases by at least $|X| > 0$. When the Rule 4 cannot be applied on any vertex in the graph, then the maximum degree of the graph is bounded by $O(k)$. Further note that because of Rule 1, there can be at most 2 parallel edges between any two vertices u and v. For the sake of convenience, we would like to deal with simple graphs and hence we do the following. For each pair of vertices u and v in G with 2 parallel edges, replace one of the edges by a path (u, a, b, v) of length 3 where a and b are two new vertices. It is easy to see that this process preserves the bound on the degree and the solution size, while resulting in a simple graph.

5.2 Bounding an Irreducible YES Instance

Now, we prove a bound on the size of an irreducible YES instance. At this point, we have a graph G where every vertex has $O(k)$ neighbors, and an approximate solution A of size $O(k)$. Note that the number of edges going across the cut $(G - A, A)$ is bounded by $O(k^2)$. Since $G - A$ is an odd cactus graph, we can find a block decomposition of this graph [6] and the corresponding *block graph* T [1]. T is a forest which contains a vertex for every block and every cut-vertex in $G - A$. There is an edge between two vertices u and v, where u corresponds to a cut-vertex and v corresponds to a block B_v in $G - A$, if $u \in B_v$ in the graph $G - A$. We call a vertex of $G - A$ an *affected vertex* if it is adjacent to a vertex in A. We call a block of $G - A$ an *affected block* if it contains a cut vertex.

Reduction 5. *Delete any pendant block in $G - A$ which is not affected.*

Correctness. It follows from the observation that such blocks are never part of any even cycle in G.

Reduction 6. *If there is a maximal path P in $G - A$ such that every vertex is of degree two and is unaffected, except for the endpoints which could either be affected vertices or be vertices of degree three or more, then replace P by a path of length two (or one) if this path is of even (or odd) length.*

Correctness. It follows from the observation that we can substitute for any such vertex by one of the endpoints, and the replacement doesn't change the parity of any cycle in G.

Definition 4. *Consider a path P of degree two vertices in T such that no vertex in P corresponds to an affected vertex or an affected block in $G - A$. We call the corresponding sequence of blocks $B_1, \ldots B_l$ a chain in $G - A$.*

It has the property that, no vertex in $\cup_{i=1}^{l} B_i$ has a neighbour in A, and every block has exactly two cut vertices. We call the two cut vertices at the two ends of a chain, the endpoints of the chain.

Reduction 7. *If there is a chain such that it consists of a single block which is an odd cycle of size greater than three, or contains at least two blocks with one of them an odd cycle, then replace this chain with a triangle which is formed by the two endpoints of the chain and a new vertex.*
If the chain contains no odd cycles, then the chain is a path in $G - A$. Furthermore, suppose that this chain contains at least three blocks. Then replace this path by a path of length two(or one) if this path is of even (or odd) length.

Correctness. The correctness follows from the fact that any vertex of the chain which is part of the solution can be simply replaced by one of the endpoints of the chain. $\square$

Lemma 13. *The number of affected vertices in $G - A$ is $O(k^2)$ and the number of affected blocks is $O(k^2)$.*

Proof. The bound on the number of affected vertices follows from the fact that $|A| = O(k)$ and every vertex in A has a degree $O(k)$. This clearly implies the second statement.

The following two lemmas give a bound on the number of vertices in the block graph T, and then on the number of vertices in $G - A$.

Lemma 14. [*] *The number of vertices in the block graph is $O(k^2)$.*

Lemma 15. [*] *There are $O(k^2)$ vertices in $G - A$.*

Using the above lemma and the fact that $|A| = O(k)$ we have the following theorem.

Theorem 4. EVENCT *parameterized by the solution size has an $O(k^2)$ vertex kernel.*

References

1. Diestel, R.: Graph Theory, Graduate Texts in Mathematics, 3rd edn., vol. 173. Springer, Heidelberg (2005), http://www.math.uni-hamburg.de/home/diestel/books/graph.theory/GraphTheoryIII.pdf
2. Downey, R.G., Fellows, M.R.: Parameterized Complexity. Springer, New York (1999)
3. Fiorini, S., Joret, G., Pietropaoli, U.: Hitting Diamonds and Growing Cacti. In: Eisenbrand, F., Shepherd, F.B. (eds.) IPCO 2010. LNCS, vol. 6080, pp. 191–204. Springer, Heidelberg (2010)
4. Flum, J., Grohe, M.: Parameterized Complexity Theory. Texts in Theoretical Computer Science. An EATCS Series. Springer, Berlin (2006)
5. Fomin, F.V., Lokshtanov, D., Misra, N., Philip, G., Saurabh, S.: Hitting forbidden minors: Approximation and kernelization. In: STACS, pp. 189–200 (2011)
6. Hopcroft, J., Tarjan, R.: Algorithm 447: efficient algorithms for graph manipulation. Commun. ACM 16, 372–378 (1973), http://doi.acm.org/10.1145/362248.362272
7. Hüffner, F.: Algorithm engineering for optimal graph bipartization. J. Graph Algorithms Appl. 13(2), 77–98 (2009)
8. Kakimura, N., Ichi Kawarabayashi, K., Kobayashi, Y.: Erdös-pósa property and its algorithmic applications: parity constraints, subset feedback set, and subset packing. In: SODA, pp. 1726–1736 (2012)
9. Lokshtanov, D., Saurabh, S., Sikdar, S.: Simpler Parameterized Algorithm for OCT. In: Fiala, J., Kratochvíl, J., Miller, M. (eds.) IWOCA 2009. LNCS, vol. 5874, pp. 380–384. Springer, Heidelberg (2009)
10. Loksthanov, D., Narayanaswamy, N.S., Raman, V., Ramanujan, M.S., Saurabh, S.: Faster Parameterized Algorithms using Linear Programming. ArXiv e-prints (March 2012)
11. Reed, B.A., Smith, K., Vetta, A.: Finding odd cycle transversals. Oper. Res. Lett. 32(4), 299–301 (2004)
12. Thomassé, S.: A quadratic kernel for feedback vertex set. In: SODA, pp. 115–119 (2009)
13. Thomassen, C.: On the presence of disjoint subgraphs of a specified type. Journal of Graph Theory 12(1), 101–111 (1988)

Bisections above Tight Lower Bounds

Matthias Mnich[1] and Rico Zenklusen[2,*]

[1] Cluster of Excellence, Saarbrücken, Germany
m.mnich@mmci.uni-saarland.de
[2] Massachusetts Institute of Technology, Cambridge, MA, USA
ricoz@math.mit.edu

Abstract. A bisection of a graph is a bipartition of its vertex set in which the number of vertices in the two parts differ by at most one, and the size of the bisection is the number of edges which go across the two parts.

Every graph with m edges has a bisection of size at least $\lceil m/2 \rceil$, and this bound is sharp for infinitely many graphs. Therefore, Gutin and Yeo considered the parameterized complexity of deciding whether an input graph with m edges has a bisection of size at least $\lceil m/2 \rceil + k$, where k is the parameter. They showed fixed-parameter tractability of this problem, and gave a kernel with $O(k^2)$ vertices.

Here, we improve the kernel size to $O(k)$ vertices. Under the Exponential Time Hypothesis, this result is best possible up to constant factors.

1 Introduction

A *bisection* of a graph is a bipartition of its vertex set in which the number of vertices in the two parts differ by at most one, and its *size* is the number of edges which go across the two parts. We are interested in finding bisections of maximum size in a given graph, which is known as the MAX-BISECTION problem. This problem is NP-hard by a simple reduction from the MAX-CUT problem. On the other hand, there is a simple randomized polynomial-time procedure [1] that finds in any m-edge graph a bisection of size at least $\lceil m/2 \rceil$, and there are graphs (such as stars) for which this bound cannot be improved. Therefore, interest arose in the study of the problem MAX-BISECTION ABOVE TIGHT LOWER BOUND (or MAX-BISECTION ATLB for short), where we seek a bisection of size at least $m/2 + k$ when given an m-edge graph G together with an integer $k \in \mathbb{N}$. The NP-hardness of MAX-BISECTION ATLB follows from the NP-hardness of MAX-BISECTION. On the positive side, Gutin and Yeo [1] showed that MAX-BISECTION ATLB is fixed-parameter tractable, that is, pairs (G, k) can be decided in time $f(k) \cdot n^{O(1)}$ for some function f dependent only on k, where n is the number of vertices of G. Fixed-parameter tractability directly implies the existence of a *kernelization* [2], which is a polynomial-time algorithm that efficiently compresses instances (G, k) to equivalent instances (G', k') (the

* Supported by NSF grants CCF-1115849 and CCF-0829878, and by ONR grants N00014-11-1-0053 and N00014-09-1-0326.

M.C. Golumbic et al. (Eds.): WG 2012, LNCS 7551, pp. 184–193, 2012.

kernel) of size $|G'| + k' \leq g(k)$ for some function g dependent only on k. Gutin and Yeo's fixed-parameter tractability result [1] is based on proving a kernel with $O(k^2)$ vertices.

Here we improve their kernel as follows.

Theorem 1. MAX-BISECTION ATLB *admits a kernel with at most* $16k$ *vertices.*

We observe that the number of vertices in our kernel is asymptotically optimal, assuming the Exponential Time Hypothesis introduced by Impagliazzo et al.[3]. The hypothesis implies that a large family of NP-complete problems cannot be solved in subexponential time, including the MAX-CUT problem. In the MAX-CUT problem the goal is to find any bipartition of the vertices (not necessarily balanced as in the bisection case) that maximizes the number of edges crossing it. The MAX-CUT problem on a graph $G = (V, E)$ can easily be reduced to the maximum bisection problem by adding $|V|$ additional isolated vertices to G to obtain G', and solving the bisection problem in G'. Clearly, the maximum bisection in G' induces a bipartition of the vertices in G that solves the MAX-CUT problem. Now notice that a kernel for the MAX-BISECTION ATLB problem with $o(k)$ vertices, would imply that we could solve the MAX-BISECTION ATLB problem in $2^{o(k)}$ time by checking all bisections of the kernel. By the above relation to the MAX-CUT problem this would yield a subexponential time algorithm for MAX-CUT, contradicting the Exponential Time Hypothesis [4].

We remark that for the MAX-CUT problem, deciding the existence of a cut with size $m/2+k$ is trivially fixed-parameter tractable, because any graph admits a cut of size $m/2 + \Omega(\sqrt{m})$ due to a classical result of Edwards [5,6]. In fact, the MAX-BISECTION problem forms an "extremal point" of a series of problems on *α-bisections*, which are cuts in which both sides of the bipartition have at least $(1/2-\alpha)n$ vertices. By a recent result of Lee et al. [7], for every $\alpha \in [0, 1/6]$ every n-vertex graph with m edges and no isolated vertices contains an α-bisection of size at least $m/2 + \alpha n$. Thus, deciding the existence of an α-bisection of size $m/2 + k$ is trivially fixed-parameter tractable for all $\alpha \in (0, 1/6]$. In this paper we prove an essentially optimal kernel for the extremal case $\alpha = 0$.

The problem MAX-BISECTION ATLB is an example of a so-called "above tight lower bound paramaterization", where the parameter k is chosen as the excess of the solution value of the given instance over a non-trivial tight lower bound on the solution value in arbitrary instances (here: $\lceil m/2 \rceil$). For many parameterizations above tight lower bound, it is often not even clear how to solve such problems in time m^k, let alone by a fixed-parameter algorithm in time $f(k) \cdot m^{O(1)}$ for some function f dependent only on k. By now, several techniques have been developed for fixed-parameter algorithms of above tight lower bound parameterizations of important computational problems, such as MAX-r-SAT [8], MAX-LIN2 [9], PERMUTATION-CSPS [10], and MAX-CUT [11]; see the survey by Gutin and Yeo [12]. Most of these techniques are based on probabilistic analysis of carefully chosen random variables, and they rarely yield kernels of linear size. Here, we introduce a new technique to establish a linear vertex-kernel for MAX-BISECTION ATLB, based on Edmonds-Gallai decompositions of graphs. We believe that this technique has the potential to find further applications in establishing linear kernels for problems parameterized above tight lower bound.

2   Preliminaries

Let G be a loopless undirected graph with vertex set $V(G)$ and edge set $E(G)$. We allow parallel edges. For each vertex $v \in V(G)$, let $N(v) = \{u \in V(G) \mid \{u, v\} \in E(G)\}$ be the set of neighbors of v. In particular, $v \notin N(v)$. For a vertex $v \in V(G)$ and a subset $U \subseteq V(G)$, we denote by $d_U(v) = |\{e \in E(G) \mid v \in e, e \setminus \{v\} \subseteq U\}|$ the *degree of v in U*, and write shorthand $d(v) = d_{V(G)}(v)$. Notice that if we have parallel edges adjacent to v, then $d(v) > |N(v)|$. Let $\sigma(G)$ denote the number of connected components of G. For sets $U, W \subseteq V(G)$, define $E(U, W) = \{\{u, w\} \in E(G) \mid u \in U, w \in W\}$ and $d(U, W) = |E(U, W)|$. For the special case of $U = \{u\}$ and $W = \{v\}$ being singleton sets, we use the shorthand $d(u, w) = d(\{u\}, \{w\})$. The subgraph of G induced by a subset $V' \subseteq V(G)$ is denoted by $G[V']$. For $v \in V$, we use the shorthand $G - v = G[V \setminus \{v\}]$.

For a graph G, a *matching* is a set M of pairwise non-adjacent edges; the vertices in $V(M) = \{v \in e \mid e \in M\}$ are *saturated by M* and vertices in $V(G) \setminus M$ are *unsaturated by M*. A matching is *perfect* for G if it saturates every vertex of G, and G is *factor-critical* if the graph $G - v$ admits a perfect matching for every $v \in V(G)$. Denote by $\nu(G)$ the cardinality of a maximum size matching in G.

For a graph G, an *Edmonds-Gallai decomposition* [13] is a tuple (X, Y, Z) such that $\{X, Y, Z\}$ forms a tripartion of $V(G)$, X is such that for every vertex $v \in X$ the size of a maximum cardinality matching in $G - v$ and G are the same, Y contains all neighbors of X in $V(G) \setminus X$, and $Z = V(G) \setminus (X \cup Y)$. Classical results on the Edmonds-Gallai decomposition imply that every connected component of $G[X]$ is factor-critical, every component of $G[Z]$ admits a perfect matching, and furthermore $\nu(G) = \frac{n - \sigma(G[X]) + |Y|}{2}$.

3   Proof of Theorem 1

In this section we prove our main result, Theorem 1.

Let G be a loopless undirected graph with $n = |V(G)|$ vertices and $m = |E(G)|$ edges, and let $k \geq 0$ be an integer. An important ingredient that we use to bound the size of the kernel is the following well-known fact, showing that large matchings lead to large bisections.

Proposition 1 ([14]). *Let G be a graph and M be a matching in G; then G has a bisection of size at least $\lceil m/2 \rceil + \lfloor |M|/2 \rfloor$ and such can be found in $O(m + n)$ time.*

Another important fact, that will prove to be useful in our reduction to obtain a small kernel, is that whenever there is a large set of vertices with the same neighbors (and same number of parallel edges to the neighbors), the problem can be reduced to a smaller one. This is a straightforward generalization of a reduction used by Gutin and Yeo [1].

Lemma 1 (straightforward extension of a result in [1]). *Let G be a loopless undirected graph with a set $I \subseteq V(G)$ of size $\lceil n/2 + j \rceil$ for some $j > 0$ such that $d(u, w) = d(v, w)$ for all $u, v \in I$ and $w \in V(G)$. Then G has a bisection of size $\lceil |E(G)|/2 \rceil + k$ if and only if the graph G' obtained from G by removing $2j$ arbitrary vertices of I has a bisection of size $\lceil E(G')/2 \rceil + k$.*

The above lemma can easily be seen to be true by observing that any balanced bipartition (V_1, V_2) of V must satisfy $|I \cap V_1| \geq j$ and $|I \cap V_2| \geq j$, due to the large size of I and the balancedness of the bipartition (V_1, V_2). One can then observe that any balanced bipartition (V_1, V_2) of V leading to a bisection of size at least $\frac{|E(G)|}{2} + k$ can be transformed into a bisection of G' of size $\frac{|E(G')|}{2} + k$ by removing any j vertices of $|I \cap V_i|$ from V_i, for $i = 1, 2$. Similarly, any bisection (V_1', V_2') of G' of size at least $\frac{|E(G')|}{2} + k$ can be completed to a bisection (V_1, V_2) of G of size $\frac{|E(G')|}{2} + k$ by adding any j vertices of $V(G) \backslash V(G')$ to V_1 to obtain V_1', and the remaining vertices of $V(G) \setminus V(G')$ to V_2 to obtain V_2'.

From now on, we assume that $\nu(G) < 2k$, since otherwise there is a bisection of size $m/2 + k$ due to Proposition 1, and such can be found efficiently through a simple randomized switching argument that can be derandomized through conditional expectations. (Details of such a derandomization in a similar setting are given by Ries and Zenklusen [15].) Furthermore, we assume that G does not contain any large set $I \subseteq V$ as defined in Lemma 1, for otherwise we could apply Lemma 1 to reduce the size of the graph.

Let (X, Y, Z) be a Gallai-Edmonds decomposition of G. As a reminder, $G[X]$ consists of factor-critical components, Y are all neighbors of X, and $G[Z]$ admits a perfect matching. Furthermore, $\nu(G) = \frac{n - \sigma(G[X]) + |Y|}{2}$.

We partition X into sets X_0, X_1, X_2, defined as

$$X_0 = \{v \in X \mid d(v) = 0\},$$
$$X_1 = \{v \in X \mid d_X(v) = 0\} \setminus X_0,$$
$$X_2 = \{v \in X \mid d_X(v) \geq 1\}.$$

Hence, $G[X_2]$ contains all connected components of $G[X]$ with more than one vertex. Notice that since these components are factor-critical, each of them has size at least 3.

Lemma 2. *We have $|X_2|/3 + |Y| + |Z|/2 < 2k$.*

Proof. Consider a maximum matching $M \subseteq E(G)$ in G. It is a well-known property of the Gallai-Edmonds decomposition [13] that M saturates all vertices in $Y \cup Z$. More precisely, M can be partitioned into $M = M_X \uplus M_Y \uplus M_Z$, where M_X are the edges of M having both endpoints in X, M_Y are the edges of M having one endpoint in Y, and M_Z are the edges of M having both endpoints in Z. Furthermore, M_Z is a perfect matching in $G[Z]$, and all edges of M_Y connect a vertex of Y with one of X. Additionally, in each connected component $G[X']$ of $G[X]$, the edges of M_X with both endpoints in X' saturate all but one vertex of X'. Observe that X_2 are precisely those vertices in X that belong to connected components of $G[X]$ of size at least 3. Hence,

$$|M_X| \geq |X_2|/3,$$

because the number of edges $|M_X|$ is minimized when all vertices in X_2 are in connected components of size precisely 3. This is the case since in a connected component of $G[X]$ of size $2p + 1 \geq 3$ (whose size must be odd since $G[X]$ is factor-critical), the ratio between number of vertices and matching edges between them is $\frac{p}{2p+1}$, and this ratio is minimized for $p = 1$.

Furthermore, $|M_Y| = |Y|$ and $|M_Z| = |Z|/2$. Thus

$$|M| = |M_X| + |M_Y| + |M_Z| \geq |X_2|/3 + |Y| + |Z|/2.$$

Since $|M| < 2k$ (by assumption there is no matching of size $\geq 2k$), the result follows.   □

Our key technique to show that $|V(G)| \leq 16k$ is a generalized version of the randomized argument used to show that a large matching leads to a large bisection. We replace the role of a matching by what we call *switching units* on $V(G)$. A switching unit is a tuple (A, B) with $A, B \subseteq V(G)$, $A \cap B = \emptyset$ and $|A| = |B|$. We will construct a *switching family* $(A_i, B_i)_i$, which is a collection of mutually disjoint switching units, i.e., $(A_i \cup B_i) \cap (A_j \cup B_j) = \emptyset$ for $i \neq j$. Any switching family can be used to define a random bisection (V_1, V_2) of $V(G)$ by randomly and independently assigning the vertices of each switching unit (A_i, B_i) to the sets V_1, V_2 as follows: with probability $1/2$ assign the vertices of A_i to V_1 ($A_i \subseteq V_1$) and the vertices of B_i to V_2 ($B_i \subseteq V_2$), otherwise set $A_i \subseteq V_2$ and $B_i \subseteq V_1$. Furthermore, for all remaining vertices $\tilde{V} = V(G) \setminus \bigcup_i (A_i \cup B_i)$, i.e., all vertices not part of any switching unit, we pick uniformly at random a bisection $(\tilde{V}_1, \tilde{V}_2)$ of $\tilde{V}$, and assign all vertices in $\tilde{V}_1$ to V_1 and all vertices of $\tilde{V}_2$ to V_2. We call the thus obtained random bisection (V_1, V_2) a random bisection corresponding to the switching family $(A_i, B_i)_i$.

To compute the expected number of edges $\mathbb{E}[d(V_1, V_2)]$ in the random bisection (V_1, V_2) corresponding to $(A_i, B_i)_i$, we observe that all edges not having both endpoints in the same switching set $A_i \cup B_i$ are in the bisection (V_1, V_2) with probability at least $1/2$. (Notice that this probability can indeed be strictly larger than $1/2$, e.g., when considering a graph consisting of two vertices connected by a single edge, then this edge will be in the bisection with probability one.) It remains to consider for each switching unit (A_i, B_i) its contribution to $d(V_1, V_2)$. Notice that this contribution is deterministic and equals $d(A_i, B_i)$.

Since we are interested in how much $\mathbb{E}[d(V_1, V_2)]$ exceeds the tight lower bound $m/2$, we introduce the *excess* $\mathsf{ex}(A_i, B_i)$ of the switching unit (A_i, B_i) as follows:

$$\mathsf{ex}(A_i, B_i) = 2d(A_i, B_i) - d(A_i \cup B_i, A_i \cup B_i)$$
$$= d(A_i, B_i) - d(A_i, A_i) - d(B_i, B_i) .$$

In words, the excess of (A_i, B_i) is the difference between the number of edges in $G[A_i \cup B_i]$ crossing the bisection (A_i, B_i) and those who do not. Using the notion of excess, we can express the expected number of edges in the random bisection (V_1, V_2) by

$$\mathbb{E}[d(V_1, V_2)] \geq \frac{m}{2} + \frac{1}{2} \sum_i \mathsf{ex}(A_i, B_i), \tag{1}$$

because every edge is in the bisection with probability at least $1/2$, except for edges e with both endpoints in one switching unit (A_i, B_i), in which case e is in the bisection if $e \in E(A_i, B_i)$ and e is not in the bisection if $e \in E(A_i, A_i) \cup E(B_i, B_i)$.

In the following, we describe a way to construct a switching family $(A_i, B_i)_i$ with a high total excess $\sum_i \mathsf{ex}(A_i, B_i)$. Let $Y = \{y_1, \ldots, y_\ell\}$. We start by constructing iteratively for each $i = 1, \ldots, \ell$ a switching unit (A_i, B_i), which might be chosen to be $(\emptyset, \emptyset)$. Assume that we already constructed switching units $(A_1, B_1), \ldots, (A_{i-1}, B_{i-1})$. Let

$$N_i = X_1 \setminus \bigcup_{j=1}^{i-1}(A_j \cup B_j),$$

where $N_1 = X_1$. Consider the partition of N_i into sets $N_i^0, N_i^1, \ldots,$ where

$$N_i^j = \{v \in N_i \mid d(y_i, v) = j\}.$$

If $N_i = N_i^j$ for some $j \in \mathbb{Z}_+$, we set $A_i = B_i = \emptyset$. Otherwise, we start by assigning y_i to A_i and we choose any element $v \in N_i \setminus N_i^0$ that we assign to B_i. Then, as long as there is an unassigned pair $(u, v) \in N_i \times N_i$ with $u \in N_i^{j_1}$ and $v \in N_i^{j_2}$, where $j_1 < j_2$, we assign u to A_i and v to B_i. Clearly, at the end of this procedure, all elements in $N_i \setminus (A_i \cup B_i)$ belong to a single group N_i^j. The key observation is that for every pair $u \in N_i^{j_1}, v \in N_i^{j_2}$ with $j_1 < j_2$ that we add, $\mathsf{ex}(A_i, B_i)$ increases by at least one unit because v has at least one more edge adjacent to y_i than u has. Furthermore, also the assignment at the start of y_i to A_i and of an arbitrarily chosen vertex $v \in N_i \setminus N_i^0$ to B_i creates an excess of at least one unit. Thus, for each $i \in \{1, \ldots, \ell\}$, the switching unit (A_i, B_i) satisfies the following properties:

(a) $\mathsf{ex}(A_i, B_i) \geq |A_i \cup B_i|/2$, since any added pair of vertices increases the excess by at least one unit, and

(b) $d(y_i, v) = d(y_i, u)$ for any $u, v \in N_i$, since otherwise another pair of vertices could have been added to the switching unit (A_i, B_i).

The switching units (A_i, B_i) with $i \in \{1, \ldots, \ell\}$ are completed by adding switching units corresponding to a perfect matching M_Z of $G[Z]$ and a maximum matching M_{X_2} in $G[X_2]$, i.e., for each $\{u, v\} \in M_Z \cup M_{X_2}$, we construct a switching unit $(\{u\}, \{v\})$. These trivial switching units together with the ones constructed above complete the construction of our switching family, which we denote by (A_i, B_i).

We first provide a lower bound for $\mathbb{E}[d(V_1, V_2)]$. Let $\tilde{X}_1 = X_1 \setminus \bigcup_{i=1}^{\ell}(A_i \cup B_i)$.

Lemma 3. *It holds that* $\mathbb{E}[d(V_1, V_2)] \geq \frac{m}{2} + \frac{1}{4}\left(|X_1 \setminus \tilde{X}_1| + |Z|\right) + \frac{1}{6}|X_2|.$

Proof. Using (1), we have $\mathbb{E}[d(V_1, V_2)] \geq \frac{m}{2} + \frac{1}{2}\sum_i \mathsf{ex}(A_i, B_i)$. We recall that the sum $\sum_i \mathsf{ex}(A_i, B_i)$ is composed of three types of terms

$$\sum_i \mathsf{ex}(A_i, B_i) = \sum_{i=1}^{\ell} \mathsf{ex}(A_i, B_i) + \sum_{\{u,v\} \in M_Z} \mathsf{ex}(\{u\}, \{v\}) + \sum_{\{u,v\} \in M_{X_2}} \mathsf{ex}(\{u\}, \{v\}),$$

$$(2)$$

where (A_i, B_i) for $i \in \{1, \ldots, \ell\}$ are the iteratively constructed switching units, the second term corresponds to switching units stemming from a perfect matching M_Z in $G[Z]$, and the third term corresponds to switching units stemming from a maximum matching M_{X_2} in $G[X_2]$.

Consider the first term of (2). By property (a), we have $\mathsf{ex}(A_i, B_i) \geq \frac{|A_i \cup B_i|}{2}$ for $i \in \{1, \ldots, \ell\}$. Since $\bigcup_{i=1}^{\ell}(A_i \cup B_i)$ contains all edges of $X_1 \setminus \tilde{X}_1$ (together with some additional vertices of Y), we obtain

$$\sum_{i=1}^{\ell} \mathsf{ex}(A_i, B_i) \geq \frac{1}{2}|X_1 \setminus \tilde{X}_1| . \tag{3}$$

Now consider the second and third term of (2). Since M_Z is a perfect matching over $G[Z]$, we have $|M_Z| \geq \frac{|Z|}{2}$. Furthermore, since each connected component of $G[X_2]$ is factor-critical and has size at least 3, we have $|M_{X_2}| \geq \frac{|X_2|}{3}$. Notice that $\mathsf{ex}(\{u\}, \{v\}) \geq 1$, and the inequality can be strict in case of parallel edges between u and v. Hence

$$\sum_{\{u,v\} \in M_Z} \mathsf{ex}(\{u\}, \{v\}) \geq \frac{|Z|}{2},$$

$$\sum_{\{u,v\} \in M_{X_2}} \mathsf{ex}(\{u\}, \{v\}) \geq \frac{|X_2|}{3} .$$

Combining the above inequalities with (3) and (2) and using $\mathbb{E}[d(V_1, V_2)] \geq \frac{m}{2} + \frac{1}{2}\sum_i \mathsf{ex}(A_i, B_i)$, the desired result is obtained. $\qquad \square$

The next step is to show that not both $\tilde{X}_1$ and X_0 can have a large size. For this we start with the following observation that follows immediately from property (b) of our iterative way to define the switching sets (A_i, B_i) for $i \in \{1, \ldots, \ell\}$.

Proposition 2. *All vertices in* $\tilde{X}_1$ *have the same neighborhood structure, i.e., for any* $u, v \in \tilde{X}_1$ *and* $w \in V(G)$, *we have* $d(u, w) = d(v, w)$.

Lemma 4. *If* $|\tilde{X}_1| \geq 2k$ *and* $|X_0| \geq 2k - 1$ *then* G *has a bisection of* G *of size at least* $m/2 + k$, *and such can be found efficiently.*

Proof. Assume $|\widetilde{X}_1| \geq 2k$ and $|X_0| \geq 2k - 1$. Consider the following switching unit (A, B). Let $v \in N(u)$ for an arbitrary $u \in \widetilde{X}_1$. Notice that the choice of u does not matter due to Proposition 2, and $v \in Y$. Observe further that $N(u) \neq \emptyset$, since any element $u \in \widetilde{X}_1 \subseteq X_1$ has at least one neighbor in Y, as otherwise it would belong to X_0. Let $A = \{v\} \cup X_0'$, where $X_0' \subseteq X_0$ is any set with $|X_0'| = 2k - 1$, and let $B = \widetilde{X}_1'$ where $\widetilde{X}_1' \subseteq \widetilde{X}_1$ is any set with $|\widetilde{X}_1'| = 2k$. Notice that since all elements of B have the same neighborhood because $B \subseteq \widetilde{X}_1$, there is an edge between any vertex of B and v.

Instead of considering a random bisection using the switching units $(A_i, B_i)_i$, consider a random bisection (V_A, V_B) corresponding to the single switching unit (A, B). Notice that $G[A \cup B]$ is a bipartite graph with bipartition $\{A, B\}$. Hence, $\mathsf{ex}(A, B)$ is equal to the number of edges in $E(G)$ with both endpoints in $A \cup B$. Since each edge of B is connected to $v \in A$, we have $\mathsf{ex}(A, B) \geq |B| = 2k$. By (1) we thus obtain

$$\mathbb{E}[d(V_A, V_B)] \geq \frac{m}{2} + \frac{\mathsf{ex}(A, B)}{2} \geq \frac{m}{2} + k .$$

A bisection of size at least $m/2 + k$ can then be found efficiently through standard derandomization arguments using conditional expectations.   □

We are now ready to combine all ingredients to obtain a kernel of size at most $16k$. To obtain the desired kernel, we first repeatedly apply Lemma 1 to reduce the given graph as long as the conditions of Lemma 1 are fulfilled, i.e., as long as there are large vertex sets with the same neighborhood structure. After that, if we can either apply Proposition 1, Lemma 4, or if the switching family $(A_i, B_i)_i$ leads to a random bisection (V_1, V_2) with $\mathbb{E}[d(V_1, V_2)] \geq \frac{m}{2} + k$, then we can obtain a large bisection with $\geq \frac{m}{2} + k$ edges. The remaining case, when none of these results leads to a large bisection, is covered by the following theorem.

Theorem 2. *Let G be a loopless graph on m edges and let $k \in \mathbb{N}$. Then either*

- *G has at most $16k$ vertices, or*
- *we can reduce G to a graph G' on $m' < m$ edges such that G has a bisection of size at least $m/2 + k$ if and only if G' has a bisection of size at least $m'/2 + k$, or*
- *G has a bisection of size at least $m/2 + k$ that we can find efficiently.*

Proof. First, suppose that G satisfies one of the following properties.

(i) If $\nu(G) \geq 2k$ then we obtain a bisection of G of size at least $\frac{m}{2} + k$ by Proposition 1.

(ii) If $|X_0| > \frac{|V(G)|}{2}$ then we can apply Lemma 1 to the vertices in X_0 to reduce the graph G, since all vertices in X_0 have the same neighborhood structure.

(iii) If $|\widetilde{X}_1| > \frac{|V(G)|}{2}$ then we can apply Lemma 1 to the vertices in $\widetilde{X}_1$, since all these vertices have the same neighborhood structure due to Proposition 2.

(iv) If $\min\{|X_0|, |\widetilde{X}_1|\} \geq 2k$ then Lemma 4 implies that there is a bisection of size at least $m/2 + k$ in G.

(v) If the random bisection (V_1, V_2) corresponding to the switching family $(A_i, B_i)_i$ satisfies $\mathbb{E}[d(V_1, V_2)] \geq \frac{m}{2} + k$ then a standard derandomization as given by Ries and Zenklusen [15] leads to a bisection of size at least $m/2 + k$ in G.

Second, suppose that G satisfies none of the conditions (i)–(v); we show that $|V(G)| \leq 16k$. By assumption, (v) does not hold, and so by Lemma 3 we have

$$\frac{m}{2} + k > \mathbb{E}[d(V_1, V_2)] \geq \frac{m}{2} + \frac{1}{4}(|X_1 \setminus \tilde{X}_1| + |Z|) + \frac{1}{6}|X_2|,$$

implying

$$4k > |X_1 \setminus \tilde{X}_1| + |Z| + \frac{2}{3}|X_2| . \tag{4}$$

Hence, we obtain

$$|V(G)| = |X_0| + |X_1| + |X_2| + |Y| + |Z|$$
$$= \underbrace{|X_1 \setminus \tilde{X}_1| + |Z| + \frac{2}{3}|X_2|}_{<4k \text{ by (4)}} + \underbrace{\frac{1}{3}|X_2| + |Y|}_{<2k \text{ by Lemma 2}} + |X_0| + |\tilde{X}_1|$$
$$< 6k + |X_0| + |\tilde{X}_1|$$
$$= 6k + \underbrace{\min\{|X_0|, |\tilde{X}_1|\}}_{<2k \text{ by (iv)}} + \underbrace{\max\{|X_0|, |\tilde{X}_1|\}}_{\leq \frac{|V|}{2} \text{ by (ii) and (iii)}}$$
$$< 8k + \frac{|V(G)|}{2} .$$

Therefore, $|V(G)| < 16k$. $\qquad\qquad\qquad\qquad\qquad\qquad\qquad\qquad\qquad\qquad\square$

Finally, as mentioned above, Theorem 1 is a direct consequence of Theorem 2.

4   Discussion

Our main result in this paper is a linear vertex-kernel for the MAX-BISECTION ATLB problem. Recently, Lee et al. [7] showed that for every $\alpha \in [0, 1/6]$, every n-vertex graph with m edges and no isolated vertices contains an α-bisection of size at least $m/2 + \alpha n$, where each side of the bipartition has at least $(1/2 - \alpha)n$ vertices. Thus, a natural problem to study is MAX-α-BISECTION ATLB for every $\alpha \in [0, 1/6]$, where we wish to decide the existence of an α-bisection of size at least $m/2 + \alpha n + k$ in a given n-vertex m-edge graph. We conjecture this problem to be fixed-parameter tractable and admit a polynomial-size kernel.

Acknowledgment. We are grateful to the referees for their helpful suggestions and comments.

References

1. Gutin, G., Yeo, A.: Note on maximal bisection above tight lower bound. Information Processing Letters 110, 966–969 (2010)
2. Downey, R.G., Fellows, M.R.: Parameterized complexity. Monographs in Computer Science (1999)
3. Impagliazzo, R., Paturi, R.: On the complexity of k-SAT. Journal of Computer and System Sciences 62, 367–375 (2001)
4. Cai, L., Juedes, D.: On the existence of subexponential parameterized algorithms. Journal of Computer and System Sciences 67, 789–807 (2003)
5. Edwards, C.S.: Some extremal properties of bipartite subgraphs. Canadian Journal of Mathematics 25, 475–485 (1973)
6. Edwards, C.S.: An improved lower bound for the number of edges in a largest bipartite subgraph. In: Recent Advances in Graph Theory (Proceedings of Second Czechoslovak Symposium, Prague, 1974), Prague, pp. 167–181 (1975)
7. Lee, C., Loh, P.S., Sudakov, B.: Bisections of graphs (2011), http://arxiv.org/abs/1109.3180/v3
8. Alon, N., Gutin, G., Kim, E.J., Szeider, S., Yeo, A.: Solving MAX-r-SAT above a tight lower bound. Algorithmica 61, 638–655 (2011)
9. Crowston, R., Fellows, M., Gutin, G., Jones, M., Rosamond, F., Thomassé, S., Yeo, A.: Simultaneously Satisfying Linear Equations Over $\mathbb{F}_2$: MaxLin2 and Max-r-Lin2 Parameterized Above Average. In: IARCS Annual Conference on Foundations of Software Technology and Theoretical Computer Science (FSTTCS 2011). Leibniz International Proceedings in Informatics (LIPIcs), vol. 13, pp. 229–240 (2011)
10. Gutin, G., van Iersel, L., Mnich, M., Yeo, A.: Every ternary permutation constraint satisfaction problem parameterized above average has a kernel with a quadratic number of variables. Journal of Computer and System Sciences 78, 151–163 (2012)
11. Crowston, R., Jones, M., Mnich, M.: Max-Cut Parameterized above the Edwards-Erdős Bound. In: Czumaj, A., Mehlhorn, K., Pitts, A., Wattenhofer, R. (eds.) ICALP 2012, part I. LNCS, vol. 7391, pp. 242–253. Springer, Heidelberg (2012)
12. Gutin, G., Yeo, A.: Constraint Satisfaction Problems Parameterized above or below Tight Bounds: A Survey. In: Bodlaender, H.L., Downey, R., Fomin, F.V., Marx, D. (eds.) Fellows Festschrift 2012. LNCS, vol. 7370, pp. 257–286. Springer, Heidelberg (2012)
13. Schrijver, A.: Combinatorial Optimization, Polyhedra and Efficiency. Springer (2003)
14. Haglin, D.J., Venkatesan, S.M.: Approximation and intractability results for the maximum cut problem and its variants. IEEE Transactions on Computing 40, 110–113 (1991)
15. Ries, B., Zenklusen, R.: A 2-approximation for the maximum satisfying bisection problem. European Journal of Operational Research 210, 169–175 (2011)

On Group Feedback Vertex Set Parameterized by the Size of the Cutset

Marek Cygan[1,*], Marcin Pilipczuk[2,**], and Michał Pilipczuk[3,***]

[1] IDSIA, University of Lugano, Switzerland
marek@idsia.ch
[2] Institute of Informatics, University of Warsaw, Poland
malcin@mimuw.edu.pl
[3] Department of Informatics, University of Bergen, Norway
michal.pilipczuk@ii.uib.no

Abstract. We study parameterized complexity of a generalization of the classical FEEDBACK VERTEX SET problem, namely the GROUP FEEDBACK VERTEX SET problem: we are given a graph G with edges labeled with group elements, and the goal is to compute the smallest set of vertices that hits all cycles of G that evaluate to a non-null element of the group. This problem generalizes not only FEEDBACK VERTEX SET, but also SUBSET FEEDBACK VERTEX SET, MULTIWAY CUT and ODD CYCLE TRANSVERSAL. Completing the results of Guillemot [Discr. Opt. 2011], we provide a fixed-parameter algorithm for the parameterization by the size of the cutset only. Our algorithm works even if the group is given as a blackbox performing group operations.

1 Introduction

The parameterized complexity is an approach for tackling NP-hard problems by designing algorithms that perform well, when the instance is in some sense simple; its difficulty is measured by an integer, called the *parameter*, additionally appended to the input. Formally, we say that a problem is *fixed-parameter tractable* (FPT), if it admits an algorithm that given input of length n and parameter k, resolves the task in time $f(k)n^c$, where f is some computable function and c is a constant independent of the parameter.

The search for fixed-parameter algorithms led to the development of a number of new techniques and gave valuable insight into structures of many classes of NP-hard problems. Among them, there is a family of so-called *graph cut* problems, where the goal is to delete as few as possible edges or vertices (depending on the variant) in order to make the graph satisfy a global separation requirement. This class is perhaps best represented by the classical FEEDBACK VERTEX SET problem (FVS) where, given an undirected graph G, we seek for a minimum set of vertices that hits all cycles of G. Other examples are MULTIWAY CUT (MWC: separate each pair from a given set of

---

* Partially supported by European Research Council (ERC) Starting Grant NEWNET 279352, NCN grant N206567140 and Foundation for Polish Science.
** Partially supported by NCN grant N206567140 and Foundation for Polish Science.
*** Partially supported by European Research Council (ERC) Grant "Rigorous Theory of Preprocessing", reference 267959.

M.C. Golumbic et al. (Eds.): WG 2012, LNCS 7551, pp. 194–205, 2012.

terminals in a graph with a minimum cutset) or ODD CYCLE TRANSVERSAL (OCT: make a graph bipartite by a minimum number of vertex deletions).

The research on the aforementioned problems had a great impact on the development of parameterized complexity. The long line of research concerning parameterized algorithms for FVS contains [1,2,3,4,10,11,12,14,15,20], leading to an algorithm working in $O(3^k n^{O(1)})$ time [7]. The search for a polynomial kernel for FVS lead to surprising applications of deep combinatorial results such as the Gallai's theorem [23], which has also been found useful in designing FPT algorithms [9]. While investigating the graph cut problems such as MWC, Márx [18] introduced the *important separator* technique, which turned out to be very robust and is now the key ingredient in parameterized algorithms for various problems such as variants of FVS [5,9] or ALMOST 2-SAT [21]. Moreover, the recent developments on MWC show applicability of linear programming in parameterized complexity, leading to the fastest currently known algorithms not only for MWC, but also ALMOST 2-SAT and OCT [8,19]. Last but not least, the research on the OCT problem resulted in the introduction of iterative compression, a simple yet powerful technique for designing parameterized algorithms [22].

Considered problem. In this paper we study a generalization of the FVS problem, namely GROUP FEEDBACK VERTEX SET[1]. Let Σ be a finite (not necessarily abelian) group, with unit element 1_Σ. We use the multiplicative convention for denoting the group operation.

Definition 1. *For a finite group Σ, a directed graph $G = (V, A)$ and a labeling function $\Lambda : A \to \Sigma$, we call (G, Λ) a Σ-labeled graph iff for each arc $(u, v) \in A$ we have $(v, u) \in A$ and $\Lambda((u, v)) = \Lambda((v, u))^{-1}$.*

We somehow abuse the notation and by $(G \setminus X, \Lambda)$ denote the Σ-labeled graph (G, Λ) with vertices of X removed, even though formally Λ has in its domain arcs that do not exist in $G \setminus X$.

For a path[2] $P = (v_1, \dots, v_\ell)$ we denote $\Lambda(P) = \Lambda((v_1, v_2)) \cdot \dots \cdot \Lambda((v_{\ell-1}, v_\ell))$. Similarly, for a cycle $C = (v_1, \dots, v_\ell, v_1)$ we denote $\Lambda(C) = \Lambda((v_1, v_2)) \cdot \dots \cdot \Lambda((v_{\ell-1}, v_\ell)) \cdot \Lambda((v_\ell, v_1))$. We call a cycle C a *non-null* cycle, iff $\Lambda(C) \neq 1_\Sigma$. Observe that if the group Σ is non-abelian, then it may happen that cyclic shifts of the same cycle yield different elements of the group; nevertheless, the notion of a non-null cycle is well-defined, as either all of them are equal to 1_Σ or none of them.

Lemma 2. *Assume that $(x_1, \dots, x_\ell, x_1)$ is a cycle in a Σ-labeled graph (G, Λ). If $\Lambda((x_1, \dots, x_\ell, x_1)) \neq 1_\Sigma$, then $\Lambda((x_2, \dots, x_\ell, x_1, x_2)) \neq 1_\Sigma$.*

Proof. Let $g_1 = \Lambda((x_1, x_2))$ and $g_2 = \Lambda((x_2, \dots, x_\ell, x_1))$. We have that $g_1 \cdot g_2 = 1_\Sigma$ iff $g_2 \cdot g_1 = 1_\Sigma$ and the lemma follows. □

In the GROUP FEEDBACK VERTEX SET problem we want to hit all non-null cycles in a Σ-labeled graph using at most k vertices.

---

[1] In this paper, we follow the notation of Guillemot [13].

[2] In this paper, all paths and cycles are simple.

> GROUP FEEDBACK VERTEX SET (GFVS)   **Parameter:** k
> **Input:** A Σ-labeled graph (G, Λ) and an integer k.
> **Question:** Does there exist a set $X \subseteq V(G)$ of at most k vertices, such that there
> are no non-null cycles in $(G \setminus X, \Lambda)$?

As observed in [13], for a graph excluding a non-null cycle we can define a consistent labeling.

Definition 3. *For a Σ-labeled graph (G, Λ) we call $\lambda : V \to \Sigma$ a consistent labeling iff for each arc $(u, v) = a \in A(G)$ we have $\lambda(v) = \lambda(u) \cdot \Lambda(a)$.*

Lemma 4 ([13]). *A Σ-labeled graph (G, Λ) has a consistent labeling iff it does not contain a non-null cycle. Moreover, there is a polynomial-time algorithm which, given (G, Λ), finds either a non-null cycle in G or a consistent labeling of G.*

Note that when analyzing the complexity of the GFVS problem, it is important how the group Σ is represented. In [13] it is assumed that Σ is given via its multiplication table as a part of the input. In this paper we assume a more general model, where operations in Σ are computed by a given blackbox. More precisely, we assume that we are given subroutines that can multiply two elements, return an inverse of an element, provide the neutral element 1_Σ, or check whether two elements are equal. The running times of our algorithms are always measured in terms of basic operations and group operations, while space complexity is measured in the number of bits and group elements stored.

As noted in [17], GFVS subsumes not only the classical FVS problem, but also OCT (with $\Sigma = \mathbb{Z}_2$) and MWC (with Σ being an arbitrary group of size not smaller than the number of terminals). We note that if Σ is given in the blackbox, GROUP FEEDBACK VERTEX SET subsumes also EDGE SUBSET FEEDBACK VERTEX SET, which is equivalent to SUBSET FEEDBACK VERTEX SET [9].

> EDGE SUBSET FEEDBACK VERTEX SET (ESFVS)   **Parameter:** k
> **Input:** An undirected graph G, a set $S \subseteq E(G)$ and an integer k.
> **Question:** Does there exist a set $X \subseteq V(G)$ of at most k vertices, such that in
> $G \setminus X$ there are no cycles with at least one edge from S?

Lemma 5. *Given an ESFVS instance (G, S, k), one can in polynomial time construct an equivalent GFVS instance (G', Λ, k) with group $\Sigma = \mathbb{Z}_2^{|S|}$.*

Proof. To construct the new GFVS instance, create the graph G' by replacing each edge of G with arcs in both direction, keep the parameter k, take $\Sigma = \mathbb{Z}_2^{|S|}$ and construct a Σ-labeling Λ by setting any $|S|$ linearly independent values of $\Lambda((u, v))$ for $uv \in S$ and $\Lambda((u, v)) = 1_\Sigma$ for $uv \notin S$. Clearly, this construction can be done in polynomial time and the operations on the group Σ can be performed by a subroutine in time polynomial in $|S|$ by representing elements of Σ as bit vectors of length $|S|$. $\square$

We note that the GROUP FEEDBACK VERTEX SET problem was also studied from the graph theoretical point of view, as, in addition to the aforementioned reductions, it also subsumes the setting of Mader's $\mathcal{S}$-paths theorem [6,16]. In particular, Kawarabayashi and Wollan proved the Erdös-Pósa property for non-null cycles in highly connected graphs, generalizing a list of previous results [16].

The study of parameterized complexity of GFVS was initiated by Guillemot [13], who presented a fixed-parameter algorithm for GFVS parameterized by $|\Sigma| + k$ running in time[3] $O^*(2^{O(k \log |\Sigma|)})$. When parameterized by k, Guillemot showed a fixed-parameter algorithm for the easier edge-deletion variant of GFVS, running in time $O^*(2^{O(k \log k)})$. Very recently, Kratsch and Wahlström presented a randomized kernelization algorithm that reduces the size of a GFVS instance to $O(k^{2|\Sigma|})$ [17].

Before we proceed to the description of our results, let us briefly sketch their motivation. The main purpose of studying the GFVS problem is to find the common points in the fixed-parameter algorithms for problems it generalizes. Precisely this approach has been presented by Guillemot in [13], where at the base of the algorithm lies a subroutine that solves a very general version of MULTIWAY CUT. When reducing various graph cut problems to GFVS, usually the size of the group depends on the number of distinguished vertices or edges in the instance, as in Lemma 5. Hence, an application of the general $O^*(2^{O(k \log |\Sigma|)})$ algorithm of Guillemot unfortunately incorporates this parameter in the running time. It appears that by a more refined combinatorial analysis, usually one can get rid of this dependence; this is the case both in SUBSET FEEDBACK VERTEX SET [9] and in MULTIWAY CUT [8,19]. This suggested that the phenomenon can be, in fact, more general.

Our result and techniques. Our main result is a fixed-parameter algorithm for GFVS parameterized by the size of the cutset only. Recall that time and space complexities refer to basic and group operations performed, and bits and group elements stored, respectively.

Theorem 6. GROUP FEEDBACK VERTEX SET *can be solved in* $O^*(2^{O(k \log k)})$ *time and polynomial space.*

Our algorithm uses a similar approach as described by Kratsch and Wahlström in [17]: in each step of iterative compression, when we are given a solution Z of size $k + 1$, we guess the values of a consistent labeling on the vertices of Z, and reduce the problem to MULTIWAY CUT. However, by a straightforward application of this approach we obtain $O^*(2^{O(k \log |\Sigma|)})$ time complexity. To reduce the dependency on $|\Sigma|$, we carefully analyze the structure of a solution, provide a few reduction rules in a spirit of the ones used in the recent algorithm for SUBSET FEEDBACK VERTEX SET [9] and, finally, for each vertex of Z we reduce the number of choices for a value of a consistent labeling to polynomial in k. Therefore, the number of reasonable consistent labelings of Z is bounded by $2^{O(k \log k)}$ and we can afford solving a MULTIWAY CUT instance for each such labeling.

Note that the bound on the running time of our algorithm matches the currently best known algorithm for SUBSET FEEDBACK VERTEX SET [9]. Therefore, we obtain the same running time as in [9] by applying a much more general framework.

---

[3] The $O^*()$ notation suppresses terms polynomial in the input size.

2   Preliminaries

Notation. We use standard graph notation. For a graph G, by $V(G)$ and $E(G)$ we denote its vertex and edge sets, respectively. In case of a directed graph G, we denote the arc set of G by $A(G)$. For $v \in V(G)$, its neighborhood $N_G(v)$ is defined as $N_G(v) = \{u : uv \in E(G)\}$, and $N_G[v] = N_G(v) \cup \{v\}$ is the closed neighborhood of v. We extend this notation to subsets of vertices: $N_G[X] = \bigcup_{v \in X} N_G[v]$ and $N_G(X) = N_G[X] \setminus X$. For a set $X \subseteq V(G)$ by $G[X]$ we denote the subgraph of G induced by X. For a set X of vertices or edges of G, by $G \setminus X$ we denote the graph with the vertices or edges of X removed; in case of vertex removal, we remove also all the incident edges. By somehow abusing the notation, we often treat the (directed) Σ-labeled graph also as an undirected graph, as the neighborhood relation in the underlying undirected graph is the same.

In the GROUP FEEDBACK VERTEX SET problem definition in [13] a set of forbidden vertices $F \subseteq V(G)$ is additionally given as a part of the input. One can easily gadget such vertices by replacing each of them by a clique of size $k + 1$ labeled with 1_Σ; therefore, for the sake of simplicity we assume that all the vertices are allowed.

3   Algorithm

In this section we prove Theorem 6. We proceed with a standard application of the iterative compression technique in Section 3.1. In each step of the iterative compression, we solve a COMPRESSION GROUP FEEDBACK VERTEX SET problem, where we are given a solution Z of size a bit too large — $k + 1$ — and we are to find a new solution disjoint with it. We first prepare the COMPRESSION GROUP FEEDBACK VERTEX SET instance by *untangling* it in Section 3.2, in the same manner as it is done in the kernelization algorithm of [17]. The main step of the algorithm is done in Section 3.3, where we provide a set of reduction rules that enable us for each vertex $v \in Z$ to limit the number of choices for a value of a consistent labeling on v to polynomial in k. Finally, we iterate over all $O^*(2^{O(k \log k)})$ remaining labelings of Z and, for each labeling, reduce the instance to MULTIWAY CUT (Section 3.4).

3.1   Iterative Compression

The first step in the proof of Theorem 6 is a standard technique in the design of parameterized algorithms, that is, iterative compression, introduced by Reed et al. [22]. Iterative compression was also the first step of the parameterized algorithm for SUBSET FEEDBACK VERTEX SET [9].

We define a *compression problem*, where the input additionally contains a feasible solution $Z \subseteq V(G)$, and we are asked whether there exists a solution of size at most k which is disjoint with Z.

COMPRESSION GROUP FEEDBACK VERTEX SET (C-GFVS) **Parameter:** $k + |Z|$
Input: A Σ-labeled graph (G, Λ), an integer k and a set $Z \subseteq V(G)$, such that $(G \setminus Z, \Lambda)$ has no non-null cycle.
Goal: Find a set $X \subseteq V(G) \setminus Z$ of at most k vertices, such that there is no non-null cycle in $(G \setminus X, \Lambda)$ or return NO, if such a set does not exist.

In Section 3.2 we prove the following lemma providing a parameterized algorithm for COMPRESSION GROUP FEEDBACK VERTEX SET.

Lemma 7. *The* COMPRESSION GROUP FEEDBACK VERTEX SET *problem can be solved in* $O^*(2^{O(|Z|(\log k + \log |Z|))} \cdot 2^k)$ *time and polynomial space.*

Armed with the aforementioned result, we can easily prove Theorem 6.

Proof (of Theorem 6). In the iterative compression approach we start with an empty solution for an empty graph, and in each of the n steps we add a single vertex both to a feasible solution and to the graph; we use Lemma 7 to compress the feasible solution after guessing which vertices of the solution of size at most $k+1$ should not be removed.

Formally, for a given instance $(G = (V, A), \Lambda, k)$ let $V = \{v_1, \ldots, v_n\}$. For $0 \le i \le n$ define $V_i = \{v_1, \ldots, v_i\}$ (in particular, $V_0 = \emptyset$) and let Λ_i be the function Λ restricted to the set of arcs $A_i = \{(u, v) \in A : u, v \in V_i\}$. Initially we set $X_0 = \emptyset$, which is a solution to the graph $(G[V_0], \Lambda_0)$. For each $i = 1, \ldots, n$ we set $Z_i = X_{i-1} \cup \{v_i\}$, which is a feasible solution to $(G[V_i], \Lambda_i)$ of size at most $k+1$. If $|Z_i| \le k$, then we set $X_i = Z_i$ and continue the inductive process. Otherwise, if $|Z_i| = k+1$, we guess by trying all possibilities a subset of vertices $Z'_i \subseteq Z_i$ that is not removed in a solution of size k to $(G[V_i], \Lambda_i)$ and use Lemma 7 for the instance $I_{Z'_i} = (G[V_i \setminus (Z_i \setminus Z'_i)], \Lambda_i, k' = |Z'_i| - 1, Z'_i)$. If for each set Z'_i the algorithm from Lemma 7 returns NO, then there is no solution for $(G[V_i], \Lambda_i)$ and, consequently, there is no solution for (G, Λ). However, if for some Z'_i the algorithm from Lemma 7 returns a set X'_i of size smaller than $|Z'_i|$, then we set $X_i = (Z_i \setminus Z'_i) \cup X'_i$. Since $|X_i| = |Z_i \setminus Z'_i| + |X'_i| < |Z_i| = k+1$, the set X_i is a solution of size at most k for the instance (G_i, Λ_i).

Finally, we observe that since $(G_n, \Lambda_n) = (G, \Lambda)$, the set X_n is a solution for the initial instance $(G = (V, A), \Lambda, k)$ of GROUP FEEDBACK VERTEX SET. The claimed bound on the running time follows from the observation that $|Z_i| \le k+1$ for each of polynomially many steps. □

At this point a reader might wonder why we do not add an assumption $|Z| \le k+1$ to the C-GFVS problem definition and parameterize the problem solely by k. The reason for this is that in Section 3.3 we will solve the C-GFVS problem recursively, sometimes decreasing the value of k without decreasing the size of Z, and to always work with a feasible instance of the C-GFVS problem we avoid adding the $|Z| \le k+1$ assumption to the problem definition.

3.2 Untangling

In order to prove Lemma 7 we use the concept of *untangling*, previously used by Kratsch and Wahlström [17]. We transform an instance of C-GFVS to ensure that each arc (u, v) with both endpoints in $V(G) \setminus Z$ is labeled 1_Σ by Λ.

Definition 8. *We call an instance $(G = (V, A), \Lambda, k, Z)$ of C-GFVS untangled, iff for each arc $(u, v) \in A$ such that $u, v \in V \setminus Z$ we have $\Lambda((u, v)) = 1_\Sigma$.*

Moreover, by untangling *a labeling Λ around vertex x with a group element g we mean changing the labeling to $\Lambda' : A \to \Sigma$, such that for $(u, v) = a \in A$, we have*

$$\Lambda'(a) = \begin{cases} g \cdot \Lambda(a) & \text{if } u = x; \\ \Lambda(a) \cdot g^{-1} & \text{if } v = x; \\ \Lambda(a) & \text{otherwise.} \end{cases}$$

Lemma 9. *Let $(G = (V, A), \Lambda)$ be a Σ-labeled graph, $x \in V$ be a vertex of G and let $g \in \Sigma$ be a group element. For any subset of vertices $X \subseteq V$ the graph $(G \setminus X, \Lambda)$ contains a non-null cycle iff $(G \setminus X, \Lambda')$ contains a non-null cycle, where Λ' is the labeling Λ untangled around the vertex x with a group element g.*

Proof. The lemma follows from the fact that for any cycle C in G we have $\Lambda(C) = 1_\Sigma$ iff $\Lambda'(C) = 1_\Sigma$.   □

In Section 3.3 we prove the following lemma.

Lemma 10. *The* COMPRESSION GROUP FEEDBACK VERTEX SET *problem for untangled instances can be solved in $O^*(2^{O(|Z|(\log k + \log |Z|))} \cdot 2^k)$ time and polynomial space.*

Having Lemmata 9 and 10 we can prove Lemma 7.

Proof (of Lemma 7). Let (G, Λ, k, Z) be an instance of C-GFVS. Since $(G \setminus Z)$ has no non-null cycle, by Lemma 4 there is a consistent labeling λ of $(G \setminus Z, \Lambda)$.

Let Λ' be a result of untangling Λ around each vertex $v \in V(G) \setminus Z$ with $\lambda(v)$. Note that, by associativity of Σ, the order in which we untangle subsequent vertices does not matter. After all the untangling operations, for an arc $a = (u, v) \in A(G)$, such that $u, v \in V(G) \setminus Z$, we have $\Lambda'(a) = (\lambda(u) \cdot \Lambda(a)) \cdot \lambda(v)^{-1} = \lambda(v) \cdot \lambda(v)^{-1} = 1_\Sigma$. Therefore, by Lemma 9 the instance (G, Λ', k, Z) is an untangled instance of C-GFVS, which is a YES-instance iff (G, Λ, k, Z) is a YES-instance. Consequently, we can use Lemma 10 and the claim follows.   □

3.3 Fixing a Labeling on Z

In this section we prove Lemma 10 using the following lemma, which we prove in Section 3.4.

Lemma 11. *Let (G, Λ, k, Z) be an untangled instances of C-GFVS. There is an algorithm which for a given function $\phi : Z \to \Sigma$, finds a set $X \subseteq V(G) \setminus Z$ of size at most k, such that there exists a consistent labeling $\lambda : V(G) \setminus X \to \Sigma$ of $(G \setminus X, \Lambda)$, where $\lambda|_Z = \phi$, or checks that such a set X does not exist; the algorithm works in $O^*(2^k)$ time and uses polynomial space.*

We could try all $(|\Sigma| + 1)^{|Z|}$ possible assignments ϕ and use the algorithm from Lemma 11. Unfortunately, since $|\Sigma|$ is not our parameter we cannot iterate over all such assignments. Therefore, the goal of this section is to show that after some preprocessing, it is enough to consider only $2^{O(|Z|(\log k + \log |Z|))}$ assignments ϕ; together with Lemma 11 this suffices to prove Lemma 10.

Definition 12. *Let (G, Λ, k, Z) be an untangled instance of* C-GFVS, *let z be a vertex in Z and by Σ_z denote the set $\Lambda(\{(z, v) \in A(G) : v \in V(G) \setminus Z\})$.*

By a flow graph $F(G, \Lambda, Z, z)$, *we denote the undirected graph (V', E'), where $V' = (V(G) \setminus Z) \cup \Sigma_z$ and $E' = \{uv : (u, v) \in A(G[V(G) \setminus Z])\} \cup \{gv : (z, v) \in A(G), v \in V(G) \setminus Z, \Lambda((z, v)) = g\}$.*

Less formally, in the flow graph we take the underlying undirected graph of $G[V(G) \setminus Z]$ and add a vertex for each group element $g \in \Sigma_z$, that is a group element for which there exists an arc from z to $V(G) \setminus Z$ labeled with g by Λ. A vertex $g \in \Sigma_z$ is adjacent to all the vertices of $V(G) \setminus Z$ for which there exists an arc going from z, labeled with g by Λ.

Lemma 13. *Let (G, Λ, k, Z) be an untangled instance of* C-GFVS. *Let H be the flow graph $F(G, \Lambda, Z, z)$ for some $z \in Z$. If for some vertex $v \in V(G) \setminus Z$, in H there are at least $k + 2$ paths from v to Σ_z that are vertex disjoint apart from v, then v belongs to every solution of* C-GFVS.

Proof. Let us assume, that v is not a part of a solution $X \subseteq V(G) \setminus Z$, where $|X| \leq k$. Then at least 2 out of the $k + 2$ paths from v to Σ_z remain in $H \setminus X$. These two paths are vertex disjoint apart from v and end in different elements of Σ_z, so they correspond to a non-null cycle in $G \setminus X$, a contradiction.   □

Definition 14. *For an untangled instance (G, Λ, k, Z) of* C-GFVS *by an* external path *we denote any path P beginning and ending in Z, but with all internal vertices belonging to $V(G) \setminus Z$. Moreover, for two distinct vertices $z_1, z_2 \in Z$ by $\Sigma(z_1, z_2)$ we denote the set of all elements $g \in \Sigma$, for which there exists an external path P from z_1 to z_2 with $\Lambda(P) = g$.*

Note that an arc (z_1, z_2) for $z_1, z_2 \in Z$ also forms an external path from z_1 to z_2.

Lemma 15. *Let (G, Λ, k, Z) be an untangled instance of* C-GFVS. *If for each $z \in Z$ and $v \in V(G) \setminus Z$ there are at most $k + 1$ vertex disjoint paths from v to Σ_z in $F(G, \Lambda, Z, z)$ and for some $z_1, z_2 \in Z$, $z_1 \neq z_2$, we have $|\Sigma(z_1, z_2)| \geq k^3(k+1)^2 + 2$, then there is no solution for (G, Λ, k, Z).*

Proof. Let us assume that $X \subseteq V(G) \setminus Z$ is a solution for (G, Λ, k, Z); in particular, $|X| \leq k$. Let $\mathcal{P}$ be a set of external paths from z_1 to z_2, containing exactly one path P for each $g \in \Sigma(z_1, z_2)$ with $\Lambda(P) = g$. Note that the only arcs with non-null labels in P are possibly the first and the last arc.

By the pigeon-hole principle, there exists a vertex $v \in X$, which belongs to at least $k^2(k + 1)^2 + 1$ paths in $\mathcal{P}$, since otherwise there would be at least two paths in $\mathcal{P}$ disjoint with X, creating a non-null closed walk disjoint with X. Note that existence of a non-null closed walk disjoint with X is a sufficient proof that X is not a solution to (G, Λ, k, Z), as it contradicts existence of a consistent labeling, guaranteed by Lemma 4.

Consider a connected component C of $G[V(G) \setminus Z]$ to which v belongs. Observe that there exists a vertex $z \in \{z_1, z_2\}$ that has at least $k(k + 1) + 1$ incident arcs going to C with pairwise different labels in Λ, since otherwise v would belong to at most $k^2(k+1)^2$ paths in $\mathcal{P}$.

Let H be the flow graph $F(G, \Lambda, Z, z)$ and let $T \subseteq \Sigma_z$ be the set of labels of arcs going from z to C; recall that $|T| > k(k+1)$. Since there is no non-null cycle in $(G \setminus X, \Lambda)$, we infer that in $H_0 = H[C \cup T] \setminus (X \cap C)$, no two vertices of T belong to the same connected component. Moreover, as C is connected in G, for each $t \in T$ there exists a path P_t with endpoints v and t in $H[C \cup T]$. Let w_t be the closest to t vertex from X on the path P_t; note that such a vertex always exists, as $v \in X$. As $|X| \leq k$ and $|T| > k(k+1)$, there exists $w \in X$ such that $w = w_t$ for at least $k + 2$ elements $t \in T$. By the definition of the vertices w_t and the fact that there are no two vertices of T in the same connected component of H_0, the subpaths of P_t from t to w_t for all t with $w = w_t$ are vertex disjoint apart from w. As there are at least $k + 2$ of them, we have a contradiction. □

We are now ready to prove Lemma 10 given Lemma 11.

Proof (of Lemma 10). If there exists a vertex v satisfying the properties of Lemma 13, we can assume that it has to be a part of the solution; therefore, we can remove the vertex from the graph and solve the problem for decremented parameter value. Hence, we assume that for each $z \in Z$ and $v \in V(G) \setminus Z$, there are at most $k+1$ vertex disjoint paths from v to Σ_z in $F(G, \Lambda, Z, z)$. We note that one can compute the number of such vertex disjoint paths in polynomial time, using a maximum flow algorithm.

By Lemma 15, if there is a pair of vertices $z_1, z_2 \in Z$ with $|\Sigma(z_1, z_2)| \geq k^3(k+1)^2 + 2$, we know that there is no solution. Observe, that one can easily verify the cardinality of $\Sigma(z_1, z_2)$, since the only non-null label arcs on paths contributing to $\Sigma(z_1, z_2)$ are the first and the last one, and we can iterate over all such arcs and check whether their endpoints are in the same connected component in $G[V(G) \setminus Z]$. Clearly, this can be done in polynomial time.

Knowing that the sets $\Sigma(z_1, z_2)$ have sizes bounded by a function of k, we can enumerate all the reasonable labelings of Z. For the sake of analysis let $G' = (Z, E')$ be an auxiliary undirected graph, where two vertices of Z are adjacent, when they are connected by an external path in $G \setminus X$, for some fixed solution $X \subseteq V(G) \setminus Z$. Let F be any spanning forest of G'. Since F has at most $|Z| - 1$ edges, we can guess F, by trying at most $|Z| \cdot |Z|^{2(|Z|-1)}$ possibilities. Let us assume, that we have guessed F correctly. Observe that for any two vertices $z_1, z_2 \in Z$, belonging to two different connected components of F, there is no path between z_1 and z_2 in $G \setminus X$. Therefore, there exists a consistent labeling of $G \setminus X$, which labels an arbitrary vertex from each connected component of F with 1_Σ. Having fixed the labeling on one vertex from each component of F, we can root the components in corresponding vertices and iteratively guess the labeling on the remaining vertices in a top-down manner. At each step we use the fact that if we have already fixed a value $\phi(z_1)$, then for each external path corresponding to an edge $z_1 z_2$ of F, there are at most $k^3(k+1)^2 + 1$ possible values of $\phi(z_2)$, since $\phi^{-1}(z_1) \cdot \phi(z_2) \in \Sigma(z_1, z_2)$. Hence, having fixed F there are at most $2^{O(|Z| \log k)}$ possible labelings ϕ of Z, as for each edge of the forest F we choose one of at most $k^3(k+1)^2 + 1$ options. As the number of choices of F is bounded by $2^{O(|Z| \log |Z|)}$, we obtain at most $2^{O(|Z|(\log k + \log |Z|))}$ labelings ϕ of Z in total, and we can use Lemma 11 for each of them. □

3.4 Reduction to Multiway Cut

In this section, we prove Lemma 11, by a reduction to MULTIWAY CUT. A similar reduction was also used recently by Kratsch and Wahlström in the kernelization algorithm for GROUP FEEDBACK VERTEX SET parameterized by k with constant $|\Sigma|$ [17]. Currently the fastest FPT algorithm for MULTIWAY CUT is due to Cygan et al. [8], and it solves the problem in $O^*(2^k)$ time and polynomial space.

MULTIWAY CUT   **Parameter:** k
Input: An undirected graph $G = (V, E)$, a set of terminals $T \subseteq V$, and a positive integer k.
Goal: Find a set $X \subseteq V \setminus T$, such that $|X| \leq k$ and no pair of terminals from the set T is contained in one connected component of the graph $G[V \setminus X]$, or return NO if such a set X does not exist.

Proof (of Lemma 11). Firstly, we check whether the given function ϕ satisfies $\phi(z_2) = \phi(z_1) \cdot \Lambda((z_1, z_2))$, for each arc $(z_1, z_2) \in G[Z]$, since otherwise there is no set X we are looking for.

Given a Σ-labeled graph (G, Λ), a set Z, an integer k, and a function $\phi : Z \to \Sigma$, we create an undirected graph $G' = (V', E')$. As the vertex set, we set $V' = (V(G) \setminus Z) \cup T$ and $T = \{g : (u, v) \in A(G), u \in Z, v \in V(G) \setminus Z, \phi(u) \cdot \Lambda((u, v)) = g\}$. Note that in the set T there are exactly these elements of Σ, which are potential values of a consistent labeling of (G, Λ) that matches ϕ on Z. As the edge set, we set $E' = \{uv : (u, v) \in A(G[V(G) \setminus Z])\} \cup \{gv : (u, v) \in A(G), u \in Z, v \in V(G) \setminus Z, \phi(u) \cdot \Lambda((u, v)) = g\}$. We show that (G', T, k) is a YES-instance of MULTIWAY CUT iff there exists a set $X \subseteq V(G) \setminus Z$, such that there exists a consistent labeling λ of $(G \setminus X, \Lambda)$ with $\lambda|_Z = \phi$.

Let X be solution for (G', T, k). We define a consistent labeling λ of $(G \setminus X, \Lambda)$. For $v \in Z$ we set $\lambda(v) = \phi(v)$. For $v \in (V(G) \setminus Z) \setminus X$, if v is reachable from a terminal $g \in T$ in $G' \setminus X$, we set $\lambda(v) = g$. If $v \in (V(G) \setminus Z) \setminus X$ is not reachable from any terminal in G', we set $\lambda(v) = 1_\Sigma$. Since each arc in $A(G[V(G) \setminus Z])$ is labeled 1_Σ by Λ, and each vertex in $V(G) \setminus Z$ is reachable from at most one terminal in $G' \setminus X$, λ is a consistent labeling of $(G \setminus X, \Lambda)$.

Let $X \subseteq V(G) \setminus Z$ be a set of vertices of G, $|X| \leq k$, such that there is a consistent labeling λ of $(G \setminus X, \Lambda)$, where $\lambda|_Z = \phi$. By the definition of edges between T and $V(G) \setminus Z$ in G', each vertex of $V(G) \setminus Z$ is reachable from at most one terminal in G', since otherwise λ would not be a consistent labeling of $(G \setminus X, \lambda)$. Therefore, X is a solution for (G', T, k).

We can now apply the algorithm for MULTIWAY CUT of [8] to the instance (G', T, k) in order to conclude the proof.   □

4 Conclusions and Open Problems

We have shown a relatively simple fixed-parameter algorithm for GROUP FEEDBACK VERTEX SET running in time $O^*(2^{O(k \log k)})$. Our algorithm works even in a robust blackbox model, that allows us to generalize the recent algorithm for SUBSET FEED-BACK VERTEX SET [9] within the same complexity bound.

We would like to note that if we represent group elements by strings consisting g and g^{-1} for $g \in \Lambda(A(G))$ (formally, we perform the computations in the free group over generators corresponding to the arcs of the graph), then after slight modifications of our algorithm we can solve the GROUP FEEDBACK VERTEX SET problem even for infinite groups for which the word problem, i.e., the problem of checking whether results of two sequences of multiplications are equal, is polynomial-time solvable. The lengths of representations of group elements created during the computation can be bounded linearly in the size of the input graph. Therefore, if a group admits a polynomial-time algorithm solving the word problem, then we can use this algorithm as the blackbox.

Both our algorithm and the algorithm for SUBSET FEEDBACK VERTEX SET of [9] seems hard to speed up to time complexity $O^*(2^{O(k)})$. Can these problems be solved in $O^*(2^{O(k)})$ time, or can we prove that such a result would violate Exponential Time Hypothesis?

Acknowledgements. We thank Stefan Kratsch and Magnus Wahlström for inspiring discussions on graph separation problems and for drawing our attention to the GROUP FEEDBACK VERTEX SET problem.

References

1. Becker, A., Bar-Yehuda, R., Geiger, D.: Randomized algorithms for the loop cutset problem. J. Artif. Intell. Res. (JAIR) 12, 219–234 (2000)
2. Bodlaender, H.L.: On disjoint cycles. Int. J. Found. Comput. Sci. 5(1), 59–68 (1994)
3. Cao, Y., Chen, J., Liu, Y.: On Feedback Vertex Set New Measure and New Structures. In: Kaplan, H. (ed.) SWAT 2010. LNCS, vol. 6139, pp. 93–104. Springer, Heidelberg (2010)
4. Chen, J., Fomin, F.V., Liu, Y., Lu, S., Villanger, Y.: Improved algorithms for feedback vertex set problems. J. Comput. Syst. Sci. 74(7), 1188–1198 (2008)
5. Chen, J., Liu, Y., Lu, S., O'Sullivan, B., Razgon, I.: A fixed-parameter algorithm for the directed feedback vertex set problem. J. ACM 55(5) (2008)
6. Chudnovsky, M., Geelen, J., Gerards, B., Goddyn, L.A., Lohman, M., Seymour, P.D.: Packing non-zero a-paths in group-labelled graphs. Combinatorica 26(5), 521–532 (2006)
7. Cygan, M., Nederlof, J., Pilipczuk, M., Pilipczuk, M., van Rooij, J.M.M., Wojtaszczyk, J.O.: Solving connectivity problems parameterized by treewidth in single exponential time. In: Ostrovsky, R. (ed.) FOCS, pp. 150–159. IEEE (2011)
8. Cygan, M., Pilipczuk, M., Pilipczuk, M., Wojtaszczyk, J.O.: On Multiway Cut Parameterized above Lower Bounds. In: Marx, D., Rossmanith, P. (eds.) IPEC 2011. LNCS, vol. 7112, pp. 1–12. Springer, Heidelberg (2012)
9. Cygan, M., Pilipczuk, M., Pilipczuk, M., Wojtaszczyk, J.O.: Subset Feedback Vertex Set Is Fixed-Parameter Tractable. In: Aceto, L., Henzinger, M., Sgall, J. (eds.) ICALP 2011, part I. LNCS, vol. 6755, pp. 449–461. Springer, Heidelberg (2011)
10. Dehne, F.K.H.A., Fellows, M.R., Langston, M.A., Rosamond, F.A., Stevens, K.: An $O(2^{O(k)})n^3$ FPT algorithm for the undirected feedback vertex set problem. Theory Comput. Syst. 41(3), 479–492 (2007)
11. Downey, R.G., Fellows, M.R.: Fixed parameter tractability and completeness. In: Complexity Theory: Current Research, pp. 191–225 (1992)

12. Downey, R.G., Fellows, M.R.: Parameterized Complexity. Springer (1999)
13. Guillemot, S.: FPT algorithms for path-transversal and cycle-transversal problems. Discrete Optimization 8(1), 61–71 (2011)
14. Guo, J., Gramm, J., Hüffner, F., Niedermeier, R., Wernicke, S.: Compression-based fixed-parameter algorithms for feedback vertex set and edge bipartization. J. Comput. Syst. Sci. 72(8), 1386–1396 (2006)
15. Kanj, I.A., Pelsmajer, M.J., Schaefer, M.: Parameterized Algorithms for Feedback Vertex Set. In: Downey, R.G., Fellows, M.R., Dehne, F. (eds.) IWPEC 2004. LNCS, vol. 3162, pp. 235–247. Springer, Heidelberg (2004)
16. Kawarabayashi, K.-I., Wollan, P.: Non-zero disjoint cycles in highly connected group labelled graphs. J. Comb. Theory, Ser. B 96(2), 296–301 (2006)
17. Kratsch, S., Wahlström, M.: Representative sets and irrelevant vertices: New tools for kernelization. CoRR, abs/1111.2195 (2011)
18. Marx, D.: Parameterized graph separation problems. Theor. Comput. Sci. 351(3), 394–406 (2006)
19. Narayanaswamy, N.S., Raman, V., Ramanujan, M.S., Saurabh, S.: LP can be a cure for parameterized problems. In: Dürr, C., Wilke, T. (eds.) STACS. LIPIcs, vol. 14, pp. 338–349. Schloss Dagstuhl - Leibniz-Zentrum fuer Informatik (2012)
20. Raman, V., Saurabh, S., Subramanian, C.R.: Faster fixed parameter tractable algorithms for finding feedback vertex sets. ACM Transactions on Algorithms 2(3), 403–415 (2006)
21. Razgon, I., O'Sullivan, B.: Almost 2-SAT is fixed-parameter tractable. J. Comput. Syst. Sci. 75(8), 435–450 (2009)
22. Reed, B.A., Smith, K., Vetta, A.: Finding odd cycle transversals. Oper. Res. Lett. 32(4), 299–301 (2004)
23. Thomassé, S.: A $4k^2$ kernel for feedback vertex set. ACM Transactions on Algorithms 6(2) (2010)

Fault Tolerant Additive Spanners

Gilad Braunschvig*, Shiri Chechik*, and David Peleg*

Department of Computer Science and Applied Mathematics,
The Weizmann Institute, Rehovot, Israel
{gilad.braunschvig,shiri.chechik,david.peleg}@weizmann.ac.il

Abstract. Graph spanners are sparse subgraphs that preserve the distances of the original graph, up to some small multiplicative factor or additive term (known as the *stretch* of the spanner). A number of algorithms are known for constructing sparse spanners with small multiplicative or additive stretch. Recently, the problem of constructing *fault-tolerant multiplicative* spanners for general graphs was given some algorithms. This paper addresses the analogous problem of constructing *fault tolerant additive* spanners for general graphs.

We establish the following general result. Given an n-vertex graph G, if H_1 is an ordinary additive spanner for G with additive stretch α, and H_2 is a fault tolerant multiplicative spanner for G, resilient against up to f edge failures, with multiplicative stretch μ, then $H = H_1 \cup H_2$ is an additive fault tolerant spanner of G, resilient against up to f edge failures, with additive stretch $O(\tilde{f}(\alpha + \mu))$ where $\tilde{f}$ is the number of failures that have actually occurred ($\tilde{f} \leq f$).

This allows us to derive a poly-time algorithm $\mathtt{Span}_{add}^{f-t}$ for constructing an additive fault tolerant spanner H of G, relying on the existence of algorithms for constructing fault tolerant multiplicative spanners and (ordinary) additive spanners. In particular, based on some known spanner construction algorithms, we show how to construct for any n-vertex graph G an additive fault tolerant spanner with additive stretch $O(\tilde{f})$ and size $O(fn^{4/3})$.

1 Introduction

1.1 Background and Motivation

The concept of spanners is a generalization of the notion of spanning trees. A spanner of a given graph is a subgraph that faithfully preserves the distances of the original graph. Two widely studied types of spanners are *multiplicative* spanners and *additive* spanners. A multiplicative spanner of the graph G is a subgraph H that preserves the distances between any two vertices in G up to a constant multiplicative factor (referred to as the stretch of the spanner), whereas an additive spanner of G preserves distances up to a constant additive term.

* Supported in part by the Israel Science Foundation (grant 894/09), the United States-Israel Binational Science Foundation (grant 2008348), the Israel Ministry of Science and Technology (infrastructures grant), and the Citi Foundation.

M.C. Golumbic et al. (Eds.): WG 2012, LNCS 7551, pp. 206–214, 2012.
© Springer-Verlag Berlin Heidelberg 2012

More formally, a subgraph $H = (V, E_H)$ is a μ-multiplicative spanner of the graph $G = (V, E_G)$ if $E_H \subseteq E_G$ and $dist(u, v, H) \leq \mu \cdot dist(u, v, G)$ for every $u, v \in V$, where $dist(u, v, G')$ for a graph G' is the distance between u and v in G'. Similarly, a subgraph $H = (V, E_H)$ is an α-additive spanner of the graph $G = (V, E_G)$ if $E_H \subseteq E_G$ and $dist(u, v, H) \leq dist(u, v, G) + \alpha$ for every $u, v \in V$.

Additive spanners provide, in some sense, a much stronger guarantee than multiplicative ones, especially when dealing with long routes, because the *penalty* in taking the alternative route offered by the spanner is not proportional to the length of the original one, but bounded by a fixed term. Clearly any graph is a 1-multiplicative spanner and a 0-additive spanner of itself, so usually we are interested in computing spanners that are compact in the number of edges.

This paper considers settings in which the underlying graph G may occasionally suffer edge failures. In such settings, we are interested in *fault tolerant spanners*, both in the case of multiplicative and in the case of additive. These are spanners that keep the locality properties even after a number of faults occur. This robustness is important in systems that are prone to local malfunctions, like for example broken links in communication networks.

We say that a subgraph $H = (V, E_H)$ is a (μ, f)-multiplicative fault tolerant spanner of the graph $G = (V, E_G)$ if for every $F = \{e_1, \ldots, e_f\} \subseteq E_G$ and $u, v \in V$, $dist(u, v, H \setminus F) \leq \mu \cdot dist(u, v, G \setminus F)$.

Analogously, we define the notion of additive fault tolerant spanners as follows. A subgraph $H = (V, E_H)$ is an (α, f)-additive fault tolerant spanner of graph $G = (V, E_G)$ if for every $F = \{e_1, \ldots, e_f\} \subseteq E_G$ and $u, v \in V$, $dist(u, v, H \setminus F) \leq dist(u, v, G \setminus F) + \alpha$.

Fault tolerant spanners were first considered by Levcopoulos, Narasimhan and Smid [11] in the context of geometric graphs (where the nodes are assumed to be in the Euclidean space and the distance between every two nodes is the Euclidean distance between them). Levcopoulos et al. [11] presented efficient constructions for fault tolerant spanners with $(1 + \epsilon)$ multiplicative stretch. The size of the spanner was later improved by Lukovszki [12] and then by Czumaj and Zhao [6].

Constructions for *multiplicative* fault tolerant spanners for general graphs that are robust to edge or vertex failures were presented in [5], later the construction for vertex failures was improved in [7]. In this paper we show a construction for *additive* fault tolerant spanners. We deal only with edge failures. Our result relies on the existence of fault tolerant multiplicative spanners and (ordinary) additive spanners and uses algorithms for constructing such spanners as subroutines.

1.2   Our Results

In this paper we prove the following general construction scheme.

Theorem 1. *Let $G = (V, E)$ be a general graph, $H_1 = (V, E_1)$ be an α-additive spanner of G and H_2 be a μ-multiplicative fault tolerant spanner of G, resilient against up to f edge failures. Then $H = H_1 \cup H_2$ is an α'-additive fault tolerant spanner resilient against up to f failures, with additive stretch $\alpha' \leq O(\tilde{f}(\mu + \alpha))$ where $\tilde{f} \leq f$ is the number of actual faults.*

Note that the stretch guarantee depends on the number of failures that have actually occurred. Hence if no failures occur, we get a stretch bound of α, independent of f, and the stretch deteriorates as the actual number of faults increases.

As a corollary, relying on existing spanner construction algorithms, we prove that for any graph $G = (V, E)$ there exists a poly-time constructible α'-additive fault tolerant spanner $H = (V, E')$, resilient against up to f edge failures, with additive stretch $\alpha' \leq O(\tilde{f})$ and size $|E'| \leq O(fn^{4/3})$.

1.3 Related Work

Graph Spanners were first introduced by Peleg and Ullman [13] as a technique for generating synchronizers. Later, spanners were used in various contexts including routing in communication networks and distributed systems [14,17], broadcasting [10], distance oracles [3,18], etc.

It is well known how to construct $(2k - 1)$-multiplicative spanners with $O(n^{1+1/k})$ edges [2]. This size-stretch tradeoff is also conjectured to be optimal.

The picture for additive spanners is far from being complete, basically there are two known constructions for additive spanners. Aingworth et al. [1] presented a construction for 2-additive spanner with $O(n^{3/2})$ edges (for further follow-up see [8,9,19,16]). Later, Baswana et al. [4] presented an efficient construction for 6-additive spanner with $O(n^{4/3})$ edges.

In lack of truly understanding the complete picture for additive spanners, many papers consider the problem of constructing spanners with either non-constant additive stretch or with both multiplicative and additive stretch (e.g., [9,19,15,4]).

In order to achieve the constants mentioned above, we make use of existing constructions of ordinary additive spanners and multiplicative fault tolerant spanners. In practice, we may use the construction for additive spanners presented in [1,4] and the construction for multiplicative fault tolerant spanners presented in [5].

2   Preliminaries

Denote by $dist(u, v, G)$ the distance between u and v in G (if there is no path from u to v in G then $dist(u, v, G) = \infty$). Denote by $SP(u, v, G)$ the shortest path between u and v in G (if there is no path from u to v in G then $SP(u, v, G) = \emptyset$, if there is more than one such path then choose one arbitrarily). For a simple path P, denote by $|P|$ the number of edges in P. For a path P in the graph and vertices x, y on this path, denote by $P[x, y]$ the subpath of P from x to y. For a graph $G = (V, E)$ and a set of edges F, denote by $G \setminus F$ the graph $G' = (V, E \setminus F)$. Throughout this paper, when talking about fault tolerant additive spanners we distinguish between f, the maximum number of faults that the spanner can tolerate while keeping its stretch promise, and $\tilde{f}$, the number of edges that actually fail. The *size* of a graph $G(V, E)$ is defined to be its number of edges, $|E|$.

3   Constructing (α, f)-Additive Fault Tolerant Spanners

3.1   The Construction

We start by describing the algorithm for constructing a fault tolerant additive spanner and continue with the analysis of the worst case additive stretch guaranteed by this construction. We rely on the existence of known algorithms Span_{add} constructing an α-additive spanner for a given graph G (for certain values of α), and Span_{mult}^{f-t} constructing a (μ, f)-multiplicative fault tolerant spanner for G (for certain values of μ) cf. [5,1,4].

Algorithm Span_{add}^{f-t}

1. Invoke Algorithm Span_{add} to generate an α-additive spanner H_1 of G
2. Invoke Algorithm Span_{mult}^{f-t} to generate a (μ, f)-multiplicative fault tolerant spanner H_2 of G
3. $H \leftarrow H_1 \cup H_2$
4. Return H

3.2   Analysis

We next analyze the additive stretch of the subgraph H constructed by Algorithm Span_{add}^{f-t}, and prove that it is bounded by a constant linear in μ, α and $\tilde{f}$, the number of actual failures.

Our analysis proceeds as follows. We inspect the shortest path P between two vertices s and t in the graph $G \setminus F$ and distinguish several key points on that path. Then we show that the additive spanner H_1 provides for each pair of these key points a fault-free detour that is not too long. In other parts along the path P we use the fault tolerant multiplicative spanner H_2 in order to progress while avoiding faults. Finally we show that the union of all of these detours provides a path in the constructed spanner H that is completely free of faults and is close in length to the shortest path P (up to an additive term).

Consider a source vertex s, a target vertex t and a set of $\tilde{f}$ edge faults $F = \{e_1, \ldots, e_{\tilde{f}}\}$ ($\tilde{f} \leq f$). Let $P = SP(s, t, G \setminus F)$ be the shortest path from s to t after the failure event. Denote by $p(v)$ the *position* of v on P, where $p(v) = 0$ if v is the first vertex on P and $p(v) = |P|$ if v is the last vertex on P. Since H is a spanner of G, every pair of vertices $w_1, w_2 \in P$ s.t. $p(w_1) < p(w_2)$, has an alternative path in H. We refer to the shortest such path $SP(w_1, w_2, H)$, as the *bypass* of w_1 and w_2 in H.

We classify the bypasses as follows. If the bypass contains an edge in F, we say that the pair (w_1, w_2) belongs to *class* (u, v) if the first faulty edge that occurs on $SP(w_1, w_2, H)$ starting from w_1 is (u, v). Note that we take into consideration the direction of the edge, i.e., for every undirected edge e we have two different classes, one for each direction. For every pair of vertices $w_1, w_2 \in P$ s.t. $p(w_1) < p(w_2)$, if $SP(w_1, w_2, H)$ does not use any edge of F, we say that the pair (w_1, w_2) is of class Φ.

Note that if the pair (w_1, w_2) is of class Φ, then

$$dist(w_1, w_2, H \setminus F) = dist(w_1, w_2, H)$$
$$\leq dist(w_1, w_2, G) + \alpha$$
$$\leq dist(w_1, w_2, G \setminus F) + \alpha \, ,$$

and therefore

$$dist(w_1, w_2, H \setminus F) \leq |P[w_1, w_2]| + \alpha \, .$$

Next, order all pairs of vertices $(w_1, w_2) \in P$ s.t. $p(w_1) < p(w_2)$ in a lexicographic order according to the value $(p(w_1), p(w_2))$.

Lemma 1. *Let x_1, x_2 and y_1, y_2 be two pairs of vertices on path P of the same class (v, u) and $p(x_1) < p(x_2) \leq p(y_1) < p(y_2)$. Then*

$$dist(x_1, y_1, H \setminus F) \leq |P[x_1, y_1]| + 2\alpha.$$

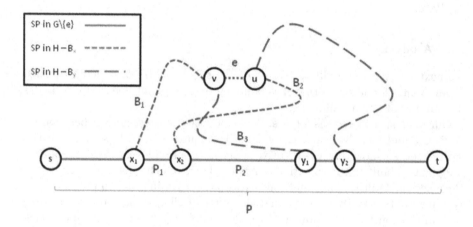

Fig. 1. Bypasses of class (v, u)

Proof. Consider the bypass $B_x = SP(x_1, x_2, H)$, $B_y = SP(y_1, y_2, H)$, and the subpaths $B_1 = B_x[x_1, v]$, $B_2 = B_x[v, x_2]$, $B_3 = B_y[y_1, v]$, $P_1 = P[x_1, x_2]$, $P_2 = P[x_2, y_1]$ (see Figure 1). By the definition of the class (v, u), the paths B_1 and B_3 do not contain any faults. Therefore,

$$dist(x_1, y_1, H \setminus F) \leq |B_1| + |B_3| \, . \tag{1}$$

Since H contains H_1 and H_1 is an additive spanner of G, $dist(w_1, w_2, H) \leq |Q| + \alpha$ for any two nodes w_1, w_2 and any path Q from w_1 to w_2 in G. In particular,

$$|B_1| + |B_2| \leq |P_1| + \alpha \tag{2}$$

and also

$$|B_3| \leq |B_2| + |P_2| + \alpha .$$ (3)

Using Inequalities (1), (2) and (3), we get that

$$
\begin{aligned}
dist(x_1, y_1, H \setminus F) &\leq |B_1| + |B_3| \\
&\leq |B_1| + |B_2| + |P_2| + \alpha \\
&\leq |P_1| + |P_2| + 2\alpha \\
&= |P[x_1, y_1]| + 2\alpha
\end{aligned}
$$

Lemma 2. *Let (x_1, x_2) be the first pair in the lexicographic order of class different than Φ and let its class be (v, u). Let (y_1, y_2) be the last pair of class (v, u) in P. Then*

$$dist(x_1, y_1, H \setminus F) \leq |P[x_1, y_1]| + 2\alpha.$$

Proof. Note that $p(x_1) \leq p(y_1)$, since (x_1, x_2) is the first pair of class (v, u). If there is only one pair of class (v, u), then the analysis is the same as if $p(x_1) = p(y_1), p(x_2) = p(y_2)$. We consider two cases. The first case is where $p(y_1) < p(x_2)$. Then the pair (x_1, y_1) is of class Φ (because it appears before the pair (x_1, x_2) in the lexicographic order and (x_1, x_2) is the first pair of class different than Φ). It follows that $dist(x_1, y_1, H \setminus F) \leq |P[x_1, y_1]| + \alpha$. The second case is where $p(x_2) \leq p(y_1)$, and then it follows from Lemma 1 that $dist(x_1, y_1, H \setminus F) \leq |P[x_1, y_1]| + 2\alpha$.

Claim. Let (x_1, x_2) be the first pair in the lexicographic order of class different than Φ and let its class be (v, u). Let (y_1, y_2) be the last pair of the class (v, u), and let s_1 be the neighbor of y_1 on the path $P[y_1, y_2]$. Then either $dist(s, s_1, H \setminus F) \leq |P[s, s_1]| + 2\alpha + \mu - 1$ or $dist(s, t, H \setminus F) \leq |P| + \alpha$.

Proof. If the class of pair (s, t) is Φ, then the bypass from s to t in H contains no failures, so $dist(s, t, H \setminus F) = dist(s, t, H) \leq |P| + \alpha$ and we are done. So now suppose the pair (s, t) is not of class Φ. Then $x_1 = s$ since otherwise $p(x_1) > p(s)$ in contradiction to the assumption that (x_1, x_2) is the first pair of class different than Φ. According to Lemma 2,

$$dist(s, y_1, H \setminus F) \leq |P[s, y_1]| + 2\alpha .$$ (4)

Since H contains H_2, which is a (μ, f)-multiplicative fault tolerant spanner of G,

$$dist(y_1, s_1, H \setminus F) \leq |P[y_1, s_1]| \cdot \mu = 1 \cdot \mu = |P[y_1, s_1]| + \mu - 1 .$$ (5)

Combining Inequalities (4) and (5), we get that

$$
\begin{aligned}
dist(s, s_1, H \setminus F) &\leq dist(s, y_1, H \setminus F) + dist(sy_1, s_1, H \setminus F) \\
&\leq |P[s, y_1]| + |P[y_1, s_1]| + 2\alpha + \mu - 1 \\
&= |P[s, s_1]| + 2\alpha + \mu - 1
\end{aligned}
$$

Lemma 3. *Let N be the number of classes on $SP(s,t,G \setminus F)$. Then*

$$dist(s,t,H \setminus F) \leq dist(s,t,G \setminus F) + N(2\alpha + \mu - 1) + \alpha .$$

Proof. We prove the lemma by induction on N. For $N = 0$, the pair (s,t) is of class Φ and the lemma holds. Assume that the lemma holds for any $n < N$. By Claim 3.2, either $dist(s,t,H \setminus F) \leq |P| + \alpha$ in which case we are done, or

$$dist(s,s_1,H \setminus F) \leq |P[s,s_1]| + 2\alpha + \mu - 1 . \tag{6}$$

Notice that the path $P[s_1,t]$ does not contain any pair of class (v,u). It follows that the number of classes on the path $P[s_1,t]$ is smaller than N, and clearly $P[s_1,t]$ is the shortest path from s_1 to t on $G \setminus F$. Therefore the induction assumption holds for the path $P[s_1,t]$, and it follows that

$$dist(s_1,t,H \setminus F) \leq dist(s_1,t,G \setminus F) + (N-1)(2\alpha + \mu - 1) + \alpha . \tag{7}$$

Combining Inequalities (6) and (7), we get that

$$
\begin{aligned}
dist(s,t,H \setminus F) &\leq dist(s,s_1,H \setminus F) + dist(s_1,t,H \setminus F) \\
&\leq dist(s,s_1,G \setminus F) + dist(s_1,t,G \setminus F) \\
&\quad + (2\alpha + \mu - 1) + (N-1)(2\alpha + \mu - 1) + \alpha \\
&= dist(s,t,G \setminus F) + N(2\alpha + \mu - 1) + \alpha .
\end{aligned}
$$

Theorem 2. *H is an (α',f)-additive fault tolerant spanner of G with $\alpha' = O(\tilde{f}(\alpha + \mu))$, and its size is $|E(H)| = |E(H_1)| + |E(H_2)|$.*

Proof. The size bound is immediate from the construction. Since there are at most $2\tilde{f}$ different classes (excluding Φ), Lemma 3 implies that $dist(s,t,H \setminus F) \leq dist(s,t,G \setminus F) + 2\tilde{f}(2\alpha + \mu - 1) + \alpha$.

A poly-time algorithm $\mathbf{Span}_{add}$ for constructing a 6-additive spanner of size $O(n^{4/3})$ for any n-vertex graph G is presented in [4]. In [5] a poly-time algorithm $\mathbf{Span}_{mult}^{f-t}$ for constructing, for any n-vertex graph, a (μ,f)-multiplicative fault tolerant spanner of size $O(fn^{1+\frac{2}{\mu+1}})$ for every odd μ and every f. Using these two results and Theorem 2, choosing $\mu = 5$, yields the following,

Corollary 1. *For every f, every graph G contains a (poly-time constructible) (α',f)-additive fault tolerant spanner of size $O(fn^{4/3})$ with $\alpha' = 32\tilde{f} + 6$.*

4   Conclusions and Open Problems

Although the concept of spanners is well established and bounds have been proven for fault tolerance in the case of multiplicative spanners, up until now there were no known constructions or lower bounds on the space and stretch of fault tolerant *additive* spanners. Hopefully this paper will open the door for more research in the field, as it leaves open several interesting problems. Our construction is relatively simple and uses previously known constructions as a *black box*. This leaves the possibility that there might exist a more sophisticated construction for fault tolerant additive spanners, with stretch that is sublinear in the number of faults f. Moreover, our analysis deals only with edge failures, and future research may focus on overcoming vertex failures. Finally, it would be interesting to consider *fault tolerant (α, β)-spanners*. For example, by simply applying our construction and analysis and using any construction for (α, β)-spanners and (μ, f)-fault tolerant multiplicative spanner as building blocks, one can present an (α', β')-spanner that is robust to f faults, where $\alpha' = \alpha^2$ and $\beta' = O(\tilde{f}(\alpha\beta + \mu))$, but this is by no means known to be the best possible.

References

1. Aingworth, D., Chekuri, C., Indyk, P., Motwani, R.: Fast estimation of diameter and shortest paths (without matrix multiplication). SIAM J. Comput. 28(4), 1167–1181 (1999)
2. Althöfer, I., Das, G., Dobkin, D., Joseph, D., Soares, J.: On sparse spanners of weighted graphs. Discrete & Computational Geometry 9, 81–100 (1993)
3. Baswana, S., Kavitha, T.: Faster algorithms for approximate distance oracles and all-pairs small stretch paths. In: Proc. IEEE Symp. on Foundations of Computer Science (FOCS), pp. 591–602 (2006)
4. Baswana, S., Kavitha, T., Mehlhorn, K., Pettie, S.: Additive spanners and (α, β)-spanners. ACM Trans. on Algo. 7, A.5 (2010)
5. Chechik, S., Langberg, M., Peleg, D., Roditty, L.: Fault-tolerant spanners for general graphs. In: Proc. 41st ACM Symp. on Theory of Computing (STOC), pp. 435–444 (2009)
6. Czumaj, A., Zhao, H.: Fault-tolerant geometric spanners. Discrete & Computational Geometry 32, 2004 (2003)
7. Dinitz, M., Krauthgamer, R.: Fault-Tolerant Spanners: Better and Simpler. In: 30th ACM Symp. on Principles of Distributed Computing (PODC), pp. 169–178 (2011)
8. Dor, D., Halperin, S., Zwick, U.: All-pairs almost shortest paths. SIAM J. Computing 29(5), 1740–1759 (2000)
9. Elkin, M., Peleg, D.: $(1 + \epsilon, \beta)$-spanner constructions for general graphs. SIAM J. Computing 33(3), 608–631 (2004)
10. Farley, A.M., Proskurowski, A., Zappala, D., Windisch, K.: Spanners and message distribution in networks. Discrete Applied Mathematics 137(2), 159–171 (2004)
11. Levcopoulos, C., Narasimhan, G., Smid, M.: Efficient algorithms for constructing fault-tolerant geometric spanners. In: Proc. 30th ACM Symp. on Theory of Computing (STOC), pp. 186–195 (1998)
12. Lukovszki, T.: New Results on Fault Tolerant Geometric Spanners. In: Dehne, F., Gupta, A., Sack, J.-R., Tamassia, R. (eds.) WADS 1999. LNCS, vol. 1663, pp. 193–204. Springer, Heidelberg (1999)
13. Peleg, D., Ullman, J.D.: An optimal synchronizer for the hypercube. SIAM J. Computing 18(2), 740–747 (1989)
14. Peleg, D., Upfal, E.: A trade-off between space and efficiency for routing tables. J. ACM 36(3), 510–530 (1989)
15. Pettie, S.: Low distortion spanners. ACM Transactions on Algorithms 6(1) (2009)

16. Roditty, L., Thorup, M., Zwick, U.: Deterministic Constructions of Approximate Distance Oracles and Spanners. In: Caires, L., Italiano, G.F., Monteiro, L., Palamidessi, C., Yung, M. (eds.) ICALP 2005. LNCS, vol. 3580, pp. 261–272. Springer, Heidelberg (2005)
17. Thorup, M., Zwick, U.: Compact routing schemes. In: Proc. 13th ACM Symp. on Parallel Algorithms and Architectures (SPAA), pp. 1–10 (2001)
18. Thorup, M., Zwick, U.: Approximate distance oracles. J. ACM 52(1), 1–24 (2005)
19. Thorup, M., Zwick, U.: Spanners and emulators with sublinear distance errors. In: 17th Symp. on Discrete Algorithms (SODA), pp. 802–809. ACM-SIAM (2006)

Multi-rooted Greedy Approximation
of Directed Steiner Trees with Applications[*]

Tomoya Hibi[**] and Toshihiro Fujito[***]

Department of Computer Science and Engineering
Toyohashi University of Technology
Tempaku, Toyohashi 441-8580 Japan
hibi@algo.cs.tut.ac.jp,
fujito@cs.tut.ac.jp

Abstract. We present a greedy algorithm for the directed Steiner tree problem (DST), where any tree rooted at any (uncovered) terminal can be a candidate for greedy choice. It will be shown that the algorithm, running in polynomial time for any constant l, outputs a directed Steiner tree of cost no larger than $2(l-1)(\ln n+1)$ times the cost of the minimum l-restricted Steiner tree. We derive from this result that 1) DST for a class of graphs, including quasi-bipartite graphs, in which the length of paths induced by Steiner vertices is bounded by some constant can be approximated within a factor of $O(\log n)$, and 2) the tree cover problem on directed graphs can also be approximated within a factor of $O(\log n)$.

1 Introduction

The Steiner tree (in graphs) problem is one of the most well-known combinatorial optimization problems with a long and rich history of being a subject for mathematical and computational studies. The problem is of fundamental importance especially in the areas of network design, network routing such as multicasting, and so on, where it is required to find a minimum cost tree, in a given edge-costed graph, spanning all the vertices specified as *terminals*. The problem is, however, one of the Karp's original NP-complete problems [9], and various approximation algorithms as well as heuristics have been developed for it. The case of undirected graphs has been and continues to be actively studied, and after the basic result of a factor 2 approximation by the minimum spanning tree based approach, the best approximation factors have been renewed several times [21,2,10,18], culminating with the recent breakthrough result with a performance ratio of $\ln(4) + \epsilon < 1.39$ [1]. It is NP-hard, on the other hand, to guarantee solutions of cost less than $96/95$ times the optimal cost [3].

---

[*] Supported in part by a Grant in Aid for Scientific Research of the Ministry of Education, Science, Sports and Culture of Japan.
[**] Currently at NTT corporation.
[***] Corresponding author. Also affiliated with Intelligent Sensing System Research Center, Toyohashi Univ. of Tech.

M.C. Golumbic et al. (Eds.): WG 2012, LNCS 7551, pp. 215–224, 2012.

The case of directed graphs in contrast has seen much less progress. The *directed Steiner tree problem* (DST), the main subject of the current paper, is to find a minimum cost subgraph T, given a directed graph $G = (V, A)$ with arc costs $c(a)$ ($\forall a \in A$), a root vertex $r \in V$, and a subset X of vertices called *terminals*, such that T contains a path starting at r and leading to every terminal, where the cost of a subgraph is defined to be the total cost of arcs in it. Those non-terminals ($\in V \setminus X$) are called *Steiner vertices*, and a (directed) tree in this paper is assumed to be the one in which every arc is directed away from its root towards a leaf. A *directed Steiner tree* ("*dst*" for short) is a tree spanning all the terminals, and DST is then equivalently defined to be the problem of computing a minimum cost dst rooted at r. The first nontrivial approximation algorithm for DST was developed by Charikar et al., achieving a performance ratio of $l^2(l-1)|X|^{1/l}$ in time $O(n^l|X|^{2l})$ for any $l > 1$ [4,5]. The algorithm thus approximates DST within a factor of $O(n^\epsilon)$ for any $\epsilon > 0$ in polynomial time and at the same time within a factor of $O(\log^3 |X|)$ in quasi-polynomial time of $n^{O(\log |X|)}$, raising a conjecture that a polylogarithmic approximation of DST might be possible. It has, in fact, been attempted to improve the Charikar et al.'s approximation bound [22,12,15], and without success, however, the $O(n^\epsilon)$ factor of Charikar et al. [4] remains as the best performance ratio known today, and the polylogarithmic approximability is still wide open for DST.

1.1   Greedy Approaches for Approximating DST

It is natural to consider DST to be a generalization of the set cover problem by representing the notion of coverage by "reachability" from the root. Here, any trees rooted at r are the subsets for covering elements and terminals are those elements to be covered. The greedy set cover algorithm repeatedly selects into a solution a most "cost-effective" subset until all the elements become covered. Here, the cost-effectiveness of a subset is measured by the ratio of its cost to the number of yet uncovered elements in it, and it is the *density* $d(T)$ of a tree T rooted at r in DST defined to be the ratio of its cost to the number of terminals in it not yet reachable from r, i.e., $d(T) = c(T)/(\#$ of terminals in T not reachable from $r)$. If it were possible to compute a rooted tree with minimum density in polynomial time, it would lead to an $O(\log n)$-approximation for DST as is for the set cover problem (or in more general, a factor of α approximation of the minimum density tree yields an $O(\alpha \log n)$ approximation of DST). It is hard to compute it exactly, however, and this is why all the greedy approaches for DST including ours have had to settle for trees with approximately lowest density in their greedy choices.

Definition 1.   – *An l-level tree is a tree in which no leaf is more than l arcs away from the root.*
– *A Steiner tree in which all the terminals are at leaves (or at the root) is called a full Steiner tree. Any Steiner tree can be decomposed into arc-disjoint full Steiner trees (full components) by splitting all the non-leaf terminals, each of them into a leaf of one tree and a root of the other tree.*
– *A Steiner tree is l-restricted if every full component in it is an l-leveled tree.*

The algorithm of Charikar et al. uses an l-level tree (in the metric closure of an original graph) of which density is a factor of at most $l - 1$ away from that of the minimum density l-level tree [4]. Zosin and Khuller show that a tree of density bounded by $D + 1$ times the minimum density (of any tree rooted at r) is polynomially computable if $V - X$ induces a tree of depth D [22]. Either algorithm considers such trees rooted at a fixed root r only. Zelikovsky's algorithm, based on a different type of density function, considers any full Steiner tree, and it computes (and adopts) a tree of which density is a factor of at most $(2 + \ln |X|)^{l-2}$ away from that of the minimum density l-restricted Steiner tree [20].

1.2 Our Approach and Contributions

The greedy algorithm designed in this paper iteratively chooses a full Steiner tree rooted at either r or any "uncovered" terminal in a way similar to Zelikovsky's [20]. It uses a density function different from Zelikovsky's and naturally from Charikar et al.'s and Zosin-Khuller's as the notion of coverage cannot be represented by "reachability from r" in our setting. A terminal in a full Steiner tree T rooted at either r or any terminal is considered "covered by T" if it is not the root of T, and the density $d(T)$ is then redefined to be the ratio of its cost to the number of yet uncovered non-root terminals in T. The main theorem (Theorem 1) of the paper states that this algorithm computes a dst in polynomial time of which cost is no larger than $2(l - 1)(\ln |X| + 1)$ times the cost of the minimum l-restricted Steiner tree. It is interesting to compare this with the Zelikovsky's [20] and Charikar et al.'s [4] algorithms; the former outputs a dst of cost no larger than $(\ln |X| + 2)^{l-1}$ times the cost of the minimum l-restricted Steiner tree, whereas the latter outputs an l-level dst of cost at most $l(l - 1)$ times the cost of the minimum l-level dst.

The main result described above does not lead to an improved performance ratio for DST per se, yet some new approximation results can be derived from it. One is the case of DST for a class of graphs G where Steiner vertices induce no path of length longer than l in G. A *quasi-bipartite* graph belongs to such a class with $l = 0$, and DST is known to be hard to approximate better than $O(\log n)$ even when inputs are restricted to quasi-bipartite graphs. It can be shown from the main theorem that our greedy algorithm approximates DST for such a case within a factor of $O(\log |X|)$ for any constant l. When combined with the $\Omega(\log^{2-\epsilon} n)$ approximation hardness of DST on general graphs [8], this separates the approximability of DST between the cases of quasi-bipartite graphs and general graphs. It has been repeatedly observed, in case of undirected graphs, that the Steiner tree problem is easier to approximate on quasi-bipartite graphs than on general graphs since [17], and it is here proven to be true in case of directed graphs.

Another application of the main theorem presented is in approximation of the *Directed Tree Cover problem (DTC)*. It is required in DTC, given an arc-costed directed graph G and a root vertex r, to compute a minimum directed tree T rooted at r such that there exists a path in T from r to every arc in G. In case of undirected graphs, the tree cover problem is known to be approximable within a factor of 2, by a simple algorithm for the uniform costs [19] and by a not so simple one for general costs [7]. The approximability of DTC, on the other hand, has remained wide open as mentioned in [11] [1]. It will be shown, by reducing general DTC to DST on bipartite graphs with terminal-Steiner bipartition, that DTC can be approximated within a factor of $O(\log n)$, again matching the known approximation lower bound of $\Omega(\log n)$ for DTC.

2   Algorithm

Let $G = (V, A)$ be a directed graph with non-negative arc cost $c(a)$ for each arc $a \in A$, a node r designated as a root, and a set $X \subseteq V$ of terminals. The greedy algorithm presented below grows a subgraph P of G in sequence, initially consisting of r and all the terminals only (no arcs), by iteratively adding trees in G rooted either at the "real" root r or at some terminals not yet covered. Here any terminal becomes "covered" whenever a tree containing it as a "non-root" is added to P by the algorithm (or equivalently, we may contract such a tree into a single vertex). The algorithm repeats this as long as uncovered terminals remain in G, and eventually ends up with a subgraph of G composed of all the trees added with all the terminals covered by some of them. As it contains a path from r to every terminal in X, a dst spanning all the terminals can be found within it, and it will be output by the algorithm.

Algorithm 1: Multi-Rooted Greedy

Input: $G = (V, A)$, root $r \in V$, and terminal set $X \subseteq V$
Output: a Steiner tree rooted at r spanning all terminals in X

1 Initialize: $P = (X \cup \{r\}, \emptyset)$ and $C = \emptyset$;
2 **while** *there remain (uncovered) terminals in $X \setminus C$* **do**
3   Compute a tree T of low density rooted at any vertex in $\{r\} \cup (X \setminus C)$;
4   Set $P = (V_P \cup V(T), A_P \cup A(T))$;
5   Letting u be the root of T, add all terminals in T but u to C by setting $C = C \cup (X(T) - u)$;
6   Reset $c(a) = 0$ for all $a \in A(T)$ and recompute the metric closure of G;
7 **end**
8 Compute and output any tree within P rooted at r spanning all terminals in X.

It remains to elaborate on how to compute a small density tree T in step 3, and we use the algorithm developed by Kortsarz, Peleg, and Charikar et al. [13,4], assuming that we are working with the metric closure of the current graph in what follows. Let $d_l^*(k, v, X)$ denote the minimum density of l-level trees T_l rooted at v containing any k terminals from X. It was shown that good approximation of the minimum density tree among l-level trees is possible when l is a constant [4]:

---

[1] In fact an $O(\log n)$-approximation of DTC was claimed in error [14].

Lemma 1 ([4]). *For any $v \in V, X \subseteq V, 1 \le k \le |X|$, and $l \ge 2$, an l-level tree T_l can be found in time $O(n^l k^{2l-1})$ such that $d(T_l) \le (l-1)d_l^*(k, v, X)$.*

Let $d_l^*(X)$ denote the minimum density of l-level trees T_l rooted at any vertex in $\{r\} \cup X$ containing any number of terminals from X. It follows immediately from Lemma 1, by running the algorithm used in it for each $1 \le k \le |X|$ and $v \in \{r\} \cup X$, that

Lemma 2. *For any $X \subseteq V$ and $l \ge 2$, an l-level tree T_l rooted at some vertex in $\{r\} \cup X$ can be found in time $O(n^l |X|^{2l+1})$ such that $d(T_l) \le (l-1)d_l^*(X)$.*

In the next section the approximation performance of Algorithm 1 is analyzed assuming that trees T are computed in step 3 by the algorithm of Lemma 2.

3   Analysis

Definition 2. *Let $P = (V_P, A_P)$ be a subgraph of G and $X_P \subseteq X$. $(P, \{r\} \cup X_P)$ is called a partial Steiner tree (PST) for (G, r, X) if*

- *$\{r\} \cup X \subseteq V_P$, and*
- *every vertex in V_P is reachable within P from some vertex in $\{r\} \cup X_P$.*

Lemma 3. *Let $P = (V_P, A_P)$ and C be a subgraph of G and a set of covered terminals, respectively, computed at any iteration of the while-loop during the execution of Algorithm 1. Then, $(P, \{r\} \cup (X \setminus C))$ with $X_P = X \setminus C$ is a PST for (G, r, X).*

Proof. Initially, $P = (\{r\} \cup X, \emptyset)$ is clearly a PST for (G, r, X). Suppose at some iteration a tree T is added to P and let u be its root. Then, all the terminals in T but u become covered (and leave X_P), but all the vertices in T are reachable from u within T. So, any vertex reachable from those newly covered terminals before addition of T becomes reachable now from u after addition of T.

For any PST $(P = (V_P, A_P), \{r\} \cup X_P)$ for (G, r, X) every vertex $v \in V_P$ is reachable from some vertex in $\{r\} \cup X_P$ within P. If v is reachable from more than one vertex in $\{r\} \cup X_P$, choose v itself if $v \in \{r\} \cup X_P$, but choose any one of them otherwise, and denote it by $r(v)$. Then, V_P is partitioned into a family of disjoint subsets, each of them consisting of vertices $v \in V_P$ with a common representative vertex $r(v) \in \{r\} \cup X_P$, and the subset v belongs to is referred to as $V(r(v))$ for any $v \in V_P$.

Fix one l-restricted Steiner tree $T_{(l)}$ for (G, r, X). Since $X_P \subseteq X$, each vertex in X_P is contained in $T_{(l)}$. Denote by $s(v)$ for $v \in X_P$ the lowest ancestor of v within $T_{(l)}$ such that $r(v) \ne r(s(v))$ (thus, $V(r(v)) \cap V(r(s(v))) = \emptyset$). As $T_{(l)}$ is rooted at r and $v \ne r$, $s(v)$ exists for any $v \in X_P$. Consider the set of $s(v)$-v paths for all $v \in X_P$, and denote it by $\mathcal{T}_0$, i.e., $\mathcal{T}_0 = \{s(v)\text{-}v \text{ path} \mid v \in X_P\}$. The following properties of the paths in $\mathcal{T}_0$ can be verified by recalling the choice of $s(v)$ for $v \in X_P$, that all the paths in $\mathcal{T}_0$ are parts of tree $T_{(l)}$ and that $V(r(u)) \cap V(r(v)) = \emptyset$ for any $u, v \in X_P$ if $u \ne v$.

Lemma 4. *The paths in T_0 possess the following properties:*

1. *On any $s(v)$-v path all the vertices but $s(v)$ come from $V(r(v)) \cup (V \setminus V_P)$.*
2. *Suppose two paths, $s(u)$-u and $s(v)$-v, in T_0 overlap.*
   (a) *If $s(u) = s(v)$, $s(u)$-u and $s(v)$-v paths overlap only in their initial segments, i.e., the subpaths starting at $s(u) = s(v)$ followed by a sequence of vertices in $V \setminus V_P$ only.*
   (b) *If $s(u) \neq s(v)$,*
      i. *either $s(u)$ is a proper ancestor of $s(v)$ in $T_{(l)}$, or the other way around, and*
      ii. *if $s(u)$ is a proper ancestor of $s(v)$ (the other case is similar), u and $s(v)$ must belong to the same set $V(r(u)) = V(r(s(v)))$, and therefore, they can overlap only in the initial segment of $s(v)$-v path consisting of $s(v)$ and vertices in $V \setminus V_P$ only.*

For any $v \in X_P$, collect all the $s(u)$-u paths in T_0 with $s(u) = s(v)$, and merge them into a single tree rooted at the common starting vertex $s(v)$ (if $s(u) \neq s(v)$ for any $u \in X_P - \{v\}$, $s(v)$-v path is such a tree by itself). Call a subtree of $T_{(l)}$ thus constructed from some paths in T_0 and rooted at $s(v)$ as $s(v)$-*tree*, and denote by T_1 the collection of all $s(v)$-trees.

Lemma 5. *T_1 satisfies the following properties:*

1. *Every vertex in X_P occurs at a leaf of exactly one tree in T_1.*
2. *No arc of $T_{(l)}$ occurs in more than two trees of T_1.*
3. *For any tree with multiple leaves in T_1, any branching occurs within the distance of $l - 1$ from the root.*

Proof. 1. This is clear from the construction of T_0 and T_1.
2. Suppose an arc (y, z) of $T_{(l)}$ is shared by three trees, $s(v_1)$-, $s(v_2)$-, and $s(v_3)$-trees, from T_1. Then, it must be the case that no two of $s(v_1)$, $s(v_2)$, and $s(v_3)$ can coincide, and that all of $s(v_1)$, $s(v_2)$, and $s(v_3)$ are ancestors of y in $T_{(l)}$. Then, $s(v_1)$-z, $s(v_2)$-z, and $s(v_3)$-z paths are all the initial segments of distinct three paths in T_0, all of them lying on the r-z path of $T_{(l)}$. There is no way, however, that they can satisfy property 2(b)ii. of Lemma 4.
3. Recall that any paths running from the root to leaves in a tree of T_1 come from T_0, and hence, any two of them must satisfy property 2(a) of Lemma 4. Recall also that $T_{(l)}$ is an l-restricted tree, and hence, the length of a consecutive run of Steiner vertices on any path is bounded by $l - 2$ in $T_{(l)}$. Therefore, any two paths starting at the same vertex must branch out within the distance of $l - 1$ from the starting vertex.

Let us assume henceforth that PST $(P, \{r\} \cup X_P)$ for (G, r, X) is the one generated during the execution of Algorithm 1 (Lemma 3); i.e., $P = (V_P, A_P)$ and C are a subgraph of G and a set of covered terminals, respectively, computed at any iteration of the while-loop, and $X_P = X \setminus C$. For any $s(v)$-tree T in T_1, we do the following operations:

1. When a path is followed from $s(v)$ to a leaf w, no branching occurs after passing the $(l-1)$st vertex u (property 3 in Lemma 5). As we are working with the metric closure of the current graph, there exists an arc (u, w) of cost no larger than that of the subpath running from u to w. So, replacing such a subpath by such an arc on any path leading to a leaf if it is longer than l, T becomes an l-level tree of no larger cost.

2. Recall that $s(v)$ is reachable from $r(s(v))$ within P, where every arc has a zero cost (due to step 6 of Algorithm 1). Hence, by connecting $r(s(v))$ directly to each child of $s(v)$ by an arc, the root of $s(v)$-tree can be replaced by $r(s(v))$ without increasing its cost nor its levels.

Let us denote by $\mathcal{T}_2$ the set of l-level trees resulting from applications of the operations above to the $s(v)$-trees in $\mathcal{T}_1$. The next lemma is a key to our main theorem and we prove it by examining properties of $\mathcal{T}_2$:

Lemma 6. *Let $T_{(l)}$ be any l-restricted Steiner tree rooted at r in G. Suppose $(P, \{r\} \cup X_P)$ is a PST for (G, r, X) generated by Algorithm 1 during its execution. Then, there exists an l-level tree T_l in G rooted at some vertex in $\{r\} \cup X_P$ such that $d(T_l) \le 2c(T_{(l)})/|X_P|$.*

Proof. Consider $\mathcal{T}_2$. Each l-level tree in it is rooted at some vertex in $\{r\} \cup X_P$ (due to operation 2), and every vertex in X_P occurs at a leaf of some tree in $\mathcal{T}_2$ (Lemma 5.1). Therefore, all the terminals in X_P can be covered by using all the trees in $\mathcal{T}_2$. The total cost of trees in $\mathcal{T}_2$ is no larger than that of those trees in $\mathcal{T}_1$. The latter can be bounded by $2c(T_{(l)})$ because of Lemma 5.2. Therefore, those trees in $\mathcal{T}_2$ can jointly cover $|X_P|$ uncovered terminals, and it costs at most $2c(T_{(l)})$ to do so. Hence, there must be a tree T_l in $\mathcal{T}_2$ of density no larger than $2c(T_{(l)})/|X_P|$. $\square$

We are now ready to bound the cost of a dst output by Algorithm 1:

Theorem 1. *Let $\mathrm{OPT}_{(l)}$ denote the cost of the minimum l-restricted Steiner tree for (G, r, X). Algorithm 1 computes a dst of cost no larger than $2(l-1)H(|X|)\mathrm{OPT}_{(l)}$, in time $O(n^l|X|^{2l+2})$, where $H(k)$ is the kth harmonic number and $H(k) = 1 + 1/2 + \cdots + 1/k$.*

Proof. The running time is dominated by that consumed in step 3, which is executed in total $O(|X|)$ times.

Suppose T is the tree computed in step 3 at any iteration of the while-loop. Assign $d(T)$ to each of the terminals newly covered by T. Total value assigned in one iteration of the while-loop coincides with the cost of T chosen during the iteration by definition of density $d(T)$. By doing this at every iteration, each terminal in X gets assigned with some density exactly once, and hence, total cost of trees chosen by Algorithm 1 can be recovered by collecting all the density values assigned to the terminals in X.

The density $d(T)$ of T can be bounded by $2(l-1)\mathrm{OPT}_{(l)}/|X \setminus C|$ according to Lemmas 2 and 6. Order the terminals in X in the order of becoming covered by Algorithm 1, and let x_i be the ith terminal covered by Algorithm 1 for

$1 \leq i \leq |X|$. As there remain at least $|X| - (i - 1)$ uncovered terminals when x_i is covered, the density x_i receives is no larger than $2(l-1)\text{OPT}_{(l)}/(|X|-(i-1))$. Therefore, the total density assigned to all the terminals in X is bounded by

$$\sum_{i=1}^{|X|} \frac{2(l-1)\text{OPT}_{(l)}}{|X| - (i-1)} = 2(l-1)\text{OPT}_{(l)} \sum_{i=1}^{|X|} \frac{1}{i} = 2(l-1)\text{OPT}_{(l)} H(|X|).$$

The output tree is a subgraph of PST P, of which cost is bounded as above, and the claim follows.

4   Applications

A graph $G = (V, A)$ is called *quasi-bipartite* (with respect to terminal set X) when the set of Steiner vertices ($= V \setminus X$) induces no arc in G. It is easy to confirm the following corollary of Theorem 1 by observing that every Steiner tree is $(l + 2)$-restricted in such special inputs as given below:

Corollary 1. *When inputs $(G = (V, A), r, X)$ are limited to those in which $V \setminus X$ induces no path of length longer than l, Algorithm 1 approximates DST within a factor of $2(l + 1)H(|X|) = O(l \log |X|)$, running in polynomial time for any constant l. In particular, when inputs are restricted to quasi-bipartite graphs, DST can be approximated within a factor of $2H(|X|) \leq 2\ln |X| + 2$.*

The set cover problem can be embedded in DST on bipartite graphs $G = (X \cup (V \setminus X), A)$. Because of $\Omega(\log n)$ lower bound for set cover approximation [16,6], it can be said that Algorithm 1 yields an optimal approximation for such special cases as given in Corollary 1 for any constant l.

Let us turn our attention to the directed tree cover problem (DTC). The set cover problem can be embedded in DTC by the almost same construction as in DST, and hence, the same approximation hardness of $\Omega(\log n)$ lower bound holds. For the upper bound, we use the following reduction:

Lemma 7. *DTC on general graphs is reducible to DST on bipartite graphs with terminal-Steiner bipartition in an approximation preserving manner.*

Proof. Let $(G = (V, A), r, c)$ be an instance of DTC. For each arc $a = (u, v) \in A$, introduce a new vertex x_a as a terminal for DST. Each arc $a = (u, v) \in A$ is replaced by three arcs, $(u, x_a), (x_a, v)$, and (v, x_a), and the costs of these arcs are set equal to $0, c(a)$, and 0, respectively. An instance $(G' = (V', A'), r, X, c')$ of DST is constructed from a DTC instance (G, r, c) in this way such that $X = \{x_a \mid a \in A\}, V' = V \cup X, A' = \{(u, x_a), (x_a, v), (v, x_a) \mid a = (u, v) \in A\}$, and $\forall a \in A, c'(u, x_a) = c'(v, x_a) = 0, c'(x_a, v) = c(a)$. It is not hard to verify that a tree cover of any cost exists in (G, r, c) if and only if a dst of the same cost exits in (G', r, X, c'). It is also clear that $G' = (V \cup X, A')$ constructed from $G = (V, A)$ in the reduction is a bipartite graph for any G.

Due to this lemma, the following optimal approximation for DTC follows from Theorem 1 as in Corollary 1:

Theorem 2. *DTC can be approximated by Algorithm 1 within a factor of* $2H(|A|) \leq 2\ln|A| + 2$.

References

1. Byrka, J., Grandoni, F., Rothvoß, T., Sanità, L.: An improved LP-based approximation for Steiner tree. In: Proc. 42nd STOC, pp. 583–592 (2010)
2. Berman, P., Ramaiyer, V.: Improved approximations for the Steiner tree problem. In: Proc. 3rd SODA, pp. 325–334 (1992)
3. Chlebík, M., Chlebíková, J.: The Steiner tree problem on graphs: Inapproximability results. Theory Comput. Syst. 406(3), 207–214 (2008)
4. Charikar, M., Chekuri, C., Cheung, T., Dai, Z., Goel, A., Guha, S., Li, M.: Approximation algorithms for directed Steiner tree problems. J. Algorithms 33, 73–91 (1999)
5. Calinescu, G., Zelikovsky, A.: The polymatroid Steiner problems. J. Comb. Opt. 9, 281–294 (2005)
6. Feige, U.: A threshold of $\ln n$ for approximating set cover. J. ACM 45(4), 634–652 (1998)
7. Fujito, T.: How to Trim an MST: A 2-Approximation Algorithm for Minimum Cost Tree Cover. In: Bugliesi, M., Preneel, B., Sassone, V., Wegener, I. (eds.) ICALP 2006. LNCS, vol. 4051, pp. 431–442. Springer, Heidelberg (2006)
8. Halperin, E., Krauthgamer, R.: Polylogarithmic inapproximability. In: Proc. 35th STOC, pp. 585–594 (2003)
9. Karp, R.M.: Reducibility among combinatorial problems. In: Complexity of Computer Computations, pp. 85–103. Plenum Press, New York (1972)
10. Karpinski, M., Zelikovsky, A.Z.: New approximation algorithms for the Steiner tree problem. J. Comb. Opt. 1, 47–65 (1997)
11. Könemann, J., Konjevod, G., Parekh, O., Sinha, A.: Improved Approximations for Tour and Tree Covers. In: Jansen, K., Khuller, S. (eds.) APPROX 2000. LNCS, vol. 1913, pp. 184–193. Springer, Heidelberg (2000)
12. Konjevod, G.: Directed Steiner trees, linear programs and randomized rounding, 8 pages (2005) (manuscript)
13. Kortsarz, G., Peleg, D.: Approximating the weight of shallow Steiner trees. Discrete Applied Mathematics 93, 265–285 (1999)
14. Nguyen, V.H.: Approximation Algorithm for the Minimum Directed Tree Cover. In: Wu, W., Daescu, O. (eds.) COCOA 2010, Part II. LNCS, vol. 6509, pp. 144–159. Springer, Heidelberg (2010)
15. Rothvoß, T.: Directed Steiner tree and the Lasserre hierarchy. ArXiv e-prints (November 2011)
16. Raz, R., Safra, S.: A sub-constant error-probability low-degree test, and a sub-constant error-probability PCP characterization of NP. In: Proc. 29th STOC, pp. 475–484 (1997)
17. Rajagopalan, S., Vazirani, V.V.: On the bidirected cut relaxation for the metric Steiner tree problem. In: Proc. 10th SODA, pp. 742–751 (1999)
18. Robins, G., Zelikovsky, A.: Tighter bounds for graph Steiner tree approximation. SIAM J. Discrete Math. 19, 122–134 (2005)

19. Savage, C.: Depth-first search and the vertex cover problem. Inform. Process. Lett. 14(5), 233–235 (1982)
20. Zelikovsky, A.: A series of approximation algorithms for the acyclic directed Steiner tree problem. Algorithmica 18, 99–110 (1997)
21. Zelikovsky, A.: An 11/6-approximation algorithm for the network Steiner problem. Algorithmica 9, 463–470 (1993)
22. Zosin, L., Khuller, S.: On directed Steiner trees. In: Proc. 13th SODA, pp. 59–63 (2002)

Approximating Infeasible 2VPI-Systems

Neele Leithäuser[1], Sven O. Krumke[2], and Maximilian Merkert[2]

[1] ITWM Fraunhofer Institut für Techno- und Wirtschaftsmathematik,
Fraunhofer-Platz 1, 67663 Kaiserslautern, Germany
neele.leithaeuser@itwm.fraunhofer.de
[2] Dept. of Mathematics, University of Kaiserslautern, Paul-Ehrlich-Str. 14,
67663 Kaiserslautern, Germany
{krumke,merkert}@mathematik.uni-kl.de

Abstract. It is a folklore result that testing whether a given system of equations with two variables per inequality (a 2VPI system) of the form $x_i - x_j = c_{ij}$ is solvable, can be done efficiently not only by Gaussian elimination but also by shortest-path computation on an associated constraint graph. However, when the system is infeasible and one wishes to delete a minimum weight set of inequalities to obtain feasibility (MINFS2$^=$), this task becomes NP-complete.

Our main result is a 2-approximation for the problem MINFS2$^=$ for the case when the constraint graph is planar using a primal-dual approach. We also give an α-approximation for the related maximization problem MAXFS2$^=$ where the goal is to maximize the weight of feasible inequalities. Here, α denotes the arboricity of the constraint graph. Our results extend to obtain constant factor approximations for the case when the domains of the variables are further restricted.

1 Introduction

The problem of checking whether a system of linear (in-)equalities admits a solution is efficiently solvable by means of linear optimization methods. However, in many applications it is known that a linear system is unsolvable. An instance $I = (X, C)$ of the maximum feasible subsystem problem (MAXFS) consists of a finite set X of variables and a set C of constraints on the variables, and the goal is to select a subset $C' \subseteq C$ of the constraints of maximum size such that the corresponding reduced system (X, C') is feasible. This problem has a wide range of applications (see e.g. [2] and the references therein) and is well-known to be NP-complete. The corresponding minimization problem MINFS asks to delete a minimum number of constraints in order to obtain a feasible system.

In this paper we consider the versions of MINFS and MAXFS where the constraints are restricted to the special form $x_i - x_j = c_{ij}$, in particular there are only two variables per inequality. Such a system is commonly referred to as a 2VPI-system. We are also given a nonnegative weight for each constraint, specifying the cost of removing it from the system. The problems MINFS2$^=$ and MAXFS2$^=$ of deleting a minimum cost set of constraints or retaining a maximum cost set of constraints, respectively, are still NP-hard to solve, even in case of unit costs for the constraints.

M.C. Golumbic et al. (Eds.): WG 2012, LNCS 7551, pp. 225–236, 2012.

With each instance $I = (X, C)$ of a 2VPI-system, one can associate a corresponding *constraint graph* G_I as follows. For each variable $x_i \in X$ there is a vertex i in G_I and for each constraint of the form $x_i - x_j = c_{ij}$ there is a directed arc a from i to j of length $\ell(a) := c_{ij}$. This arc may be traversed in both directions, where the length for traversing in direction from j to i is given by $-c_{ij}$. It is then easy to see that the system given in $I = (X, C)$ is feasible if and only if G_I does not contain a negative length (undirected) cycle. Thus, the consistency of such a system can be tested by an all-pairs shortest path computation in $O(mn)$ time, where $n = |X|$ and $m = |C|$ denote the number of variables and constraints, respectively.

This graph-theoretic view on a 2VPI-system which we are going to take in this paper means that the problem MinFs2$^=$ is equivalent to the following problem: Delete a minimum cost set of the arcs of a given graph such that the resulting graph does not contain any negative length cycle.

MaxFs and MinFs have been studied extensively in terms of complexity and approximability, see, e.g., [2,10,4,13] and the references therein, and also many heuristic algorithms have been proposed (e.g., [3,1]), but MaxFs2$^=$ and MinFs2$^=$ have not been studied specifically so far in literature.

We present the first constant factor approximation algorithm for MaxFs2$^=$ when restricted to planar graphs. We also give an exact pseudo-polynomial time algorithm for series-parallel graphs and a polynomial time algorithm for extension-parallel graphs. Moreover we derive new hardness results for generalized versions of the problems, where the variables have values in some domain other than $\mathbb{R}$ or the operations are replaced by the group operations in some group.

Our paper is organized as follows: Section 3 contains our new hardness results which among other things show that MinFs2$^=$ and MaxFs2$^=$ are still NP-hard to solve, even if the corresponding constraint graph is planar. In Section 4 we give a factor 2-approximation for MinFs2$^=$ on planar graphs. Our algorithm is based on the primal-dual framework by Goemans and Williamson [8] and their work on feedback vertex problems in [9]. In fact, although we apply a very similar technique as Goemans and Williamson do in [9] for minimum feedback vertex problems in planar graphs, we obtain an improved performance guarantee of 2 compared to 3 in [9] (their best result is a factor of $\frac{9}{4}$).[1] In Section 5 we obtain approximation algorithms for the maximization version MaxFs2$^=$. In Section 6 we extend our results to the case of additional restrictions on the domains of the variables.

2   Preliminaries and Problem Definition

An instance I of MaxFs2$^=$ (and MinFs2$^=$) is given by a finite set X of variables and a finite set $\mathcal{E}$ of equations of the form $x_i - x_j = c_{ij}$ over the variables in X. Also, for each equation $e \in \mathcal{E}$ we are given a nonnegative weight w_e.

---

[1] We stress that, unfortunately, our results do not imply improved approximation results for the feedback vertex problem.

We call a subset $\mathcal{E}' \subseteq \mathcal{E}$ of the equations *consistent* if the included equations can be satisfied simultaneously, otherwise $\mathcal{E}'$ is termed *inconsistent*. The goal in MAXFS2$^=$ is to find a consistent subset $\mathcal{E}'$ of the equations of maximum weight $w(\mathcal{E}') = \sum_{e \in \mathcal{E}'} w_e$. Similarly, MINFS2$^=$ asks to remove a minimum weight subset of the equations to obtain a consistent system.

With each instance I of MAXFS2$^=$ (and MINFS2$^=$), the associated *constraint graph* G_I contains a vertex i for each variable $x_i \in X$ and for each constraint $x_i - x_j = c_{ij}$ a directed arc a from i to j of length $\ell(a) := c_{ij}$. As mentioned before, this arc may be traversed in both directions, where the length for traversing in direction from j to i is given by $-c_{ij}$. In all what follows we identify a constraint e with the corresponding arc in G_I. As a consequence, in addition to its length, any arc e has an associated weight $w_e \geq 0$.

Since we allow traversing arcs against their directions at the corresponding negative length, G_I contains a negative length (undirected) cycle if and only if G_I contains a cycle of positive length. Thus, we call any (undirected) cycle in G_I of nonzero length an *inconsistent cycle*. By the observations made in the introduction, a subset of the equations is consistent if and only if the corresponding subgraph of G_I does not contain an inconsistent cycle.

Thus, one can state MAXFS2$^=$ (MINFS2$^=$) equivalently in graph theoretic terms as follows: Given a directed graph $G = (V, E)$ with lengths and weights on the arcs, find a maximum weight subset of the arcs (delete a minimum weight subset of the arcs) such that the resulting subgraph does not contain any inconsistent cycle. We call such a subgraph a *consistent subgraph*. In the sequel we assume that all graphs are (weakly) connected, since otherwise we can consider the problem on each connected component separately.

3   Complexity of MinFs2$^=$ and MaxFs2$^=$

Although MAXFS2$^=$ and MINFS2$^=$ were defined in such a way that each equality contains exactly two variables, in the following we will nevertheless consider equations with only one variable, since this apparently more general case can easily be reduced to MAXFS2$^=$ and MINFS2$^=$, respectively: Introduce a new variable x_0 and add it with appropriate sign to every equation that has only one variable, e.g., an equation $x_i = c$ is replaced by $x_i - x_0 = c$. This problem has an optimal solution x^* with $x_0^* = 0$: Add $-\hat{x}_0$ to every component of an arbitrary given solution $\hat{x}$ to obtain a new solution with $x_0 = 0$ and equally many (exactly the same) satisfied equations.

Theorem 1. MAXFS2$^=$ *is APX-hard.*

Proof. We provide an L-reduction from the APX-complete optimization problem MAX-2-SAT: Given an instance I_1 of MAX-2-SAT with variables x_i, $i = 1, \ldots, n$ and clauses $y_{c_{j1}} \lor y_{c_{j2}}, j = 1, \ldots, m$ with exactly two literals, $y_{c_{jl}} \in \{x_{c_{jl}}, \overline{x}_{c_{jl}}\}$, construct an instance I_2 of MAXFS2$^=$ as follows: For every clause $y_{c_{j1}} \lor y_{c_{j2}}$, $j \in \{1, \ldots, m\}$ create the following 14 equations:

$$y_{c_{j1}} - \overline{y}_{c_{j2}} = 0$$
$$y_{c_{j1}} - \overline{y}_{c_{j2}} = 1$$
$$(\text{where } \overline{\overline{x}}_{c_{jl}} := x_{c_{jl}})$$
$$x_{c_{j1}} = 1, \quad x_{c_{j2}} = 1, \quad \overline{x}_{c_{j1}} = 1, \quad \overline{x}_{c_{j2}} = 1$$
$$x_{c_{j1}} = 0, \quad x_{c_{j2}} = 0, \quad \overline{x}_{c_{j1}} = 0, \quad \overline{x}_{c_{j2}} = 0$$
$$x_{c_{j1}} - \overline{x}_{c_{j1}} = 1, \quad x_{c_{j2}} - \overline{x}_{c_{j2}} = 1$$
$$x_{c_{j1}} - \overline{x}_{c_{j1}} = -1, \quad x_{c_{j2}} - \overline{x}_{c_{j2}} = -1$$

This transformation is not cost preserving, but it is not too hard to verify that it is in fact an L-reduction: There exists a solution x for the instance I_2 of MaxFs2$^=$ which satisfies k_x equations if and only if there exists a solution x^{SAT} for I_1 that satisfies at least $k_{x^{SAT}} = k_x - 6m$ clauses. Moreover, $\text{OPT}(I_2) \leq 7m = 14 \cdot \frac{m}{2} \leq 14 \cdot \text{OPT}(I_1)$.

Note that for the definition of MaxFs2$^=$ we do not need to have a multiplication on the domains and also no algorithm presented in this paper relies on the fact that there is a multiplication on $\mathbb{R}$. In fact a group is perfectly sufficient. For these algebraic versions we have a similar complexity result, which essentially uses the same construction as in Theorem 1 if the group has an element of order greater than 2 and otherwise uses a reduction from MaxCut:

Theorem 2. *For every nontrivial group $(G,+)$, MaxFs2$^=$, where all variables must attain values in G, is APX-hard.* □

A simple greedy method shows that furthermore MaxFs2$^=$ can be approximated within the constant factor $|G|$ for any finite group. The algorithm iteratively chooses a variable x_i. Let U denote the equations which are (currently) unary with respect to x_i, i.e., which currently contain no variable other than x_i. It then assigns a value to x_i which maximizes the number of satisfied equations in U (if U is empty, an arbitrary value is assigned to x_i) and updates the remaining equations by substituting the chosen value for x_i. It is easy to see that this provides a $|G|$-approximation.

By a reduction from PARTITION we can prove the following result:

Theorem 3. MaxFs2$^=$ *is NP-hard, even if the constraint graph is series-parallel.* □

Theorem 4. MaxFs2$^=$ *is strongly NP-hard, even if all weights are unit weights and the constraint graph is planar.*

Proof. Due to lack of space, we only sketch the proof. We provide a reduction from the strongly NP-complete problem Rectilinear Steiner Tree Problem MRST ([6, ND13]). We are given a set $P \subset \mathbb{Z} \times \mathbb{Z}$ of points in the plane and an integer k. The task is to decide whether there is a finite set $Q \subset \mathbb{Z} \times \mathbb{Z}$ such that there is a spanning tree of total weight k or less for the vertex set $V \cup Q$, where the weight of an edge $\{(x_1, y_1), (x_2, y_2)\}$ is measured with respect to the rectilinear metric $|x_1 - x_2| + |y_1 - y_2|$.

As common, we call the points in P terminals. Given an arbitrary instance I_1 from MRST with a terminal set $P = \{v_1, \ldots, v_n\} \subset \mathbb{Z} \times \mathbb{Z}$ where $v_i = (x_i, y_i)$, we compute

$$x_{\min} := \min_{i=1,\ldots,n} x_i \qquad\qquad y_{\min} := \min_{i=1,\ldots,n} y_i$$

$$x_{\max} := \max_{i=1,\ldots,n} x_i \qquad\qquad y_{\max} := \max_{i=1,\ldots,n} y_i$$

and set up a grid-graph $G^* = (V, E^*)$ with nodes $V = \{v_{x,y} : x \in \mathbb{Z} \cap [x_{\min}, x_{\max}], y \in \mathbb{Z} \cap [y_{\min}, y_{\max}]\}$. The edges connect vertices that are horizontally or vertically adjacent, that is, $E^* = \{e = [(x_1, y_1), (x_2, y_2)] : |x_1 - x_2| = 1, |y_1 - y_2| = 1\}$. Since the original reduction in [6] constructs only terminal points within a grid of maximal extension in $O(n^3)$, we can assume that instance I_1 also induces a grid which is polynomially bounded in n and hence only polynomially many nodes have to be introduced in G^*.

We now construct an instance I_2 of MINFS2$^=$ on the plane dual graph $G = (G^*)^*$ of G^* as follows: In G the vertices in G^* correspond to faces and the edges incident to a vertex in G^* form the boundary of the corresponding face in G. Assign a value to every node of G^* as follows:

$$c(v_i) := \begin{cases} i, & \text{if } i \in \{1, \ldots, n-1\} \\ -\frac{1}{2}n(n-1), & \text{if } i = n \\ 0, & \text{otherwise.} \end{cases}$$

It can then be shown that we can always assign lengths to the edges of G such that the clockwise length of the face corresponding to vertex v_i equals $c(v_i)$.

The graph G corresponds in fact to an instance of MINFS2$^=$. Observe that for no strict subset of the terminals their (face) values add up to zero. It can now be seen that the feasible solutions for the instance I_2 of MINFS2$^=$ correspond exactly to Steiner trees for I_1 and that I_2 has a consistent subset of at least $m - k$ edges if and only if for I_1, there is a Steiner Tree with at most k edges.

4 Approximation Algorithm for MinFs2$^=$ on Planar Graphs

The problem MINFS2$^=$ is a special case of the well-known HITTING SET Problem. In the general HITTING SET Problem, we are given a collection $\mathcal{C}$ of subsets of a finite set E and weights $w_e \geq 0$ for all $e \in E$. The goal is to find a minimum weight subset $A \subset E$ such that A contains at least one element from each subset in $\mathcal{C}$. In the case of MINFS2$^=$, the collection $\mathcal{C}$ is the set of all inconsistent cycles and the finite set of ground elements is the set of all arcs in the constraint graph with their respective weights.

Goemans and Williamson [8] gave a general primal-dual framework for designing approximation algorithms for HITTING SET. The problem can be formulated as an Integer Linear Program whose linear relaxation and dual read as follows:

$$\min \sum_{e \in E} w_e x_e \qquad\qquad \max \sum_{C \in \mathcal{C}} y_C$$

$$\text{s.t.} \sum_{e \in C} x_e \geq 1 \quad \forall C \in \mathcal{C} \qquad\qquad \text{s.t.} \sum_{C \in \mathcal{C} : e \in C} y_C \leq w_e \quad \forall e \in E$$

$$x_e \geq 0 \quad \forall e \in E \qquad\qquad\qquad y_C \geq 0 \quad \forall C \in \mathcal{C}$$

The primal-dual-method constructs simultaneously a feasible solution x for the Integer Linear Program and a feasible solution y for the dual of the LP-relaxation. The algorithm starts with the trivial dual feasible solution $y = 0$ and uses a violation oracle VIOLATION, which outputs for a given subset S of the ground set E a subset of the sets not hit by S. It then increases the dual variables of all the sets in VIOLATION(S) until one of the dual packing constraints becomes tight for some element, which then gets added to the solution. At the end, a cleanup step is performed. Algorithm 1 shows the translation of the general algorithm for MINFS2$^=$. We assume without loss of generality that $w_e > 0$ for all $e \in E$, since edges of zero weight can be removed from the graph at the beginning without adding to the solution cost.

A key theorem from Goemans and Williamson [8] is the following:

Theorem 5 ([8]). *The primal-dual algorithm 1 delivers a solution for* HITTING SET *of weight at most* $\gamma \sum_{C \in \mathcal{C}} y_C \leq \gamma OPT$ *if for any infeasible* $\overline{E}$ *and any minimal augmentation* F *of* $\overline{E}$, *the collection* $\mathcal{V}(\overline{E})$ *returned by the violation oracle on input* $\overline{E}$ *satisfies:*

$$\sum_{C \in \mathcal{V}(\overline{E})} |C \cap F| \leq \gamma |\mathcal{V}(\overline{E})|.$$

A *minimal augmentation* is a feasible solution F containing $\overline{E}$ which is inclusionwise minimal, that is, for any $e \in F$ it holds that $F \setminus \{e\}$ is infeasible.

In order to apply the result of Theorem 5 we need to design an appropriate violation oracle. The techniques will be similar to that in [9], where Goemans and Williamson construct a primal-dual 3-approximation for feedback problems on planar graphs. We will use the following intuitive definitions:

Definition 1 (cf. [9]). *Let G be a directed plane graph (i.e., G is planar and has been embedded in the plane, so it makes sense to talk about faces of G). A face of G whose boundary forms an inconsistent cycle is called a inconsistent face. Every cycle C of G divides the plane into two regions, the interior (the one with finite diameter) and the exterior. We define $f(C)$ to be the set of all faces in the interior of C. We say that a cycle C_1 contains a cycle C_2 if $f(C_1) \supseteq f(C_2)$ and write $C_2 \subseteq_f C_1$.*

The relation "$\subseteq_f$" defines a partial order on the set $\mathcal{C}$ of inconsistent cycles of G. The inclusionwise minimal inconsistent cycles with respect to this partial order are of particular interest. We will abuse notation slightly and call them the minimal inconsistent cycles.

Two cycles C_1 and C_2 cross if none of the sets $f(C_1) \cap f(C_2)$, $f(C_1) \setminus f(C_2)$, $f(C_2) \setminus f(C_1)$ is empty. A family of cycles is called laminar if no two of its cycles

---

Algorithm 1 Primal-Dual Approximation Algorithm for MINFS2$^=$

1: **Input:** Graph $G = (V, E)$ with arc lengths and weights $w_e \geq 0$.
2: **Output:** An edge set $A \subseteq E$ such that the subgraph $(V, E \setminus A)$ does not contain any inconsistent cycles, i.e., an edge set A such that $C \cap A \neq \emptyset$ for each inconsistent cycle C.

3: $y = 0$ {dual solution}
4: $A = \emptyset$ {primal solution, initially empty}
5: Set $x_e = 0$ for all edges $e \in E$.
6: $l = 0$ {iteration count}
7: **while** there is an inconsistent cycle in $(V, E \setminus A)$ **do**
8:    $\mathcal{V} = \text{VIOLATION}(A)$
9:    Increase y_C uniformly for all $C \in \mathcal{V}$ until $\exists e_l \notin A : \sum_{C : e_l \in C} y_C = w_{e_l}$
10:    $A = A \cup \{e_l\}$
11: **end while**
12: **for all** $j = l, \ldots, 1$ **do**
13:    **if** $A \setminus e_l$ is feasible **then**
14:       $A = A \setminus \{e_l\}$
15:    **end if**
16: **end for**

---

are crossing, i.e., any two cycles either do not share an interior face or one of them contains the other.

The definition of our violation oracle is the first and also the most important point where we make use of the planarity of the constraint graph $G := G_I$: Define VIOLATION(A) to be the set of all minimal inconsistent cycles in the graph obtained by deleting all edges in A from G. As we will show, the minimal inconsistent cycles are all inconsistent faces and therefore it is clear that VIOLATION can be computed in polynomial time.

Choose an orientation for the plane and orient the boundaries of the faces of G accordingly. Summing up all (the lengths of) those boundaries we see that every edge contributes to exactly two summands and with different sign, so the result must be 0. Also, if we only sum up all the boundaries of the faces inside some cycle C, all inner boundaries cancel, leaving only the edges of C.

To put this into a formula, define $\ell(C)$ to be the length of the cycle C (=sum of the signed lengths of its edges) with clockwise orientation. If C is the boundary of a face F, we also define $\ell(F) := \ell(C)$, except for the exterior face, where we set $\ell(F) := -\ell(C)$. This gives us

$$\sum_{F: \text{ is a face of } G} \ell(F) = 0 \tag{1}$$

and

$$\ell(C) = \sum_{F \in f(C)} \ell(F) \quad \text{for all cycles } C \text{ of } G. \tag{2}$$

Lemma 1. *If there is an inconsistent cycle C in G, there are (at least) two faces corresponding to inconsistent cycles, one of which lies in the interior of C and the other one in the exterior of C.*

Proof. We first show that the existence of one inconsistent face implies that there are actually two inconsistent faces. To see this use Equation (1). If there were only one inconsistent face, its boundary would be the only nonzero summand, which means that the sum is nonzero contradicting (1).

Let $C \in \mathcal{C}$ be an inconsistent cycle in G. Consider the subgraph G_{int} of G, which only contains C and all edges and vertices in the interior of C. By construction, C is the boundary of an inconsistent face of G_{int} (namely the exterior face) and by the above claim G_{int} must have another inconsistent face, which lies in the interior of C. But this is, of course, also an inconsistent face in G.

The existence of an inconsistent face in the exterior of C follows analogously if we consider the subgraph G_{ext} of G, which only contains C and all edges and vertices in the exterior of C, instead of G_{int}.

Proposition 1. *Every minimal cycle in $\mathcal{C}$ with respect to $\subseteq_f$ is the boundary of a face of G and, therefore, face-minimal inconsistent cycles do not cross.*

Proof. Follows immediately from Lemma 1.

Lemma 2. *A graph is consistent if and only if all face-minimal cycles are consistent.*

Proof. Immediately from Proposition 1.

Inspired by Lemma 2, given a partial solution A, our violation oracle returns the collection of face-minimal inconsistent cycles in $G = (V, E \setminus A)$. Observe that this violation oracle can be implemented to run in polynomial time, since there is only a linear number of faces, which can be checked exhaustively. Our main ingredient for the analysis of the primal-dual algorithm for MINFS2$^=$ is the following:

Theorem 6. *Let G be a planar graph and let $\mathcal{M}$ be a collection of face-minimal inconsistent cycles. Then, for any minimal solution A we have*

$$\sum_{C \in \mathcal{M}} |A \cap C| \leq 2|A| \leq 2|\mathcal{M}|.$$

In order to prove the theorem, we need a number of auxiliary results. Let A be an inclusionwise minimal solution for MINFS2$^=$. By the minimality of A, for every $e \in A$ there must be an inconsistent cycle C_e such that $C_e \cap A = \{e\}$. In fact, if the intersection were empty or contained more than $\{e\}$ for every inconsistent cycle, then we could remove e from A because it either hits only cycles which are already being hit or it does not hit any cycle. We call such a cycle C_e with $C_e \cap A = \{e\}$ a *witness cycle* of e.

Due to lack of space, the proof of the following lemma is deferred to the full version of the paper:

Lemma 3. *Let $A \subset E$ be an inclusionwise minimal solution. Then, there exists a laminar family of witness cycles $C_e \in \mathcal{C}$, $e \in A$.* □

We are now ready to complete the proof of Theorem 6. We first see from Lemma 2 that the violation oracle $\mathcal{V}$, which returns all face-minimal inconsistent cycles is valid in the sense that if G contains an inconsistent cycle, $\mathcal{V}$ will be non-empty.

Let $\mathcal{F} = \{C_e | e \in A\}$ be a laminar family of witness cycles for A, which exists by Lemma 3. Since $\mathcal{F}$ is laminar, we can construct a forest with node set $\mathcal{F}$ representing the partial order induced by "$\subseteq_f$" restricted to the elements of $\mathcal{F}$. We add a root node r, connect it to all maximal elements of $\mathcal{F}$, and denote the resulting tree by T. For the analysis we now assign an edge $e \in A$ to the node of T corresponding to its witness cycle; furthermore, every element C of $\mathcal{M}$ can be assigned to (the node of T corresponding to) the smallest witness cycle that contains C; if there is no such witness cycle, C is assigned to the root node r. Let $\mathcal{M}_e$ denote the set of elements of $\mathcal{M}$ which are assigned to node C_e. Note that $\mathcal{M}_e$ might be empty for some nodes of T, but never for leaves. This is because by Lemma 1, there must be some inconsistent face inside every witness cycle.

We wish to bound $\sum_{C \in \mathcal{M}_e} |A \cap C|$ from above: For a cycle $C \in \mathcal{M}_e$, we know that $|A \cap C|$ can only contain the edge e and edges assigned to the children of C_e, since no element of $\mathcal{M}$ crosses any element of $\mathcal{F}$ and every witness cycle contains exactly one element of A (and therefore separates its inside faces from all elements of A but the aforementioned ones). By definition of T, the number of those candidate edges is $\deg_T(C_e)$, i.e., the degree of node C_e in the tree T. Since every edge touches at most two inconsistent faces, each edge can only appear once in $\sum_{C \in \mathcal{M}_e} |A \cap C|$, so

$$\sum_{C \in \mathcal{M}_e} |A \cap C| \le \deg_T(C_e).$$

If $\mathcal{M}_e = \emptyset$, we can use 0 as a trivial better bound. Summing up over all $e \in A$, we obtain:

$$\sum_{C \in \mathcal{M}} |A \cap C| = \sum_{e \in A} \sum_{C \in \mathcal{M}_e} |A \cap C| \le \sum_{e \in A : \mathcal{M}_e \neq \emptyset} \deg_T(C_e)$$

The average vertex-degree of a tree with n nodes is $\frac{2n-2}{n}$; we would like to use this, but the sum on the right-hand side does not contain all vertex-degrees of T, but only some of them. Here the key observation is that all the leaves of T appear in the sum; and the absence of any node with degree ≥ 2 can only decrease the average vertex-degree, which is below 2. A special case is the root node: It may have degree 1 but $\mathcal{M}_r$ can be empty. Taking this into account, we get

$$\sum_{e \in A : \mathcal{M}_e \neq \emptyset} \deg_T(C_e) \le 2|\{e \in A : \mathcal{M}_e \neq \emptyset\}| - 1 \le 2|\mathcal{M}| - 1.$$

This completes the proof of Theorem 6. Together with Theorem 5, we thus obtain:

Corollary 1. *There exists a polynomial time 2-approximation algorithm for* MINFS2$^=$ *when the associated constraint graph is planar.*   □

5   Approximation Algorithms for MaxFs2$^=$

It is easy to obtain a 2-approximation algorithm for MAXFS2$^\leq$, that is, for the case when all constraints are inequalities of the form $x_i - x_j \leq c_{ij}$. Numbering the vertices in the constraint graph G_I from 1 to n, one splits the arcs into two groups, one that contains those arcs going from smaller to higher numbers and one that contains the other arcs going from higher to smaller numbers. Thus, the arc set of G_I is partitioned into two parts each of which forms an acyclic subgraph (and hence is consistent in the case of inequalities). One of the subgraphs contains at least one half of the total weight of the arcs and is, thus, a 2-approximation to MAXFS2$^\leq$.

Unfortunately, this approach fails for MAXFS2$^=$. On the other hand, if the constraint graph associated with an instance of MAXFS2$^=$ is a forest, it is trivially consistent. This observation can be used to obtain an approximation for MAXFS2$^=$. Recall that the *arboricity* of a graph is the minimum number of forests in which the edge set of a graph can be partitioned. This value can be computed in polynomial time as shown in [5,12]. Moreover, a simple planar graph with n vertices can be edge-partitioned into three forests in $O(n)$ time [14].

Suppose first that the constraint graph does not contain parallel edges. Then, computing a decomposition of the constraint graph G into the minimum number of forests and then selecting the one with the largest weight by averaging yields an approximation with a factor α, where α is the arboricity of G. Now, suppose that G does contain parallel edges. Let e_1 and e_2 be such two parallel edges. If they have the same length ℓ, we can collapse them into one edge with length ℓ and weight $w(e_1) + w(e_2)$; otherwise no solution at all can contain both e_1 and e_2, since they form an inconsistent cycle. In this case, we remove the lighter edge. Thus, we can eliminate all parallel edges without affecting solution quality.

Theorem 7. MAXFS2$^=$ *can be approximated within a factor of α, where α is the arboricity of the reduced graph, where parallel edges have been processed as above.*   □

An immediate corollary is the following:

Corollary 2. *Choosing a maximum weight spanning tree of the constraint graph yields an approximation within a factor of α.*   □

6   Extensions to Bounded Domains

In this section, we replace the domain $\mathbb{R}$ for each variable x_i by an interval $[a_i, b_i]$. We then have additional box constraints $a_i \leq x_i \leq b_i$, which are not optional, but need all to be satisfied. We call the extension of MINFS2$^=$ and MAXFS2$^=$ to interval domains MINFS2$^=$-D and MAXFS2$^=$-D respectively. Note that we can test whether all box constraints can be satisfied simultaneously in polynomial time by a a shortest-path computation.

Analogously to inconsistent cycles, we define (domain) inconsistent paths: A *(domain) inconsistent path* $p = v_1, \ldots, v_k$ is a simple path with the property that either

$$a_{v_1} + \sum_{i=1}^{k-1} \ell(v_i, v_{i+1}) > b_{v_k} \quad \text{or} \quad b_{v_1} + \sum_{i=1}^{k-1} \ell(v_i, v_{i+1}) < a_{v_k}.$$

Obviously, an inconsistent path or an inconsistent cycle render the whole system invalid. In fact, we can show that it is also sufficient to destroy all inconsistent paths and cycles in order to have a domain feasible system.

Due to lack of space, most of the proofs of the following results will be deferred to the full version of the paper.

Theorem 8. *The problem of hitting all inconsistent paths by a subset of edges of minimum weight is NP-complete in general graphs. It remains strongly NP-complete on planar graphs, even if no vertex is incident to more than three arcs.*

Theorem 9. *The problems* MINFS2=-D *and* MAXFS2=-D *are polynomial solvable on trees. The result still holds if we have domains of the form* $[a_v, b_v] \cap \mathbb{Z}$.

Theorem 10. MINFS2=-D *on graphs without inconsistent cycles can be approximated within a factor of* $O(\log n)$, *where* n *is the number of vertices. In the special case of planar graphs, there is a constant approximation factor.*

Proof. Given a graph without inconsistent cycles, each path between two nodes i and j has the same length d_{ij}. We can therefore determine which node pairs are connected by inconsistent paths. Consequently, we can regard the problem as a multicut problem with the conflicting nodes as terminal pairs and use the approximation algorithm from [7] which provide the stated approximation factors. □

Using the result of Theorem 9 with the techniques of Section 5 we obtain:

Theorem 11. MAXFS2=-D *has an* α*-approximation guarantee on general graphs, where* α *is again the arboricity of the reduced graph.* MINFS2=-D *has a constant approximation guarantee on planar graphs.*

References

1. Amaldi, E., Bruglieri, M., Casale, G.: A two-phase relaxation-based heuristic for the maximum feasible subsystem problem. Computers and Operations Research 35, 1465–1482 (2008)
2. Amaldi, E., Kann, V.: The complexity and approximability of finding maximum feasible subsystems of linear relations. Theoretical Computer Science 147(1-2), 181–210 (1995)

3. Chinneck, J.: Fast heuristics for the maximum feasible subsystem problem. IN-FORMS Journal on Computing 13(3), 211–223 (2001)
4. Elbassioni, K., Raman, R., Ray, S., Sitters, R.A.: On the approximability of the maximum feasible subsystem problem with 0/1-coefficients. In: Proceedings of the 20th Annual ACM-SIAM Symposium on Discrete Algorithms, pp. 1210–1219 (2009)
5. Gabow, H.N., Westermann, H.H.: Forests, frames, and games: algorithms for matroid sums and applications. Algorithmica 7(1), 465–497 (1992)
6. Garey, M.R., Johnson, D.S.: The rectilinear steiner tree problem is NP-complete. SIAM Journal on Applied Mathematics, 826–834 (1977)
7. Garg, N., Vazirani, V.V., Yannakakis, M.: Approximate max-flow min-(multi) cut theorems and their applications. SIAM Journal on Computing 25, 235 (1996)
8. Goemans, M.X., Williamson, D.P.: The primal-dual method for approximation algorithms and its application to network design problems. In: [11], pp. 144–191. PWS Publishing Company (1997)
9. Goemans, M.X., Williamson, D.P.: Primal-dual approximation algorithms for feedback problems in planar graphs. Combinatorica 18(1), 37–59 (1998)
10. Greer, R.: Trees and hills: Methodology for maximizing functions of systems of linear relations. Annals of Discrete Mathematics 22 (1984)
11. Hochbaum, D.S. (ed.): Approximation algorithms for NP-hard problems. PWS Publishing Company, 20 Park Plaza, Boston, MA 02116–4324 (1997)
12. Nash-Williams, C.: Decomposition of finite graphs into forests. Journal of the London Mathematical Society 1(1), 12 (1964)
13. Pfetsch, M.: Branch-and-cut for the maximum feasible subsystem problem. SIAM Journal on Optimization 19(1), 21–38 (2008)
14. Schnyder, W.: Embedding planar graphs on the grid. In: Proceedings of the First Annual ACM-SIAM Symposium on Discrete Algorithms, pp. 138–148 (1990)

Hydras: Directed Hypergraphs
and Horn Formulas[*]

Robert H. Sloan[1], Despina Stasi[1], and György Turán[1,2]

[1] University of Illinois at Chicago
[2] Hungarian Academy of Sciences and University of Szeged,
Research Group on Artificial Intelligence

Abstract. We consider a graph parameter, the *hydra number*, arising
from an optimization problem for Horn formulas in propositional logic.
The hydra number of a graph $G = (V, E)$ is the minimal number of hyperarcs of the form $u, v \rightarrow w$ required in a directed hypergraph $H = (V, F)$,
such that for every pair (u, v), the set of vertices reachable in H from
$\{u, v\}$ is the entire vertex set V if $(u, v) \in E$, and it is $\{u, v\}$ otherwise.
Here reachability is defined by the standard forward chaining or marking
algorithm.

Various bounds are given for the hydra number. We show that the
hydra number of a graph can be upper bounded by the number of
edges plus the path cover number of its line graph, and this is a sharp
bound for some graphs. On the other hand, we construct graphs with
hydra number equal to the number of edges, but having arbitrarily
large path cover number. Furthermore we characterize trees with low
hydra number, give bounds for the hydra number of complete binary
trees, discuss a related optimization problem and formulate several open
problems.

1 Introduction

We consider a problem concerning the minimal number of hyperarcs in directed
hypergraphs with prescribed reachability properties. In this paper, a directed
hypergraph $H = (V, F)$ has size-3 hyperarcs of the form $u, v \rightarrow w$ where u, v is
called the *body* (or tail) and w is called the *head* of the hyperarc. Reachability
is defined by a marking procedure known as *forward chaining*. A vertex $w \in V$
is *reachable* from a set $S \subset V$ if the following process marks w: start by marking
vertices in S, and as long as there is a hyperarc $a, b \rightarrow c$ such that both a and b
are marked, mark c as well.

Given an undirected graph $G = (V, E)$, we would like to find the minimal
number of hyperarcs in a directed hypergraph $H = (V, F)$, such that for every
pair $(u, v) \in E$, the set of vertices reachable from $\{u, v\}$ in H is the whole vertex
set V if $(u, v) \in E$, and it is $\{u, v\}$ otherwise. In other words, given a set of

---

[*] An earlier version of this paper entitled "Hydra formulas and directed hypergraphs:
A preliminary report" appears in the online proceedings of the International Symposium on Artificial Intelligence and Mathematics 2012.

M.C. Golumbic et al. (Eds.): WG 2012, LNCS 7551, pp. 237–248, 2012.
© Springer-Verlag Berlin Heidelberg 2012

bodies, we look for the minimal total number of heads assigned to these bodies such that every body can reach every vertex. The minimum is called the *hydra number*[1] of G, denoted by $h(G)$.

The problem is a combinatorial reformulation of a special case of the *minimization problem for propositional Horn formulas*. Horn formulas are a basic knowledge representation formalism. Horn minimization is the problem of finding a shortest possible Horn formula equivalent to a given formula. There are approximation algorithms, computational hardness and inapproximability results for this problem [1,2,3,4]. Special cases correspond to the well studied transitive reduction and minimum equivalent digraph problems for directed graphs. Estimating the size of a minimal formula is not well understood even in rather simple cases. A *hydra formula* φ is a definite Horn formula with clauses of size 3 such that every body occurring in the formula occurs with all possible heads. The minimal number of clauses needed to represent φ is the hydra number of the undirected graph G corresponding to the bodies in φ.

Besides being a natural subproblem of Horn minimization, the hydra minimization problem may also be of interest for the following reason. The Horn *body minimization* problem is the problem of finding, given a definite Horn formula, an equivalent Horn formula with the minimal number of distinct bodies. There are efficient algorithms for this problem [5,6,7,8]. Thus one possible approach to Horn minimization is to find an equivalent formula with the minimal number of bodies and then to select as few heads as possible from the set of heads assigned to the bodies. This approach is indeed used in an approximate Horn minimization algorithm [2]. Hydras are a natural test case for this approach.

The paper is organized as follows. Section 2 contains some background, including a discussion of the motivating Horn minimization problem. The rest of the paper presents various results on hydra numbers.

It is easy to see that $|E(G)| \leq h(G) \leq 2|E(G)|$ for every graph G on at least three vertices. Graphs satisfying the lower bound are called *single-headed*. In Section 3 we give some sufficient and some necessary conditions for single-headedness. In Section 4 we show that the hydra number is related to the path cover number of the line graph (Theorem 13, Theorem 14). In Section 5 it is shown that single-headed trees are precisely the stars and that trees with hydra number $|E(G)| + 1$ are precisely the non-star caterpillars (Theorem 16). In Section 6 we show that the hydra number of a complete binary tree is between $\frac{13}{12}|E(G)|$ and $\lceil \frac{8}{7}|E(G)| \rceil$ (Theorem 19).

In Section 7 we consider the related problem of finding the minimal number of hyperarcs for which every k-tuple of vertices is good, and we give almost matching lower and upper bounds. We conclude the paper by mentioning several open problems.

Due to space limitations, the proof of Theorem 19 is omitted. Further results on hydra numbers will appear in [9].

---

[1] In Greek mythology the Lernaean Hydra is a beast possessing many heads.

2   Background

A *definite Horn clause* is a disjunction of literals where exactly one literal is unnegated. Such a disjunction can also be viewed as an implication, for example the clause $\bar{x} \vee \bar{y} \vee z$ is equivalent to the implication $x, y \to z$. The tuple x, y is the *body* and the variable z is the *head* of the clause. The size of a clause is the number of its literals. A definite d-Horn formula is a conjunction of definite Horn clauses of size d. A clause C is an implicate of a formula φ if every truth assignment satisfying φ satisfies C as well. The implicate C is a prime implicate if none of its proper subclauses is an implicate.

Implication between a definite Horn formula φ and a definite Horn clause C can be decided by *forward chaining*: mark every variable in the body of C, and while there is a clause in φ with all its body variables marked, mark the head of that clause as well. Then φ implies C iff the head of C gets marked.

Definition 1. *A definite 3-Horn formula φ is a* hydra *formula, or a hydra, if for every clause $x, y \to z$ in φ and every variable u, the clause $x, y \to u$ also belongs to φ.*

For example, $(x, y \to z) \wedge (x, y \to u) \wedge (x, z \to y) \wedge (x, z \to u)$ is a hydra[2].

In the following proposition we note that every prime implicate of a hydra is a clause occurring in the hydra itself (this is not true for definite 3-Horn formulas in general). Thus minimization for hydras amounts to selecting a minimal number of clauses from the hydra that are equivalent to the original formula.

Proposition 2. *Every prime implicate of a hydra belongs to the hydra.*

Proof. First note that all prime implicates of a definite Horn formula are definite Horn clauses [10]. Let us consider a hydra φ and a definite Horn clause C. If the body of C is of size 1, or it is of size 2 but it does not occur as a body in φ then forward chaining cannot mark any further variables, thus C cannot be an implicate. If the body of C has size at least 3 then it must contain a body x, y occurring in φ, otherwise, again, forward chaining cannot mark any further variables. But then the clause $x, y \to head(C)$ occurs in φ and so C is not prime. $\qquad\square$

A definite Horn formula may also be viewed as a directed hypergraph of the type described in the introduction, and the two descriptions of forward chaining are equivalent. The *closure* $cl_H(S)$ of a set of vertices S with respect to H is the set of vertices marked by forward chaining started from S. A set of vertices is *good* if its closure is the set of all vertices.

For completeness, we restate the main notions used in this paper.

Definition 3. *A directed 3-hypergraph $H = (V, F)$ represents an undirected graph $G = (V, E)$ if*

---

[2] Redundant clauses like $x, y \to x$ are omitted for simplicity.

    i. $(u, v) \in E$ implies $cl_H(u, v) = V$,
    ii. $(u, v) \notin E$ implies $cl_H(u, v) = \{u, v\}$.

Definition 4. *The* hydra number *$h(G)$ of an undirected graph $G = (V, E)$ is*

$$\min\{|F| \; : \; H = (V, F) \; represents \, G\}.$$

Proposition 2 implies that the minimal formula size of a hydra φ and the hydra number of the undirected graph G formed by the bodies in φ are the same.

Remark 5. For the rest of the paper we assume that every variable in a hydra occurs in some body, or, equivalently, that graphs contain no isolated vertices. The removal of a variable occurring only as a head decreases minimal formula size by one, and, similarly, the removal of an isolated vertex decreases the hydra number by one.

For the remainder of the paper we use hypergraph terminology.

3   The Hydra Number of Graphs

In this section we note some simple properties of the hydra number.

Proposition 6. *For every graph $G = (V, E)$ with at least three vertices*

$$|E(G)| \leq h(G) \leq 2|E(G)|.$$

Proof. For the upper bound construct a hypergraph of size $2|E(G)|$ by first ordering the edges of G, and then using each edge as the body of two hyperarcs whose heads are the two endpoints of the next edge in G. For the lower bound, note that each edge of G must be a body of at least one hyperarc. $\square$

Equality holds in the upper bound when G is a matching. Graphs satisfying the lower bound are of particular interest as they represent 'most compressible' hydras.

Definition 7. *A graph G is* single-headed *if $h(G) = |E(G)|$.*

A graph is single-headed iff there is a hypergraph $H = (V, F)$ such that every edge of G has *exactly* one head assigned to it, every hyperarc body in H is an edge of G and every edge of G is good in H. Cycles, for example, are single-headed, as shown by the directed hypergraph

$$(v_1, v_2 \to v_3), (v_2, v_3 \to v_4), \ldots, (v_{k-1}, v_k \to v_1). \tag{1}$$

Adding edges to the cycle preserves single-headedness. For example, the graph obtained by adding edge (v_i, v_j) is represented by the directed hypergraph obtained from (1) by adding the hyperarc $v_i, v_j \to v_{i+1}$, where $i+1$ is meant modulo m. Thus we obtain the following.

Proposition 8. *Hamiltonian graphs are single-headed.*

We will discuss stronger forms of this statement in the next section. Matchings, on the other hand, satisfy the upper bound in Proposition 6. Indeed, every edge must occur as the body of at least two hyperarcs as otherwise forward chaining cannot mark any further vertices.

We call a body u, v single-headed (resp., multi-headed) with respect to a directed hypergraph H representing a graph G, if it is the body of exactly one (resp., more than one) hyperarc of H.

Remark 9. Assume that the directed hypergraph $H = (V, F)$ represents the graph $G = (V, E)$ and $|V| \geq 4$. If $u, v \to w \in F$ and u, v is single-headed in H then w must be a neighbor of u or v. Indeed, otherwise $cl_H(u, v) = \{u, v, w\} \subset V$. This is a fact which we use numerous times in our proofs without referring to it explicitly.

The following proposition generalizes the argument proving Proposition 8.

Proposition 10. *Let G be a connected graph and let G' be a connected spanning subgraph of G. Then*

$$h(G) \leq h(G') + |E(G)| - |E(G')|.$$

If G' is single-headed then G is also single-headed.

Proof. Let H' be a directed hypergraph of size $h(G')$ representing G'. Since G' is a connected spanning subgraph of G, for every edge $(u, v) \in E(G) \smallsetminus E(G')$ there is an edge $(v, w) \in E(G')$. The directed hypergraph H representing G obtained from H' by adding the hyperarc $u, v \to w$ to H' for each edge $(u, v) \in E(G) \smallsetminus E(G')$ satisfies the requirements. The second statement follows trivially. □

A second proposition gives a sufficient condition for single-headedness based on single-headedness of a non-spanning subgraph.

Proposition 11. *Let G be a connected graph and $(u, v) \notin E(G)$. Construct the graph $\hat{G}$ with vertex set $V(\hat{G}) = V(G) \cup \{w\}$ and edge set $E(\hat{G}) = E(G) \cup \{(u, v), (v, w)\}$, for some $w \notin V(G)$. If G is single-headed then $\hat{G}$ is single-headed.*

Proof. Let H be a directed hypergraph representing G and containing exactly $|E(G)|$ hyperarcs. Construct $\hat{H}$ from H by adding hyperarcs $u, v \to w$ and $v, w \to z$, where z is a neighbor of v in G guaranteed to exist by the connectivity of G. Since all pairs in $E(G)$ reach both u and v in H (and in $\hat{H}$), hyperarc $u, v \to w$ ensures all pairs in $E(G)$ can reach in $\hat{H}$ the new variable w as well. On the other hand, hyperarc $v, w \to z$ ensures that the new pairs (u, v) and (v, w) can reach all other variables. Finally, there are $|E(\hat{G})|$ hyperarcs in H. □

Next we see a general sufficient condition for a graph *not* to be single-headed.

Proposition 12. *Let G be the union of two disjoint subgraphs $G_1 = (V_1, E_1)$ and $G_2 = (V_2, E_2)$, connected by a cut-edge. If both G_1, G_2 contain at least two vertices then G is not single-headed.*

Proof. Assume that G is single-headed and let H be a directed hypergraph demonstrating this. Let $u \in G_1$, $v \in G_2$, and (u, v) be the cut-edge. There is exactly one hyperarc of the form $u, v \to z$ in H. If z is in G_1 (resp., in G_2) then forward chaining started from z, u (resp., z, v) cannot mark any vertices in G_2 (resp., G_1) other than v (resp., u).   □

4   Line Graphs

In this section we consider a graph parameter that can be used to prove bounds on the hydra number. The *line graph* $L(G)$ of G has vertex set $V(L(G)) = E(G)$ and edge set $E(L(G)) = \{(e, f) | e \neq f \in E(G) \text{ and } e \cap f \neq \emptyset\}$. A (vertex-disjoint) *path cover* of G is a set of vertex-disjoint paths such that every vertex $v \in V$ is in exactly one path. The *path cover number* of G is the smallest integer k such that G has a path cover containing k paths.

In Proposition 8 we noted that hamiltonian graphs are single-headed. This can be extended to show that hamiltonicity of the line graph is also sufficient for single-headedness. Note that hamiltonicity of the line graph is a strictly weaker condition than hamiltonicity. Hamiltonicity of the graph is easily seen to imply hamiltonicity of the line graph, and a triangle with a pendant edge shows that the converse fails. Furthermore, the path cover number of the line graph of any spanning connected subgraph gives a general upper bound for the hydra number.

Theorem 13. *Let G be a connected graph and G' be a connected spanning subgraph of G. Then the following statements are true:*

i. If $L(G')$ is hamiltonian then G is single-headed.
ii. If $L(G')$ has a path cover of size k then $h(G) \leq |E(G)| + k$.

Proof. By Proposition 10 it is sufficient to prove the bounds for G'.

For i, let C be a hamiltonian cycle in $L(G')$. Direct the edges of C so that $\vec{C}$ is a directed hamiltonian cycle. The directed hypergraph H satisfying the requirements is constructed by adding a hyperarc $u, v \to w$ for each directed edge $(e, f) \in \vec{C}$, where $e = (u, v)$ and $f = (v, w)$.

For ii, let $\{P_i\}_1^k$ be the minimum path cover of $L(G')$ and let l_i be the number of vertices of the path P_i. Direct the edges of each path P_i so that $\vec{P_i}$ is a directed path. Let $e_i = (x_i, y_i)$ and $f_i = (u_i, v_i)$ be the first and last edges in $\vec{P_i}$, respectively (if $\vec{P_i}$ is a single vertex then $e_i = f_i$).

We construct a directed hypergraph H representing G' and satisfying the requirements as follows. First, for each path $\vec{P_i}$ of at least 2 vertices we add $l_i - 1$ hyperarcs: for each directed edge $(e, f) \in \vec{P_i}$, where $e = (u, v)$ and $f = (v, w)$, add a hyperarc $u, v \to w$ to H.

If $k = 1$ then we complete the construction of H by adding two hyperarcs, $u_1, v_1 \to x_1$ and $u_1, v_1 \to y_1$. If $k > 1$ then we complete the construction by adding the $2k$ hyperarcs

$$(u_k, v_k \to x_1), (u_k, v_k \to y_1) \text{ and } (u_i, v_i \to x_{i+1}), (u_i, v_i \to y_{i+1}),$$

for $1 \le i \le k - 1$.   □

The condition of Theorem 13(i) is sufficient but not necessary for a graph G to be single-headed. In fact there exist single-headed graphs such that the line graph of any of the connected spanning subgraphs has a large path cover number.

Theorem 14. *There is a family of single-headed graphs G_k with $\Theta(k)$ edges such that for every spanning, connected subgraph $G' \subseteq G_k$, $L(G')$ has path cover number $\Theta(k)$.*

Proof. Consider the sequence of graphs $\{G_k : k \ge 1\}$ constructed from an $8k$-cycle, with vertices $v_0, \ldots, v_{8k-1}$, and pendant edges $x_i v_{4i}$ and $y_i v_{4k+4i}$ for $0 \le i \le k - 1$. Add a vertex z_i and the edges (x_i, y_i), (y_i, z_i), for each i, $0 \le i \le k - 1$, corresponding to the construction in Proposition 11.

By Proposition 11, G_k is single-headed, since a cycle with attached pendant edges has a hamiltonian line graph. We will show that for an arbitrary connected spanning subgraph $G' \subseteq G_k$ the path cover number of $L(G')$ is at least $k/4$.

Define D_i to be the set of vertices in the ith diagonal of $L(G')$, namely $x_i v_{4i}$, $x_i y_i$, $y_i z_i$, and $y_i v_{4k+4i}$. Consider an arbitrary path cover $S = \{P_j : 1 \le j \le s\}$ of the vertices of $L(G')$.

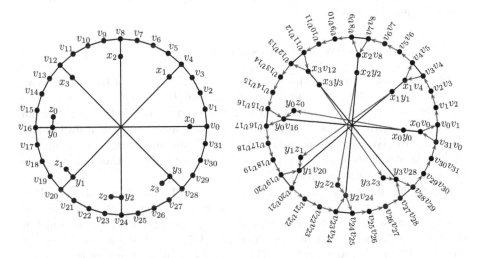

Fig. 1. Single-headed graph G_k for $k = 4$ from Theorem 14 (left) and its line graph (right)

Lemma 15. *Let $D_i' = D_i \cap V(L(G'))$, and let $G[D_i']$ be the subgraph of $L(G')$ induced by D_i'. If $G[D_i']$ does not contain an endpoint of a path in S, then $D_i' = D_i$ and one path in S covers all vertices in D_i.*

Proof. Suppose that $G[D_i']$ does not contain an endpoint of a path in S, and assume for contradiction that $e \in D_i \backslash D_i'$. Since G' is both spanning and connected, it must contain the edge (y_i, z_i), and so $e \neq y_i z_i$. Also $e \notin \{x_i y_i, y_i v_{4k+4i}\}$, else $y_i z_i$ would be a degree-1 vertex in $L(G')$, and thus it would be an endpoint of a path in S. Furthermore $e \neq x_i v_{4i}$, otherwise S must have a path endpoint in the triangle $\{x_i y_i, y_i z_i, y_i v_{4k+4i}\}$. Thus D_i is contained in the vertex set of $L(G')$, and due to the structure of the diagonal and the assumption that no path endpoints of S fall in D_i, all vertices in the diagonal are covered by exactly one path P of S. □

Define X_i to include all vertices in D_i along with the cycle vertices $v_{4i-3}v_{4i-2}$, $v_{4i-2}v_{4i-1}$, $v_{4i-1}v_{4i}$, $v_{4i}v_{4i+1}$, $v_{4i+1}v_{4i+2}$, $v_{4i+2}v_{4i+3}$, and their antipodes on the circle $v_{4k+4i-3}v_{4k+4i-2}$, $v_{4k+4i-2}v_{4k+4i-1}$, $v_{4k+4i-1}v_{4k+4i}$, $v_{4k+4i}v_{4k+4i+1}$, $v_{4k+4i+1}v_{4k+4i+2}$, $v_{4k+4i+2}v_{4k+4i+3}$.

Let $X_i' = X_i \cap V(L(G'))$. We claim that the subgraph $G[X_i']$ induced by the vertex set X_i' contains at least one endpoint of a path in S. Suppose not. By Lemma 15 all vertices in D_i are in $L(G')$. A case analysis shows that all other vertices in X_i must be present, otherwise a degree-1 vertex is introduced in $G[X_i']$ or G' is not both spanning and connected. Thus there must be a path P in S going through all the vertices of X_i. A further case analysis shows that this is not possible.

$G[X_i']$, which is a contradiction.

There are $k/2$ disjoint sets X_i' and so there are at least $k/4$ paths in S. □

A more involved case analysis gives at least two endpoints of paths of S in X_i', and so at least $k/2$ paths in S.

5   Trees with Low Hydra Number

In this section we begin the discussion of the hydra number of trees, with trees having low hydra numbers, that is, hydra number $|E(T)|$ or $|E(T)| + 1$.

A *star* is a tree that contains no length-3 path. A *caterpillar* is a tree for which deleting all vertices of degree one and their incident edges from the tree gives a path. We call this path the spine of T, and note that it is unique. A useful characterization of caterpillars is that they do not contain the subgraph in Fig. 2 [11] (see also [12, p.88]).

Caterpillars have been instrumental in [13], where finding maximal caterpillars starting from the leaves of the tree was the basis for a polynomial algorithm used to find a minimum hamiltonian completion of the line graph of a tree (which is the same as finding a minimum path cover). A linear algorithm was later put forth by [14] for the same problem. For general graphs the problem is NP-hard. Furthermore, [15] proves that finding a hamiltonian path is NP-complete even for line graphs.

Fig. 2. The forbidden subgraph for caterpillars

Stars are the only trees that are single-headed, and caterpillars are the only non-star trees that can attain $h(T) = |E(T)| + 1$.

Theorem 16. *Let T be a tree. Then*

i. $h(T) = |E(T)|$ *if and only if T is a star.*
ii. $h(T) = |E(T)| + 1$ *if and only if T is a non-star caterpillar.*

We first show that a tree that is not a star cannot be single-headed.

Lemma 17. *If T is a tree that is not a star, then $h(T) \geq |E(T)| + 1$.*

Proof. Since T is not a star, it contains a path of length three. The middle edge is a cut-edge between two components of at least two vertices, hence we can apply Proposition 12.   □

In fact a hypergraph that represents a non-caterpillar tree requires even more hyperarcs.

Lemma 18. *If T is a tree that is not a caterpillar then $h(T) > |E(T)| + 1$.*

Proof. A non-caterpillar tree T contains the subgraph in Fig. 2. Let us call the central vertex of that forbidden subgraph u.

Assume for contradiction that H is a hypergraph with $|E(T)| + 1$ hyperarcs that represents T. Let the two-headed body of H be α.

We claim α must have a head in every non-singleton subtree attached to u that does not contain both vertices of α. Suppose not. Let v be a neighbor of u, and let T_v be a non-singleton subtree of T not containing any heads of α, and

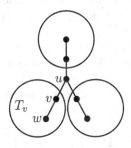

Fig. 3. Part of the non-caterpillar tree T from the proof of Lemma 18

also not containing both vertices of α. Finally let $w \in V(T_v)$ be a neighbor of v. (See Fig. 3.) Body u, v must have a head that is a neighbor of v in T_v: among the vertices in T_v, only v itself can be a head to a body completely outside T_v; so if u, v has only heads outside of T_v, then u, v cannot reach w. Body u, v must also have a head outside T_v, because otherwise only vertices in T_v and u would be reachable from u, v in H. So u, v must be α, which contradicts α having no heads in T_v.

Since there are at least three non-singleton subtrees attached to u, it must be that two of those subtrees each contain one head of α, and the third subtree contains both vertices of α. The two heads of α must not be adjacent to α, because they are in different subtrees. Those two heads also cannot be adjacent to each other. Therefore, the only vertices reachable from α in H are α's two heads and α itself.   □

Proof (of Theorem 16). We need to prove the upper bounds. The single-headedness of stars is easily seen directly, or follows from Theorem 13(i). For T a caterpillar, the upper bound follows from Theorem 13(ii) as the line graph of a caterpillar contains a hamiltonian path.   □

6   Complete Binary Trees

In this section we obtain upper and lower bounds for $h(G)$ when G is a complete binary tree. A *complete binary tree* of depth d, denoted B_d, is a tree with $d+1$ levels, where every node on levels 1 through d has exactly 2 children. B_d has $2^{d+1} - 1$ vertices and $2^{d+1} - 2$ edges.

Theorem 19. *For $d \geq 3$ it holds that*

$$\frac{13}{12} |E(B_d)| \leq h(B_d) \leq \left\lceil \frac{8}{7} |E(B_d)| \right\rceil.$$

Proof omitted due to space constraints.

7   Minimal Directed Hypergraphs with All k-Tuples Good

In this section we consider a problem related to hydra numbers. Given n and a number k ($2 \leq k \leq n - 1$), let $f(n, k)$ be the minimal number of hyperarcs in an n-vertex hypergraph H such that every k-element subset of the vertices is good for H. The case $k = 2$ is just the hydra number of complete graphs and so $f(n, 2) = \binom{n}{2}$.

We use Turán's theorem from extremal graph theory (see, e.g. [12]). The *Turán graph* $T(n, k - 1)$ is formed by dividing n vertices into $k - 1$ parts as evenly as possible (i.e., into parts of size $\lfloor n/(k - 1) \rfloor$ and $\lceil n/(k - 1) \rceil$) and connecting two vertices iff they are in different parts. The number of edges of $T(n, k - 1)$ is denoted by $t(n, k - 1)$. If $k - 1$ divides n then

$$t(n, k - 1) = \left(1 - \frac{1}{k - 1} \right) \frac{n^2}{2}.$$

Turán's theorem states that if an n-vertex graph contains no k-clique then it has at most $t(n, k - 1)$ edges and the only extremal graph is $T(n, k - 1)$. Switching to complements it follows that if an n-vertex graph has no empty subgraph on k vertices then it has at least $\binom{n}{2} - t(n, k - 1)$ edges.

Theorem 20. *If $k \leq (n/2) + 1$ then*

$$\binom{n}{2} - t(n, k - 1) \leq f(n, k) \leq \binom{n}{2} - t(n, k - 1) + (k - 1).$$

Proof. Suppose H is a 3-uniform directed hypergraph with all k-tuples good. Then every k-element set S of vertices must contain at least one body of a hyperarc in H, otherwise forward chaining started from S cannot mark any vertices. Thus the undirected graph formed by the bodies in H contains no empty subgraph on k vertices, and the lower bound follows by Turán's theorem.

For the upper bound we construct a directed hypergraph based on the complement of $T(n, k-1)$ over the vertex set $\{x_1, \ldots, x_n\}$, consisting of $k-1$ cliques of size differing by at most 1. Assume that each clique has size at least 3. In each clique do the following. Pick a hamiltonian path, direct it, and introduce hyperarcs as in (1) (with the exception of the last edge closing the cycle). For every other edge (u, v), introduce a hyperarc $u, v \to w$ where w is a vertex on the hamiltonian path that is adjacent to u or v. For each edge e closing a hamiltonian cycle, add *two* hyperarcs with body e, and heads the endpoints of the first edge on the hamiltonian path of the next clique (where 'next' assumes an arbitrary cyclic ordering of the cliques). For cliques of size 2 the single edge in the clique plays the role of the unassigned edge and the construction is similar.   □

8   Open Problems

We list only a few of the related open problems. As computing hydra numbers is a special case of Horn minimization, it would be interesting to determine the computational complexity of computing hydra numbers and recognizing single-headed graphs. What is the maximal hydra number among connected n-vertex graphs? Can the path cover number of the line graph be used to get a lower bound for the hydra number of trees?

Acknowledgement. This material is based upon work supported by the National Science Foundation under Grant No. CCF-0916708. We would like to thank Daniel Uber for useful discussions.

References

1. Ausiello, G., D'Atri, A., Saccà, D.: Minimal representation of directed hypergraphs. SIAM J. Comput. 15(2), 418–431 (1986)

2. Bhattacharya, A., DasGupta, B., Mubayi, D., Turán, G.: On Approximate Horn Formula Minimization. In: Abramsky, S., Gavoille, C., Kirchner, C., Meyer auf der Heide, F., Spirakis, P.G. (eds.) ICALP 2010, Part I. LNCS, vol. 6198, pp. 438–450. Springer, Heidelberg (2010)
3. Boros, E., Gruber, A.: Hardness results for approximate pure Horn CNF formulae minimization. In: International Symposium on AI and Mathematics, ISAIM (2012)
4. Hammer, P.L., Kogan, A.: Optimal compression of propositional Horn knowledge bases: complexity and approximation. Artificial Intelligence 46, 131–145 (1993)
5. Maier, D.: The Theory of Relational Databases. Computer Science Press (1983)
6. Guigues, J., Duquenne, V.: Familles minimales d'implications informatives résultant d'un tableau de données binaires. Mathématiques et Sciences Humaines 95, 5–18 (1986)
7. Angluin, D., Frazier, M., Pitt, L.: Learning conjunctions of Horn clauses. Machine Learning 9, 147–164 (1992)
8. Arias, M., Balcázar, J.L.: Construction and learnability of canonical Horn formulas. Machine Learning 85, 273–297 (2011)
9. Stasi, D.: Combinatorial Problems in Graph Drawing and Knowledge Representation. PhD thesis, University of Illinois at Chicago (Forthcoming Summer 2012)
10. Hammer, P.L., Kogan, A.: Horn functions and their DNFs. Inf. Process. Lett. 44, 23–29 (1992)
11. Harary, F., Schwenk, A.: Trees with hamiltonian squares. Mathematika 18, 138–140 (1971)
12. West, D.B.: Introduction to Graph Theory, 2nd edn. Prentice Hall (2001)
13. Raychaudhuri, A.: The total interval number of a tree and the Hamiltonian completion number of its line graph. Inf. Process. Lett. 56, 299–306 (1995)
14. Agnetis, A., Detti, P., Meloni, C., Pacciarelli, D.: A linear algorithm for the Hamiltonian completion number of the line graph of a tree. Inf. Process. Lett. 79, 17–24 (2001)
15. Bertossi, A.A.: The Edge Hamiltonian Path Problem is NP-complete. Inf. Process. Lett. 13(4/5), 157–159 (1981)

Minimum Weight Dynamo and Fast Opinion Spreading

(Extended Abstract)

Sara Brunetti[1], Gennaro Cordasco[2], Luisa Gargano[3], Elena Lodi[1],
and Walter Quattrociocchi[1]

[1] Department of Mathematics and Computer Science, University of Siena, Italy
[2] Department of Psychology, Second University of Naples, Italy
[3] Department of Computer Science, University of Salerno, Italy

Abstract. We consider the following multi–level opinion spreading model on networks. Initially, each node gets a weight from the set $\{0, \ldots, k-1\}$, where such a weight stands for the individuals conviction of a new idea or product. Then, by proceeding to rounds, each node updates its weight according to the weights of its neighbors. We are interested in the initial assignments of weights leading each node to get the value $k - 1$ –e.g. unanimous maximum level acceptance– within a given number of rounds. We determine lower bounds on the sum of the initial weights of the nodes under the irreversible simple majority rules, where a node increases its weight if and only if the majority of its neighbors have a weight that is higher than its own one. Moreover, we provide constructive tight upper bounds for some class of regular topologies: rings, tori, and cliques.

Keywords: multicolored dynamos, information spreading, linear threshold models.

1 Introduction

New opinions and behaviors usually spread gradually through social networks. In 1966 a classical study showed how doctors' willingness to prescribe a new antibiotic diffused through professional contacts. A similar pattern can be detected in a variety of innovations: Initially a few innovators adopt, then people in contact with the innovators get interested and then adopt, and so forth until eventually the innovation spreads throughout the society. A classical question is then how many innovators are needed, and how they need to be disposed, in order to get a fast unanimous adoption [17].

In the wide set of the information spreading models, the first computational study about information diffusion [9] used the *linear threshold model* where the threshold triggering the adoption of a new idea to a node is given by the majority of its active neighbors.

Recently, information spreading has been intensively studied also in the context of *viral marketing*, which uses social networks to achieve marketing objectives through self-replicating viral processes, analogous to the spread of viruses. The goal here is to create a marketing message that can initially convince a selected set of people and then spread to the whole network in a short period of time [6]. One problem in viral marketing is the *target set selection problem* which asks for identifying the minimal

M.C. Golumbic et al. (Eds.): WG 2012, LNCS 7551, pp. 249–261, 2012.
© Springer-Verlag Berlin Heidelberg 2012

number of nodes which can activate, under some conditions, the whole network [7]. The target set selection problem has been proved to be NP-hard through a reduction to the node cover problem [10]. Recently, inapproximability results of opinion spreading problems have been presented in [5].

In this paper, we consider the following novel opinion spreading model. Initially, each node is assigned a weight from the set $\{0, \ldots, k-1\}$; where the weight of a node represents the level of acceptance of the opinion by the actor represented by the node itself. Then, the process proceeds in synchronous rounds where each node updates its weight depending on the weights of its neighbors. We are interested in the initial assignments of weights leading to the all–$(k-1)$ configuration within a given number of rounds. The goal is to minimize the sum of the initial weights of the nodes.

Essentially, we want everyone to completely accept the new opinion within a given time bound while minimizing the initial convincing effort (sum of the initial node weights).

We notice that we are interested in the case in which the spreading is essentially a one-way process: once an agent has adopted an opinion (or behavior, innovation, ...), she sticks with it. These are usually referred as *irreversible* spreading processes.

Dynamic Monopolies and Opinion Spreading. In a different scenario, spreading processes have been studied under the name of dynamic monopolies. Monopolies were initially introduced to deal with faulty nodes in distributed computing systems. A monopoly in a graph is a subset M of nodes such that each other node of the graph has a prescribed number of neighbors belonging to M. The problem of finding monopolies in graphs has been widely studied, see for example [1], [12], and [13] for connections with minimum dominating set problem.

Dynamic monopolies or shortly *dynamo* were introduced by Peleg [15]. A strong relationship between opinion spreading problems, such as the target set selection, and dynamic monopolies exists. Indeed, they can be used to model the irreversible spread of opinions in social networks.

Dynamic monopolies have been intensively studied with respect to the bounds of the size of the monopolies, the time needed to converge into a fixed point, and topologies over which the interaction takes place [2], [3], [8], [11], [14], [16].

Our Results: Weighted Opinion Spreading. We model the opinion spreading process considered in this paper, by means of weighted dynamos.

We extend the setting of dynamos from 2 possible weights (denoting whether a node has accepted the opinion or not) to k levels of opinion acceptance (a different extension has been studied in [4]). Initially, each node has a weight (which represents the node initial level of acceptance of the opinion) in the set $\{0, \ldots, k-1\}$. Then, each node updates its weight by increasing it of one unit if the weights of the simple majority of its neighbors is larger than its own. We call k-dynamos, the initial weight assignments which lead each node in the network to have maximum weight $k-1$. We are interested in the minimum weight (i.e. the sum of the weight initially assigned to the nodes) of a k-dynamo. We focus on both the weight and the time (e.g., number of rounds needed to reach the final configuration); namely, we study k-dynamos of minimum weight which converge into at most t rounds.

Paper Organization. In Section 2, we formalize the model and fix the notation. In Section 3, we determine lower bounds on the weight of k-dynamos which converge into at most t rounds. Section 4 provides tight constructive upper bounds for rings, tori and cliques. In the last section, we conclude and state a few open problems.

2 The Model

Let $G = (V, E)$ be an undirected connected graph. For each $v \in V$, we denote by $N(v) = \{u \in V \mid \{u, v\} \in E\}$ the neighborhood of v and by $d(v) = |N(v)|$ its cardinality (i.e., the degree of v).

We assume the nodes of G to be weighted by the set $[k] = \{0, 1, \dots, k-1\}$ of the first $k \geq 2$ integers. For each $v \in V$ we denote by $c_v \in [k]$ the weight assigned to a given node v.

Definition 1. *A configuration C on G is a partition of V into k sets $\{V_0, V_1, \dots, V_{k-1}\}$, where $V_j = \{v \in V \mid c_v = j\}$ is the set of nodes of weight j. The weight $w(C)$ of C is the weighted sum of its nodes*

$$w(C) = \sum_{j=0}^{k-1} j \times |V_j| = \sum_{v \in V} c_v.$$

Consider the following node weighting game played on G using the set of weights $[k]$ and a threshold value λ (for some $0 < \lambda \leq 1$):

> In the initial configuration, each node has a weight in $[k]$. Then node weights are updated in synchronous rounds (i.e., round i depends on round $i-1$ only). Let $c_v(i)$ denote the weight of node v at the end of round $i \geq 0$; during round $i \geq 1$, each node updates its weight according to the weight of its neighbors at round $i - 1$. Specifically, each node v
> - first computes the number $n^+(v) = |\{u \in N(v) \mid c_u(i-1) > c_v(i-1)\}|$ of neighbors having a weight larger than its current one $c_v(i-1)$;
> - then, it applies the following *irreversible rule:*
>
> $$c_v(i) = \begin{cases} c_v(i-1) + 1 & \text{if } n^+(v) \geq \lceil \lambda d(v) \rceil \\ c_v(i-1) & \text{otherwise} \end{cases}$$

We denote the initial configuration by C^0 and the configuration at round i by C^i.

We are interested into initial configurations that converge to the unanimous all-$(k-1)$s configuration – i.e., there exists a round t^* such that for each $i \geq t^*$ and for each node v, it holds $c_v(i) = k - 1$. Such configurations are named k-weights dynamic monopoly (henceforth k-*dynamo*).

A (k, t)-*dynamo* is a k-dynamo which reaches its final configuration within t rounds, that is, $c_v(i) = k - 1$ for each node $v \in V$ and $i \geq t$. An example of (k, t)-dynamo, with $\lambda = 1/2$, is depicted in Figure 1. Given a graph G, a set of weights $[k]$, a threshold λ, and an integer $t > 0$, we aim for a minimum weight (k, t)-dynamo.

Definition 2. *A (k, t)-dynamo on a graph G with threshold λ is optimal if its weight is minimal among all the (k, t)-dynamos for the graph G with threshold λ.*

3   Time Bounded Dynamos

In this section we provide a lower bound on the weight of a (k, t)–dynamo and study the minimum value of t for which an optimal (k, t)–dynamo coincides with a k–dynamo.

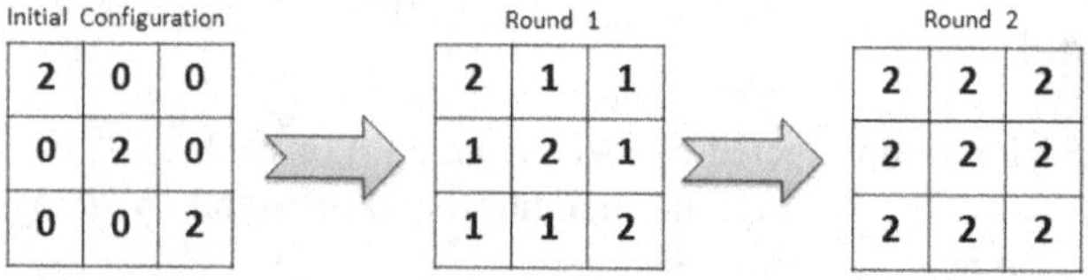

Initial Configuration				Round 1				Round 2		
2	0	0		2	1	1		2	2	2
0	2	0		1	2	1		2	2	2
0	0	2		1	1	2		2	2	2

Fig. 1. A $(3, 2)$-dynamo on a 3×3 Tori $(\lambda = 1/2)$: Starting from the initial configuration (left), two rounds are needed to reach the final all-(2)s configuration

3.1   Preliminary Results

Definition 3. *Consider an undirected connected graph $G = (V, E)$. Let $k \geq 2$ and $t \geq 1$ be integers and $0 < \lambda \leq 1$. An initial configuration $\mathcal{C}$ for G is called (k, t)-simple-monotone if V can be partitioned into $t + 1$ sets $X_{-s}, X_{-s+1}, \ldots X_{k-1}$ (here $s = t - k + 1$) where $X_{k-1} \neq \emptyset$, and for each $v \in X_i$*
  (i) $c_v(0) = \max(i, 0)$;
  (ii) v has at least $\lceil \lambda d(v) \rceil$ neighbours in $\bigcup_{j=i+1}^{k-1} X_j$.

Lemma 1. *Any (k, t)-simple-monotone configuration for an undirected connected graph G is a (k, t)-dynamo for G.*

Proof. We show that for each $i = -s, -s + 1, \ldots, k - 1$ (here $s = t - k + 1$) and $j = 0, \ldots, t$ and for each $u \in X_i$

$$c_u(j) = \begin{cases} \min(j + i, k - 1) & \text{if } j + i > 0 \\ 0 & \text{otherwise.} \end{cases}$$

We prove this statement by induction on i from $k - 1$ back to $-s$. For $i = k - 1$ the nodes in X_{k-1} have weight $k - 1$ from the initial configuration and the statement is trivially true for each round j.

Assume now that the statement is true for any $r > i$. For each $u \in X_i$, we know that u has at least $\lceil \lambda d(v) \rceil$ neighbours which belong to $\bigcup_{r=i+1}^{k-1} X_r$. By induction, each of this neighbor nodes, for each round j has a weight greater or equal to $\min(j+i+1, k-1)$ if $j + (i + 1) > 0$.

Hence, u preserves its weight $c_u(j) = max(i, 0) = 0$ until it increases its weight at each round j such that $j + (i+1) > 1$ (i.e. $j+i > 0$) and $c_u(j) < k-1$; as a result each node in X_i has weight $\min(j + i, k - 1)$ whenever $j + i > 0$, for each $j = 0, 1, \ldots, t$.

The Lemma follows since at round t, $i + j = i + t \geq -s + t = k - 1 > 0$. Hence, all the nodes will have weight $\min(i + t, k - 1) = k - 1$.   □

Lemma 2. *Let $G = (V, E)$ be an undirected connected graph. There exists an optimal (k, t)-dynamo for G which is a (k, t)-simple-monotone configuration for G.*

Proof. Let $\mathcal{C}$ be an optimal (k, t)-dynamo. Define a new configuration $\mathcal{C}'$ as follows: Let $s = t - k + 1$, for $i = k - 1, k - 2, \ldots, -s$, let X_i be the set of nodes that, starting with configuration $\mathcal{C}$, reaches permanently the weight $k - 1$ at round $k - 1 - i$, that is,

$$X_i = \{u \in V \mid c_u(k - 2 - i) \neq k - 1, \text{ and } c_u(j) = k - 1 \text{ for each } j \geq k - 1 - i\}.$$

In $\mathcal{C}'$, for each $u \in X_i$ set $c'_u(0) = \max(i, 0)$.

Notice that since $\mathcal{C}$ is a k-dynamo which converges into t rounds, $\{X_{-s}, X_{-s+1}, \ldots, X_{k-1}\}$ is a partition of V and $X_{k-1} \neq \emptyset$. We now show that $w(\mathcal{C}') \leq w(\mathcal{C})$ and $\mathcal{C}'$ is a (k, t)-simple-monotone configuration for G. Clearly,

(a) for each index $i \leq 0$, and for each $u \in X_i$, $c_u(0) \geq c'_u(0) = 0$;
(b) for each $i > 0$ and for each $u \in X_i$ we have $c_u(0) \geq c'_u(0) = i$ (otherwise u cannot reach the final weight $k - 1$ by round $k - 1 - i$, since the weight of a node increases by at most 1 at each round).

By using (a) and (b) above we have that $w(\mathcal{C}') \leq w(\mathcal{C})$. It remains to show that $\mathcal{C}'$ is a (k, t)-simple-monotone configuration for G. By construction, $\mathcal{C}'$ satisfies point (i) of Definition 3. Moreover, for each $u \in X_i$, we know that u in the configuration $\mathcal{C}$ reaches the weight $k - 1$ at round $k - 1 - i$. Hence at least $\lceil \lambda d(v) \rceil$ of its neighbors have weight $k - 1$ at round $k - 1 - i - 1 = k - 1 - (i + 1)$, that is at least $\lceil \lambda d(v) \rceil$ of its neighbors belong to $\bigcup_{j=i+1}^{k-1} X_j$. Hence, point (ii) of Definition 3 also holds. □

3.2 A Lower Bound

Theorem 1. *Consider an undirected connected graph $G = (V, E)$ and let $k \geq 2$ and $t \geq 1$ be integers. Any (k, t)-dynamo $\mathcal{C}$, with $\lambda = 1/2$, has weight*

$$w(\mathcal{C}) \geq \begin{cases} \frac{|V|}{2\rho(\ell+s+1)+1} \times (k-1 + \rho\ell(\ell+1)) \\ \quad \text{where } \ell = \left\lfloor \frac{\sqrt{(2\rho s+\rho+1)^2+4\rho(k-1)}-(2\rho s+\rho+1)}{2\rho} \right\rfloor & \text{if } t \geq k-1 \\ \frac{|V|}{2\rho(\ell+s+1)+1} \times (k-1 + \rho(\ell(\ell+1) - s(s+1))) \\ \quad \text{where } \ell = \left\lfloor \frac{\sqrt{4\rho(t+1)+(\rho-1)^2}-(2\rho s+\rho+1)}{2\rho} \right\rfloor & \text{otherwise,} \end{cases}$$

where ρ is the ratio between the maximum and the minimum degree of the nodes in V and $s = t - k + 1$.

Proof. By Lemma 2 we can restrict our attention to (k, t)-simple-monotone configurations for G. Therefore, the set V can be partitioned into $t + 1$ subsets $X_{-s}, X_{-s+1}, \ldots, X_{k-1}$ where $s = t - k + 1$ and for $i = -s, -s + 1, \ldots, k - 1$, X_i denotes the set of nodes whose weight at round j is $\max(0, \min(j + i, k - 1))$. Henceforth, we denote the size of X_i by x_i and the sum of the degree of nodes in $A \subseteq V$ by $d(A)$.

In order to prove the theorem, we first show that, for each $i = -s, -s+1, \ldots, k-2$, it holds

$$x_i \leq 2\rho x_{k-1}. \tag{1}$$

Let $E(A, B) = |\{e = (u, v) \in E : u \in A \text{ and } v \in B\}|$ denote the number of edges between a node in A and one in B. Each node $v \in X_i$ must increase its weight for each round r such that $0 < r + i < k - 1$; hence, at round $r = \max(-i + 1, 0)$, node v must have at least $\lceil d(v)/2 \rceil$ neighbors which belong to $\bigcup_{j=i+1}^{k-1} X_j$. Overall the number of edges between X_i and $\bigcup_{j=i+1}^{k-1} X_j$ satisfies

$$E\left(X_i, \bigcup_{j=i+1}^{k-1} X_j\right) \geq \frac{d(X_i)}{2} \geq \frac{|X_i|d_{min}}{2} = \frac{x_i d_{min}}{2}, \qquad (2)$$

where d_{min} represents the minimum degree of a node in G. Moreover, for each $i = -s, -s+1, \ldots, k-2$, the number of edges between X_i and $\bigcup_{j=i+1}^{k-1} X_j$ is

$$E\left(X_i, \bigcup_{j=i+1}^{k-1} X_j\right) \leq \sum_{j=i+1}^{k-1} d(X_j) - 2E\left(X_{i+1}, \bigcup_{j=i+1}^{k-1} X_j\right) - 2E\left(X_{i+2}, \bigcup_{j=i+2}^{k-1} X_j\right) - \ldots$$

$$\ldots - 2E\left(X_{k-2}, X_{k-2} \cup X_{k-1}\right) - 2E\left(X_{k-1}, X_{k-1}\right)$$

$$\leq \sum_{j=i+1}^{k-1} d(X_j) - 2\left[E\left(X_{i+1}, \bigcup_{j=i+2}^{k-1} X_j\right) + E\left(X_{i+2}, \bigcup_{j=i+3}^{k-1} X_j\right) + \right.$$

$$\left. \ldots + E(X_{k-2}, X_{k-1})\right]$$

$$\leq \sum_{j=i+1}^{k-1} d(X_j) - 2\left[d(X_{i+1})/2 + d(X_{i+2})/2 + \ldots + d(X_{k-2})/2\right]$$

$$= d(X_{k-1}) \leq d_{max}|X_{k-1}| = d_{max}x_{k-1},$$

where d_{max} is the maximum node degree of a node in G. By this and (2), recalling that $\rho = d_{max}/d_{min}$, we get (1).

Define now $y_i = x_i/x_{k-1}$. By (1), $0 \leq y_i \leq 2\rho$. Our goal is to minimize the weight function $w(\mathcal{C}) = \sum_{j=1}^{k-1} j x_j = x_{k-1}\left((k-1) + \sum_{j=1}^{k-2} j y_j\right)$ with $|V| = \sum_{j=-s}^{k-1} x_j = x_{k-1}\left(1 + \sum_{j=-s}^{k-2} y_j\right)$. Hence, $x_{k-1} = \frac{|V|}{1 + \sum_{j=-s}^{k-2} y_j}$ and we can write

$$w(C) = |V| \times \frac{k - 1 + \sum_{j=1}^{k-2} j y_j}{1 + \sum_{j=-s}^{k-2} y_j}. \qquad (3)$$

We distinguish now two cases depending on whether $t \geq k - 1$ or $t < k - 1$.

Case I ($t \geq k - 1$): In this case, it is possible to show that the rightmost term of (3) is minimized when

$$y_i = \begin{cases} 2\rho & \text{if } -s \leq i \leq \ell \\ 0 & \text{if } \ell < i \leq k - 2, \end{cases} \qquad (4)$$

where $\ell = \left\lfloor \frac{\sqrt{(2\rho s + \rho + 1)^2 + 4\rho(k-1)} - (2\rho s + \rho + 1)}{2\rho} \right\rfloor$ is the floor of the positive root of the equation $\rho i^2 + (2\rho s + \rho + 1)i - (k+1)$.

Let $f(y_{-s}, y_{-s+1}, \ldots, y_{k-2}) = \frac{k-1+\sum_{j=1}^{k-2} j y_j}{1+\sum_{j=-s}^{k-2} y_j}$. This function is decreasing in y_i for each $-s \leq i \leq 0$. Hence, since $0 \leq y_j \leq 2\rho$ for each j,

$$f(y_{-s}, y_{-s+1}, \ldots, y_0, y_1, \ldots, y_{k-2}) \geq f(2\rho, 2\rho, \ldots, 2\rho, y_1, \ldots, y_{k-2}).$$

Moreover, we show that the following two inequalities hold:

$$
\begin{aligned}
f(2\rho, 2\rho, \ldots, 2\rho, y_1, \ldots, y_\ell, \ldots, y_{k-2}) &\geq f(2\rho, 2\rho, \ldots, 2\rho, y_2, \ldots, y_\ell, \ldots, y_{k-2}) \\
&\geq f(2\rho, 2\rho, \ldots, 2\rho, y_3, \ldots, y_\ell, \ldots, y_{k-2}) \\
&\geq \cdots \\
&\geq f(2\rho, 2\rho, \ldots, 2\rho, y_{\ell+1}, \cdots y_{k-2}) \qquad (5)
\end{aligned}
$$

$$
\begin{aligned}
f(2\rho, 2\rho, \ldots, 2\rho, y_{\ell+1}, \cdots y_{k-2}) &\geq f(2\rho, 2\rho, \ldots, 2\rho, y_{\ell+1}, \cdots y_{k-3}, 0) \\
&\geq f(2\rho, 2\rho, \ldots, 2\rho, y_{\ell+1}, \cdots y_{k-4}, 0, 0) \\
&\geq \cdots \\
&\geq f(2\rho, 2\rho, \ldots, 2\rho, 0, 0, \ldots, 0). \qquad (6)
\end{aligned}
$$

We first prove (5). Each inequality in (5) is obtained by considering the following one for some $i \leq \ell$ (recalling that ℓ is the floor of the positive root of the equation $\rho i^2 + (2\rho s + \rho + 1)i - (k+1)$)

$$f(2\rho, \ldots, 2\rho, y_i, \ldots, y_{k-2}) = \frac{A+iy_i}{B+y_i} \geq \frac{A+2\rho i}{B+2\rho} = f(2\rho, \ldots, 2\rho, y_{i+1}, \ldots, y_{k-2}) \quad (7)$$

where $A = k-1 + \sum_{j=i+1}^{k-2} j y_j + \rho i(i-1)$ and $B = 1 + \sum_{j=i+1}^{k-2} y_j + 2\rho(i+s)$.

We notice that (7) is satisfied whenever $y_i(A - iB) \leq 2\rho(A - iB)$ and that for $i \leq \ell$

$$
\begin{aligned}
A - iB &= k-1 + \sum_{j=i+1}^{k-2} j y_j + \rho i(i-1) - i\left(1 + \sum_{j=i+1}^{k-2} y_j + 2\rho(i+s)\right) \\
&= k-1 + \sum_{j=i+1}^{k-2} (j-i) y_j + \rho i^2 - \rho i - i - 2\rho i^2 - 2\rho i s \\
&\geq -\rho i^2 - (2\rho s + \rho + 1)i + k - 1 \geq 0.
\end{aligned}
$$

Hence, (7) and consequently (5) are satisfied. In order to get (6), we show that for each $i > \ell$

$$
\begin{aligned}
f(2\rho, \ldots, 2\rho, y_{\ell+1}, \ldots, y_i, 0, \ldots, 0) &= \frac{C+iy_i}{D+y_i} \\
&\geq \frac{C}{D} = f(2\rho, \ldots, 2\rho, y_{\ell+1}, \ldots, y_{i-1}, 0, \ldots, 0) \, (8)
\end{aligned}
$$

where $C = k-1 + \sum_{j=\ell+1}^{i-1} j y_j + \rho \ell(\ell+1)$ and $D = 1 + \sum_{j=\ell+1}^{i-1} y_j + 2\rho(s+\ell+1)$.

Since (8) is satisfied whenever $y_i(C - iD) \leq 0$ and since now $i > \ell$ we get

$$C - iD = k - 1 + \sum_{j=\ell+1}^{i-1} jy_j + \rho\ell(\ell+1) - i\left(1 + \sum_{j=\ell+1}^{i-1} y_j + 2\rho(s+\ell+1)\right)$$

$$\leq k - 1 + \rho\ell^2 + \rho\ell - (\ell+1) - 2\rho(\ell+1)s - 2\rho(\ell+1)\ell - 2\rho(\ell+1)$$

$$= -\rho\ell^2 - (2\rho s + 3\rho + 1)\ell + k - 2\rho s - 2\rho - 2 \leq 0.$$

Hence, (8) and consequently (6) are satisfied. Summarizing, we have that the minimizing values are

$$x_i = \begin{cases} \frac{|V|}{1+\sum_{j=-s}^{k-2} y_j} = \frac{|V|}{2\rho(\ell+s+1)+1}, & \text{for } i = k-1 \\ 2\rho x_{k-1} = \frac{2\rho|V|}{2\rho(\ell+s+1)+1} & \text{for } i = -s, -s+1, \ldots, \ell \\ 0 & \text{otherwise.} \end{cases}$$

Therefore,

$$\sum_{j=1}^{k-1} jx_j = \frac{|V|}{2\rho(\ell+s+1)+1}\left(k-1+2\rho\sum_{j=1}^{\ell} j\right) = \frac{|V|}{2\rho(\ell+s+1)+1} \times (k-1+\rho\ell(\ell+1)),$$

and we can conclude that $w(\mathcal{C}) \geq \frac{|V|}{2\rho(\ell+s+1)+1} \times (k - 1 + \rho\ell(\ell+1))$, when $t \geq k-1$.

Case II ($t < k - 1$): The proof of this case is left to the reader. $\qquad\square$

Corollary 1. *Consider an undirected connected d-regular graph $G = (V, E)$. Let $k \geq 2$ and $t \geq 1$ be integers. Any (k,t)-dynamo $\mathcal{C}$, with $\lambda = 1/2$, has weight*

$$w(\mathcal{C}) \geq \begin{cases} \frac{|V|}{2\ell+2s+3} \times (k-1+\ell(\ell+1)) \text{ where } \ell = \lfloor\sqrt{t+1+s^2+s}\rfloor-(s+1) & \text{if } t \geq k-1 \\ \frac{|V|}{2\ell+2s+3} \times (k-1+\ell(\ell+1) - s(s+1)) \text{ where } \ell = \lfloor\sqrt{t+1}\rfloor-(s+1) & \text{otherwise,} \end{cases}$$

where $s = t - k + 1$.

We are now able to answer the question: *Which is the smallest value of t such that the optimal dynamo contains only two weights?* By analyzing the value of ℓ in the case $t \geq k - 1$ we have that whenever $t > \frac{k(2\rho+1)-2\rho-4}{2\rho}$ then $\ell = 0$, hence only the weights 0 and $k - 1$ will appear in the optimal configuration. When $\rho = 1$ (i.e., on regular graphs) one has $t > \frac{3}{2}k - 3$.

Remark 1. Our result generalizes the one in [8] with $k = 2$. Indeed, when $t \geq k-1 = 1$ by the above consideration we get $t > \frac{3}{2}k - 3 = 0$ and $\ell = 0$. Hence, $w(\mathcal{C}) \geq \frac{|V|}{2s+3} \times (k - 1) = \frac{|V|}{2t+1}$.

Theorem 2. *Let $G = (V, E)$ be an undirected connected graph, if t is sufficiently large, then:*

(i) *any optimal (k,t)-dynamo contains only the weights 0 and $k - 1$;*

(ii) *let $k \geq 2$ be an integer and C_2 a 2–dynamo on G. Let C_k be obtained from C_2 by replacing the weight 1 with the weight $k-1$. If C_2 is an optimal 2-dynamo then C_k is an optimal k-dynamo. Moreover, $w(C_k) = w(C_2) \times (k-1)$ and $t(C_k) = t(C_2) + k - 2$ (where $t(C)$ is the time needed to reach the final configuration).*

Proof omitted.

4 Building (k, t)-Dynamo

In this section we provide several optimal (or almost optimal) (k, t)-dynamo constructions for Rings and Tori ($\lambda = 1/2$) and Cliques (any λ).

4.1 Rings

A n-node ring $\mathcal{R}_n$ consists of n nodes and $n - 1$ edges, where for $i = 0, 1, \ldots, n - 1$ each node v_i is connected with $v_{(i-1) \bmod n}$ and $v_{(i+1) \bmod n}$.

A necessary condition for $C(\mathcal{R}_n, k)$ to be a k-dynamo ($\lambda \leq 1/2$) is that at least one node of $\mathcal{R}_n$ is weighted by $k - 1$. This condition is also sufficient.

Theorem 3. *An optimal k-dynamo ($\lambda \leq 1/2$) $C(\mathcal{R}_n, k)$ has weight $w(C(\mathcal{R}_n, k)) = (k - 1)$, and it reaches its final configuration within $t = k - 2 + \lceil \frac{n-1}{2} \rceil$ rounds.*

A (k, t)-dynamo ($\lambda = 1/2$) for a ring $\mathcal{R}_n$ is obtained by the following partition of V which defines the initial configuration (see Figure 2) $C(\mathcal{R}_n, k, t)$: for $i = 0, 1, \ldots, n$,

$$\forall v_i \in \mathcal{R}_n, \quad v_i \in \begin{cases} X_{k-1} & \text{if } j = 0 \\ X_{\ell+1-j} & \text{if } 1 \leq j \leq \ell + s + 1 \\ X_{j-\ell-2s-2} & \text{if } \ell + s + 2 \leq j \leq 2\ell + 2s + 2 \end{cases}$$

where $s = t - k + 1$, $j = i \bmod (2\ell + 2s + 3)$ and $\ell = \lfloor \sqrt{t+1+s^2+s} \rfloor - (s+1)$ if $t \geq k - 1$ and $\ell = \lfloor \sqrt{t+1} \rfloor - (s+1)$ otherwise.

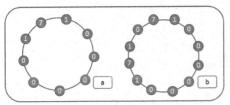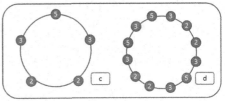

Fig. 2. (k, t)**-dynamos on Rings**: (a) $C(\mathcal{R}_9, 8, 9)$, a (8,9)-dynamo on $\mathcal{R}_9$ ($\ell = 1$), in this particular case $n = 2\ell + 2s + 3$; (b) $C(\mathcal{R}_{12}, 8, 9)$ a (8,9)-dynamo on $\mathcal{R}_{12}$ ($\ell = 1$); (c) $C(\mathcal{R}_5, 6, 3)$, a (6,3)-dynamo on $\mathcal{R}_5$ ($\ell = 3$), in this particular case $n = 2\ell + 2s + 3$; (d) $C(\mathcal{R}_{12}, 6, 3)$, a (6,3)-dynamo on $\mathcal{R}_{12}$ ($\ell = 3$).

Theorem 4. *(i) The configuration $\mathcal{C}(\mathcal{R}_n, k, t)$ is a (k, t)-dynamo for any value of n, $\lambda = 1/2$, $k \geq 2$ and $t \geq 1$. (ii) The weight of $\mathcal{C}(\mathcal{R}_n, k, t)$ is*

$$w(\mathcal{C}(\mathcal{R}_n, k, t)) \leq \begin{cases} \left\lceil \frac{n}{2\ell+2s+3} \right\rceil (k-1+\ell(\ell+1)) & \text{if } t \geq k-1 \\ \quad \text{where } \ell = \lfloor \sqrt{t+1+s^2+s} \rfloor - (s+1) \\ \left\lceil \frac{n}{2\ell+2s+3} \right\rceil (k-1+\ell(\ell+1) - s(s+1)) & \text{otherwise} \\ \quad \text{where } \ell = \lfloor \sqrt{t+1} \rfloor - (s+1) \end{cases}$$

Proof. (i) By construction $\mathcal{C}(\mathcal{R}_n, k, t)$ is (k, t)-simple-monotone, hence by Lemma 1, $\mathcal{C}(\mathcal{R}_n, k, t)$ is a (k, t)-dynamo. (ii) There are two cases to consider: if $t \geq k - 1$, then starting from v_0 each set of $2\ell + 2s + 3$ nodes weights $k - 1 + 2\sum_{i=1}^{\ell} i = k - 1 + \ell(\ell+1)$. Then the weight of $\mathcal{C}(\mathcal{R}_n, k, t)$ is smaller than the weight of $\mathcal{C}(\mathcal{R}_{\bar{n}}, k, t)$ where $\bar{n} = \lceil \frac{n}{2\ell+2s+3} \rceil \times (2\ell + 2s + 3)$. Hence, $w(\mathcal{C}(\mathcal{R}_n, k, t)) \leq w(\mathcal{C}(\mathcal{R}_{\bar{n}}, k, t)) = \lceil \frac{n}{2\ell+2s+3} \rceil (k-1+\ell(\ell+1))$. Similarly for $t < k - 1$. $\square$

By Corollary 1 and Theorem 4 we have the following Corollary.

Corollary 2. *When $n/(2\ell+2s+3)$ is integer, $\mathcal{C}(\mathcal{R}_n, k, t)$ is an optimal (k, t)-dynamo.*

4.2 Tori

A $n \times m$-node tori $\mathcal{T}_{n,m}$ consists of $n \times m$ nodes and $2(n \times m)$ edges, where for $i = 0, 1, \ldots, n - 1$ and $j = 0, 1, \ldots, m - 1$, each node $v_{i,j}$ is connected with four nodes: $v_{i,(j-1) \bmod m}$, $v_{i,(j+1) \bmod m}$, $v_{(i-1) \bmod n,j}$ and $v_{(i+1) \bmod n,j}$.

A (k, t)-dynamo $(\lambda = 1/2)$ for $\mathcal{T}_{2\ell+2s+3, 2\ell+2s+3}$ is obtained by weighting diagonals with the same order defined for dynamos on rings. Specifically, the configuration $\mathcal{C}(\mathcal{T}_{2\ell+2s+3, 2\ell+2s+3}, k, t)$ is defined by the partition of V described as follows, let $D_i = \{v_{a,b} : i = (b - a) \bmod (2\ell + 2s + 3)\}$ denote the i-th diagonal of $\mathcal{T}_{2\ell+2s+3, 2\ell+2s+3}$, for $i = 0, 1, \ldots, 2\ell + 2s + 2$,

$$\forall v \in D_i, \quad v \in \begin{cases} X_{k-1} & \text{if } i = 0 \\ X_{\ell+1-i} & \text{if } 1 \leq i \leq \ell+s+1 \\ X_{i-\ell-2s-2} & \text{if } \ell+s+2 \leq i \leq 2\ell+2s+2, \end{cases}$$

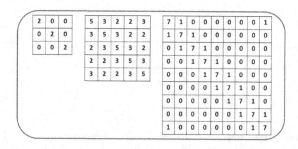

Fig. 3. (k, t)-dynamos on Tori: (left) $\mathcal{C}(\mathcal{T}_{3,3}, 3, 2)$, a $(3,2)$-dynamo on $\mathcal{T}_{3,3}$ $(\ell=0)$; (middle) $\mathcal{C}(\mathcal{T}_{5,5}, 6, 3)$, a $(6,3)$-dynamo on $\mathcal{T}_{5,5}$ $(\ell=3)$; (right) $\mathcal{C}(\mathcal{T}_{9,9}, 8, 9)$ a $(8,9)$-dynamo on $\mathcal{T}_{9,9}$ $(\ell=1)$

where $s = t - k + 1$, $\ell = \lfloor \sqrt{t+1+s^2+s} \rfloor - (s+1)$ if $t \geq k - 1$ and $\ell = \lfloor \sqrt{t+1} \rfloor - (s+1)$ otherwise. Some examples are depicted in Figure 3.

Theorem 5. *The configuration* $C(\mathcal{T}_{2\ell+2s+3,2\ell+2s+3}, k, t)$ *is an optimal* (k,t)*-dynamo for any* $k \geq 2$, $t \geq 1$ *and* $\lambda = 1/2$.

Proof. Let $C = C(\mathcal{T}_{2\ell+2s+3,2\ell+2s+3}, k, t)$. By construction C is (k,t)-simple-monotone, hence by Lemma 1, it is a (k,t)-dynamo. To show its optimality we distinguish two cases. If $t \geq k - 1$, each row (resp. each column) corresponds to $C(\mathcal{R}_{2\ell+2s+3}, k, t)$ and its weight is $k - 1 + \ell(\ell+1)$. Overall, $w(C) = (2\ell + 2s + 3) \times (k - 1 + \ell(\ell+1))$ that matches the bound in Corollary 1. Similarly for $t < k - 1$. $\qquad\square$

A (k,t)-dynamo for $\mathcal{T}_{n,m}$ is obtained by building a grid $\lceil \frac{n}{2\ell+2s+3} \rceil \times \lceil \frac{m}{2\ell+2s+3} \rceil$, where each cell is filled with a configuration $C(\mathcal{T}_{2\ell+2s+3,2\ell+2s+3}, k, t)$ defined above. Then, the exceeding part is removed and the last row and the last column are updated. In particular, for each column (resp. row), if the removed part contains a $k - 1$, then the element in the last row (resp. column) is given the value $k - 1$ (see Figure 4). We call this configuration $C(\mathcal{T}_{n,m}, k, t)$.

Theorem 6.
(i) $C(\mathcal{T}_{n,m}, k, t)$ *is a* (k,t)*-dynamo for any value of* n, $m, \lambda = 1/2$, $k \geq 2$ *and* $t \geq 1$.
(ii) The weight of $C(\mathcal{T}_{n,m}, k, t)$ *is*

$$w(C(\mathcal{T}_{n,m}, k, t)) \leq \begin{cases} \lceil \frac{n}{2\ell+2s+3} \rceil \lceil \frac{m}{2\ell+2s+3} \rceil (2\ell+2s+3)(k-1+\ell(\ell+1)) & \text{if } t \geq k-1 \\ \quad \text{where } \ell = \lfloor \sqrt{t+1+s^2+s} \rfloor - (s+1) \\ \lceil \frac{n}{2\ell+2s+3} \rceil \lceil \frac{m}{2\ell+2s+3} \rceil (2\ell+2s+3)(k-1+\ell(\ell+1)-s(s+1)) & \text{otherwise} \\ \quad \text{where } \ell = \lfloor \sqrt{t+1} \rfloor - (s+1). \end{cases}$$

Proof. (i) By construction $C(\mathcal{T}_{n,m}, k, t)$ is (k,t)-simple-monotone (cfr. Figure 4), hence by Lemma 1, $C(\mathcal{T}_{n,m}, k, t)$ is a (k,t)-dynamo.
(ii) The grid contains $\lceil \frac{n}{2\ell+2s+3} \rceil \times \lceil \frac{m}{2\ell+2s+3} \rceil$ cells. If $t \geq k - 1$, each cell has weight $w(C(\mathcal{T}_{2\ell+2s+3,2\ell+2s+3}, k, t)) = (2\ell + 2s + 3) \times (k - 1 + \ell(\ell+1))$.

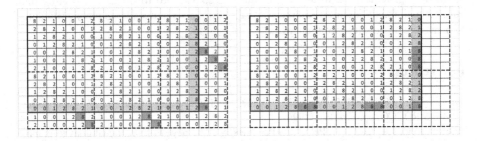

Fig. 4. $C(\mathcal{T}_{12,18}, 9, 8)$, a $(9,8)$-dynamo on $\mathcal{T}_{12,18}$ ($\ell = 2$): (left) a grid 2×3 is filled with 6 configuration $C(\mathcal{T}_{7,7}, 9, 8)$; (right) The exceeding parts i.e., the last two rows and the last three columns are removed. Finally the last row and the last column are updated in order to obtain a configuration that satisfies Lemma 1.

Moreover, the nodes that change their weight take the weight of a removed element. Hence, the weight of $\mathcal{C}(\mathcal{T}_{n,m}, k, t)$ is upper bounded by the weight of the full grid which is $\lceil \frac{n}{2\ell+2s+3} \rceil \times \lceil \frac{m}{2\ell+2s+3} \rceil \times w(\mathcal{C}(\mathcal{T}_{2\ell+2s+3,2\ell+2s+3}, k, t))$. Similarly for $t < k - 1$. $\square$

By Corollary 1 and Theorem 6 we have the following Corollary.

Corollary 3. *If both n and m are multiples of $2\ell + 2s + 3$, $\mathcal{C}(\mathcal{T}_{n,m}, k, t)$ is an optimal (k, t)-dynamo.*

4.3 Cliques

Let K_n be the clique on n nodes. A necessary condition for a k-dynamo $\mathcal{C}(K_n, k)$ is that $\lceil \lambda(n - 1) \rceil$ nodes are weighted by $k - 1$. The condition is also sufficient and if the remaining $\lfloor \lambda(n - 1) \rfloor$ nodes are weighted by 0, the k-dynamo is optimal and reaches its final configuration within $t = k - 1$ rounds. So, when $t \geq k - 1$ the optimal configuration is obtained by weighting $\lceil \lambda(n - 1) \rceil$ nodes by $k - 1$ and the remaining nodes by 0. For $t < k - 1$, an optimal (k,t)-dynamo is obtained by assigning weight $k - t - 1$ to all the non-$k - 1$ weighted nodes. Clearly this configuration is optimal, if we assign a weight smaller than $k - t - 1$ to a node v, then v can not reach the weight $k - 1$ within t rounds. Therefore:

Theorem 7. *Let K_n be the clique on n nodes. An optimal (k,t)-dynamo $\mathcal{C}(K_n, k, t)$ has weight $w(\mathcal{C}(K_n, k, t)) = (k - 1) \times \lceil \lambda(n - 1) \rceil + \max(k - t - 1, 0) \times \lfloor \lambda(n - 1) \rfloor$.*

5 Conclusion and Open Problems

We studied multivalued dynamos with respect to both weight and time. We derived lower bounds on the weight of (k, t)-dynamo and provided constructive tight upper bounds for rings, tori and cliques. Several questions remain open: In particular, different updating rules could also be investigated, as for instance reversible rules. Construction based on different graphs would also be interesting.

Acknowledgments. We would like to thank Ugo Vaccaro for many stimulating discussions and the anonymous referees whose helpful comments allowed to significantly improve the presentation of their work.

References

1. Bermond, J.C., Bond, J., Peleg, D., Perennes, S.: Tight bounds on the size of 2-monopolies. In: SIROCCO, pp. 170–179 (1996)
2. Bermond, J.C., Bond, J., Peleg, D., Perennes, S.: The power of small coalitions in graphs. Discrete Applied Mathematics 127(3), 399–414 (2003)
3. Bermond, J.C., Gargano, L., Rescigno, A.A., Vaccaro, U.: Fast Gossiping by Short Messages. SIAM J. on Computing 27(4), 917–941 (1998)
4. Brunetti, S., Lodi, E., Quattrociocchi, W.: Dynamic monopolies in colored tori. In: IPDPS Workshops, pp. 626–631 (2011)

5. Chen, N.: On the approximability of influence in social networks. SIAM J. Discret. Math. 23, 1400–1415 (2009)
6. Domingos, P., Richardson, M.: Mining the network value of customers. In: KDD 2001: Proceedings of the Seventh ACM SIGKDD International Conference on Knowledge Discovery and Data Mining, New York, NY, USA, pp. 57–66 (2001)
7. Easley, D., Kleinberg, J.: Networks, Crowds, and Markets: Reasoning About a Highly Connected World. Cambridge University Press (2010)
8. Flocchini, P., Královič, R., Ružička, P., Roncato, A., Santoro, N.: On time versus size for monotone dynamic monopolies in regular topologies. J. of Discrete Algorithms 1(2), 129–150 (2003)
9. Granovetter, M.: Economic action and social structure: the problem of embeddedness. American Journal of Sociology 91 (1985)
10. Kempe, D., Kleinberg, J., Tardos, E.: Maximizing the spread of influence through a social network. In: KDD 2003: Proceedings of the Ninth ACM SIGKDD International Conference on Knowledge Discovery and Data Mining, pp. 137–146. ACM, New York (2003)
11. Kulich, T.: Dynamic monopolies with randomized starting configuration. Theor. Comput. Sci. 412(45), 6371–6381 (2011)
12. Linial, N., Peleg, D., Rabinovich, Y., Saks, M.: Sphere packing and local majorities in graphs. In: ISTCS, pp. 141–149. IEEE Computer Soc. Press (1993)
13. Mishra, S., Rao, S.B.: Minimum monopoly in regular and tree graphs. Electronic Notes in Discrete Mathematics 15(0), 126 (2003)
14. Nayak, A., Pagli, L., Santoro, N.: Efficient construction of catastrophic patterns for vlsi reconfigurable arrays with bidirectional links. In: ICCI, pp. 79–83 (1992)
15. Peleg, D.: Size bounds for dynamic monopolies. Discrete Applied Mathematics 86(2-3), 263–273 (1998)
16. Peleg, D.: Local majorities, coalitions and monopolies in graphs: a review. Theor. Comput. Sci. 282(2), 231–257 (2002)
17. Valente, T.W.: Network models of the diffusion of innovations. Hampton Press (1995)

Immediate versus Eventual Conversion: Comparing Geodetic and Hull Numbers in P_3-Convexity

Carmen Cecilia Centeno[1], Lucia Draque Penso[2], Dieter Rautenbach[2], and Vinícius Gusmão Pereira de Sá[1]

[1] Instituto de Matemática, NCE, and COPPE,
Universidade Federal do Rio de Janeiro, Rio de Janeiro, RJ, Brazil
carmen@cos.ufrj.br, vigusmao@dcc.ufrj.br
[2] Institut für Optimierung und Operations Research,
Universität Ulm, Ulm, Germany
lucia.penso@uni-ulm.de, dieter.rautenbach@uni-ulm.de

Abstract. We study the graphs G for which the hull number $h(G)$ and the geodetic number $g(G)$ with respect to P_3-convexity coincide. These two parameters correspond to the minimum cardinality of a set U of vertices of G such that the simple expansion process that iteratively adds to U, all vertices outside of U that have two neighbors in U, produces the whole vertex set of G either eventually or after one iteration, respectively. We establish numerous structural properties of the graphs G with $h(G) = g(G)$, which allow the constructive characterization as well as the efficient recognition of all triangle-free such graphs. Furthermore, we characterize the graphs G that satisfy $h(H) = g(H)$ for every induced subgraph H of G in terms of forbidden induced subgraphs.

Keywords: Hull number, geodetic number, P_3-convexity, irreversible 2-threshold processes, triangle-free graphs.

1 Introduction

As one of the most elementary models of the spreading a property within a network — like sharing an idea or disseminating a virus — one can consider a graph G, a set U of vertices of G that initially possess the property, and an iterative process whereby new vertices u enter U whenever sufficiently many neighbors of u are already in U. The simplest choice for "sufficiently many" that results in interesting effects is 2. This choice leads to the *irreversible 2-threshold processes* considered by Dreyer and Roberts [5]. Similar models were studied in various contexts such as statistical physics, social networks, marketing, and distributed computing under different names such as bootstrap percolation, influence dynamics, local majority processes, irreversible dynamic monopolies, catastrophic fault patterns and many others [1–3, 5, 7–10].

From the point of view of discrete convexity, the above spreading process is nothing but the formation of the convex hull of the set U of vertices of G with

M.C. Golumbic et al. (Eds.): WG 2012, LNCS 7551, pp. 262–273, 2012.
© Springer-Verlag Berlin Heidelberg 2012

respect to the so-called P_3-*convexity* in G, where a set C of vertices of G is considered to be P_3-*convex* if no vertex of G outside of C has two neighbors in C. A P_3-*hull set* of G is a set of vertices whose P_3-convex hull equals the whole vertex set of G, and the minimum cardinality of a P_3-hull set of G is the P_3-*hull number* $h(G)$ of G.

Closely related to the notion of hull sets and the hull number of a graph are geodetic sets and the geodetic number. A P_3-*geodetic set* of graph G is a set of vertices such that every vertex u of G either belongs to the set or has two neighbors in the set. The minimum cardinality of a P_3-geodetic set of G is the P_3-*geodetic number* $g(G)$ of G. Different types of graph convexities have been considered in the literature, and the definitions of hull and geodetic sets change accordingly. For the special case of P_3-convexity, the P_3-geodetic number coincides with the well-studied 2-domination number [6].

In view of the iterative spreading process considered above, a hull set *eventually* distributes the property within the entire network, whereas a geodetic set spreads the property within the entire network *in exactly one iteration*. In [11] we considered spreading processes with arbitrary deadlines between 1 and ∞. Clearly, every geodetic set is a hull set, which implies

$$h(G) \leq g(G) \tag{1}$$

for every graph G. Furthermore, both parameters are computationally hard in general, and efficient algorithms are only known for quite restricted graph classes [4, 6].

In the present paper we study graphs that satisfy (1) with equality. After summarizing useful notation and terminology, we collect numerous structural properties of such graphs in Section 2. Based on these properties, we construct a large subclass of those graphs in Section 3, comprising all triangle-free such graphs. In Section 4 we derive an efficient algorithm for the recognition of the triangle-free graphs that satisfy (1) with equality. In Section 5 we give a complete characterization in terms of forbidden induced subgraphs of the class of all graphs for which (1) holds with equality for every induced subgraph. Finally, we conclude with some open problems in Section 6.

1.1   Notation and Terminology

We consider finite and simple graphs and digraphs, and use standard terminology. For a graph G, the vertex set is denoted $V(G)$ and the edge set is denoted $E(G)$. For a vertex u of a graph G, the neighborhood of u in G is denoted $N_G(u)$ and the degree of u in G is denoted $d_G(u)$. A vertex of a graph whose removal increases the number of components is a cut vertex. A set C of vertices of G is P_3-*convex* exactly if no vertex of G outside C has two neighbors in C. The P_3-*convexity of* G is the collection $\mathcal{C}(G)$ of all P_3-convex sets. Since we only consider P_3-convexity, we will omit the prefix "P_3-" from now on.

For a set U of G, let the *interval* $I_G(U)$ *of* U *in* G be the set $U \cup \{u \in V(G) \setminus U \mid |N_G(u) \cap U| \geq 2\}$, and let $H_G(U)$ denote the convex hull of U in G, that is, $H_G(U)$ is the unique smallest set in $\mathcal{C}(G)$ containing U. Within this notation, U is a geodetic set of G if $I_G(U) = V(G)$, and U is a hull set of G if $H_G(U) = V(G)$. The inequality (1) follows from the immediate observation that $I_G(U) \subseteq H_G(U)$ for every set U of vertices of some graph G.

If U is a hull set of G, then there is an acyclic orientation D of a spanning subgraph of G such that the in-degree $d_D^-(u)$ is 0 for every vertex u in U and 2 for every vertex u in $V(G) \setminus U$. We call D a hull proof for U in G.

Since the hull number and the geodetic number are both additive with respect to the components of G, we consider the set of graphs

$$\mathcal{H} = \{G \mid G \text{ is a connected graph with } h(G) = g(G)\}.$$

2   Structural Properties of Graphs in $\mathcal{H}$

We collect some structural properties of the graphs in $\mathcal{H}$ in the form of lemmas which will be required to prove our main results in the next section. The proofs of many lemmas in this section, however, were omitted due to space limitations and left to an extended version of this paper.

Let G be a fixed graph in $\mathcal{H}$. Let W be a geodetic set of G of minimum order and let $B = V(G) \setminus W$. By definition, every vertex in B has at least two neighbors in W. Therefore, G has a spanning bipartite subgraph G_0 with bipartition $V(G_0) = W \cup B$ such that every vertex in B has degree exactly 2 in G_0. Let E_1 denote the set of edges in $E(G) \setminus E(G_0)$ between vertices in the same component of G_0 and let E_2 denote the set of edges in $E(G) \setminus E(G_0)$ between vertices in distinct components of G_0. Note that, by construction, W is a geodetic set of G_0. Since $|W| = g(G) = h(G) \leq h(G_0) \leq g(G_0) \leq |W|$, we obtain $h(G_0) = g(G_0) = |W|$, that is, G_0 has no geodetic set and no hull set of order less than $|W|$. Thus, if C is a component of G_0, then $W \cap V(C)$ is a minimum geodetic set of C as well as a minimum hull set of C.

Lemma 1. *Let C be a component of G_0.*

(i) *No two vertices in C are incident with edges in E_2.*
(ii) *If some vertex u in C is incident with at least two edges in E_2, then u belongs to B and u is a cut vertex of C.*

Proof. (i) We consider different cases. If two vertices w and w' in $V(C) \cap W$ are incident with edges in E_2, then let $P : w_1 b_1 \ldots w_{l-1} b_{l-1} w_l$ be a shortest path in C between $w = w_1$ and $w' = w_l$. The set $(W \setminus \{w_1, \ldots, w_l\}) \cup \{b_1, \ldots, b_{l-1}\}$ is a hull set of G, which is a contradiction.

If a vertex w in $V(C) \cap W$ and a vertex b in $V(C) \cap B$ are incident with edges in E_2, then let $P : w_1 b_1 \ldots w_l b_l$ be a shortest path in C between $w = w_1$ and $b = b_l$. Note that b has a neighbor in G_0 that does not belong to P. Therefore, the set $(W \setminus \{w_1, \ldots, w_l\}) \cup \{b_1, \ldots, b_{l-1}\}$ is a hull set of G, which is a contradiction.

Finally, if two vertices b and b' in $V(C) \cap B$ are incident with edges in E_2, then let $P : b_1 w_1 \ldots b_{l-1} w_{l-1} b_l$ be a shortest path in C between $b = b_1$ and $b' = b_l$. Note that b and b' both have neighbors in G_0 that do not belong to P.

Therefore, the set $(W \setminus \{w_1, \ldots, w_{l-1}\}) \cup \{b_2, \ldots, b_{l-1}\}$ is a hull set of G, which is a contradiction.

(ii) If a vertex w in $V(C) \cap W$ is incident with at least two edges in E_2, then $W \setminus \{w\}$ is a hull set of G, which is a contradiction.

If a vertex b in $V(C) \cap B$ that is not a cut vertex of C is incident with at least two edges in E_2, then let $P : w_1 b_1 \ldots w_{l-1} b_{l-1} w_l$ be a path in C avoiding b between the two neighbors w_1 and w_l of b in G_0. The set $(W \setminus \{w_1, \ldots, w_l\}) \cup \{b_1, \ldots, b_{l-1}\}$ is a hull set of G, which is a contradiction and completes the proof.

Lemma 2. *If G_0 is not connected, no two vertices in W that belong to the same component of G_0 are adjacent.*

Proof. For contradiction, we assume that ww' is an edge of G where w and w' are vertices in W that belong to the same component C of G_0. Since G is connected, there is an edge uv in E_2 with $u \in V(C)$ and $v \in V(G) \setminus V(C)$.

First, we assume that u belongs to W. Let $P : w_1 b_1 \ldots w_{l-1} b_{l-1} w_l$ be a shortest path in C between $u = w_1$ and a vertex w_l in $\{w, w'\}$. Note that $l = 1$ is possible. The set $(W \setminus \{w_1, \ldots, w_l\}) \cup \{b_1, \ldots, b_{l-1}\}$ is a hull set of G, which is a contradiction.

Next, we assume that u belongs to B. Let $P : b_1 w_1 \ldots b_l w_l$ be a shortest path in C between $u = b_1$ and a vertex w_l in $\{w, w'\}$. Note that $l = 1$ is possible. Furthermore, note that b_1 has a neighbor in G_0 that does not belong to P. The set $(W \setminus \{w_1, \ldots, w_l\}) \cup \{b_2, \ldots, b_l\}$ is a hull set of G, which is a contradiction and completes the proof.

Lemma 3. *If G_0 is not connected and C is a component of G_0, then there are no two vertices w in $V(C) \cap W$ and b in $V(C) \cap B$ such that $wb \in E(G) \setminus E(G_0)$.*

Proof. For contradiction, we assume that wb is an edge of G where w in W and b in B belong to the same component C of G_0. Since G is connected, there is an edge uv in E_2 with $u \in V(C)$ and $v \in V(G) \setminus V(C)$.

First, we assume that $u \in W$. Let P be a shortest path in C between u and a vertex u' in $\{w, b\}$. If $u' = w$, then let $P : w_1 b_1 \ldots b_{l-1} w_l$ where $u = w_1$ and $w = w_l$. Note that $l = 1$ is possible. In this case the set $(W \setminus \{w_1, \ldots, w_l\}) \cup \{b_1, \ldots, b_{l-1}\}$ is a hull set of G, which is a contradiction. If $u' = b$, then let $P : w_1 b_1 \ldots b_{l-1} w_l b_l$ where $u = w_1$ and $b = b_l$. Note that $l = 1$ is possible. Furthermore, note that b has a neighbor in G_0 that does not belong to P. In this case, the set $(W \setminus \{w_1, \ldots, w_l\}) \cup \{b_1, \ldots, b_{l-1}\}$ is a hull set of G, which is a contradiction.

Next, we assume that $u = b$. Let $P : b_1 w_1 \ldots b_l w_l$ be a shortest path in C between $b = b_1$ and $w = w_l$. Note that the edge bw does not belong to C, hence $l \geq 2$. Furthermore, note that b has a neighbor in G_0 that does not belong to P. In this case, the set $(W \setminus \{w_1, \ldots, w_l\}) \cup \{b_2, \ldots, b_l\}$ is a hull set of G, which is a contradiction.

Finally, we assume that $u \in B \setminus \{b\}$. Let P be a shortest path in C between u and a vertex u' in $\{w, b\}$. If $u' = w$, then let $P : b_1 w_1 \ldots b_l w_l$, where $u = b_1$ and $w = w_l$. Note that $l = 1$ is possible. Furthermore, note that w is the unique neighbor of b in P, and that u has a neighbor in G_0 that does not belong to P. In this case, the set $(W \setminus \{w_1, \ldots, w_l\}) \cup \{b_2, \ldots, b_l\}$ is a hull set of G, which is a contradiction. If $u' = b$, then let $P : b_1 w_1 \ldots w_{l-1} b_l$, where $u = b_1$ and $b = b_l$. In this case, the set $(W \setminus \{w_1, \ldots, w_{l-1}\}) \cup \{b_2, \ldots, b_{l-1}\}$ is a hull set of G, which is a contradiction and completes the proof.

Lemma 4. *Let G_0 be disconnected and let b and b' be two vertices in B that belong to the same component C of G_0 satisfying $bb' \in E_1$.*

  (i) *Neither b nor b' is incident with an edge in E_2.*
  (ii) *If some vertex w in $V(C) \cap W$ is incident with an edge in E_2 and $P :$ $w_1 b_1 \ldots w_l b_l$ is a path in C between $w = w_1$ and a vertex b_l in $\{b, b'\}$, then w_l is adjacent to both b and b', and C contains no path between b and b' that does not contain w_l.*
  (iii) *If some vertex b'' in $(V(C) \cap B) \setminus \{b, b'\}$ is incident with an edge in E_2 and $P : b_1 w_1 \ldots w_{l-1} b_l$ is a path in C between $b'' = b_1$ and a vertex b_l in $\{b, b'\}$, then w_{l-1} is adjacent to both b and b' and C contains no path between b and b' that does not contain w_{l-1}.*

Lemma 5. *If C is a component of G_0, then there are no two vertices w and w' of C that belong to W and two edges e and e' that belong to $E(G) \setminus E(G_0)$ such that w is incident with e, w' is incident with e', and e' is distinct from ww'.*

Lemma 6. *If C is a component of G_0, then there are no two edges wb and wb' that belong to $E(G) \setminus E(G_0)$ with $w \in W \cap V(C)$ and $b, b' \in B \cap V(C)$.*

Lemma 7. *If G_0 is connected and G is triangle-free, then there are no two edges ww' and bb' in G with $w, w' \in W$ and $b, b' \in B$.*

Lemma 8. *If G_0 is connected and G is triangle-free, then there are no two edges wb and $b'b''$ in G with $w \in W$ and $b, b', b'' \in B$.*

Lemma 9. *If G_0 is connected and G is triangle-free, then there are no two distinct edges bb' and $b''b'''$ in G with $b, b', b'', b''' \in B$.*

3   Constructing All Triangle-Free Graphs in $\mathcal{H}$

Let $\mathcal{G}_0$ denote the set of all bipartite graphs G_0 with a fixed bipartition $V(G_0) = B \cup W$ such that every vertex in B has degree exactly 2.

We consider four distinct operations that can be applied to a graph G_0 from $\mathcal{G}_0$.

  – **Operation $\mathcal{O}_1$**
    Add one arbitrary edge to G_0.
  – **Operation $\mathcal{O}_1'$**
    Select two vertices w_1 and w_2 from W and arbitrarily add new edges between vertices in $\{w_1, w_2\} \cup (N_{G_0}(w_1) \cap N_{G_0}(w_2))$.

- **Operation** $\mathcal{O}_2$
  Add one arbitrary edge between vertices in distinct components of G_0.
- **Operation** $\mathcal{O}_3$
  Choose a non-empty subset X of B such that all vertices in X are cut vertices of G_0 and no two vertices in X lie in the same component of G_0. Add arbitrary edges between vertices in X so that X induces a connected subgraph of the resulting graph. For every component C of G_0 that does not contain a vertex from X, add one arbitrary edge between a vertex in C and a vertex in X.

Let $\mathcal{G}_1$ denote the set of graphs that are obtained by applying operation $\mathcal{O}_1$ once to a connected graph G_0 in $\mathcal{G}_0$. Let $\mathcal{G}_1'$ denote the set of graphs that are obtained by applying operation $\mathcal{O}_1'$ once to a connected graph G_0 in $\mathcal{G}_0$. Let $\mathcal{G}_2$ denote the set of graphs that are obtained by applying operation $\mathcal{O}_2$ once to a graph G_0 in $\mathcal{G}_0$ that has exactly two components. Let $\mathcal{G}_3$ denote the set of graphs that are obtained by applying operation $\mathcal{O}_3$ once to a graph G_0 in $\mathcal{G}_0$ that has at least three components. Note that $\mathcal{O}_3$ can only be applied if G_0 has at least one cut vertex that belongs to B.

Finally, let

$$\mathcal{G} = \mathcal{G}_1 \cup \mathcal{G}_1' \cup \mathcal{G}_2 \cup \mathcal{G}_3. \tag{2}$$

Since the operation $\mathcal{O}_1'$ allows that no edges are added, the set $\mathcal{G}_1'$ contains all connected graphs in $\mathcal{G}_0$.

Theorem 1. $\mathcal{G} \subseteq \mathcal{H}$.

Proof. Let G be a graph in $\mathcal{G}$ that is obtained by applying some operation to a graph G_0 in $\mathcal{G}_0$. Let $V(G_0) = B \cup W$ be the fixed bipartition of G_0. Since every vertex in B has two neighbors in W, the partite set W is a geodetic set of G and therefore $g(G) \leq |W|$. By (1), it suffices to show that $h(G) \geq |W|$ to conclude the proof. For contradiction, we assume that U is a hull set of G with $|U| < |W|$. Let D be a hull proof for U in G.

The proof naturally splits into four cases according to which of the four sets $\mathcal{G}_1$, $\mathcal{G}_1'$, $\mathcal{G}_2$, and $\mathcal{G}_3$ the graph G belongs to. Due to space limitation, we give the details of the proof only for one case.

Case 1. $G \in \mathcal{G}_1$.

Let $W_1 = W \setminus U$ and $B_0 = B \cap U$. Note that, by the above assumption, $|W_1| > |B_0| \geq 0$.

We claim that there is at most one vertex w in W_1 for which the set $N_{G_0}(w)$ contains exactly one vertex of B_0 and that for every other vertex w' in W_1, the set $N_{G_0}(w')$ contains at least two vertices of B_0. In other words, there is a vertex w^* in W_1 such that

$$|N_{G_0}(w) \cap B_0| \geq \begin{cases} 1, \ w = w^*, \\ 2, \ w \in W_1 \setminus \{w^*\}. \end{cases} \tag{3}$$

Let w be a vertex in W_1. Since $|W_1| > |B_0|$, we may assume that $N_{G_0}(w)$ contains at most one vertex from B_0. Let x and y denote the two in-neighbors of w in D. Let e denote the edge added by operation $\mathcal{O}_1$.

If x belongs to W, then e is the edge xw. Hence $y \in B$ and $d_G(y) = 2$. Therefore $y \in B_0$, that is, $y \in N_{G_0}(w) \cap B_0$. Furthermore, for every other vertex w' in $W_1 \setminus \{w\}$, its two in-neighbors x' and y' in D both belong to B and are not incident with e. Hence $d_G(x') = d_G(y') = 2$ and therefore $x', y' \in B_0$, that is, $x', y' \in N_{G_0}(w) \cap B_0$. Hence, we may assume that x and y both belong to B.

If e is the edge wx, then we obtain as above that $y \in N_{G_0}(w) \cap B_0$. Hence x does not belong to B_0. This implies that the two edges of G_0 incident with x are both oriented towards x in D. For every other vertex w' in $W_1 \setminus \{w\}$, it follows that its two in-neighbors x' and y' in D satisfy $x', y' \in N_{G_0}(w) \cap B_0$. Hence, we may assume that e is neither wx nor wy.

Since $N_{G_0}(w)$ contains at most one element from B_0, we may assume that e is incident with x and oriented towards x in D. This implies that $y \in N_{G_0}(w) \cap B_0$. Furthermore, for every other vertex w' in $W_1 \setminus \{w\}$, its two in-neighbors x' and y' in D both belong to B, and, if they are incident with e, then e is not oriented towards them in D. This implies that x' and y' belong to B_0, that is, $x', y' \in N_{G_0}(w') \cap B_0$.

Altogether, the existence of a vertex w^* in W_1 with (3) follows. If m denotes the number of edges in G_0 between W_1 and B_0, then (3) implies $m \geq 2(|W_1| - 1) + 1$. Furthermore, every vertex in B has degree 2 in G_0 and therefore $m \leq 2|B_0|$. Thus, $2|W_1| - 1 \leq 2|B_0|$. Since both cardinalities are integers, we obtain $|W_1| \leq |B_0|$, hence

$$|U| = |W \cap U| + |B \cap U| \geq |W \cap U| + |W \setminus U| = |W|,$$

which is a contradiction. This completes the proof.

In conjunction, the results in Sections 2 and Theorem 1 allow for a complete constructive characterization of the triangle-free graphs in $\mathcal{H}$.

Corollary 1. *If $\mathcal{T}$ denotes the set of all triangle-free graphs, then $\mathcal{G} \cap \mathcal{T} = \mathcal{H} \cap \mathcal{T}$.*

Proof. Theorem 1 implies $\mathcal{G} \cap \mathcal{T} \subseteq \mathcal{H} \cap \mathcal{T}$. For the converse inclusion, let G be a triangle-free graph in $\mathcal{H}$. Similarly as in Section 2, let W be a minimum geodetic set of G, let $B = V(G) \setminus W$, and let G_0 be a spanning bipartite subgraph of G with bipartition $V(G_0) = W \cup B$ such that every vertex in B has degree exactly 2 in G_0. Let E_1 denote the set of edges in $E(G) \setminus E(G_0)$ between vertices in the same component of G_0 and let E_2 denote the set of edges in $E(G) \setminus E(G_0)$ between vertices in distinct components of G_0.

First, we assume that G_0 is connected. In this case, $E_1 = E(G) \setminus E(G_0)$. For contradiction, we assume that E_1 contains two edges e and e'. By Lemmas 5 and 6, the edges e and e' are not both incident with vertices in W. We may therefore assume that e connects two vertices from B. Now, since G is triangle-free, Lemmas 7, 8, and 9 imply a contradiction. Hence E_1 contains at most one edge, which implies $G \in \mathcal{G}_1 \cup \mathcal{G}_1'$.

Next, we assume that G_0 is disconnected. By Lemmas 2 and 3, all vertices incident with edges in E_1 belong to B. For contradiction, we assume that E_1 is not empty. Let $bb' \in E_1$, where b and b' belong to some component C of G_0. Since G is connected but G_0 is not, some vertex of C is incident with an edge f from E_2. By Lemma 4, the edge f is not incident with b or b'. Furthermore, by Lemma 4 (ii) and (iii), G necessarily contains a triangle, which is a contradiction. Hence E_1 is empty. Now Lemma 1 immediately implies $G \in \mathcal{G}_2 \cup \mathcal{G}_3$, which completes the proof.

Corollary 1 implies several restrictions on the cycle structure of a triangle-free graph G in $\mathcal{H}$. Let G_0 with bipartition $B \cup W$ be the underlying graph in $\mathcal{G}_0$. Clearly, all cycles of G that are also cycles of G_0 are of even length and alternate between B and W. Furthermore, at most one of the vertices from B in such a cycle can have degree more than 2 in G. If G_0 is connected, the cycles of G are either such cycles of G_0 or they contain the unique edges in $E(G) \setminus E(G_0)$. If G_0 has two components, then G arises from G_0 by adding a bridge and all cycles of G are also cycles of G_0. Finally, if G_0 has at least three components and X is as described in $\mathcal{O}_3$, then X induces an arbitrary connected triangle-free graph in G, that is, the cycle structure of $G[X]$ can be quite complicated. Nevertheless, all cycles in $G[X]$ contain only vertices of degree at least 4 in G. All further cycles of G are totally contained within one component of G_0 and contain at least one vertex from B that has degree 2 in G.

4   Recognizing All Triangle-Free Graphs in $\mathcal{H}$

By Corollary 1, the structure of the triangle-free graphs in $\mathcal{H}$ is quite restricted. In fact, it is not difficult to recognize these graphs in polynomial time. This section is devoted to the details of a corresponding algorithm.

Let G be a given connected triangle-free input graph. By Corollary 1, the graph G belongs to $\mathcal{H}$ if and only if either G belongs to $\mathcal{G}_0 \cup \mathcal{G}_1 \cup \mathcal{G}_2$ or G belongs to $\mathcal{G}_3$.

Lemma 10. *It can be checked in polynomial time whether $G \in \mathcal{G}_0 \cup \mathcal{G}_1 \cup \mathcal{G}_2$.*

Proof. By definition, the graph G belongs to $\mathcal{G}_0 \cup \mathcal{G}_1 \cup \mathcal{G}_2$ if and only if deleting at most one edge from G results in a graph in $\mathcal{G}_0$ with at most two components. Since the graphs in $\mathcal{G}_0$ can obviously be recognized in linear time, it suffices to check whether $G \in \mathcal{G}_0$ and to consider each edge e of G in turn and check whether $G - e \in \mathcal{G}_0$. Since the graphs in $\mathcal{G}_0 \cup \mathcal{G}_1 \cup \mathcal{G}_2$ have a linear number of edges, all this can be done in quadratic time. This completes the proof.

In view of Lemma 10, we may assume from now on that G does not belong $\mathcal{G}_0 \cup \mathcal{G}_1 \cup \mathcal{G}_2$. The following lemma is an immediate consequence of the definition of operation $\mathcal{O}_3$.

Lemma 11. *If G belongs to $\mathcal{G}_3$, then there is a vertex x of G of degree at least three and two edges $e_l = xy_l$ and $e_r = xy_r$ of G incident with x such that, in the* *graph G' that arises by deleting from G all edges incident with x except for e_l and e_r, the component $C(x, e_l, e_r)$ of G' that contains x has the following properties:*

   (i) x is a cut vertex of $C(x, e_l, e_r)$;
  (ii) $C(x, e_l, e_r)$ has a unique bipartition with partite sets $B_l \cup \{x\} \cup B_r$ and $W_l \cup W_r$;
 (iii) Every vertex in $B_l \cup \{x\} \cup B_r$ has degree 2 in $C(x, e_l, e_r)$;
 (iv) $B_l \cup W_l$ and $B_r \cup W_r$ are the vertex sets of the two components of $C(x, e_l, e_r) - x$ such that $y_l \in W_l$ and $y_r \in W_r$;
  (v) None of the deleted edges connects x to a vertex from $V(C(x, e_l, e_r)) \setminus \{x\}$;
 (vi) W_l and W_r both contain a vertex of odd degree.

Proof. Choosing as x one of the vertices from the non-empty set X in the definition of $\mathcal{O}_3$ and choosing as e_1 and e_2 the two edges of G_0 incident with x, the properties (i) to (v) follow immediately. Note that $C(x, e_l, e_r)$ is the component of G_0 that contains x. For property (vi), observe that the number of edges of $C(x, e_l, e_r)$ between $B_l \cup \{x\}$ and W_l is exactly $2|B_l| + 1$, that is, it is an odd number, which implies that not all vertices of W_l can be of even degree. A similar argument applies to W_r. This completes the proof.

The key observation for the completion of the algorithm is the following lemma, which states that the properties from Lemma 11 uniquely characterize the elements of X.

Lemma 12. *If G belongs to $\mathcal{G}_3$ and a vertex x of G of degree at least three and two edges $e_l = xy_l$ and $e_r = xy_r$ of G incident with x are such that properties (i) to (vi) from Lemma 11 hold, then*

  (i) G is obtained by applying operation $\mathcal{O}_3$ to a graph G_0 in $\mathcal{G}_0$ with at least three components such that x belongs to the set X used by operation $\mathcal{O}_3$ and
 (ii) $C(x, e_l, e_r)$ defined as in Lemma 11 is the component of G_0 that contains x.

We proceed to the main result in this section.

Theorem 2. *For a given triangle-free graph G, it can be checked in polynomial time whether $h(G) = g(G)$ holds.*

Proof. Clearly, we can consider each component of G separately and may therefore assume that G is connected. Let n denote the order of G. By Lemma 10, we can check in $O(n^2)$ time whether G belongs $\mathcal{G}_0 \cup \mathcal{G}_1 \cup \mathcal{G}_2$. If this is the case, then Corollary 1 implies $h(G) = g(G)$. Hence, we may assume that G does not belong to $\mathcal{G}_0 \cup \mathcal{G}_1 \cup \mathcal{G}_2$. Note that there are $O(n^3)$ choices for a vertex x of G and two incident edges e_l and e_r of G. Furthermore, note that for every individual choice of the triple (x, e_l, e_r), the properties (i) to (vi) from Lemma 11 can be checked in $O(n)$ time. Therefore, by Lemmas 11 and 12, in $O(n^4)$ time, we can

  – either determine that no choice of (x, e_l, e_r) satisfies the conclusion of Lemma 11, which, by Corollary 1, implies $h(G) \neq g(G)$,

– or find a suitable triple (x, e_l, e_r) and reduce the instance G to a smaller instance $G^- = G - V(C(x, e_l, e_r))$.

Since the order of G^- is at least three less than n, this leads to an overall running time of $O(n^5)$. This completes the proof.

5   Forbidden Induced Subgraphs

It is an easy exercise to prove $h(G) = g(G)$ whenever G is a path, a cycle, or a star.

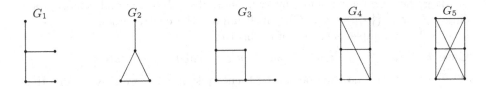

Fig. 1. The five graphs $G_1, \ldots, G_5$

Theorem 3. *If G is a graph, then $h(H) = g(H)$ for every induced subgraph H of G if and only if G is $\{G_1, \ldots, G_5\}$-free.*

Proof. Since $3 = h(G_1) = h(G_3) < g(G_1) = g(G_3) = 4$ and $2 = h(G_2) = h(G_4) = h(G_5) < g(G_2) = g(G_4) = g(G_5) = 3$, the "only if"-part of the statement follows. In order to prove the "if"-part, we may assume, for contradiction, that G is a connected $\{G_1, \ldots, G_5\}$-free graph with $h(G) < g(G)$. We consider different cases.

Case 1. *G contains a triangle $T : abca$.*

Since G is G_2-free, no vertex has exactly one neighbor on T.

If some vertex has no neighbor on T, then, by symmetry, we may assume that there are two vertices u and v of G such that uva is a path and u has no neighbor on T. Since G is G_2-free, we may assume that v is adjacent to b. Now u, v, a, and b induce G_2, which is a contradiction. Hence every vertex has at least one neighbor on T. This implies that the vertex set of G can be partitioned as

$$V(G) = \{a, b, c\} \cup N(\{a, b\}) \cup N(\{a, c\}) \cup N(\{b, c\}) \cup N(\{a, b, c\}),$$

where $N(S) = \{u \in V(G) \setminus \{a, b, c\} \mid N_G(u) \cap \{a, b, c\} = S\}$.

If two vertices, say u and v, in $N(\{a, b\})$ are adjacent, then u, v, a, and c induce G_2, which is a contradiction. Hence, by symmetry, each of the three sets $N(\{a, b\})$, $N(\{a, c\})$, and $N(\{b, c\})$ is independent. If some vertex u in $N(\{a, b\})$ is not adjacent to some vertex v in $N(\{a, c\})$, then u, v, a, and b induce G_2, which is a contradiction. Hence, by symmetry, there are all possible edges between every two of the three sets $N(\{a, b\})$, $N(\{a, c\})$, and $N(\{b, c\})$. If some vertex u in $N(\{a, b\})$ is not adjacent to some vertex v in $N(\{a, b, c\})$, then u, v, a, and c induce G_2, which is a contradiction. Hence, by symmetry, there are all possible edges between the two sets $N(\{a, b\}) \cup N(\{a, c\}) \cup N(\{b, c\})$ and $N(\{a, b, c\})$. If $N(\{a, b\})$ contains exactly one vertex, say u, then $I_G(\{u, c\}) = V(G)$, which implies the contradiction $2 \le h(G) \le g(G) \le 2$. Hence, by symmetry, none of the three sets $N(\{a, b\})$, $N(\{a, c\})$, and $N(\{b, c\})$ contains exactly one vertex. If there are two vertices in $N(\{a, b\})$, say u_1 and u_2, and two vertices in $N(\{b, c\})$, say v_1 and v_2, then u_1, u_2, v_1, v_2, a, and c induce G_5, which is a contradiction. Hence no two of the three sets $N(\{a, b\})$, $N(\{a, c\})$, and $N(\{b, c\})$ contain at least two vertices.

Altogether, we may assume, by symmetry, that $N(\{a, c\})$ and $N(\{b, c\})$ are empty. Now $I_G(\{a, b\}) = V(G)$, which implies the contradiction $2 \le h(G) \le g(G) \le 2$ and completes the proof in this case.

Case 2. *G contains no triangle but a cycle of length four $C : abcda$.*

If some vertex has no neighbor on C, then, by symmetry, we may assume that there are two vertices u and v of G such that uva is a path. Since G is triangle-free, v is not adjacent to b or d. Hence u, v, a, b, and d induce G_1, which is a contradiction. Hence every vertex has at least one neighbor on C. Since G is triangle-free, this implies that the vertex set of G can be partitioned as

$$V(G) = \{a, b, c, d\} \cup N(\{a\}) \cup N(\{b\}) \cup N(\{c\}) \cup N(\{d\}) \cup N(\{a, c\}) \cup N(\{b, d\}),$$

where $N(S) = \{u \in V(G) \setminus \{a, b, c, d\} \mid N_G(u) \cap \{a, b, c, d\} = S\}$.

If there is a vertex u in $N(\{a\})$ and a vertex v in $N(\{c\})$, then u and v are adjacent, because G is G_3-free. Now u, v, a, b, and d induce G_1, which is a contradiction. Hence, by symmetry, we may assume that $N(\{c\}) \cup N(\{d\})$ is empty. If there is a vertex u in $N(\{b\})$ and a vertex v in $N(\{a, c\})$, then u and v are not adjacent, because G is G_4-free. Now u, v, a, b, and d induce G_1, which is a contradiction. Hence, by symmetry, one of the two sets $N(\{b\})$ and $N(\{a, c\})$ is empty and one of the two sets $N(\{a\})$ and $N(\{b, d\})$ is empty. If there is a vertex u in $N(\{a, c\})$ and a vertex v in $N(\{b, d\})$, then u, v, a, b, c, and d induce either G_4 or G_5, which is a contradiction. Hence, by symmetry, we may assume that $N(\{b, d\})$ is empty.

Since G is G_1-free, there are all possible edges between the two sets $N(\{a\})$ and $N(\{b\})$.

Since G is G_1-free, both of the sets $N(\{a\})$ and $N(\{b\})$ contain at most one vertex.

Since G is G_1-free, there is no edge between $N(\{a\})$ and $N(\{a, c\})$.

If both of the sets $N(\{a\})$ and $N(\{b\})$ are not empty, then G is a graph of order 6 with $h(G) = g(G) = 3$, which is a contradiction. Hence, by symmetry, we may assume that

$$V(G) = \{a, b, c, d\} \cup N(\{a\}) \cup N(\{a, c\}).$$

If $N(\{a\})$ is empty, then $I_G(\{a, c\}) = V(G)$, which implies the contradiction $2 \le h(G) \le g(G) \le 2$. Hence $N(\{a\})$ contains exactly one vertex, say u, and

$I_G(\{a, c, u\}) = V(G)$, which implies $g(G) \leq 3$. If $H_G(U) = V(G)$ for some set U of vertices of G, then $u \in U$. In view of the structure of G, it follows easily that $h(G) \geq 3$, which implies the contradiction $h(G) = g(G)$ and completes the proof in this case.

Case 3. *G does not contain a triangle or a cycle of length four.*

If G contains no vertex of degree at least 3, then G is a path or a cycle, which implies the contradiction $h(G) = g(G)$. Hence, we may assume G contains a vertex of degree at least 3. Since G is G_1-free, G is a star, which implies the contradiction $h(G) = g(G)$ and completes the proof.

6   Conclusion

Several open problems/tasks are immediate.

- Give a constructive characterization of all graphs in $\mathcal{H}$.
- Describe an efficient algorithm to recognize all graphs in $\mathcal{H}$.

Acknowledgments. This work was supported in part by the DAAD/CAPES Probral project *"Cycles, Convexity and Searching in Graphs"* grant number 50724218.

References

1. Balister, P., Bollobás, B., Johnson, J.R., Walters, M.: Random majority percolation. Random Struct. Algorithms 36, 315–340 (2010)
2. Balogh, J., Bollobás, B.: Sharp thresholds in Bootstrap percolation. Physica A 326, 305–312 (2003)
3. Bermond, J.-C., Bond, J., Peleg, D., Perennes, S.: The power of small coalitions in graphs. Discrete Appl. Math. 127, 399–414 (2003)
4. Centeno, C.C., Dourado, M.C., Penso, L.D., Rautenbach, D., Szwarcfiter, J.L.: Irreversible Conversion of Graphs. Theor. Comput. Sci. 412, 3693–3700 (2011)
5. Dreyer Jr., P.A., Roberts, F.S.: Irreversible k-threshold processes: Graph-theoretical threshold models of the spread of disease and of opinion. Discrete Appl. Math. 157, 1615–1627 (2009)
6. Haynes, T.W., Hedetniemi, S.T., Slater, P.J.: Fundamentals of domination in graphs. Marcel Dekker (1998)
7. Kempe, D., Kleinberg, J., Tardos, É.: Maximizing the spread of influence through a social network. In: Proceedings of the 9th ACM SIGKDD International Conference on Knowledge Discovery and Data Mining, pp. 137–146 (2003)
8. Mustafa, N.H., Pekeč, A.: Listen to your neighbors: How (not) to reach a consensus. SIAM J. Discrete. Math. 17, 634–660 (2004)
9. Nayak, A., Ren, J., Santoro, N.: An improved testing scheme for catastrophic fault patterns. Inf. Process. Lett. 73, 199–206 (2000)
10. Peleg, D.: Local majorities, coalitions and monopolies in graphs: A review. Theor. Comput. Sci. 282, 231–257 (2002)
11. Rautenbach, D., dos Santos, V., Schäfer, P.M.: Irreversible conversion processes with deadlines (2010) (manuscript)

Bend-Bounded Path Intersection Graphs: Sausages, Noodles, and Waffles on a Grill[*]

Steven Chaplick[1,**], Vít Jelínek[2], Jan Kratochvíl[3], and Tomáš Vyskočil[3]

[1] Department of Mathematics, Wilfrid Laurier University, Waterloo, CA
chaplick@cs.toronto.edu
[2] Computer Science Institute, Faculty of Mathematics and Physics,
Charles University, Prague
jelinek@iuuk.mff.cuni.cz
[3] Department of Applied Mathematics, Faculty of Mathematics and Physics,
Charles University, Prague
{honza,whisky}@kam.mff.cuni.cz

Abstract. In this paper we study properties of intersection graphs of k-bend paths in the rectangular grid. A k-bend path is a path with at most k 90 degree turns. The class of graphs representable by intersections of k-bend paths is denoted by B_k-VPG. We show here that for every fixed k, B_k-VPG $\subsetneq B_{k+1}$-VPG and that recognition of graphs from B_k-VPG is NP-complete even when the input graph is given by a B_{k+1}-VPG representation. We also show that the class B_k-VPG (for $k \geq 1$) is in no inclusion relation with the class of intersection graphs of straight line segments in the plane.

1 Introduction

In this paper we continue the study of Vertex-intersection graphs of Paths in Grids[1] (VPG graphs) started by Asinowski et. al [1,2]. A *VPG representation* of a graph G is a collection of paths of the rectangular grid where the paths represent the vertices of G in such a way that two vertices of G are adjacent if and only if the corresponding paths share at least one vertex.

VPG representations arise naturally when studying circuit layout problems and layout optimization [15] where layouts are modelled as paths (wires) on grids. One approach to minimize the cost or difficulty of production involves minimizing the number of times the wires bend [3,13]. Thus the research has been focused on VPG representations parameterized by the number of times each path is allowed to *bend* (these representations are also the focus of [1,2]). In particular, a *k-bend path* is a path in the grid which contains at most k

---

[*] Research support by Czech research grants CE-ITI GAČR P202/12/6061 and GraDR EUROGIGA GAČR GIG/11/E023.
[**] This author was also supported by the Natural Sciences and Engineering Research Council of Canada and the University of Toronto.
[1] The grids to which we refer are always rectangular.

M.C. Golumbic et al. (Eds.): WG 2012, LNCS 7551, pp. 274–285, 2012.
© Springer-Verlag Berlin Heidelberg 2012

bends where a *bend* is when two consecutive edges on the path have different horizontal/vertical orientation. In this sense a B_k-VPG representation of a graph G is a VPG representation of G where each path is a k-bend path. A graph is B_k-VPG if it has a B_k-VPG representation.

Several relationships between VPG graphs and traditional graph classes (i.e., circle graphs, circular arc graphs, interval graphs, planar graphs, segment (SEG) graphs, and string (STRING) graphs) were observed in [1,2]. For example, the equivalence between string graphs (the intersection graphs of curves in the plane) and VPG graphs is formally proven in [2], but it was known as folklore result [6]. Additionally, the base case of this family of graph classes (namely, B_0-VPG) is a special case of segment graphs (the intersection graphs of line segments in the plane). Specifically, B_0-VPG is more well known as the 2-DIR[2]. The recognition problem for the VPG = string graph class is known to be NP-Hard by [9] and in NP by [14]. Similarly, it is NP-Complete to recognize 2-DIR = B_0-VPG graphs [11]. However, the recognition status of B_k-VPG for every $k > 0$ was given as an open problem from [2] (all cases were conjectured to be NP-Complete). We confirm this conjecture by proving a stronger result. Namely, we demonstrate that deciding whether a B_{k+1}-VPG graph is a B_k-VPG graph is NP-Complete (for any fixed $k > 0$) – see Section 4.

Furthermore, in [1,2] it is shown that B_0-VPG $\subsetneq B_1$-VPG $\subsetneq$ VPG and it was conjectured that B_k-VPG $\subsetneq B_{k+1}$-VPG for every $k > 0$. We confirm this conjecture constructively – see Section 3.

Finally, we consider the relationship between the B_k-VPG graph classes and segment graphs. In particular, we show that SEG and B_k-VPG are incomparable through the following pair of results (the latter of which is somewhat surprising): (1) There is a B_1-VPG graph which is not a SEG graph; (2) For every k, there is a 3-DIR graph which has no B_k-VPG representation.

The paper is organized as follows. In Section 2 we introduce the Noodle-Forcing Lemma, which is the key to restricting the topological structure of VPG representations[3]. In Section 3 we introduce the "sausage" structure which is the crucial gadget that we use for the hardness reduction and which by itself shows that B_k-VPG is strict subset of B_{k+1}-VPG[4]. We also demonstrate the incomparability of B_k-VPG and SEG in Section 3. The NP-hardness reduction is presented in Section 4. We end the paper with some remarks and open problems.

2   Noodle-Forcing Lemma

In this section, we present the key lemma of this paper (see Lemma 1). Essentially, we prove that, for "proper" representations R of a graph G, there is a graph G' where G is an induced subgraph of G' and R is "sub-representation"

---

[2] Note: a k-DIR graph is an intersection graph of straight line segments in the plane with at most k distinct directions (slopes).

[3] This was inspired by the order forcing lemma of [12].

[4] This gadget is named due to its VPG representation resembling sausage links.

of every representation of G' (i.e., all representations of G' require the part corresponding to G to have the "topological structure" of R). We begin this section with several definitions.

Let $G = (V, E)$ be a graph. A *representation* of G is a collection $R = \{R(v),\ v \in V\}$ of piecewise linear curves in the plane, such that $R(u) \cap R(v)$ is nonempty iff uv is an edge of G.

An *intersection point* of a representation R is a point in the plane that belongs to (at least) two distinct curves of R. Let $\text{In}(R)$ denote the set of intersection points of R.

A representation is *proper* if

1. each $R(v)$ is a simple curve, i.e., it does not intersect itself,
2. R has only finitely many intersection points (in particular no two curves may overlap) and finitely many bends, and
3. each intersection point p belongs to exactly two curves of R, and the two curves cross in p (in particular, the curves may not touch, and an endpoint of a curve may not belong to another curve).

Let R be a proper representation of $G = (V, E)$, let R' be another (not necessarily proper) representation of G, and let ϕ be a mapping from $\text{In}(R)$ to $\text{In}(R')$. We say that ϕ is *order-preserving* if it is injective and has the property that for every $v \in V$, if $p_1, p_2, \ldots, p_k$ are all the distinct intersection points on $R(v)$, then $\phi(p_1), \ldots, \phi(p_k)$ all belong to $R'(v)$ and they appear on $R'(v)$ in the same relative order as the points $p_1, \ldots, p_k$ on $R(v)$. (If $R'(v)$ visits the point $\phi(p_i)$ more than once, we may select one visit of each $\phi(p_i)$, such that the selected visits occur in the correct order $\phi(p_1), \ldots, \phi(p_k)$.)

For a set P of points in the plane, the *ε-neighborhood of P*, denoted by $\mathcal{N}_\varepsilon(P)$, is the set of points that have distance less than ε from P.

Lemma 1 (Noodle-Forcing Lemma). *Let $G = (V, E)$ be a graph with a proper representation $R = \{R(v),\ v \in V\}$. Then there exists a graph $G' = (V', E')$ containing G as an induced subgraph, which has a proper representation $R' = \{R'(v),\ v \in V'\}$ such that $R(v) = R'(v)$ for every $v \in V$, and $R'(w)$ is a horizontal or vertical segment for $w \in V' \setminus V$. Moreover, for any $\varepsilon > 0$, any (not necessarily proper) representation of G' can be transformed by a homeomorphism of the plane and by circular inversion into a representation $R^\varepsilon = \{R^\varepsilon(v),\ v \in V'\}$ with these properties:*

1. *for every vertex $v \in V$, the curve $R^\varepsilon(v)$ is contained in the ε-neighborhood of $R(v)$, and $R(v)$ is contained in the ε-neighborhood of $R^\varepsilon(v)$.*
2. *there is an order-preserving mapping $\phi\colon \text{In}(R) \to \text{In}(R^\varepsilon)$, with the additional property that for every $p \in \text{In}(R)$, the point $\phi(p)$ coincides with the point p.*

Due to space limitations, we only sketch the proof of the lemma. Suppose R is a proper representation of a graph G. The main idea is to overlay the representation R with a sufficiently fine grid-like configuration C of short horizontal and vertical segments, so that the position of a curve $R(v) \in R$ is well approximated by the set of segments of C that are intersected by $R(v)$. We refer to this step as 'grilling' of the representation R, since the segments of C form a structure resembling a grill.

We let R' be the representation $R \cup C$ and G' be the graph represented by R'. Moreover, let G_C be the graph whose intersection representation is C. The configuration of C has the property that any representation C' of the graph G_C can be transformed into the representation C by a homeomorphism and a circular inversion, followed possibly by a truncation of some of the curves of C'. In particular, any representation of G' can be transformed by a homeomorphism and a circular inversion into a representation R'' that essentially contains a copy of C. The segments of C then constrain the relative positions of the curves representing the vertices of G in R''.

This allows us to argue that the curve $R''(v) \in R''$ representing a vertex v of G can be deformed to be arbitrarily close to the corresponding curve $R(v)$ of R, and conversely, every point of $R(v)$ is close to a point of $R''(v)$. In fact, for every ε, we may deform $R''(v)$ into a curve $R^{\varepsilon}(v)$ which is confined to the the the ε-neighborhood of the original curve $R(v)$, without affecting the intersections between this curve and the curves of C'. We call the ε-neighborhood of $R(v)$ *the noodle* of $R(v)$, denoted by $N(v)$.

It now remains to provide the order-preserving mapping ϕ. Suppose that $R(u)$ and $R(v)$ are two curves of R that cross at a point p. Assuming ε is small enough, $N(u)$ and $N(v)$ intersect in a parallelogram-shaped region surrounding the point p. We call this region *the zone* of p. We may assume that distinct intersection points of R have disjoint zones.

Assume from now on that all the curves of R and R^{ε} have a prescribed orientation, i.e., a fixed beginning and end. Suppose that a curve $R(v)$ contains k intersection points $p_1, \dots, p_k$ appearing in this order, with zones $P_1, \dots, P_k$. We may assume that the two endpoints of $R^{\varepsilon}(v)$ are ε-close to the corresponding endpoints of $R(v)$, otherwise we only consider a truncated part of $R^{\varepsilon}(v)$ that has this property. Following the curve $R^{\varepsilon}(v)$ from beginning to end, we eventually encounter all the zones $P_1, \dots, P_k$. Of course, a given zone P_i may be intersected several times by $R^{\varepsilon}(v)$, since the curve $R^{\varepsilon}(v)$ may be folded inside $N(v)$ in a complicated way. For every zone P_i, we fix the first occurrence when $R^{\varepsilon}(v)$ enters inside P_i and then exits through the opposite side of P_i. The subcurve of $R^{\varepsilon}(v)$ inside P_i that corresponds to this occurrence will be called *the representative of $R^{\varepsilon}(v)$ inside P_i*, and denoted by $r_i(v)$. Note that the representatives appear on $R^{\varepsilon}(v)$ in the 'correct' order, i.e., $r_1(v), r_2(v), \dots, r_k(v)$.

We now define the order-preserving mapping ϕ. Let $p \in \text{In}(R)$ be an intersection point of two curves $R(u)$ and $R(v)$, and let P be its zone. Let $r(u)$ and $r(v)$ be the representatives of $R^{\varepsilon}(u)$ and $R^{\varepsilon}(v)$ inside this zone, and let p'' be an arbitrary intersection of $r(u)$ and $r(v)$. We then put $\phi(p) = p''$. This mapping is order-preserving by the construction of the representatives. Deforming the curves of R^{ε} inside each zone, we may even assume that p'' coincides with p. This completes the sketch of proof of the Noodle-Forcing Lemma.

3   Relations between Classes

With the Noodle-Forcing Lemma, we can prove our separation results.

Theorem 1. *For any $k \geq 1$, there is a graph G' that has a proper representation using k-bend axis-parallel curves, but has no representation using $(k-1)$-bend axis-parallel curves.*

Proof. Consider the graph K_2 consisting of a single edge uv, with a representation R in which both u and v are represented by weakly increasing k-bend staircase curves that have $k+1$ common intersections $p_1, \ldots, p_{k+1}$, in left-to-right order, see Fig.1. We refer to this representation as a *sausage* due to it resembling sausage links.

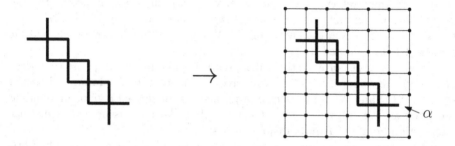

Fig. 1. The sausage representation for $k = 3$ and its grilled version

We now grill the sausage (i.e., we apply the Noodle-Forcing Lemma to K_2 and R) to obtain a graph G' with a k-bend representation R'. We claim that G' has no $(k-1)$-bend representation. Assume for contradiction that there is a $(k-1)$-bend representation R'' of G'. Lemma 1 then implies that there is an order-preserving mapping $\phi: \text{In}(R) \to \text{In}(R'')$. Let $s_i(u)$ be the subcurve of $R''(u)$ between the points $\phi(p_i)$ and $\phi(p_{i+1})$, and similarly for $s_i(v)$ and $R''(v)$. Consider, for each $i = 1, \ldots, k$, the union $c_i = s_i(u) \cup s_i(v)$. We know from Lemma 1 that $s_i(u)$ and $s_i(v)$ cannot completely overlap, and therefore the closed curve c_i must surround at least one nonempty bounded region of the plane. Therefore c_i contains at least two bends different from $\phi(p_i)$ and $\phi(p_{i+1})$. We conclude that $R''(u)$ and $R''(v)$ together have at least $2k$ bends, a contradiction.

A straightforward consequence is the following.

Corollary 1. *For every k, B_k-VPG $\subsetneq B_{k+1}$-VPG.*

Because two straight-line segments in the plane cross at most once, the Noodle-Forcing Lemma also implies the following.

Corollary 2. *For every $k \geq 1$, B_k-VPG $\not\subseteq$ SEG.*

This raises a natural question: Is there some k such that every SEG graph is contained in B_k-VPG? The following theorem answers it negatively.

Theorem 2. *For every k, there is a graph which belongs to 3-DIR but not to B_k-VPG.*

Proof. We fix an arbitrary k. Consider, for an integer n, a representation $R \equiv R(n)$ formed by $3n$ segments, where n of them are horizontal, n are vertical and n have a slope of 45 degrees. Suppose that every two segments of R with different slopes intersect, and their intersections form the regular pattern depicted in Figure 2 (with a little bit of creative fantasy this pattern resembles a waffle, especially when viewed under a linear transformation).

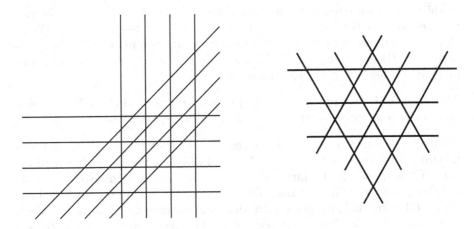

Fig. 2. The 'waffle' representation R from Theorem 2 and its transformed representation

Note that the representation R forms $\Omega(n^2)$ empty internal triangular faces bounded by segments of R, and the boundaries of these faces intersect in at most a single point. Suppose that n is large enough, so that there are more than $3kn$ such triangular faces. Let G be the graph represented by R.

The representation R is proper, so we can apply the Noodle-Forcing Lemma to R and G, obtaining a graph G' together with its 3-DIR representation R'. We claim that G' has no B_k-VPG representation.

Suppose for contradiction that there is a B_k-VPG representation R'' of G'. We will show that the $3n$ curves of R'' that represent the vertices of G contain together more than $3kn$ bends.

From the Noodle-Forcing Lemma, we deduce that there exists an order-preserving mapping $\phi: \text{In}(R_n) \to \text{In}(R''_n)$. Let T be a triangular face of the representation R. The boundary of T consists of three intersection points $p, q, r \in \text{In}(R)$ and three subcurves a, b, c. The three intersection points $\phi(p)$, $\phi(q)$ and $\phi(r)$ determine the corresponding subcurves a'', b'' and c'' in R''.

The Noodle-Forcing Lemma implies that there is a homeomorphism h which sends a'', b'', and c'' into small neighborhoods of a, b and c, respectively. This shows that each of the three curves a'', b'' and c'' contains a point that does not belong to any of the other two curves. This in turn shows that at least one of the three curves is not a segment, i.e., it has a bend in its interior.

Since the triangular faces of R have non-overlapping boundaries, and since ϕ is order-preserving, we see that for each triangular face of R there is at least one bend in R'' belonging to a curve representing a vertex of G. Since G has $3n$ vertices and R determines more than $3kn$ triangular faces, we conclude that at least one curve of R'' has more than k bends, a contradiction.

4   Hardness Results

In this section we strengthen the separation result of Corollary 1 by showing that not only are the classes B_k-VPG and B_{k+1}-VPG different, but providing a B_{k+1}-VPG representation does not help in deciding B_k-VPG membership. This also settles the conjecture on NP-hardness of recognition of these classes stated in [2], in a considerably stronger form than it was asked.

Theorem 3. *For every $k \geq 0$, deciding membership in B_k-VPG is NP-complete even if the input graph is given with a B_{k+1}-VPG representation.*

Proof. It is not difficult to see that recognition of B_k-VPG is in NP and therefore we will be concerned in showing NP-hardness only. We use the NP-hardness reduction developed in [11] for showing that recognizing grid intersection graphs is NP-complete. Grid intersection graphs are intersection graphs of vertical and horizontal segments in the plane with additional restriction that no two segments of the same direction share a common point. Thus these graphs are formally close but not equal to B_0-VPG graphs (where paths of the same direction are allowed to overlap). However, bipartite B_0-VPG graphs are exactly grid intersection graphs. This follows from a result of Bellantoni et al. [4] who proved that bipartite intersection graphs of axes parallel rectangles are exactly grid intersection graphs.

The reduction in [11] constructs, given a Boolean formula Φ, a graph G_Φ which is a grid intersection graph if and only if Φ is satisfiable. In arguing about this, a representation by vertical and horizontal segments is described for a general layout of G_Φ for which it is also shown how to represent its parts corresponding to the clauses of the formula, referred to as *clause gadgets*, if at least one literal is true. The clause gadget is reprinted with a generous approval of the author in Fig. 3, while Fig. 4 shows the grid intersection representations of satisfied clauses, and Fig. 5 shows the problem when all literals are false. In Fig. 6, we show that in the case of all false literals, the clause gadget can be represented by grid paths with at most 1 bend each. It follows that $G_\Phi \in B_1$-VPG and a 1-bend representation can be constructed in polynomial time. Thus, recognition of B_0-VPG is NP-complete even if the input graph is given with a B_1-VPG representation.

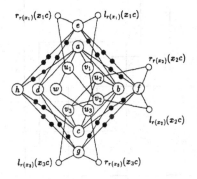

Fig. 3. The clause gadget reprinted from [11]

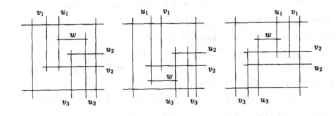

Fig. 4. The representations of satisfied clauses reprinted from [11]

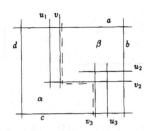

Fig. 5. The problem preventing the representation of an unsatisfied clause reprinted from [11]

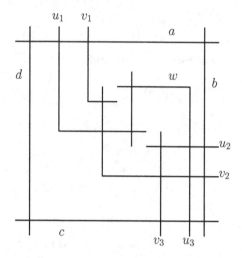

Fig. 6. The representation of an unsatisfied clause gadget via curves with one bend

We use a similar approach for arbitrary $k > 0$ with a help of the Noodle-Forcing Lemma. We grill the same representation R of K_2 as in the proof of Theorem 1. We call the resulting graph $P(u)$ where u is one of the vertices of the K_2, the one whose curve in R is ending in a boundary cell denoted by α in the schematic Fig. 1. We call this graph the *pin* since it follows from Lemma 1 that it has a B_k-VPG representation such that the bounding paths of the cell α wrap around the grill and the last segment of $R(u)$ extends arbitrarily far (see the schematic Fig. 7). We will refer to this extending segment as the *tip* of the pin. It is crucial to observe that in any B_k-VPG representation R' of $P(u)$ all bends of $R'(u)$ are consumed between the crossing points with the curve representing the other vertex of K_2 and hence the part of $R'(u)$ that lies in the α cell of R' is necessarily straight.

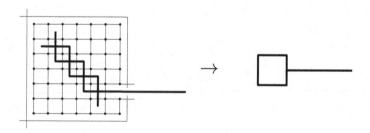

Fig. 7. Construction of a pin

Next we combine two pins together to form a *clothespin*. The construction is illustrated in the schematic Fig. 8. We start with a K_4 whose edges are subdivided by one vertex each. Every STRING representation of this graph contains

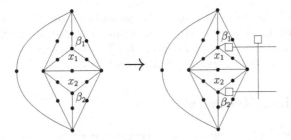

Fig. 8. Construction of a clothespin

4 basic regions which correspond to the faces of a drawing of the K_4 (this is true for every 3-connected planar graph and it is seen by contracting the curves corresponding to the degree 2 vertices, the argument going back to Sinden [15]). We add two vertices x_1, x_2 that are connected by paths of length 2 to the boundary vertices of two triangles, say β_1 and β_2. The curves representing x_1 and x_2 must lie entirely inside the corresponding regions. Then we add two pins, say $P(u_1)$ and $P(u_2)$, connect the vertices of the boundary of α_i to x_i by paths of length 2 and make u_i adjacent to a vertex on the boundary of β_i (for $i = 1, 2$). Finally, we add a third pin $P(u_3)$ and make u_3 adjacent to u_1 and u_2. We denote the resulting graph by $CP(u)$.

It is easy to check that the clothespin has a B_k-VPG representation $\check{R}$ such that the tips of $\check{R}(u_1)$ and $\check{R}(u_2)$ are parallel and extend arbitrarily far from the rest of the representation, as indicated in Fig. 8.

On the other hand, in any B_k-VPG representation R' of $CP(u)$, if a curve crosses $R'(u_1)$ and $R'(u_2)$ and no other path of $R'(CP(u))$, then it must cross the tips of $R'(u_1)$ and $R'(u_2)$. This follows from the fact that for $i = 1, 2$, $R'(x_i)$ must lie in α_i (to be able to reach all its bounding curves), and hence, by circle inversion, all bends of $R'(u_i)$ are trapped inside β_i. If a curve crosses both $R'(u_1)$ and $R'(u_2)$, it must cross them outside $\beta_1 \cup \beta_2$, and hence it only may cross their tips.

Now we are ready to describe the construction of G'_Φ. We take G_Φ as constructed in [11] replace every vertex u by a clothespin $CP(u)$, and whenever $uv \in E(G_\Phi)$, we add edges $u_i v_j, i, j = 1, 2$. Now we claim that $G'_\Phi \in B_k$-VPG if and only if Φ is satisfiable, while $G'_\Phi \in B_{k+1}$-VPG is always true.

On one hand, if $G'_\Phi \in B_k$-VPG and R' is a B_k-VPG representation of G'_Φ, then the tips of $R'(u_1), u \in V(G_\Phi)$ form a 2-DIR representation of G_Φ ($R'(u_1)$ and $R'(v_1)$ may only intersect in their tips) and Φ is satisfiable.

On the other hand, if Φ is satisfiable, we represent G_Φ as a grid intersection graph and replace every segment of the representation by a clothespin with slim parallel tips and the body of the pin tiny enough so that does not intersect anything else in the representation. Similarly, if Φ is not satisfiable, we modify a 1-bend representation of G_Φ by replacing the paths of the representation by clothespins with 1-bend on the tips, thus obtaining a B_{k+1}-VPG representation of G_Φ. The representation consists of a large part inherited from the representation of G_Φ and tiny parts representing the heads of the pins, but these can be made all of the same constant size and thus providing only a constant ratio refinement of the representation of G_Φ. The representation is thus still of linear size and can be constructed in polynomial time.

5   Concluding Remarks

In this paper we have affirmatively settled two main conjectures of Asinowski et al [2] regarding VPG graphs. We have also demonstrated the relationship between B_k-VPG graphs and segment graphs.

The first conjecture that we settled claimed that B_k-VPG is a strict subset of B_{k+1}-VPG for all k. We have proven this constructively. Previously only the following separation was known: B_0-VPG $\subsetneq B_1$-VPG $\subsetneq$ VPG.

The second conjecture claimed that the B_k-VPG recognition problem is NP-Complete for all k. We have actually proven a stronger result; namely, that the B_k-VPG recognition problem is NP-Complete for all k even when the input graph is a B_{k+1}-VPG graph. Previously only the NP-Completeness of B_0-VPG (from 2-DIR [11]) and VPG (from STRING [9,14]) were known.

Finally due to the close relationship between VPG graphs and segment graphs (i.e., since B_0-VPG = 2-DIR, and SEG $\subsetneq$ STRING = VPG) we have considered the relationship between these classes. In particular, we have shown that:

- There is no k such that 3-DIR is contained in B_k-VPG (i.e., SEG is not contained in B_k-VPG for any k).
- B_1-VPG is not contained in SEG.

Thus, to obtain polynomial time recognition algorithms, one would need to restrict attention to specific cases with (potentially) useful structure. In this respect, in [8], certain subclasses of B_0-VPG graphs have been characterized and shown to admit polynomial time recognition; namely split, chordal claw-free, and chordal bull-free B_0-VPG graphs are discussed in [8]. Additionally, in [5], B_0-VPG chordal and 2-row B_0-VPG[5] have been shown to have polynomial time recognition algorithms. In particular, we conjecture that applying similar restrictions to the B_k-VPG graph class will also yield polynomial time recognition algorithms. It is interesting to note that since our separating examples are not chordal it is also open whether B_k-VPG chordal $\subsetneq B_{k+1}$-VPG chordal.

References

1. Asinowski, A., Cohen, E., Golumbic, M.C., Limouzy, V., Lipshteyn, M., Stern, M.: String graphs of k-bend paths on a grid. Electronic Notes in Discrete Mathematics 37, 141–146 (2011)
2. Asinowski, A., Cohen, E., Golumbic, M.C., Limouzy, V., Lipshteyn, M., Stern, M.: Vertex Intersection Graphs of Paths on a Grid. Journal of Graph Algorithms and Applications 16(2), 129–150 (2012)

---

[5] Where the VPG representation has at most two rows.

3. Bandy, M., Sarrafzadeh, M.: Stretching a knock-knee layout for multilayer wiring. IEEE Trans. Computing 39, 148–151 (1990)
4. Bellantoni, S., Ben-Arroyo Hartman, I., Przytycka, T.M., Whitesides, S.: Grid intersection graphs and boxicity. Discrete Mathematics 114, 41–49 (1993)
5. Chaplick, S., Cohen, E., Stacho, J.: Recognizing Some Subclasses of Vertex Intersection Graphs of 0-Bend Paths in a Grid. In: Kolman, P., Kratochvíl, J. (eds.) WG 2011. LNCS, vol. 6986, pp. 319–330. Springer, Heidelberg (2011)
6. Coury, M.D., Hell, P., Kratochvíl, J., Vyskočil, T.: Faithful Representations of Graphs by Islands in the Extended Grid. In: López-Ortiz, A. (ed.) LATIN 2010. LNCS, vol. 6034, pp. 131–142. Springer, Heidelberg (2010)
7. Di Battista, G., Eades, P., Tamassia, R., Tollis, I.G.: Graph Drawing. Prentice-Hall (1999)
8. Golumbic, M.C., Ries, B.: On the intersection graphs of orthogonal line segments in the plane: characterizations of some subclasses of chordal graphs. To Appear in Graphs and Combinatorics
9. Kratochvíl, J.: String graphs II, Recognizing string graphs is NP-hard. J. Comb. Theory, Ser. B 52, 67–78 (1991)
10. Kratochvíl, J., Matoušek, J.: String graphs requiring exponential representations. J. Comb. Theory, Ser. B 53, 1–4 (1991)
11. Kratochvíl, J.: A Special Planar Satisfiability Problem and a Consequence of Its NP-completeness. Discrete Applied Mathematics 52, 233–252 (1994)
12. Kratochvíl, J., Matoušek, J.: Intersection Graphs of Segments. J. Comb. Theory, Ser. B 62, 289–315 (1994)
13. Molitor, P.: A survey on wiring. EIK Journal of Information Processing and Cybernetics 27, 3–19 (1991)
14. Schaefer, M., Sedgwick, E., Stefankovic, D.: Recognizing string graphs in NP. J. Comput. Syst. Sci. 67, 365–380 (2003)
15. Sinden, F.: Topology of thin film circuits. Bell System Tech. J. 45, 1639–1662 (1966)

On the Recognition of k-Equistable Graphs[*]

Vadim E. Levit[1], Martin Milanič[2,3], and David Tankus[1]

[1] Department of Computer Science and Mathematics
Ariel University Center of Samaria, Ariel 40700, Israel
{levitv,davidta}@ariel.ac.il
[2] University of Primorska, UP IAM, Muzejski trg 2, SI6000 Koper, Slovenia
[3] University of Primorska, UP FAMNIT, Glagoljaška 8, SI6000 Koper, Slovenia
martin.milanic@upr.si

Abstract. A graph $G = (V, E)$ is called *equistable* if there exist a positive integer t and a weight function $w : V \longrightarrow \mathbb{N}$ such that $S \subseteq V$ is a maximal stable set of G if and only if $w(S) = t$. The function w, if exists, is called an *equistable function* of G. No combinatorial characterization of equistable graphs is known, and the complexity status of recognizing equistable graphs is open. It is not even known whether recognizing equistable graphs is in NP.

Let k be a positive integer. An equistable graph $G = (V, E)$ is said to be k-*equistable* if it admits an equistable function which is bounded by k. For every constant k, we present a polynomial time algorithm which decides whether an input graph is k-equistable.

Keywords: maximal stable set, equistable graph, polynomial time algorithm.

1 Introduction and Preliminaries

All graphs considered in this paper will be finite, simple and undirected. For a graph G, we denote by $V = V(G)$ the vertex set of G, and by $E = E(G)$ its edge set. Let $n = |V|$ and $m = |E|$. A set $S \subseteq V$ is *stable* if its members are pairwise non-adjacent. A stable set is *maximal* if it is not a subset of another stable set. A graph $G = (V, E)$ is called *equistable* if there exist a positive integer t and a weight function $w : V \longrightarrow \mathbb{N}$ such that a set $S \subseteq V$ is a maximal stable set of G if and only if $w(S) = \sum_{v \in S} w(v) = t$ [33]. The function w, if exists, is called an *equistable function* of G, while the pair (w, t) is called an *equistable structure*. No combinatorial characterization of equistable graphs is known, and the complexity status of recognizing equistable graphs is open.

---

[*] MM is supported in part by "Agencija za raziskovalno dejavnost Republike Slovenije", research program P1–0285 and research projects J1–4010, J1–4021 and N1–0011. Research was partly done during a visit of the second author at the Department of Computer Science and Mathematics at the Ariel University Center of Samaria in the frame of a Slovenian Research Agency project MU-PROM/11-007. The second author thanks the Department for its hospitality and support.

M.C. Golumbic et al. (Eds.): WG 2012, LNCS 7551, pp. 286–296, 2012.
© Springer-Verlag Berlin Heidelberg 2012

It is not even known whether recognizing equistable graphs is in NP. There is an exponential time algorithm based on linear programming to recognize equistable graphs [19]. Deciding whether a given weight function on the vertices of a graph G is an equistable function is co-NP-complete [29]. Our current level of (non-)understanding the structure of equistable graphs provides ample motivation for further investigation of their structural properties and complexity aspects, initiated for general equistable graphs in [23] and continued for particular graph classes in [18–20, 23, 27, 34] and for general equistable graphs and related classes in [28, 29].

Mahadev et al. introduced in [23] a subclass of equistable graphs, called *strongly equistable* graphs, using the following notation and definitions. The set of all maximal stable sets of a graph $G = (V, E)$ is denoted $\mathcal{S}(G)$, and the set of all other nonempty subsets of V is denoted by $\mathcal{T}(G)$. A graph $G = (V, E)$ is said to be *strongly equistable* if for each $T \in \mathcal{T}(G)$ and each $\gamma \leq 1$ there exists a weight function $w : V \longrightarrow \mathbb{R}^+$ such that $w(S) = 1$ for all $S \in \mathcal{S}(G)$, and $w(T) \neq \gamma$.

Theorem 1. *[23] All strongly equistable graphs are equistable.*

Conjecture 1. *[23] All equistable graphs are strongly equistable.*

This conjecture is known to hold for perfect graphs [23], for series-parallel graphs [18], for AT-free graphs [27], for simplicial graphs [20], for very well covered graphs [20], for line graphs [20], and for various product graphs [27].

Although no combinatorial characterization of equistable graphs is known, there are some necessary and sufficient conditions of a combinatorial flavor for a graph to be equistable. Following [31, 32], we say that a graph G is a *triangle graph* if for every maximal stable set S in G and every edge uv in $G - S$ there is a vertex $s \in S$ such that $\{u, v, s\}$ induces a triangle in G.

Theorem 2. *[27] Every equistable graph is a triangle graph.*

A graph $G = (V, E)$ is a *general partition graph* if there exists a set U and an assignment of non-empty subsets $U_v \subseteq U$ to the vertices $v \in V$ such that two vertices x and y are adjacent if and only if $U_x \cap U_y \neq \emptyset$, and for every maximal stable set S of G, the set $\{U_x : x \in S\}$ is a partition of U. General partition graphs arise in the geometric setting of lattice polygon triangulations [11] and were studied in a series of papers [1, 8–10, 12, 17, 20, 27]. It is proved in [24] that all general partition graphs are triangle graphs.

A *strong clique* in a graph G is a set of pairwise adjacent vertices that meets all maximal stable sets. The following two theorems relate strong cliques and equistable graphs.

Theorem 3. *[23] Every equistable graph with a strong clique is strongly equistable.*

Theorem 4. *[24] Let G be a graph. The following are equivalent:*

(i) G is a general partition graph.
(ii) Every edge of G is contained in a strong clique.

Theorem 5. *(Jim Orlin (personal communication, 2009), see [27] for a proof)*
All general partition graphs are equistable.

Conjecture 2. *(Jim Orlin (personal communication, 2009)) All equistable*
graphs are general partition graphs.

Theorems 3, 4 and 5 imply the following.

Theorem 6. *All general partition graphs are strongly equistable.*

The inclusion relations between the considered graph classes read as follows:

$$\text{general partition graphs} \subseteq \text{strongly equistable graphs}$$
$$\subseteq \text{equistable graphs} \subset \text{triangle graphs}.$$

Let k be a positive integer. An equistable graph $G = (V, E)$ is said to be k-
equistable if it admits an equistable function $w : V \longrightarrow [k] := \{1, \ldots, k\}$. Such
a weight function is called a k-*equistable function*, and the corresponding structure (w, t) is a k-*equistable structure*. In this paper, we present a polynomial
time algorithm which decides for any constant k whether an input graph is k-
equistable. If the answer is affirmative, then the algorithm finds a k-equistable
function of the input graph. The algorithm, presented in Section 3, relies on the
notion of the twin equivalence relation, which we define and analyze in Section 2.
We conclude the paper with some remarks in Section 4.

2   The Twin Equivalence Relation

We say that vertices u and v of a graph G are *twins* if they have exactly the
same set of neighbors, other than u and v. We denote the twin relation by $\sim_t$:

$$u \sim_t v \quad \text{if and only if} \quad N(u) \setminus \{v\} = N(v) \setminus \{u\}.$$

Note that twins may be either adjacent or non-adjacent. Adjacent twins are *true*
twins and non-adjacent twins are *false twins*.

We now list some easily verified properties of the twin relation. We start with
the following basic observation.

Lemma 7. *Let $G = (V, E)$ be a graph. The twin relation is an equivalence*
relation, and every equivalence class is either a clique or a stable set.

An equivalence class of the twin relation will be referred to as a *twin class*. Twin
classes that are cliques will be referred to *true twin classes*, and the remaining
classes will be referred as *false twin classes*. The set of all twin classes will be
denoted by $\Pi(G)$ and referred to as the *twin partition* of G. The number of
twin classes of G will be denoted by $\pi(G) = |\Pi(G)|$. The twin relation was

also considered in the literature under the name *similarity* [13]. For further applications of the notion of twin classes, see [2, 6, 15].[1]

Two disjoint sets of vertices X and Y in a graph G *see* each other if every vertex of X is adjacent to every vertex of Y, and they *miss* each other if every vertex of X is non-adjacent to every vertex of Y. A vertex x *sees* a set $Y \subseteq V(G) \setminus \{x\}$ if the singleton $\{x\}$ sees Y, and similarly x *misses* Y if $\{x\}$ misses Y.

Lemma 8. *Every two distinct twin classes either see each other or miss each other.*

For later use, we also mention a straightforward consequence of the homogeneous structure of the twin classes specified by the last two lemmas:

Corollary 9. *For every graph G and for every two permutations $(u_1, \ldots, u_n)$ and $(v_1, \ldots, v_n)$ of $V(G)$ such that $u_i \sim_t v_i$ for every $1 \leq i \leq n$, the mapping $\varphi : V(G) \to V(G)$ given by $\varphi(u_i) = v_i$ for all i is an automorphism of G.*

The following *quotient graph* of G, denoted $\mathcal{Q}(G)$, is well defined: Its vertex set is $\Pi(G)$, and two twin classes are adjacent if and only if they see each other. Given a graph G, it is possible to find in linear time the twin partition $\Pi(G)$, the quotient graph $\mathcal{Q}(G)$ and $\pi(G)$, using any of the linear time algorithms for modular decomposition [7, 26, 36].

3   A Recognition Algorithm for k-Equistable Graphs

In this section, we show that k-equistable graphs can be recognized in polynomial time. We adopt the usual simplifying assumption that addition and comparison of two numbers can be carried out in $O(1)$ time. Recall that an equistable graph $G = (V, E)$ is said to be *k-equistable* if it admits an equistable function $w : V \longrightarrow [k] := \{1, \ldots, k\}$. For a graph G, we denote by $\kappa(G)$ the minimum integer k such that G is k-equistable. In particular $\kappa(G)$ is finite if and only if G is equistable.

Theorem 10. *For every fixed k, there is an $O\left(n^{2k}\right)$ algorithm for recognizing graphs with $\kappa(G) \leq k$. In case of a positive instance, the algorithm also produces a k-equistable structure of G.*

The main idea behind the algorithm is to examine the space of all k^n weight functions $w : V \to [k]$. First, based on the notion of twin classes, we develop some necessary conditions that every k-equistable graph and every k-equistable function of it must satisfy. Exploiting the structural properties of the twin classes, we then partition the weight functions satisfying the necessary conditions into

---

[1] Besides the twin relation, the similarly defined *true twin relation* and *false twin relation* were also considered in the literature [35]. Equivalence classes with respect to true twin relation appeared in the literature under various names such as maximal clique modules [3, 4, 25], maxmods [5] and critical cliques [21]. See also [14, 30]. Equivalence classes of the false twin relation appeared in the literature under various names such as maximal independent-set modules [16], or similarity classes [22].

polynomially many equivalence classes such that for every equivalence class, either all or none of the functions from the class are equitable. Finally, we show how to efficiently test whether a representative function from each equivalence class is equitable.

3.1   Necessary Conditions

Lemma 11. *For every equitable function w of an equitable graph G, every two vertices x and y of the same weight are twins.*

Proof. Suppose for a contradiction that there exist an equitable function w of G, and two vertices x and y such that $w(x) = w(y)$ and $N(x) \setminus \{y\} \neq N(y) \setminus \{x\}$. Without loss of generality, we may assume that there exists a vertex $z \in (N(x) \setminus \{y\}) \setminus (N(y) \setminus \{x\})$. Clearly, $z \neq x$. Let S be a maximal stable set in G containing y and z, and let $T = (S \setminus \{y\}) \cup \{x\}$. Then $w(T) = w(S)$, which is a contradiction since T is not stable.   □

The following are immediate consequences of Lemmas 7 and 11.

Corollary 12. *For every equitable function w of G and for every i, the set $W_i = \{x \in V : w(x) = i\}$ is either a clique or a stable set in G.*

Corollary 13. *For every equitable graph G, $\pi(G) \leq \kappa(G)$.*

Hence, in what follows, we will assume that the input graph G satisfies $\pi(G) \leq k$, since otherwise G is not k-equitable.

The following lemma is a partial converse of Lemma 11.

Lemma 14. *For every equitable function w of an equitable graph G and for every two true twins x and y, it holds that $w(x) = w(y)$.*

Proof. Suppose for a contradiction that in some equitable function w of G, there exist two true twins, x and y, such that $w(x) \neq w(y)$. Let S be any maximal stable set in G containing x. Then $y \notin S$, and $S' = (S \setminus \{x\}) \cup \{y\}$ is a maximal stable set in G. Since $w(x) \neq w(y)$, it follows that $w(S') \neq w(S)$, which contradicts the fact that w is an equitable function of G.   □

Remark 1. *If x and y are false twins in an equitable graph G, and w is an equitable function of the graph, then it is possible that $w(x) \neq w(y)$. An example of the above is the 4-cycle with vertices v_1, v_2, v_3, v_4 in the cyclic order, and the equitable structure of it given by $w(v_1) = w(v_3) = 2$, $w(v_2) = 1$, $w(v_4) = 3$ and $t = 4$. Moreover, it is easy to see that every equitable function of this graph assigns different weights to at least one pair of false twins.*

By Lemmas 11 and 14, in our search for a k-equitable function of G, we can restrict our attention to functions $w : V \to [k]$ satisfying the following two properties:

1. If $w(x) = w(y)$ then x and y are twins.
2. If x and y are true twins then $w(x) = w(y)$.

A weight function $w : V \to [k]$ satisfying these two properties will be called a *candidate* weight function of (G, k).

3.2 Witnesses and Equivalent Weight Functions

We now describe a useful way of partitioning the set of candidate weight functions of (G, k) into polynomially many equivalence classes. This partition relies on the notion of *witness*, defined as follows: A *witness* of (G, k) is a $2k$-tuple $\omega = (C_1, \ldots, C_k, n_1, \ldots, n_k)$ such that:

1. For every $1 \le i \le k$, either $C_i = \emptyset$ or C_i is a twin class of G.
2. For every $1 \le i \le k$, it holds that $n_i \in \{0, 1, \ldots, |C_i|\}$, with $n_i = |C_i|$ if C_i is a clique in G.
3. For every twin class C of G, it holds that $\sum \{n_i : 1 \le i \le k$ and $C_i = C\} = |C|$.

Given a candidate weight function w of (G, k), we can associate to it a unique witness of (G, k)

$$\omega(w) = (C_1, \ldots, C_k, n_1, \ldots, n_k)$$

(called *witness of w*) as follows: for every $1 \le i \le k$, set $n_i = |\{x \in V : w(x) = i\}|$ and

$$C_i = \begin{cases} \text{the unique twin class of } G \text{ containing vertices of weight } i, \text{ if } n_i > 0; \\ \emptyset, \hspace{6cm} \text{otherwise.} \end{cases}$$

Using Properties 1 and 2 of candidate weight functions, it can be easily verified that $\omega(w)$ is well defined, and a witness of (G, k).

Conversely, for every witness $\omega = (C_1, \ldots, C_k, n_1, \ldots, n_k)$ of (G, k) there exists a candidate weight function w such that $\omega(w) = \omega$: For every twin class C of G, let $I = \{i \in [k] : C_i = C\}$. By property 3 of witnesses, there exists a partition $\{W_i : i \in I_C\}$ of C indexed over I such that $|W_i| = n_i$. Moreover, the set $\{I_C : C \in \Pi(G)\}$ forms a partition of $[k]$, and the set $\{W_i : i \in [k]\}$ forms a partition of V. Hence, ω is the witness of the candidate weight function $w : V \to [k]$, defined by $w(x) = i$ if and only if $x \in W_i$.

Let us now partition the set of all candidate weight functions of (G, k) into equivalence classes as follows: we say that two functions w and w' are *equivalent* if and only if $\omega(w) = \omega(w')$. It follows from Corollary 9 that for every two equivalent functions w and w', function w is an equistable function of G if and only if w' is. Thus, we can examine the space of all candidate weight functions of (G, k) by working with equivalence classes represented by witnesses of (G, k). We say that a witness ω of (G, k) is *equistable* if there exists a k-equistable function w of G such that $\omega(w) = \omega$. (Equivalently: every candidate weight function w such that $\omega(w) = \omega$ is an equistable function of G.)

3.3 Testing a Witness for Equistability

We now derive a polynomially testable necessary and sufficient condition for a witness ω of (G, k) to be equistable. Let $\omega = (C_1, \ldots, C_k, n_1, \ldots, n_k)$ be a witness of (G, k). Fix an arbitrary candidate weight function w such that $\omega(w) = \omega$. For $i \in [k]$, let $W_i = \{x \in V : w(x) = i\}$. (Notice that $|W_i| = n_i$.) Recall that w is an equistable function of G if and only if there exists an integer t such that a subset $S \subseteq V$ is maximal stable set in G if and only if $w(S) = t$. Using similar ideas as above, let us now argue that this can be tested in polynomial time, avoiding an explicit generation of all 2^n subsets of V.

We say that a vector $x \in \mathbb{Z}^k$ is *dominated by* ω if $0 \le x_i \le n_i$ for every $i \in [k]$. Let us denote by $X(\omega)$ the set of all vectors dominated by ω. To every subset $S \subseteq V$, we can associate a vector $x^S \in X(\omega)$ by setting $(x^S)_i = |S \cap W_i|$ for all $i \in [k]$. And conversely, for every vector $x \in X(\omega)$, there exists a subset S of V such that $x^S = x$: just put in S, for every $i \in [k]$, exactly x_i vertices from W_i.

Now, let us partition the power set of V according to the following equivalence relation: two subsets $S, S' \subseteq V$ are *equivalent* if and only if $x^S = x^{S'}$. It follows from Properties 1 and 2 that for every two equivalent sets S and S', S is a maximal stable set in G if and only if S' is; moreover $w(S) = w(S') = \sum_{i=1}^{k} i x_i$ where $x = x^S$. Therefore, to determine whether w is an equistable function of G, it suffices to test the defining property for an arbitrary collection of sets $\{S_x : x \in X(\omega)\}$ where for every vector $x \in X(\omega)$, $S_x \subseteq V$ is an arbitrary set such that $x^{S_x} = x$. Let us also define the weight of $x \in X(\omega)$ as $w(x) := \sum_{i=1}^{k} i x_i$.

We say that a vector $x \in X(\omega)$ is a *maximal stable set vector* if S_x is a maximal stable set of G. This happens if and only if the following conditions hold:

1. For all $i \in [k]$ such that $x_i > 0$, it holds that

$$x_i = \begin{cases} 1, & \text{if the unique twin class of } G \text{ containing } W_i \text{ is a clique;} \\ n_i, & \text{otherwise.} \end{cases}$$

2. The set $\{C \in \Pi(G) : (\exists i \in [k])(x_i > 0 \text{ and } W_i \subseteq C)\}$ is a maximal stable set of the quotient graph $\mathcal{Q}(G)$.

Hence, it is possible to test whether x is a maximal stable set vector in $O(k^2)$ time.

Lemma 15. *Function w is an equistable function of G if and only if there exists an integer t such that for every vector $x \in X(\omega)$, it holds that $w(x) = t$ if and only if x is a maximal stable set vector.*

Notice that $|X(\omega)| = O(n^k)$. Hence verifying whether w is an equistable function can be done in time $O(k^2 n^k)$.

3.4   The Algorithm

In summary, the following algorithm tests whether G is k-equistable, returning a k-equistable structure of G in case of a positive instance.

The correctness of the algorithm follows from the above discussion in this section. Let us analyze its running time. The twin partition $\Pi(G)$, the quotient graph $\mathcal{Q}(G)$ and $\pi(G)$ can be computed in time $O(n + m)$, using any of the linear time algorithms for modular decomposition [7, 26, 36]. Computing a maximal stable set S of G can be done in time $O(n + m)$ by a straightforward

---

Algorithm 1. Recognizing graphs with $\kappa(G) \leq k$

Input: A graph $G = (V, E)$;
Output: A k-equistable structure of G if one exists, NO, otherwise.

1 compute the twin partition $\Pi(G)$, the quotient graph $\mathcal{Q}(G)$, and $\pi(G)$;
2 **if** $\pi(G) > k$ **then**
   | // G is not k-equistable because it has too many twin classes
3 | return NO;

4 compute an arbitrary maximal stable set S of G;
5 compute the set of all witnesses of (G, k);
6 **for** *every witness* $\omega = (C_1, \ldots, C_k, n_1, \ldots, n_k)$ *of* (G, k) **do**
   | // compute the weight t of S with respect to the candidate weight
   | functions represented by ω
7 | $t \leftarrow \sum\{i \,:\, C_i$ is a clique and $S \cap C_i \neq \emptyset\} +$
   | $\sum\{in_i \,:\, C_i$ is a stable set and $S \cap C_i \neq \emptyset\}$;
8 | compute the set $X(\omega)$ of all vectors dominated by ω;
9 | **for** *every vector* $x \in X(\omega)$ **do**
10 | | $w(x) \leftarrow \sum_{i=1}^{k} ix_i$;
11 | | **if** x *is a maximal stable set vector* **then**
12 | | | **if** $w(x) \neq t$ **then**
   | | | | // witness ω is not equistable
13 | | | | go to the next iteration of the **for** loop in line 6;
14 | | **else if** $w(x) = t$ **then**
   | | | // witness ω is not equistable
15 | | | go to the next iteration of the **for** loop in line 6;
   | // witness ω is equistable
16 | let w be an arbitrary candidate weight function w such that $\omega(w) = \omega$;
17 | return (w, t);

   // no equistable witness was found
18 return NO;

---

algorithm. Computing the set of all $O((kn)^k)$ witnesses of (G, k) can be done in time $O(k(kn)^k)$ (for each of the $O((kn)^k)$ $2k$-tuples $\omega = (C_1, \ldots, C_k, n_1, \ldots, n_k)$ such that for every $1 \leq i \leq k$, either $C_i = \emptyset$ or C_i is a twin class of G, and $n_i \in \{0, 1, \ldots, |C_i|\}$, we can test in $O(k)$ time whether ω is a witness of (G, k)). The **for** loop in line 6 will be executed at most $O((kn)^k)$ times. Within each execution, the set $X(\omega)$ of all $O(n^k)$ vectors dominated by ω can be computed in time $O(kn^k)$. The **for** loop in line 9 will be executed at most $O(n^k)$ times, incurring a total time complexity of $O(k^2 n^k)$. Line 17 can be carried out in $O(n)$ time.

Hence, the overall time complexity is $O\big(n + m + k(kn)^k + (kn)^k \big(k(kn)^k + k^2 n^k\big)\big) = O\big(k^{2k+1} n^{2k}\big)$. For a fixed k this complexity is polynomial in n, namely, $O(n^{2k})$.

As an application of Theorem 10, we remark that if a k-equistable structure (w, t) of an equistable graph G is given, then the MAXIMUM WEIGHT STABLE SET and the WEIGHTED INDEPENDENT DOMINATION problems for G can be solved by dynamic programming in time $O(nt)$, which, for fixed k, is of the order $O(n^2)$ [29].

4   Concluding Remarks

No combinatorial characterization of equistable graphs is known, and the complexity status of recognizing equistable graphs is open. In this paper, we introduced the parameter $\kappa(G)$ of equistable graphs as the smallest k such that G admits an equistable weight function bounded by k. Exploiting the relationship between equistable weight functions and the twin classes, we showed that for every k, equistable graphs with $\kappa(G) \leq k$ (that is, the k-equistable graphs) can be recognized in $O\left(k^{2k+1}n^{2k}\right)$ time.

It seems natural to consider the parameterized version of the problem.

Question 1. *Is there an FPT algorithm for recognizing k-equistable graphs?*

Another natural question is to ask for explicit characterizations of k-equistable graphs for small values of k. Such characterizations might lead to faster recognition algorithms. For example:

- Corollary 12 implies that a graph is 1-equistable if and only if it is either complete or edgeless.
- For $k = 2$, the class of 2-equistable graphs can be characterized in terms of 8 forbidden induced subgraphs. Moreover, a graph is 2-equistable if and only if it is either a complete graph, an edgeless graph, a complete graph minus an edge, the disjoint union of a complete graph with K_1 or K_2, or the disjoint union of an edgeless graph with K_2.

We postpone the proofs of these characterizations of 2-equistable graphs together with a further investigation of k-equistable graphs to the journal version of the paper.

References

1. Anbeek, C., DeTemple, D., McAvaney, K.L., Robertson, J.M.: When are chordal graphs also partition graphs? Australas. J. Combin. 16, 285–293 (1997)
2. Bagheri Gh., B., Jannesari, M., Omoomi, B.: Uniquely dimensional graphs, arXiv:1205.0327v1
3. Berry, A., Bordat, J.-P.: Separability generalizes Dirac's theorem. Discrete Appl. Math. 84, 43–53 (1998)
4. Berry, A., Sigayret, A.: Representing a concept lattice by a graph. Discrete Appl. Math. 144, 27–42 (2004)
5. Berry, A., SanJuan, E., Sigayret, A.: Generalized domination in closure systems. Discrete Appl. Math. 154, 1064–1084 (2006)

6. Bui-Xuan, B.-M., Suchý, O., Telle, J.A., Vatshelle, M.: Feedback vertex set on graphs of low cliquewidth. European J. of Combinatorics (2011) (accepted for publication)
7. Cournier, A., Habib, M.: A New Linear Algorithm of Modular Decomposition. In: Tison, S. (ed.) CAAP 1994. LNCS, vol. 787, pp. 68–84. Springer, Heidelberg (1994)
8. DeTemple, D., Dineen, M.J., Robertson, J.M., McAvaney, K.L.: Recent examples in the theory of partition graphs. Discrete Math. 113, 255–258 (1993)
9. DeTemple, D., Harary, F., Robertson, J.M.: Partition graphs. Soochow J. Math. 13, 121–129 (1987)
10. DeTemple, D., Robertson, J.M.: Constructions and the realization problem for partition graphs. J. Combin. Inform. System Sci. 13, 50–63 (1988)
11. DeTemple, D., Robertson, J.M.: Graphs associated with triangulations of lattice polygons. J. Austral. Math. Soc. Ser. A 47, 391–398 (1989)
12. DeTemple, D., Robertson, J.M., Harary, F.: Existential partition graphs. J. Combin. Inform. System Sci. 9, 193–196 (1984)
13. Feder, T., Hell, P.: On realizations of point determining graphs, and obstructions to full homomorphisms. Discrete Math. 308, 1639–1652 (2008)
14. de Figueiredo, C.M.H., Meidanis, J., de Mello, C.P.: A linear-time algorithm for proper interval graph recognition. Inform. Process. Lett. 56, 179–184 (1995)
15. Habib, M., Paul, C.: A simple linear time algorithm for cograph recognition. Discrete Applied Math. 145, 183–197 (2005)
16. Heggernes, P., Meister, D., Papadopoulos, C.: Graphs of linear clique-width at most 3. Theoret. Comput. Sci. 412, 5466–5486 (2011)
17. Kloks, T., Lee, C.-M., Liu, J., Müller, H.: On the Recognition of General Partition Graphs. In: Bodlaender, H.L. (ed.) WG 2003. LNCS, vol. 2880, pp. 273–283. Springer, Heidelberg (2003)
18. Korach, E., Peled, U.N.: Equistable series-parallel graphs. Stability in Graphs and Related Topics. Discrete Appl. Math. 132, 149–162 (2003)
19. Korach, E., Peled, U.N., Rotics, U.: Equistable distance-hereditary graphs. Discrete Appl. Math. 156, 462–477 (2008)
20. Levit, V.E., Milanič, M.: Equistable simplicial, very well-covered, and line graphs (2011) (submitted for publication)
21. Lin, G.-H., Jiang, T., Kearney, P.E.: Phylogenetic k-Root and Steiner k-Root. In: Lee, D.T., Teng, S.-H. (eds.) ISAAC 2000. LNCS, vol. 1969, pp. 539–551. Springer, Heidelberg (2000)
22. Lozin, V., Milanič, M.: On the maximum independent set problem in subclasses of planar graphs. Journal of Graph Algorithms and Applications 14, 269–286 (2010)
23. Mahadev, N.V.R., Peled, U.N., Sun, F.: Equistable graphs. J. Graph Theory 18, 281–299 (1994)
24. McAvaney, K.L., Robertson, J.M., DeTemple, D.: A characterization and hereditary properties for partition graphs. Discrete Math. 113, 131–142 (1993)
25. McConnell, R.M.: Linear-time recognition of circular-arc graphs. Algorithmica 37, 93–147 (2003)
26. McConnell, R.M., Spinrad, J.P.: Modular decomposition and transitive orientation. Discrete Math. 201, 189–241 (1999)
27. Miklavič, Š., Milanič, M.: Equistable graphs, general partition graphs, triangle graphs, and graph products. Discrete Appl. Math. 159, 1148–1159 (2011)
28. Milanič, M., Rudolf, G.: Structural results for equistable graphs and related graph classes. RUTCOR Research Report, 25-2009
29. Milanič, M., Orlin, J., Rudolf, G.: Complexity results for equistable graphs and related classes. Ann. Oper. Res. 188, 359–370 (2011)

30. Nikoletseas, S., Raptopoulos, C., Spirakis, P.G.: Maximum cliques in graphs with small intersection number and random intersection graphs. arXiv:1204.4054v1
31. Orlovich, Y.L., Blazewicz, J., Dolgui, A., Finke, G., Gordon, V.S.: On the complexity of the independent set problem in triangle graphs. Discrete Math. 311, 1670–1680 (2011)
32. Orlovich, Y.L., Zverovich, I.E.: Independent domination in triangle graphs. 6th Czech-Slovak International Symposium on Combinatorics, Graph Theory, Algorithms and Applications. In: 6th Czech-Slovak International Symposium on Combinatorics, Graph Theory, Algorithms and Applications, Electron. Notes Discrete Math., vol. 28, pp. 341–348 (2007)
33. Payan, C.: A class of threshold and domishold graphs: equistable and equidominating graphs. Discrete Math. 29(1), 47–52 (1980)
34. Peled, U.N., Rotics, U.: Equistable chordal graphs. Stability in Graphs and Related Topics. Discrete Appl. Math. 132, 203–210 (2003)
35. Roberts, F.S.: Indifference graphs. In: Harary, F. (ed.) Proof Techniques in Graph Theory, pp. 139–146. Academic Press, New York (1969)
36. Tedder, M., Corneil, D., Habib, M., Paul, C.: Simpler Linear-Time Modular Decomposition Via Recursive Factorizing Permutations. In: Aceto, L., Damgård, I., Goldberg, L.A., Halldórsson, M.M., Ingólfsdóttir, A., Walukiewicz, I. (eds.) ICALP 2008, Part I. LNCS, vol. 5125, pp. 634–645. Springer, Heidelberg (2008)

Maximum Induced Multicliques and Complete Multipartite Subgraphs in Polygon-Circle Graphs and Circle Graphs

Fanica Gavril

Computer Science Dept., Technion, Haifa 32000, Israel
gavrilf@013.net.il

Abstract. A graph is a *multiclique* if its connected components are cliques. A graph is a *complete multipartite graph* if it is the complement of a multiclique. A graph is a *multiclique-multipartite graph* if its vertex set has a partition U, W such that $G(U)$ is complete multipartite, $G(W)$ is a multiclique and every two vertices $u \in U$, $v \in W$ are adjacent. We describe a polynomial time algorithm to find in polygon-circle graphs a maximum induced complete multipartite subgraph containing an induced $K_{2,2}$. In addition, we describe polynomial time algorithms to find maximum induced multicliques and multiclique-multipartite subgraphs in circle graphs. These problems have applications for clustering of proteins by PPI criteria.

Keywords: polygon-circle graph, circle graph, induced multiclique, induced complete multipartite subgraph, Protein-Protein-Interaction.

1   Introduction

We consider only finite graphs $G(V,E)$ with no parallel edges and no self-loops, where V is the set of vertices and E the set of edges. For $U \subseteq V$, $G(U)$ is the subgraph induced by U. We denote $N(v) = \{ u \mid u \text{ adjacent to } v \}$ and $N[v] = N(v) \cup \{v\}$. The *complement* of a graph G is denoted coG. A graph is a *multiclique* if its connected components are cliques. A graph is a *complete multipartite graph* if its vertex set has a partition into independent sets such that every two vertices in different independent sets are adjacent, that is, its complement is a multiclique. A graph $G(V,E)$ is a *multiclique-multipartite graph* if its vertex set has a partition U, W such that $G(U)$ is complete multipartite, $G(W)$ is a multiclique and every two vertices $u \in U$, $v \in W$ are adjacent.

A graph G is an *intersection graph* of a family S of subsets of a set if there is a one-to-one correspondence between the vertices of G and the subsets in S such that two vertices are adjacent if and only if their corresponding subsets in S intersect [19]. Intersection graphs of intervals on a line are called *interval graphs* [19]. *Polygon-circle graphs* [15] are intersection graphs of families of convex polygons inscribed in a circle. *Circle graphs* are intersection graph of families of chords in a circle [3,6]. A transitively orientable graph is called a *comparability graph* [12,19]; a vertex is a *source* if all its edges are outgoing and is a *sink* if they are incoming.

M.C. Golumbic et al. (Eds.): WG 2012, LNCS 7551, pp. 297–307, 2012.

A *dissociation set of a graph* is a vertex set which induces a subgraph whose connected components are edges or single vertices. In a bipartite graph, finding a maximum induced multiclique is NP-complete since it is the problem of finding a maximum dissociation set, while finding a maximum induced complete multipartite subgraph is polynomial [22].

A graph is *weakly-chordal* if it has no holes or antiholes with five or more vertices. Cameron and Hell [1] described for these graphs a polynomial time algorithm for maximum weight dissociation sets, using the algorithm in [20] for maximum weight independent sets in weakly-chordal graphs.

In the present paper we describe a polynomial time algorithm to find in polygon-circle graphs a maximum induced complete multipartite subgraph containing an induced $K_{2,2}$. This algorithm can also be applied to the *circle n-gon graphs* and the *circle trapezoid graphs*, analyzed in [11]. In addition, we describe polynomial time algorithms to find maximum induced multicliques and multiclique-multipartite subgraphs in circle graphs. These problems are NP-complete for general graphs [5]. The partition of all the vertices of a graph into a given number of independent sets and cliques, with various restrictions on mutual interconnections, was discussed in [2,13]. Note that the recognition problem of polygon-circle graphs is NP-complete [18].

Gavril [10] described a polynomial time algorithm for maximum induced bicliques in polygon-circle graphs, using separation by chords. This algorithm can be extended to find maximum induced multicliques with a constant number k of cliques, by considering all combinations of k chords in the circle.

The above problems have applications when a given set of entities related by some property, must be clustered into cliques and independent sets by some strongly connected vs. non-connected or similarity vs. dissimilarity criteria. For example, in Protein-Protein-Interaction (PPI) problems, the proteins must be clustered into strongly interacting groups, with weak or no interaction between the groups. The criteria for clustering proteins are lock-and-key criteria [17], complementary domains criteria [21], domain-domain interaction criteria [14] or interacting motifs criteria [16].

In Section 2 we describe a representation of polygon-circle graphs on a line. In Section 3 we describe an algorithm for maximum induced complete multipartite subgraphs containing an induced $K_{2,2}$ in polygon-circle graphs. In Sections 4,5 we describe algorithms for maximum induced multiclique and multiclique-multipartite subgraphs in circle graphs.

2   Representation of Polygon-Circle Graphs on a Line

Consider an intersection representation of G by polygons on a circle CR and let Z be a point on CR distinct from any corner point (Figure 1(a)). For every polygon with more than one chord we delete its chord facing Z, that is, its chord delimiting the arc containing Z. The intersection relationship does not change since two intersecting polygons have two pairs of crossing chords. Now, we open CR at Z, straighten CR into a line L (Figure 1(b)), and transform every chord into a semicircle arc above L through the chord's endpoints on L. The intersection relationship does not change since two chords in *CR* are not crossing if and only if their corresponding semicircle arcs are not intersecting. The remaining boundary of every polygon becomes a sequence of semicircle arcs with their endpoints on *L*, called *polygon-filament*. The reverse process is also true, thus a graph is a polygon-circle graph if and only if it is the intersection graph of a family of polygon-filaments on a line.

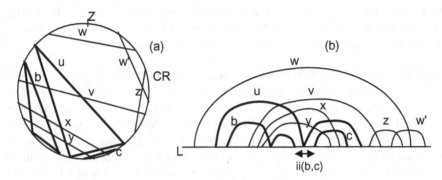

Fig. 1. A family of polygons in a circle and its representation as polygon-filaments on a line

These graphs are a subfamily of the interval-filament graphs defined by Gavril [7,8,9]. For a vertex v, we denote by $a(v)$ its corresponding polygon-filament and by $i(v)$ the interval on L delimiting $a(v)$. For two vertices u,v having $i(u) \cap i(v) = \phi$, we denote by $ii(u,v)$ the interval between $i(u)$ and $i(v)$. For a pair $p=(u,v)$ of adjacent vertices we denote $i(p)=i(u) \cup i(v)$, $N[p]=N[u] \cup N[v]$ and $U(p)=\{ x \mid i(x) \subseteq i(p) \}$. For a point X on L we denote $V_X=\{ v \mid X \in i(v) \}$. The *endpoints* of a polygon-filament $a(v)$, are the endpoints of its arcs. The *interval i of an arc* is the interval between its two endpoints. We denote by $R(G)$ the intersection representation of a polygon-circle graph G by polygon-filaments. The edges of $coG(V_X)$ represent containment of intervals of non-intersecting polygon-filaments, since the interval of every vertex of $G(V_X)$ contains X. We orient the edges u,v of $coG(V_X)$ from u to v, whenever $i(u) \subset i(v)$; all edges of $coG(V_X)$ become oriented. This orientation is acyclic and transitive since $i(u) \subset i(v) \subset i(w)$ implies $i(u) \subset i(w)$. The properties of families of polygon-filaments are the following:

Property 1. Two semicircle arcs of distinct polygon-filaments on L do not intersect (even when their polygon-filaments intersect) if and only if they have disjoint intervals or the interval of one arc appears between the endpoints of the other. This, because chords in *CR* corresponding to two non-intersecting arcs are non-crossing.

Property 2. Two polygon-filaments b, c do not intersect if and only if they have disjoint intervals, or the interval of one $i(b)$, is contained between the two endpoints of an arc of the other c (b, u in Figure 1(b)).

Lemma 1. In G, $R(G)$, for every pair $p=(u,v)$ of adjacent or identical vertices, there are no edges between a vertex $x \in U(p)$ and a vertex $w \in V - (N[p] \cup U(p))$ (Figure 1(b)).

Proof. Consider a pair $p=(u,v)$ of adjacent vertices and a vertex $w\in V-(N[p]\cup U(p))$. If $i(p)\cap i(w)=\phi$ then there are no edges between w and the vertices of $U(p)$. Otherwise, the interval $i(p)=i(u)\cup i(v)$ is contained between the endpoints of an arc of $a(w)$ and so is the interval of every $x\in U(p)$. Hence, by Property 2, $a(x)$ and $a(w)$ cannot intersect.

Lemma 2. In G, $R(G)$, consider four polygon-filaments b,c,x,y such that b,c have disjoint intervals, x,y are non-intersecting and both x,y intersect both b,c (Figure 1(b)). If there exists a polygon-filament u which intersects both x,y and does not intersect b,c, then u has an endpoint in $ii(b,c)$ and $i(b)\subset i(u)$ or $i(c)\subset i(u)$ or both.

Proof. Consider four such polygon-filaments b,c,x,y. Let X be the middle point of $ii(b,c)$. Both $i(x),i(y)$ must contain X, hence one of them must contain the other, say $i(y)\subset i(x)$. Consider a polygon-filament u which intersects both x,y and does not intersect b,c. If u has no endpoints in $ii(b,c)$, then it has no endpoints in $i(b)\cup ii(b,c)\cup i(c)$. Since $X\in i(y)$, the interval $i(y)$ must appear between the endpoints of $i(x)$ in $i(b)$ and $i(c)$. Therefore $i(y)\subset[i(b)\cup ii(b,c)\cup i(c)]$ implying that y and u cannot intersect. Thus, u has an endpoint in $ii(b,c)$. If $i(u)\subset ii(b,c)$, then by the above argument c has an endpoint in $ii(b,u)\subset ii(b,c)$ which is a contradiction. Therefore u has an endpoint in $ii(b,c)$ and $i(b)\subset i(u)$ or $i(c)\subset i(u)$.

Lemma 2 proves that in CR there are no three non-intersecting polygons b,c,u, each facing the other two, and two non-intersecting polygons x,y, intersecting b,c,u.

Theorem 3. In a representation $R(G)$ by polygon-filaments of a polygon-circle graph G, for every point $X\in L$, $G(V_X)$ is a weakly-chordal cocomparability graph.

Proof. As described earlier, we orient the edges of the comparability graph $coG(V_X)$ by containment of intervals to obtain an acyclic transitive orientation. Cocomparability graphs are perfect having no odd holes or antiholes with five or more vertices. Also [4], the cocomparability graphs cannot have holes with six or more vertices, since such holes contain asteroidal triples. Hence, $G(V_X)$ has no holes with five or more vertices.

Assume that $G(V_X)$ has an even antihole $h=\{v_1, v_2,..., v_{2k}\}$ with six or more vertices. The hole coh is transitively oriented by the orientation of $coG(V_X)$. W.l.o.g. assume that the two edges of every v_{2i-1} are incoming and the two edges of every v_{2i} are outgoing. Let $arc(v)$ denote the arc of $a(v)$ containing X.

Let us prove that for two adjacent vertices v_{2i+1}, v_{2j+1} of h, the intervals $i(arc(v_{2i+1}))$, $i(arc(v_{2j+1}))$ are intersecting but are not contained one into another. Assume that $i(arc(v_{2i+1}))\subset i(arc(v_{2j+1}))$ and $2i<2i+1<2j+1$ (when $2i+1=1$, we take $2k$ for $2i$), otherwise we renumber the vertices. Hence, $i(v_{2i})\subset i(arc(v_{2i+1}))\subset i(arc(v_{2j+1}))$, implying, by Property 2, that $a(v_{2i})$ cannot intersect $a(v_{2j+1})$, contradicting the fact that v_{2i}, v_{2j+1} are adjacent.

Let us prove that for every three $i(arc(v_{2i-1}))$, $i(arc(v_{2i+1}))$, $i(arc(v_{2i+3}))$, we have $i(arc(v_{2i+1}))\subset[i(arc(v_{2i-1}))\cup i(arc(v_{2i+3}))]$ (when $2i+1=1$, we take $2k-1$ for $2i-1$). Since every two of the three intervals are intersecting but are not contained one into another, one of the three is contained in the union of the two others. Assume that $i(arc(v_{2i+3})) \subseteq [i(arc(v_{2i-1})) \cup i(arc(v_{2i+1}))]$. By above, neither $i(arc(v_{2i+1}))$ nor $i(arc(v_{2i-1}))$ can contain $i(arc(v_{2i+3}))$. Thus $[i(arc(v_{2i-1})) \cap i(arc(v_{2i+1}))] \subseteq i(arc(v_{2i+3}))$ since the three intervals contain X. Then, $X \in i(v_{2i}) \subseteq [i(arc(v_{2i-1})) \cap i(arc(v_{2i+1}))] \subseteq i(arc(v_{2i+3}))$ and by Property 2, $a(v_{2i})$ cannot intersect $a(v_{2i+3})$, contradicting the fact that v_{2i}, v_{2i+3} are adjacent.

Hence, the left (right) endpoint of $i(arc(v_{2i+1}))$ appears on L between the left (right, respectively) endpoints of $i(arc(v_{2i-1}))$ and $i(arc(v_{2i+3}))$. Assume that the left endpoints of $i(arc(v_1))$, $i(arc(v_3))$, $i(arc(v_5))$ appear from left to right on L in the order: left endpoint of $i(arc(v_1))$, left endpoint of $i(arc(v_3))$, left endpoint of $i(arc(v_5))$. Then, by induction, we obtain that the left endpoints and the right endpoints of $i(arc(v_1))$, $i(arc(v_3)),\ldots,i(arc(v_{2k-1}))$ appear on L in this order from left to right. Hence, $i(v_{2k}) \subseteq i(arc(v_1)) \cap i(arc(v_{2k-1})) \subseteq i(arc(v_3))$ and by Property 2, $a(v_{2k})$ cannot intersect $a(v_3)$, contradicting the fact that v_3, v_{2k} are adjacent. Therefore, $G(V_X)$ has no holes and antiholes with five or more vertices, and is weakly-chordal.

3   Algorithm for Complete Multipartite Subgraphs Containing an Induced $K_{2,2}$, in Polygon-Circle Graphs

Consider a polygon-filament representation $R(G)$ of a polygon-circle graph $G(V,E)$.

Lemma 4. For two non-adjacent vertices u,v having $i(u) \subseteq i(v)$ let

$$V(u,v) = \{ w \mid i(u) \subseteq i(w) \subseteq i(v), w \notin N(u) \cup N(v)\}.$$

Then, every vertex $z \in N(u) \cap N(v)$ is adjacent to every vertex $w \in V(u,v)$ (Figure 2).

Proof. Assume that there are two non-adjacent vertices $z \in N(u) \cap N(v)$ and $w \in V(u,v)$. If $i(w) \subseteq i(z)$ then $i(u) \subseteq i(w) \subseteq i(z)$ and $a(z)$ cannot intersect $a(u)$. If $i(z) \subseteq i(w)$ then $i(z) \subseteq i(w) \subseteq i(v)$ and $a(z)$ cannot intersect $a(v)$. If $i(w) \cap i(z) = \phi$, then $i(u) \cap i(z) = \phi$, since $i(u) \subseteq i(w)$. All three cases contradict the fact that z is adjacent to both u and v but not to w.

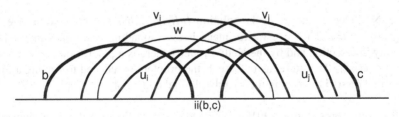

Fig. 2. For $w \notin N(u_i) \cup N(v_i)$ and $i(u_i) \subseteq i(w) \subseteq i(v_i)$, every vertex $v_j \in N(u_i) \cap N(v_i)$ is adjacent to w

The algorithm to find a maximum induced complete multipartite subgraph of a polygon-circle graph G solves separately the following two cases:

Case 1: The solution $B(IND_1,...,IND_k,E)$, where every IND_j is an independent set, contains two vertices with disjoint intervals. W.l.o.g. assume that IND_1 has two vertices b,c having $i(b) \cap i(c) = \phi$, and their intervals are minimal in IND_1 (Figure 2). Consider some IND_j, $2 \le j \le k$: for every $x \in IND_j$, $i(x)$ contains the middle point X of $ii(b,c)$; hence $IND_j \subseteq V_X$. In addition, every two vertices $u,v \in IND_j$, being non-adjacent and their intervals containing the point X, fulfill $i(u) \subset i(v)$. Thus, IND_j cannot contain two vertices both with minimal or both with maximal intervals, since the interval of one must be contained into the other. Let u_j, v_j be the unique vertices of IND_j with minimal and maximal intervals; we may have $u_j = v_j$. Then, by Lemma 4 (Figure 2), $IND_j \subseteq V(u_j,v_j) \cup \{u_j,v_j\}$ and for every $s \ne j$ $IND_s \subseteq N(u_j) \cap N(v_j)$; we assign the weight $|IND_j|$ to the pair u_j,v_j. The vertices of every pair u_j,v_j, $2 \le j \le k$, are not adjacent while every two vertices in different pairs are adjacent. Thus, the set of pairs u_j,v_j, $2 \le j \le k$, forms a weighted dissociation set of the complement $coG(V_X)$ of $G(V_X)$.

Since B contains an induced $K_{2,2}$, some IND_j, $2 \le j \le k$, contains at least two vertices. Hence by Lemma 2, the polygon-filament of every vertex $d \in IND_1$, $d \ne b,c$, has an endpoint in $ii(b,c)$ and $i(b) \subset i(d)$ or $i(c) \subset i(d)$ or both. The above implies that $IND_1 - \{b,c\}$ is a subset of

$$S(b,c) = \{ d \mid d \notin N(b) \cup N(c), a(d) \text{ has an endpoint in } ii(b,c) \text{ and } i(b) \subset i(d) \text{ or } i(c) \subset i(d)\}$$

and for every $2 \le j \le k$, IND_j is a subset of $V_X(b,c) = V_X \cap N(b) \cap N(c)$. The two sets $S(b,c) \cup \{b,c\}$ and $V_X(b,c)$ are disjoint. Therefore, the family IND_j, $1 \le j \le k$, is defined by the family of pairs $\{(b,c)\} \cup \{(u_j,v_j) \mid 2 \le j \le k\}$ fulfilling that the vertices of every pair are not adjacent while, every two vertices in different pairs are adjacent. Thus, for Case 1, the algorithm considers every two non-adjacent vertices b,c fulfilling $i(b) \cap i(c) = \phi$, and the middle point X of $ii(b,c)$. The algorithm finds by the algorithm in [7] a maximum independent set in $S(b,c)$ to obtain $IND_1 - \{b,c\}$, for every pair u,v of non-adjacent vertices in $V_X(b,c)$ finds a maximum independent set IND in $V(u,v)$ and assigns the weight $|IND \cup \{u,v\}|$ to the pair u,v. Now, the algorithm finds a maximum weight dissociation set in $coG(V_X(b,c))$, by the algorithm in [1,20], since by Theorem 3, $coG(V_X(b,c))$ is weakly-chordal. By the above explanation, the independent sets corresponding in G to this maximum weight dissociation set together with IND_1 form a maximum induced complete multipartite subgraph of G.

Case 2: No independent set in the solution $B(IND_1,...,IND_k,E)$ has two vertices b,c having $i(b) \cap i(c) = \phi$. Then every two intervals corresponding to vertices in B have a non-empty intersection and by the Helly property there is a point X on L contained in all these intervals. Therefore, as in Case 1, the problem is reduced to finding a maximum weight dissociation set in the weakly-chordal comparability graph $coG(V_X)$.

The algorithm works in time $O(|V|^5 + |V|^2 F(|V|))$, where $F(|V|)$ is the time required to find a maximum weight dissociation set in a weakly-chordal comparability graph.

In the special case when B contains no induced $K_{2,2}$ implying that every IND_j, $2 \le j \le k$, contains one vertex, the problem is to find a maximum induced subgraph with a vertex partition into an independent set IND and a clique C, completely interconnected. If IND

contains at least three vertices b,c,d, with mutually disjoint intervals, the above algorithm cannot be applied, and the problem remains open. Note that in such a case, by Lemma 2, $N(b) \cap N(c) \cap N(d)$ is a clique.

4 Algorithm for Multicliques in Circle Graphs

Consider a polygon-filament representation $R(G)$ of a circle graph $G(V,E)$: a vertex is represented by one semicircle. For a clique C let $i(C) = \cup_{w \in C} i(w)$ and $a(C) = \cup_{w \in C} a(w)$. The pair of (not necessarily distinct) vertices $u,v \in C$ to which the endpoints of $i(C)$ belong, fulfils $i(C) = i(u) \cup i(v)$; we say that the pair $p = (u,v)$ *delimits* $i(C)$ (Figure 3). By Lemma 1, there are no edges between vertices in $U(p)$ and vertices in $V - (N[p] \cup U(p))$. Let H be the graph whose vertices are pairs of adjacent or identical vertices of G, two pairs p,q being connected by an edge if and only if they have a vertex in common, or two vertices one in p one in q are adjacent. The graph H is an intersection graph, in which every vertex $p = (u,v)$ is represented by the union $a(p)$ of the two intersecting polygon-filaments $a(u)$ and $a(v)$. Let $E2$ be the oriented edge subset $\{q \rightarrow p\}$ of the edge set of coH given by the relation $i(q) \subset i(p)$ and $a(q) \cap a(p) = \phi$, this orientation of $E2$ is transitive. By Lemma 1, for an edge $q \rightarrow p$ in $E2$, there are no edges in E between $U(q)$ and $U(p) - (N[q] \cup U(q))$.

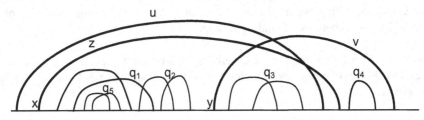

Fig. 3. For the pair $p = (u,v)$ we have $C_p = \{u,z,v\}$, $P_M(p) = \{q_1,q_3,q_4,q_5\}$, sinks are q_1, q_3, q_4, and $M(p) = C_p \cup M(q_1) \cup M(q_3) \cup M(q_4)$; in the interval $i_{j,p} = [x,y]$, $s_1 = q_1$ is the unique sink of $coH(P_M \cap W(i_{j,p}), E2)$

For a multiclique M, let P_M be the vertex set of H corresponding to the pairs delimiting the cliques in M; P_M is an independent set of H. For a pair $p(u,v) \in P_M$ (Figure 3), let $P_M(p) = \{ q \mid q \in P_M, i(q) \subset i(p) \}$ and let $M(p)$ be the partial multiclique of M defined by $P_M(p)$ in the subgraph $G(U(p))$. $M(p)$ is composed of a clique $C_p \subseteq N[u] \cap N[v] \cap U(p)$ and of the cliques defined by $P_M(p) - \{p\}$. Every pair q in $P_M(p) - \{p\}$ fulfils $i(q) \subset i(p)$ and $a(q) \cap a(p) = \phi$, implying that $q \rightarrow p \in E2$ and p is a sink of $P_M(p)$ in $coH(P_M, E2)$. Similarly, for every two pairs s,q in $P_M(p) - \{p\}$ either $s \rightarrow q \in E2$ or $i(s) \cap i(q) = \phi$. Let $q_1,...,q_k$ be the sinks of $P_M(p) - \{p\}$ in the transitive orientation of $coH(P_M, E2)$. Then, $M(p) = C_p \cup M(q_1) \cup ... \cup M(q_k)$ and for every $q \in M(q_1) \cup ... \cup M(q_k)$, $i(q)$ appears between consecutive endpoints of $a(C_p)$. When M is a maximum induced multiclique, $M(p)$ is a maximum induced multiclique of $G(U(p))$, otherwise we can replace $M(p)$ by a maximum one. We assign to p the weight $weight(p) = |M(p)| = |C_p| + |M(q_1)| + ... + |M(q_k)|$. Consider an interval $i_{j,p}$ between two consecutive endpoints of $a(C_p)$. Let $W(i_{j,p})=\{ q \mid i(q) \subseteq i_{j,p}\}$; the polygon-filaments corresponding to the vertices of $W(i_{j,p})$ do not intersect the polygon-filaments corresponding to the vertices of C_p or of the cliques of M delimited by intervals of $a(C_p)$ disjoint from $i_{j,p}$. Consider the weighted interval graph $I(W(i_{j,p}))$ in which every vertex q in $W(i_{j,p})$ is represented by $i(q)$ with $weight(q)$. Let $s_1,...,s_r$ be the sinks of $coH(P_M \cap W(i_{j,p}),E2)$. When M is a maximum induced multiclique, $s_1,...,s_r$ is a maximum weight independent set of the interval graph $I(W(i_{j,p}))$, otherwise we could obtain a larger induced multiclique by replacing $s_1,...,s_r$ by a maximum weight independent set.

The algorithm works as follows: Using the topological ordering defined by the transitive orientation of $E2$ on coH, we go from sources to sinks on $E2$ and construct for every pair $p=(u,v)$ a maximum induced multiclique $M(p)$ of $G(U(p))$, using the maximum induced multicliques $M(q)$ of the pairs q having $q \rightarrow p \in E2$. For a given p, we must find a clique $C_p \subseteq N[u] \cap N[v] \cap U(p)$, and a maximum weight independent set IND which is the union of maximum weight independent sets (with sinks $s_1,...,s_r$) in the intervals between consecutive endpoints of $a(C_p)$ such that $|C_p|+weight(s_1)+...+weight(s_r)$ is maximum. Then, $C_p \cup M(s_1) \cup .. \cup M(s_r)$ is a maximum induced multiclique of $G(U(p))$.

By the Helly property of intervals on a line, every clique in $N[u] \cap N[v] \cap U(p)$ is contained in a vertex set V_X for a point X in $i(u) \cap i(v)$: we must consider every subinterval between consecutive endpoints of polygon-filaments, in $i(u) \cap i(v)$, in each a point X, and for each X we must construct C_p and IND. For the semicircle $a(u)$ of a vertex u, let l_u, r_u denote its left and right endpoints.

Lemma 5. In a circle graph G, the vertices of a clique C_p, $p=(u,v)$, $C_p \subseteq V_X$, are represented by semicircles whose endpoints at the left and the right of X are in the same order.

Proof. Assume that the endpoints of $a(x)$, $a(y)$ representing $x,y \in C_p$ are in order $l_x \triangleleft l_y$ at the left of X and in $r_y \triangleleft r_x$ at the right of X. Then, the endpoints of $a(y)$ are contained in $i(x)$ and $a(x),a(y)$ cannot intersect, contradicting the fact that $x,y \in C_p$.

For a pair $p=(u,v)$ and a point X in $i(u) \cap i(v)$ we denote by $l_u=l_1< l_2 <,...,< l_s=l_v \triangleleft X$ the left endpoints of the semicircles representing the vertices in $U(p) \cap V_X$. For a vertex w_i whose left endpoint of $i(w_i)$ is l_i, we denote the right endpoint by r_i; note that $X \triangleleft r_u \triangleleft r_i \triangleleft r_v$. We now go on the left endpoints from left to right and for every i we find a maximum multiclique $M(p,1,i)$ within the intervals $[l_1,l_i] \cup [r_1,r_i]$; let $C_{p,i}$ denote its clique containing u and w_i. Assume that we found such a solution for $1,..,i-1$ and we want to find one for i and w_i. We consider every $l_j \triangleleft l_i$, such that w_j, w_i are adjacent, hence by Lemma 5 $r_j \triangleleft r_i$, and we evaluate a maximum weight independent set $IM(p,j,i)$ in the weighted interval graph $I(W([l_j,l_i]) \cup W([r_j,r_i]))$ in which every vertex q is represented by $i(q)$ with $weight(q)$. By Lemma 5, every vertex w_k of the clique $C_{p,j}$ in the partial solution $M(p,1,j)$ has $l_k \leq l_j \triangleleft l_i$ and $r_k \leq r_j \triangleleft r_i$, hence w_k is adjacent to w_i implying that $C_{p,j} \cup \{w_i\}$ is a clique, Among all j, we take the solution with maximum $|M(p,1,j)|+weight(IM(p,j,i))+1$ and assign it to i as $M(p,1,i)$; by induction

$\{w_i\} \cup M(p,1,j) \cup \{ M(q) \mid q \in IM(p,j,i)\}$ is a maximum induced multiclique $M(p,1,i)$. For the final solution, we find a maximum weight independent set in the interval graph defined by all the pairs of adjacent vertices in G.

The algorithm works in time $O(|V|^6)$. Polygon-circle graphs do not fulfill Lemma 5 and their problem remains open.

5   Algorithm for Multipartite-Multiclique Subgraphs in Circle Graphs

Consider a circle graph $G(V,E)$ represented as an intersection graph of chords in a circle CR.

Let $M(U,W)$ be an induced multiclique-multipartite subgraph of $G(V,E)$ where $M(U)$ is complete multipartite, $M(W)$ is a multiclique and every two vertices $u \in U$, $v \in W$ are adjacent. Consider a vertex u in U with chord $X_u Y_u$ (Figure 4). Let x_v, x_z be the endpoints of chords of vertices v,z in W closest to X_u. Let y_w, y_s be the endpoints of chords of vertices w,s in W closest to Y_u. We denote all arcs counterclockwise. Every vertex in W is adjacent to $u \in U$, hence its chord intersects the chord $X_u Y_u$ and has its endpoints in the disjoint arcs $x_v y_w$ and $y_s x_z$. The vertices v,z can be identical or must have intersecting chords; similarly for w,s. Let $Z(v,z,w,s)$ be the set of vertices of G whose chords have the endpoints one in each arc $x_v y_w$ and $y_s x_z$. Thus $W \subseteq Z(v,z,w,s)$. Let $Q(v,z,w,s)$ be the set of vertices of G whose chords have the endpoints one in each arc $x_z x_v$ and $y_w y_s$.

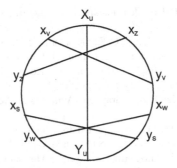

Fig. 4. The chords on CR of the vertices $u \in U$ and $v,z,y,s \in W$; the pairs v,z and y,s are in different cliques of W.

Case 1: Assume that $M(W)$ has at least two cliques. This implies that the chords of v,z do not intersect the chords of w,s. Since the chord of every vertex in U intersects the chords of all the vertices in W, it has its endpoints one in each arc $x_z x_v$ and $y_w y_s$. Hence, $U \subseteq Q(v,z,w,s)$. The algorithm works as follows: We consider every two pairs v,z and w,s of adjacent (or identical) vertices with no interconnecting edges (Figure 4). Let $Z(v,z,w,s)$ and $Q(v,z,w,s)$ be defined as above. By the algorithms in Sections 3,4, we find a maximum induced multiclique $G(W)$ in $G(Z(v,z,w,s))$, and a maximum induced multipartite subgraph $G(U)$ in $G(Q(v,z,w,s))$. Among these pairs v,z and w,s we chose the induced multipartite-multiclique subgraph with a maximum number of vertices.

Case 2: Assume that $M(W)$ has only one clique C. Hence $v=s$, $z=w$ and v,z are adjacent. Therefore, among the four arcs defined by the chords of v,z on CR, there is a pair of opposite arcs such that the chords corresponding to the vertices of U have the endpoints one in each arc of this pair. The algorithm works as follows: For every pair v,z of adjacent vertices and for every pair of their opposite arcs we find a maximum clique $G(W)$ for the pair v,z, and a maximum induced multipartite subgraph $G(U)$ for their opposite pairs of arcs. To find $G(U)$: we use the algorithm in Section 3, to cover the case that it contains an induced $K_{2,2}$; we use an algorithm to find a maximum independent set in a permutation graph to cover the case that $G(U)$ has no induced $K_{2,2}$ and thus has only one independent set. Among all pairs v,z of adjacent vertices, we chose the induced multipartite-multiclique subgraph with maximum number of vertices. Note that this covers the case that $M(U,W)$ has one independent set and one clique, unsolved in Section 4.

The algorithm works in time $O(|V|^4 F_1(V)+|V|^4 F_2(V))$ where $F_1(V)$ is the time required to find a maximum induced multiclique in $G(Z(v,z,w,s))$, and $F_2(V)$ the time required to find a maximum induced multipartite subgraph in $G(Q(v,z,w,s))$.

References

1. Cameron, K., Hell, P.: Independent Packings in Structured Graphs. Math. Program. Ser. B 105, 201–213 (2006)
2. Cameron, K., Eschen, E.M., Hoang, C.T., Sritharan, R.: The Complexity of the List Partition Problem of Graphs. SIAM J. Discrete Math. 21, 900–929 (2007)
3. Even, S., Itai, A.: Queues, Stacks and Graphs. In: Kohavi, Z., Paz, A. (eds.) Theory of Machines and Computations, pp. 71–86. Academic Press, New York (1971)
4. Gallai, T.: Transitiv Orientirbare Graphen. Acta Math. Acad. Sci. Hungar 18, 25–26 (1967)
5. Garey, M.R., Johnson, D.S.: Computers and Intractability: A Guide to the Theory NP-Completeness. W. H. Freeman and Co., San Francisco (1979)
6. Gavril, F.: Algorithms for a Maximum Clique and a Maximum Independent Set of a Circle Graph. Networks 3, 261–273 (1973)
7. Gavril, F.: Maximum Weight Independent Sets and Cliques in Intersection Graphs of Filaments. Inform. Proc. Lett. 73, 181–188 (2000)
8. Gavril, F.: 3D-Interval-Filament Gaphs. Discrete Appl. Math. 155, 2625–2636 (2007)
9. Gavril, F.: Algorithms on Subtree Filament Graphs. In: Lipshteyn, M., Levit, V.E., McConnell, R.M. (eds.) Graph Theory, Computational Intelligence and Thought. LNCS, vol. 5420, pp. 27–35. Springer, Heidelberg (2009)
10. Gavril, F.: Algorithms for Induced Biclique Optimization Problems. Inform. Proc. Lett. 111, 469–473 (2011)
11. Gavril, F.: Minimum Weight Feedback Vertex Sets in Circle n-Gon Graphs and Circle Trapezoid Graphs. Discrete Math., Algorithms and Applications 3, 323–336 (2011)

12. Golumbic, M.C., Rotem, D., Urrutia, J.: Comparability Graphs and Intersection Graphs. Discrete Math. 43, 37–46 (1983)
13. Hell, P., Klein, S., Nogueira, L.T., Protti, F.: Partitioning Chordal Graphs into Independent Sets and Cliques. Discrete Appl. Math. 141, 185–194 (2004)
14. Holm, L.: Evaluation of Different Domain-based Methods in Protein Interaction Prediction. Biochem. Biophys. Res. Commun. 390, 357–362 (2009)
15. Kratochvil, J., Kostochka, A.: Covering and Coloring Polygon-Circle Graphs. Discrete Math. 163, 299–305 (1997)
16. Liu, X., Li, J., Wang, L.: Modeling Protein Interacting Groups by Quasi-Multicliques: Complexity, Algorithm and Application. IEEE/ACM Trans. on Comp. Biology and Bioinf. 7, 354–364 (2010)
17. Morrison, J.L., Breitling, R., Higham, D.J., Gilbert, D.R.: A Lock-and-Key Model for Protein-Protein Interactions. Bioinformatics 22, 2012–2019 (2006)
18. Pergel, M.: Recognition of Polygon-Circle Graphs and Graphs of Interval Filaments Is NP-Complete. In: Brandstädt, A., Kratsch, D., Müller, H. (eds.) WG 2007. LNCS, vol. 4769, pp. 238–247. Springer, Heidelberg (2007)
19. Spinrad, J.P.: Efficient Graph Representations. Fields Institute Monographs. The American Mathematical Society, USA (2003)
20. Spinrad, J.P., Sritharan, R.: Algorithms for weakly triangulated graphs. DAM 19, 181–191 (1995)
21. Thomas, A., Cannings, R., Monk, N.A.M., Cannings, C.: On the Structure of Protein-Protein Interaction Networks. Biochem. Soc. Trans. 31, 1491–1496 (2003)
22. Yannakakis, M.: Node-Deletion Problems in Bipartite Graphs. SIAM J. Computing 10, 310–327 (1981)

Parameterized Domination in Circle Graphs[*]

Nicolas Bousquet[1], Daniel Gonçalves[1], George B. Mertzios[2],
Christophe Paul[1], Ignasi Sau[1], and Stéphan Thomassé[3]

[1] AlGCo Project-Team, CNRS, LIRMM, Montpellier, France
{FirstName.FamilyName}@lirmm.fr
[2] School of Engineering and Computing Sciences, Durham University, U.K.
george.mertzios@durham.ac.uk
[3] Laboratoire LIP (U. Lyon, CNRS, ENS Lyon, INRIA, UCBL), Lyon, France
stephan.thomasse@ens-lyon.fr

Abstract. A *circle graph* is the intersection graph of a set of chords in a circle. Keil [*Discrete Applied Mathematics*, 42(1):51-63, 1993] proved that DOMINATING SET, CONNECTED DOMINATING SET, and TOTAL DOMINATING SET are NP-complete in circle graphs. To the best of our knowledge, nothing was known about the parameterized complexity of these problems in circle graphs. In this paper we prove the following results, which contribute in this direction:

- DOMINATING SET, INDEPENDENT DOMINATING SET, CONNECTED DOMINATING SET, TOTAL DOMINATING SET, and ACYCLIC DOMINATING SET are $W[1]$-hard in circle graphs, parameterized by the size of the solution.
- Whereas both CONNECTED DOMINATING SET and ACYCLIC DOMINATING SET are $W[1]$-hard in circle graphs, it turns out that CONNECTED ACYCLIC DOMINATING SET is polynomial-time solvable in circle graphs.
- If T is a *given* tree, deciding whether a circle graph has a dominating set isomorphic to T is NP-complete when T is in the input, and FPT when parameterized by $|V(T)|$. We prove that the FPT algorithm is subexponential.

Keywords: circle graphs, domination problems, parameterized complexity, parameterized algorithms, dynamic programming, constrained domination.

1   Introduction

A *circle graph* is the intersection graph of a set of chords in a circle (see Fig. 1 for an example of a circle graph G together with a circle representation of it). The class of circle graphs has been extensively studied in the literature, due in part to its applications to sorting [12] and VLSI design [27]. Many problems which are NP-hard in general graphs turn out to be solvable in polynomial time

[*] The third author was partially supported by EPSRC Grant EP/G043434/1. The other authors were partially supported by AGAPE (ANR-09-BLAN-0159) and GRATOS (ANR-09-JCJC-0041) projects (France).

M.C. Golumbic et al. (Eds.): WG 2012, LNCS 7551, pp. 308–319, 2012.
© Springer-Verlag Berlin Heidelberg 2012

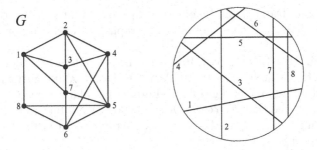

Fig. 1. A circle graph G on 8 vertices together with a circle representation of it

when restricted to circle graphs. For instance, this is the case of MAXIMUM CLIQUE and MAXIMUM INDEPENDENT SET [17], TREEWIDTH [24], MINIMUM FEEDBACK VERTEX SET [18], RECOGNITION [19,28], DOMINATING CLIQUE [22], or 3-COLORABILITY [30].

But still a few problems remain NP-complete in circle graphs, like k-COLORABILITY for $k \geq 4$ [29], HAMILTONIAN CYCLE [8], or MINIMUM CLIQUE COVER [23]. In this article we study a variety of domination problems in circle graphs, from a parameterized complexity perspective. A *dominating set* in a graph $G = (V, E)$ is a subset $S \subseteq V$ such that every vertex in $V \setminus S$ has at least one neighbor in S. Some extra conditions can be imposed to a dominating set. For instance, if $S \subseteq V$ is a dominating set and $G[S]$ is connected (resp. acyclic, an independent set, a graph without isolated vertices, a tree, a path), then S is called a *connected* (resp. *acyclic, independent, total, tree, path*) *dominating set*. In the example of Fig. 1, vertices 1 and 5 (resp. 3, 4, and 6) induce an independent (resp. connected) dominating set. The corresponding minimization problems are defined in the natural way. Given a set of graphs $\mathcal{G}$, the MINIMUM $\mathcal{G}$-DOMINATING SET problem consists in, given a graph G, finding a dominating set $S \subseteq V(G)$ of G of minimum cardinality such that $G[S]$ is isomorphic to some graph in $\mathcal{G}$. Throughout the article, we may omit the word "MINIMUM" when referring to a specific problem.

For an introduction to parameterized complexity theory, see for instance [10, 14, 26]. A decision problem with input size n and parameter k having an algorithm which solves it in time $f(k) \cdot n^{\mathcal{O}(1)}$ (for some computable function f depending only on k) is called *fixed-parameter tractable*, or FPT for short. The parameterized problems which are $W[i]$-hard for some $i \geq 1$ are not likely to be FPT [10,14,26]. A parameterized problem is in XP if it can be solved in time $f(k) \cdot n^{g(k)}$, for some (unrestricted) functions f and g. The parameterized versions of the above domination problems when parameterized by the cardinality of a solution are also defined naturally.

Previous Work. DOMINATING SET is one of the most prominent classical graph-theoretic NP-complete problems [16], and has been studied intensively in the literature. Keil [22] proved that DOMINATING SET, CONNECTED DOMINATING SET, and TOTAL DOMINATING SET are NP-complete when restricted to circle graphs, and Damian and Pemmaraju [9] proved that INDEPENDENT DOM-INATING SET is also NP-complete in circle graphs, answering an open question from Keil [22].

Hedetniemi, Hedetniemi, and Rall [20] introduced acyclic domination in graphs. In particular, they proved that ACYCLIC DOMINATING SET can be solved in polynomial time in interval graphs and proper circular-arc graphs. Xu, Kang, and Shan [31] proved that ACYCLIC DOMINATING SET is linear-time solvable in bipartite permutation graphs. The complexity status of ACYCLIC DOMINATING SET in circle graphs was unknown.

In the theory of parameterized complexity [10,14,26], DOMINATING SET also plays a fundamental role, being the paradigm of a $W[2]$-hard problem. For some graph classes, like planar graphs, DOMINATING SET remains NP-complete [16] but becomes FPT when parameterized by the size of the solution [2]. Other more recent examples can be found in H-minor-free graphs [3] and claw-free graphs [7].

The parameterized complexity of domination problems has been also studied in geometric graphs, like k-polygon graphs [11], multiple-interval graphs and their complements [13,21], k-gap interval graphs [15], or graphs defined by the intersection of unit squares, unit disks, or line segments [25]. But to the best of our knowledge, the parameterized complexity of the aforementioned domination problems in circle graphs was open.

Our Contribution. In this paper we prove the following results, which settle the parameterized complexity of a number of domination problems in circle graphs:

- In Section 2, we prove that DOMINATING SET, CONNECTED DOMINAT-ING SET, TOTAL DOMINATING SET, INDEPENDENT DOMINATING SET, and ACYCLIC DOMINATING SET are $W[1]$-hard in circle graphs, parameterized by the size of the solution. Note that ACYCLIC DOMINATING SET was not even known to be NP-hard in circle graphs. The reductions are from k-COLORED CLIQUE in general graphs.
- Whereas both CONNECTED DOMINATING SET and ACYCLIC DOMINATING SET are $W[1]$-hard in circle graphs, it turns out that CONNECTED ACYCLIC DOMINATING SET is polynomial-time solvable in circle graphs. This is proved in Section 3.
- Furthermore, if T is a *given* tree, we prove in Section 3 that the problem of deciding whether a circle graph has a dominating set isomorphic to T is NP-complete but FPT when parameterized by $|V(T)|$. The NP-completeness reduction is from 3-PARTITION, and we prove that the running time of the FPT algorithm is subexponential. As a corollary of this algorithm, we also deduce that if T has bounded degree, then deciding whether a circle graph has a dominating set isomorphic to T can be solved in polynomial time.

Due to lack of space, the proofs marked with '[⋆]' have been omitted in this extended abstract, and can be found in [5].

Further Research. Some interesting questions remain open. We proved that several domination problems are $W[1]$-hard in circle graphs. Are they $W[1]$-complete, or may they also be $W[2]$-hard? On the other hand, we proved that finding a dominating set isomorphic to a tree can be done in polynomial time. It could be interesting to generalize this result to dominating sets isomorphic to a connected graph of fixed treewidth. Finally, even if DOMINATING SET parameterized by treewidth is FPT in general graphs due to Courcelle's theorem [6], it is not plausible that it has a polynomial kernel in general graphs [4]. It may be the case that the problem admits a polynomial kernel parameterized by treewidth (or by vertex cover) when restricted to circle graphs.

2 $W[1]$-Hardness Results

In this section we prove hardness results for a number of domination problems in circle graphs. In a representation of a circle graph, we will always consider the circle oriented anticlockwise. Given three points a, b, c in the circle, by $a < b < c$ we mean that starting from a and moving anticlockwise along the circle, b comes before c. In a circle representation, we say that two chords with endpoints (a, b) and (c, d) are *parallel twins* if $a < c < d < b$, and there is no other endpoint of a chord between a and c, nor between d and b. Note that for any pair of parallel twins (a, b) and (c, d), we can slide c (resp. d) arbitrarily close to a (resp. b) without modifying the circle representation.

We start with the main result of this section.

Theorem 1. DOMINATING SET *is $W[1]$-hard in circle graphs, when parameterized by the size of the solution.*

Proof: The reduction is from the k-COLORED CLIQUE problem: given a graph $G = (V, E)$ and a coloring of V using k colors, the question is whether there is clique of size k in G containing exactly one vertex from each color. This problem is $W[1]$-hard when parameterized by k [13]. It can be easily seen that we may assume that all color classes are independent sets of the same size. We shall reduce the k-COLORED CLIQUE problem to the problem of finding a dominating set of size at most $k(k + 1)/2$ in circle graphs. Let k be an integer and let G be a k-colored graph on kn vertices such that n vertices are colored with color i for all $1 \leq i \leq k$. For every $1 \leq i \leq k$, we denote by x_j^i the vertices of color i, with $1 \leq j \leq n$. Let us prove that G has a k-colored clique of size k if and only if the following circle graph C has a dominating set of size at most $k(k + 1)/2$. We choose an arbitrary point of the circle as the *origin*. The circle graph C is defined as follows:

- We divide the circle into k disjoint open intervals $]s_i, s_i'[$ for $1 \leq i \leq k$, called *sections*. Each section is divided into $k + 1$ disjoint intervals $]c_{ij}, c_{ij}'[$ for $1 \leq j \leq k + 1$, called *clusters* (see Fig. 2 for an illustration). Each cluster has n particular points denoted by $1, \ldots, n$ following the order of the circle. These intervals are constructed in such a way that the origin is not in a section.

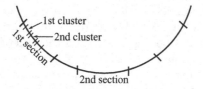

Fig. 2. Sections and clusters in the reduction of Theorem 1

- Sections are numbered from 1 to k following the anticlockwise order from the origin. Similarly, the clusters inside each section are numbered from 1 to $k + 1$.
- For each $1 \leq i \leq k, 1 \leq j \leq k + 1$, we add a chord with endpoints c_{ij} and c'_{ij}, which we call the *extremal chord* of the j-cluster of the i-th section.
- For each $1 \leq i \leq k$ and $1 \leq j \leq k$, we add chords between the j-th and the $(j + 1)$-th clusters of the i-th section as follows. For each $0 \leq l \leq n$, we add two parallel twin chords, each having one endpoint in the interval $]l, l + 1[$ of the j-th cluster, and the other endpoint in the interval $]l, l + 1[$ of the $(j + 1)$-th cluster. These chords are called *inner chords* (see Fig. 3(a) for an illustration). We note that the endpoints of the inner chords inside each interval can be chosen arbitrarily. The interval $]0, 1[$ is the interval between c_{ij} and the point 1, and similarly $]n, n + 1[$ is the interval between the point n and c'_{ij}.
- We also add chords between the first and the last clusters of each section. For each $1 \leq i \leq k$ and $1 \leq l \leq n$, we add a chord joining the point l of the first cluster and the point l of the last cluster of the i-th section. For each $1 \leq i \leq k$, these chords are called the *i-th memory chords*.
- Extremal, inner, and memory chords will ensure some structure on the solution. On the other hand, the following chords will simulate the behavior of the original graph. In fact, the n particular points in each cluster of the i-th section will simulate the behavior of the n vertices of color i in G. Let $i < j$. The chords from the i-th section to the j-th section are between the j-th cluster of the i-th section and the $(i + 1)$-th cluster of the j-th section. Between this pair of clusters, we add a chord joining the point h (in the i-th section) and the point l (in the j-th section) if and only if $x_h^i x_l^j \in E(G)$. We say that such a chord is called *associated* with an edge of the graph G, and such chords are called *outer chords*. In other words, there is an outer chord in C if the corresponding vertices are connected in G.

Intuitively, the idea of the above construction is as follows. For each $1 \leq i \leq k$, among the $k + 1$ clusters in the i-th section, the first and the last one do not contain endpoints of outer chords, and are only used for technical reasons (as discussed below). The remaining $k - 1$ clusters in the i-th section capture the edges of G between vertices of color i and vertices of the remaining $k - 1$ colors. Namely, for any two distinct colors i and j, there is a cluster in the i-th section and a cluster in the j-th section such that the outer chords between these two

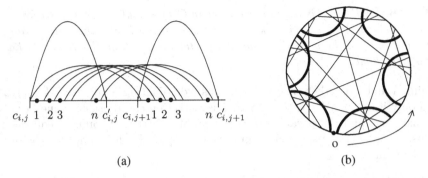

Fig. 3. (a) Representation of the chords between the j-th and the $(j+1)$-th cluster of the i-th section. The higher chords are extremal chords. The others are inner chords and have to be replaced by two parallel twin chords. (b) The general form of a solution. Thick chords are memory chords and the other ones are outer chords. The origin is depicted with a small "o".

clusters correspond to the edges in G between colors i and j. The rest of the proof is structured along a series of claims.

Claim 1. [⋆] *If there exists a k-colored clique in G, then there exists a dominating set of size $k(k+1)/2$ in C.*

In the following we will state some properties about the dominating sets in C of size $k(k+1)/2$.

Claim 2. [⋆] *A dominating set in C has size at least $k(k+1)/2$, and a dominating set of this size has exactly one endpoint in each cluster.*

Claim 3. [⋆] *A dominating set of size $k(k+1)/2$ in C contains no inner nor extremal chord.*

By Claim 3, a dominating set in C of size $k(k+1)/2$ contains only memory and outer chords. Thus, the unique (by Claim 2) endpoint of the dominating set in each cluster is one of the points $\{1,\ldots,n\}$, and we call it the *value* of a cluster. Fig. 3(b) illustrates the general form of a solution.

Claim 4. [⋆] *Assume that C contains a dominating set of size $k(k+1)/2$. Then, in a given section, the value of a cluster does not increase between consecutive clusters.*

Claim 5. [⋆] *Assume that C contains a dominating set of size $k(k+1)/2$. Then, for each $1 \leq i \leq k$, all the clusters of the i-th section have the same value.*

The *value* of a section is the value of the clusters in this section (note that it is well-defined by Claim 5). The *vertex associated with the i-th section* is the vertex x_k^i if the value of the i-th section is k.

Claim 6. [$\star$] *If there is a dominating set in C of size $k(k+1)/2$, then for each pair (i,j) with $1 \le i < j \le k$, the vertex associated with the i-th section is adjacent in G to the vertex associated with the j-th section. Therefore, G has a k-colored clique.*

Claims 1 and 6 together ensure that C has a dominating set of size $k(k+1)/2$ if and only if G has a k-colored clique. The reduction can be easily done in polynomial time, and the parameters of the problems are polynomially equivalent. Thus, DOMINATING SET in circle graphs is $W[1]$-hard. This completes the proof of Theorem 1.   □

Note that in the construction of Theorem 1, if there is a dominating set of size $k(k+1)/2$ in C, it is necessarily connected (see the form of the solution in Fig. 3(b)). Indeed, the memory chords ensure the connectivity between all the chords with one endpoint in a section. Since there is a chord between each pair of sections, the dominating set is connected. Note also that a connected dominating set is also a total dominating set, as it contains no isolated vertices. Therefore, we obtain the following corollary.

Corollary 1. CONNECTED DOMINATING SET *and* TOTAL DOMINATING SET *are $W[1]$-hard in circle graphs, when parameterized by the size of the solution.*

In the following hardness result, we use a completely different reduction from k-COLORED CLIQUE.

Theorem 2. [$\star$] INDEPENDENT DOMINATING SET *is $W[1]$-hard in circle graphs.*

The construction of Theorem 2 can be appropriately modified to deal with the case when the dominating set is required to induce an acyclic subgraph.

Theorem 3. [$\star$] ACYCLIC DOMINATING SET *is $W[1]$-hard in circle graphs.*

3   Tree Dominating Sets

In this section we focus on finding dominating sets in a circle graph which induce graphs isomorphic to trees. Namely, in Theorem 4 we give a polynomial-time algorithm to find a dominating set isomorphic to *some* tree. We prove in Theorem 5 that finding a dominating set isomorphic to a *given* tree is NP-complete. In Theorem 6 we modify the algorithm of Theorem 4 to find a dominating set isomorphic to a *given* tree T in FPT time, the parameter being the size of T. By carefully analyzing its running time, we prove that this FPT algorithm runs in *subexponential* time. It also follows from this analysis that if the given tree T has bounded degree (in particular, if it is a path), then the problem of find a dominating set isomorphic to T can be solved in polynomial time. Note that, in contrast with Theorem 4 below, Theorem 3 in Section 2 states that, if $\mathcal{F}$ is the set of all forests, then $\mathcal{F}$-DOMINATING SET is $W[1]$-hard in circle graphs.

Theorem 4. *Let $\mathcal{T}$ be the set of all trees. Then $\mathcal{T}$-Domination Set can be solved in polynomial time in circle graphs. In other words,* Connected Acyclic Dominating Set *can be solved in polynomial time in circle graphs.*

Proof: Let C be a circle graph on n vertices, and let $\mathcal{C}$ be a circle representation of C. We denote by $\mathcal{P}$ the set of intersections of the circle and the chords in this representation. The elements of $\mathcal{P}$ are called *points*. W.l.o.g., we can assume that only one chord intersects a given point. Given two points $a, b \in \mathcal{P}$, the *interval* $[a, b]$ is the interval from a to b in the anticlockwise order. Given four (non-necessarily distinct) points $a, b, c, d \in \mathcal{P}$, with $a \leq c \leq d \leq b$, by the *region* $ab - cd$ we mean the union of the two intervals $[a, c]$ and $[d, b]$. Note that these two intervals can be obtained by "subtracting" the interval $[c, d]$ from the interval $[a, b]$; this is why we use the notation $ab - cd$.

In the following, by *size* of a set of chords we mean the number of vertices of C in this set. We say that a forest F of C *spans* a region $ab - cd$ if each of a, b, c, and d is an endpoint of some chord in F, and each endpoint of a chord of F is either in $[a, c]$ or in $[d, b]$. A forest F is *split* by a region $ab - cd$ if for each connected component of F there is exactly one chord with one endpoint in $[a, c]$ and one endpoint in $[d, b]$. Given a region $ab - cd$, a forest F is $(ab - cd)$-*dominating* if all the chords of C with both endpoints either in the interval $[a, c]$ or in the interval $[d, b]$ are dominated by F. A forest is *valid* for a region $ab - cd$ if it spans $ab - cd$, is split by $ab - cd$, and is $(ab - cd)$-dominating.

Note that an $(ab - cd)$-dominating forest with several connected components might not dominate some chord going from $[a, c]$ to $[d, b]$. This is not the case if F is connected, as stated in the following claim.

Claim 7. [⋆] *Let T be a valid tree for a region $ab - cd$. Then all the chords of C with both endpoints in $[a, c] \cup [d, b]$ are dominated by T.*

We now state two properties that will be useful in the algorithm. Their correctness is proved below.

T1. Let F_1 and F_2 be two valid forests for two regions $ab - cd$ and $ef - gh$, respectively, such that $a \leq c \leq e \leq g \leq h \leq f \leq d \leq b$. If there is no chord with both endpoints either in $[c, e]$ or in $[f, d]$, then $F_1 \cup F_2$ is valid for $ab - gh$ (see Fig. 4).

T2. Let F_1 and F_2 be two valid forests for two regions $ab - cd$ and $ef - gh$, respectively (F_2 being possibly empty), and let uv be a chord such that $u \leq a \leq c \leq e \leq g \leq v \leq h \leq f \leq d \leq b$, and such that there is no chord with both endpoints either in $[u, a]$, or in $[g, v]$, or in $[v, h]$, or in $[b, u]$. Then $F_1 \cup F_2 \cup \{uv\}$ is a tree which is valid for $df - ce$. When F_2 is empty, we consider that e, f, g, h correspond to the point v. (see Fig. 4).

Roughly speaking, the intuitive idea behind this two properties is to reduce the length of the circle in which we still have to do some computation (that is, outside the valid regions). Again, the proof is structured along a series of claims. Before verifying the correctness of Properties **T1** and **T2**, let us first state a useful general fact.

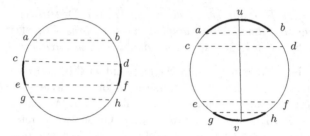

Fig. 4. On the left (resp. right), regions corresponding to Property **T1** (resp. Property **T2**). Full lines correspond to real chords of C, dashed lines correspond to the limit of regions. Bold intervals correspond to intervals with no chord of C with both endpoints in the interval.

Claim 8. [⋆] *Let $ab - cd$ be a region and let F be a valid forest for $ab - cd$. The chords with one endpoint in $[c, d]$ and one endpoint in $[d, c]$ are dominated by F.*

Claim 9. [⋆] *Properties* **T1** *and* **T2** *are correct.*

For a region $ab - cd$, we denote by $v_{ab,cd}^{f}$ (resp. $v_{ab,cd}^{t}$) the least integer l for which there is a valid forest (resp. tree) of size l for $ab - cd$. If there is no valid forest (resp. tree) for $ab - cd$, we set $v_{ab,cd}^{f} = +\infty$ (resp. $v_{ab,cd}^{t} = +\infty$). Let us now describe our algorithm based on dynamic programming. With each region $ab - cd$, we associate two integers $v_{ab,cd}^{1}$ and $v_{ab,cd}^{2}$. Algorithm 1 below calculates these two values for each region. We next show that $v_{ab,cd}^{1} = v_{ab,cd}^{f}$ and $v_{ab,cd}^{2} = v_{ab,cd}^{t}$, and that Algorithm 1 correctly computes the result in polynomial time.

Algorithm 1. Dynamic programming for computing a dominating tree

---

for each region $ab - cd$ **do** $v_{ab,cd}^{1} \leftarrow \infty$; $v_{ab,cd}^{2} \leftarrow \infty$
for each chord ab of the circle graph **do** $v_{ab,ab}^{1} \leftarrow 1$; $v_{ab,ab}^{2} \leftarrow 1$
for $j = 2$ to n **do**
    if there are two regions $ab - cd$ and $ef - gh$ such that $v_{ab,cd}^{1} = j_1$ and $v_{ef,gh}^{1} = j_2$
    with $j_1 + j_2 = j$ satisfying Property **T1**, with $v_{ab,gh}^{1} = +\infty$ **then**
        $v_{ab,gh}^{1} \leftarrow j$
    if there is a region $ab - cd$ and a chord uv such that $v_{ab,cd}^{1} = j - 1$ satisfying
    Property **T2** with an empty second forest **then**
        if $v_{dv,cv}^{1} = +\infty$ **then**
            $v_{dv,cv}^{1} \leftarrow j$
        if $v_{dv,cv}^{2} = +\infty$ **then**
            $v_{dv,cv}^{2} \leftarrow j$
    if there are two regions $ab - cd$ and $ef - gh$ and a chord uv such that $v_{ab,cd}^{1} = j_1$
    and $v_{ef,gh}^{1} = j_2$ with $j_1 + j_2 = j - 1$ satisfying Property **T2 then**
        if $v_{df,ce}^{1} = +\infty$ **then**
            $v_{df,ce}^{1} \leftarrow j$
        if $v_{df,ce}^{2} = +\infty$ **then**
            $v_{df,ce}^{2} \leftarrow j$

Claim 10. [⋆] *For any region* $ab - cd$, $v^1_{ab,cd} = v^f_{ab,cd}$ *and* $v^2_{ab,cd} = v^t_{ab,cd}$.

Let us now explain how we can verify if there is a dominating set in C isomorphic to some tree of a given size.

Claim 11. [⋆] *Let* k *be a positive integer. There is a dominating tree of size at most* k *in* C *if and only if there is a region* $ab - cd$ *such that* $v^t_{ab,cd} \le k$ *and such that there is no chord strictly contained in* $[b, a]$ *nor in* $[c, d]$.

By Claims 10 and 11, it follows that Algorithm 1 computes the regions for which there is a valid tree of any size from 1 to n. Given a region $ab - cd$ with $v^t_{ab,cd} \le k$, we just have to verify that there are no chords in the intervals $[b, a]$ and $[c, d]$, which can clearly be done in polynomial time. One can easily check that Algorithm 1 runs in time $\mathcal{O}(n^{10})$, but we did not make any attempt to improve its time complexity. □

It turns out that when we seek a dominating set isomorphic to a *given* tree T, the problem is NP-complete. The reduction is from the 3-PARTITION problem, which consists in deciding whether a given multiset of integers can be partitioned into triples such that the three integers in each triple have the same sum.

Theorem 5. [⋆] *Let* T *be a given tree. Then* {T}-DOMINATING SET *is NP-complete in circle graphs when* T *is part of the input.*

Finally, we show that {T}-DOMINATING SET in circle graphs can be solved by a subexponential FPT algorithm, when parameterized by $|V(T)|$.

The algorithm of Theorem 6 below goes along the same lines of Algorithm 1 given in the proof of Theorem 4. The main difference is that in the proof of Theorem 4, when Properties **T1** or **T2** are satisfied, we can directly apply them and still obtain a forest or a tree. But when looking for a given tree T, when we make the union of two forests, we have to make sure that the union of these two forests is still a subforest of T, and that we can correctly complete it to obtain the desired tree T. For obtaining that, we will apply two new properties corresponding to Properties **T1** or **T2**, whenever it is possible to create forests which are induced by the children of the same vertex of T.

Let us give some more intuition on the algorithm. We consider the tree T rooted at an arbitrary vertex r. Let v be a vertex of T, and let $w_1, \ldots, w_l$ be the children of v. We define $T(v)$ as the forest $T[w_1] \cup T[w_2] \ldots \cup T[w_l]$, where $T[w_i]$ is the subtree of the rooted tree T induced by w_i and the descendants of w_i. Roughly speaking, the idea of the algorithm is to exhaustively seek a dominating set isomorphic to any possible subforest of $F(v)$ for every vertex v in T, and then try to grow it until hopefully obtaining the target tree T. Note that if a vertex v of T has k children, there are a priori 2^k possible subsets of children of y, which define 2^k possible types of subforests in $F(v)$. But the key point in order to obtain a subexponential algorithm is that if some of the trees in $F(v)$ are isomorphic, some of the choices of subsets of subforests will give rise to the same tree. In order to avoid this redundancy, for each vertex v of T, we partition the trees in $F(v)$ into isomorphism classes, and then the choices within each isomorphism class reduce to choosing the multiplicity of this tree. Note that carrying out this partition into isomorphism classes can be done in polynomial time (in the size of T) for each vertex of T, using the fact that one can test whether two rooted trees T_1 and T_2 with t vertices are isomorphic in $\mathcal{O}(t)$ time [1]. The details can be found in [5].

From our analysis, it also follows that if T has bounded degree (in particular, if it is a path), then $\{T\}$-Dominating Set can be solved in polynomial time in circle graphs.

Theorem 6. [⋆] *Let T be a given tree. There exists an* FPT *algorithm to solve* $\{T\}$-Dominating Set *in a circle graph on n vertices, when parameterized by* $t = |V(T)|$, *running in time* $2^{\mathcal{O}\left(t \cdot \frac{\log \log t}{\log t}\right)} \cdot n^{\mathcal{O}(1)} = 2^{o(t)} \cdot n^{\mathcal{O}(1)}$. *Furthermore, if T has bounded degree, then $\{T\}$-Dominating Set can be solved in polynomial time in circle graphs.*

Acknowledgment. We would like to thank Sylvain Guillemot for stimulating discussions that motivated some of the research carried out in this paper.

References

1. Aho, A.V., Hopcroft, J.E., Ullman, J.D.: The Design and Analysis of Computer Algorithms. Addison-Wesley (1974)
2. Alber, J., Bodlaender, H.L., Fernau, H., Kloks, T., Niedermeier, R.: Fixed Parameter Algorithms for Dominated Set and Related Problems on Planar Graphs. Algorithmica 33(4), 461–493 (2002)
3. Alon, N., Gutner, S.: Kernels for the Dominating Set Problem on Graphs with an Excluded Minor. Electronic Colloquium on Computational Complexity (ECCC) 15(066) (2008)
4. Bodlaender, H.L., Downey, R.G., Fellows, M.R., Hermelin, D.: On problems without polynomial kernels. Journal of Computer and System Sciences 75(8), 423–434 (2009)
5. Bousquet, N., Gonçalves, D., Mertzios, G.B., Paul, C., Sau, I., Thomassé, S.: Parameterized Domination in Circle Graphs. Manuscript available at http://arxiv.org/abs/1205.3728 (2012)
6. Courcelle, B.: The Monadic Second-Order Logic of Graphs: Definable Sets of Finite Graphs. In: van Leeuwen, J. (ed.) WG 1988. LNCS, vol. 344, pp. 30–53. Springer, Heidelberg (1989)
7. Cygan, M., Philip, G., Pilipczuk, M., Pilipczuk, M., Wojtaszczyk, J.O.: Dominating set is fixed parameter tractable in claw-free graphs. Theoretical Computer Science 412(50), 6982–7000 (2011)
8. Damaschke, P.: The Hamiltonian Circuit Problem for Circle Graphs is NP-Complete. Information Processing Letters 32(1), 1–2 (1989)
9. Damian-Iordache, M., Pemmaraju, S.V.: Hardness of Approximating Independent Domination in Circle Graphs. In: Aggarwal, A.K., Pandu Rangan, C. (eds.) ISAAC 1999. LNCS, vol. 1741, pp. 56–69. Springer, Heidelberg (1999)
10. Downey, R.G., Fellows, M.R.: Parameterized Complexity. Springer, New York (1999)

11. Elmallah, E.S., Stewart, L.K.: Independence and domination in polygon graphs. Discrete Applied Mathematics 44(1-3), 65–77 (1993)
12. Even, S., Itai, A.: Queues, stacks and graphs. In: Press, A. (ed.) Theory of Machines and Computations, pp. 71–86 (1971)
13. Fellows, M.R., Hermelin, D., Rosamond, F.A., Vialette, S.: On the parameterized complexity of multiple-interval graph problems. Theoretical Computer Science 410(1), 53–61 (2009)
14. Flum, J., Grohe, M.: Parameterized Complexity Theory. Springer (2006)
15. Fomin, F., Gaspers, S., Golovach, P., Suchan, K., Szeider, S., Jan Van Leeuwen, E., Vatshelle, M., Villanger, Y.: k-Gap Interval Graphs. In: Fernández-Baca, D. (ed.) LATIN 2012. LNCS, vol. 7256, pp. 350–361. Springer, Heidelberg (2012), http://arxiv.org/abs/1112.3244
16. Garey, M., Johnson, D.: Computers and Intractability. W.H. Freeman, San Francisco (1979)
17. Gavril, F.: Algorithms for a maximum clique and a maximum independent set of a circle graph. Networks 3, 261–273 (1973)
18. Gavril, F.: Minimum weight feedback vertex sets in circle graphs. Information Processing Letters 107(1), 1–6 (2008)
19. Gioan, E., Paul, C., Tedder, M., Corneil, D.: Circle Graph Recognition in Time $O(n + m) \cdot \alpha(n + m)$. Manuscript available at http://arxiv.org/abs/1104.3284 (2011)
20. Hedetniemi, S.M., Hedetniemi, S.T., Rall, D.F.: Acyclic domination. Discrete Mathematics 222(1-3), 151–165 (2000)
21. Jiang, M., Zhang, Y.: Parameterized Complexity in Multiple-Interval Graphs: Domination. In: Marx, D., Rossmanith, P. (eds.) IPEC 2011. LNCS, vol. 7112, pp. 27–40. Springer, Heidelberg (2012)
22. Keil, J.M.: The complexity of domination problems in circle graphs. Discrete Applied Mathematics 42(1), 51–63 (1993)
23. Keil, J.M., Stewart, L.: Approximating the minimum clique cover and other hard problems in subtree filament graphs. Discrete Applied Mathematics 154(14), 1983–1995 (2006)
24. Kloks, T.: Treewidth of circle graphs. International Journal of Foundations of Computer Science 7(2), 111–120 (1996)
25. Marx, D.: Parameterized Complexity of Independence and Domination on Geometric Graphs. In: Bodlaender, H.L., Langston, M.A. (eds.) IWPEC 2006. LNCS, vol. 4169, pp. 154–165. Springer, Heidelberg (2006)
26. Niedermeier, R.: Invitation to Fixed-Parameter Algorithms. Oxford University Press (2006)
27. Sherwani, N.A.: Algorithms for VLSI Physical Design Automation. Kluwer Academic Press (1992)
28. Spinrad, J.: Recognition of circle graphs. Journal of Algorithms 16(2), 264–282 (1994)
29. Unger, W.: On the k-Colouring of Circle-Graphs. In: Cori, R., Wirsing, M. (eds.) STACS 1988. LNCS, vol. 294, pp. 61–72. Springer, Heidelberg (1988)
30. Unger, W.: The Complexity of Colouring Circle Graphs (Extended Abstract). In: Finkel, A., Jantzen, M. (eds.) STACS 1992. LNCS, vol. 577, pp. 389–400. Springer, Heidelberg (1992)
31. Xu, G., Kang, L., Shan, E.: Acyclic domination on bipartite permutation graphs. Information Processing Letters 99(4), 139–144 (2006)

How to Eliminate a Graph[*]

Petr A. Golovach[1], Pinar Heggernes[2], Pim van 't Hof[2], Fredrik Manne[2],
Daniël Paulusma[1], and Michał Pilipczuk[2]

[1] School of Engineering and Computing Sciences, Durham University, UK
{petr.golovach,daniel.paulusma}@durham.ac.uk
[2] Department of Informatics, University of Bergen, Norway
{pinar.heggernes,pim.vanthof,fredrik.manne,michal.pilipczuk}@ii.uib.no

Abstract. Vertex elimination is a graph operation that turns the neighborhood of a vertex into a clique and removes the vertex itself. It has widely known applications within sparse matrix computations. We define the ELIMINATION problem as follows: given two graphs G and H, decide whether H can be obtained from G by $|V(G)| - |V(H)|$ vertex eliminations. We study the parameterized complexity of the ELIMINATION problem. We show that ELIMINATION is $W[1]$-hard when parameterized by $|V(H)|$, even if both input graphs are split graphs, and $W[2]$-hard when parameterized by $|V(G)| - |V(H)|$, even if H is a complete graph. On the positive side, we show that ELIMINATION admits a kernel with at most $5|V(H)|$ vertices in the case when G is connected and H is a complete graph, which is in sharp contrast to the $W[1]$-hardness of the related CLIQUE problem. We also study the case when either G or H is tree. The computational complexity of the problem depends on which graph is assumed to be a tree: we show that ELIMINATION can be solved in polynomial time when H is a tree, whereas it remains NP-complete when G is a tree.

1 Introduction

Consider the problem of choosing a set S of resilient communication hubs in a network, such that if any subset of the hubs should stop functioning then all the remaining hubs in S can still communicate. Such a set is attractive if the probability of a hub failure is high, or if the network is dynamic and hubs can leave the network. We can formulate this as a graph problem in the following way. Given a graph G and an integer k, is there a set S of k vertices, such that if any subset of S is removed from G, then every pair of remaining vertices in S are still connected via paths in the modified graph. Obviously, choosing S to be a clique of size k would solve the problem, but only allowing for cliques is overly restrictive. A necessary and sufficient condition on S is that for each pair $u, v \in S$, either u and v are adjacent or there is a path between u and v in G not containing any vertex of S except u and v. Thus we can view the described problem as a relaxation of the well-known CLIQUE problem.

[*] This work is supported by EPSRC (EP/G043434/1) and Royal Society (JP100692), and by the Research Council of Norway (197548/F20).

M.C. Golumbic et al. (Eds.): WG 2012, LNCS 7551, pp. 320–331, 2012.
© Springer-Verlag Berlin Heidelberg 2012

The above problem can be stated in terms of a well-known graph operation related to Gaussian elimination: vertex elimination [14]. The *elimination* of a vertex v from a graph G is the operation that adds edges to G such that the neighbors of v form a clique, and then removes v from the resulting graph. With this operation, the above problem can be defined as follows: find a set S of size k such that eliminating all vertices of $V(G) \setminus S$ leaves S as a clique. In fact, we state a more general problem: the ELIMINATION problem takes as input two graphs G and H, and asks whether a graph isomorphic to H can be obtained by the elimination of $|V(G)| - |V(H)|$ vertices from G. If this is possible, then we say that H is an *elimination* of G.

The vertex elimination operation described above has long known applications within linear algebra, and it simulates in graphs the elimination of a variable from subsequent rows during Gaussian elimination of symmetric matrices [14]. The resulting *Elimination Game* [14] repeatedly chooses a vertex and eliminates it from the graph until the graph becomes empty. The amount of edges added during the process, called the *fill-in*, is crucial for sparse matrix computations, and a vast amount of results have appeared on this subject during the last 40 years; see e.g., [6, 7, 14, 17]. Our problem ELIMINATION is equivalent to stopping Elimination Game after $|V(G)| - |V(H)|$ steps to see whether the resulting graph at that point is isomorphic to H. A crucial aspect of Elimination Game is the order in which the vertices are chosen, as this influences the fill-in. Note however that, for our problem, only the set of $|V(G)| - |V(H)|$ vertices chosen to be eliminated is important, and not the order in which they are eliminated.

Graph modification problems resulting from operations like vertex deletion, edge deletion, edge contraction, and local complementation are well studied, especially within fixed-parameter tractability; see e.g., [1, 3, 5, 8, 9, 11–13, 15, 19]. Given the wide use of the vertex elimination operation, we find it surprising that the ELIMINATION problem does not seem to have been studied before. The only related study we are aware of is by Samdal [18], who generated all eliminations of the $n \times n$ grids for $n \leq 7$.

Our Contribution. In this paper we study the computational complexity of ELIMINATION. In particular, we show that ELIMINATION is $W[1]$-hard when parameterized by $|V(H)|$ even when both input graphs are split graphs, and $W[2]$-hard when parameterized by $|V(G)| - |V(H)|$ even when H is a complete graph. On the positive side, for the case when H is complete, we show that ELIMINATION is fixed-parameter tractable when parameterized by $|V(H)|$, and has a kernel with at most $5|V(H)|$ vertices on connected graphs, which contrasts the hardness of the CLIQUE problem. We also study the cases when one of the input graphs is a tree. It turns out that the complexity of the problem changes completely depending on which input graph is a tree; we show that if G is a tree then the problem remains NP-complete, whereas if H is a tree then it can be solved in polynomial time. The mentioned kernel result is obtained by proving a combinatorial theorem on the maximum number of leaves in a spanning tree of a graph, similar to a proof by Kleitman and West [10]. We find this a contribution of independent interest.

Notation. All graphs in this paper are undirected, finite, and simple. Let $G = (V, E)$ be a graph. We sometimes use $V(G)$ and $E(G)$ to denote V and E, respectively. The *neighborhood* of a vertex $v \in V$ is the set of its neighbors $N_G(v) = \{w \in V \mid vw \in E\}$, and the *closed neighborhood* of v is the set $N_G[v] = N_G(v) \cup \{v\}$. The *degree* of v is $d_G(v) = |N_G(v)|$. For any subset $A \subseteq V$, we define $N_G[A] = \bigcup_{a \in A} N_G[a]$, $N_G(A) = N_G[A] \setminus A$, and $d_G(A) = |N_G(A)|$. For any subset $A \subseteq V$, $G[A]$ denotes the subgraph of G induced by A. For a subgraph H of G, we write $G \setminus H$ to denote the graph obtained from G by deleting all the vertices of H from G, i.e., $G \setminus H = G[V(G) \setminus V(H)]$.

A *clique* is a set of vertices that are all pairwise adjacent. A vertex v is *simplicial* if $N_G(v)$ is a clique. A graph G is *complete* if $V(G)$ is a clique. The complete graph on k vertices is denoted by K_k. An *independent* set is a set of vertices that are pairwise non-adjacent. If G is a bipartite graph, where (A, B) is a partition of V into two independent sets, then we denote it as $G = (A, B, E)$ and we call (A, B) a *bipartition* of G. A graph is a *split graph* it its vertex set can be partitioned into a clique and an independent set. A vertex is a *cut-vertex* if the removal of the vertex leaves the graph with more connected components than before.

A parameterized problem Q belongs to the class XP if each instance (I, k) can be solved in $f(k)|I|^{g(k)}$ time for some functions f and g that depend only on the *parameter k*, and $|I|$ denotes the size of I. If a problem belongs to XP, then it can be solved in polynomial time for every fixed k. If a parameterized problem can be solved by an algorithm with running time $f(k)|I|^{O(1)}$, then we say the problem is *fixed-parameter tractable*. The class of all fixed-parameter tractable problems is denoted FPT. Between FPT and XP is a hierarchy of parameterized complexity classes, FPT $\subseteq$ W[1] $\subseteq$ W[2] $\subseteq \cdots \subseteq$ W[P] $\subseteq$ XP, where hardness for one of the W-classes is considered to be strong evidence of intractability with respect to the class FPT. A parameterized problem is said to admit a *kernel* if there is a polynomial-time algorithm that transforms each instance of the problem into an *equivalent* instance whose size and parameter value are bounded from above by $g(k)$ for some (possibly exponential) function g. We refer to the textbook by Downey and Fellows [5] for formal background on parameterized complexity.

In this extended abstract, proofs of some theorems and lemmas, which are marked with the symbol ♠, have been omitted due to page restrictions.

2 Preliminaries and Hardness of ELIMINATION

We start this section with an observation that provides a characterization of graphs that have some fixed graph H as an elimination. Our proofs heavily rely on this observation.

Observation 1 ([17]). *Let G and H be two graphs, where $V(H) = \{u_1, \ldots, u_h\}$. Then H is an elimination of G if and only if there exists a set $S = \{v_1, \ldots, v_h\}$ of h vertices in G that satisfies the following: $u_i u_j \in E(H)$ if and only if $v_i v_j \in E(G)$ or there is a path in G between v_i and v_j whose internal vertices are all in $V(G) \setminus S$, for $1 \leq i < j \leq h$.*

For two input graphs G and H that form an instance of ELIMINATION, we let n denote the number of vertices in G. If G and H form a yes-instance, we say that a subset $X \subseteq V(G)$ is a *solution* if H is the resulting graph when all vertices in X are eliminated. By Observation 1, the vertices in X can be eliminated in any order. A vertex which is not eliminated is said to be *saved*. The set $S = V(G) \setminus X$ of saved vertices is called a *witness*.

Since we can check in polynomial time whether a set $S \subseteq V(G)$ of $|V(H)|$ vertices is a witness, Observation 1 immediately implies the following result.

Corollary 1. ELIMINATION *is in* XP *when parameterized by* $|V(H)|$.

Corollary 1 naturally raises the question whether ELIMINATION is FPT when parameterized by $|V(H)|$. The following theorem shows that this is highly unlikely.

Theorem 1 (♠). ELIMINATION *is* $W[1]$-*hard when parameterized by* $|V(H)|$, *even if both* G *and* H *are split graphs.*

Since ELIMINATION is unlikely to be FPT in general as a result of Theorem 1, it is natural to ask whether certain restrictions on G or H make the problem tractable. In Section 3, we restrict H to be a complete graph; note that due to Theorem 1, restricting H to be a split graph does not suffice to guarantee tractability. In Section 4, we study the variant where either G or H is a tree.

Another possible way of achieving tractability is to investigate a different parameterization of the problem. For instance, instead of choosing the size of the witness as the parameter, we can parameterize ELIMINATION by the size of the solution, i.e., the number of eliminated vertices. The next theorem shows that the problem remains intractable with this parameter.

Theorem 2 (♠). ELIMINATION *is* $W[2]$-*hard when parameterized by* $|V(G)| - |V(H)|$, *even if* H *is a complete graph.*

We point out that the reductions used in the proofs of Theorems 1 and 2 immediately imply that the unparameterized version of ELIMINATION is NP-complete, even if both G and H are split graphs, or if H is a complete graph.

3   Eliminating to a Complete Graph

In this section, we consider a special case of the ELIMINATION problem when H is a complete graph. This corresponds exactly to the problem described in the first paragraph of Section 1. We define the problem CLIQUE ELIMINATION, which takes as input a graph G on n vertices and an integer k, and asks whether the complete graph K_k is an elimination of G. Since CLIQUE ELIMINATION is $W[2]$-hard when parameterized by $|V(G)| - k$ due to Theorem 2, we choose k as the parameter throughout this section.

If G contains a tree T with k leaves as a subgraph, then K_k is an elimination of G, as the leaves of T can serve as a witness. It is easy to observe that G contains a tree with k leaves as a subgraph if and only if G contains $K_{1,k}$, i.e., a star with k leaves, as a minor. Moreover, by Observation 1, for any fixed graph H, the property that H is an elimination of a graph G can be expressed in monadic second-order logic. Since graphs that exclude $K_{1,k}$ as a minor have bounded treewidth [16], Courcelle's Theorem [4] implies that CLIQUE ELIMINATION is FPT when parameterized by k.

Even though fixed-parameter tractability of CLIQUE ELIMINATION is already established, two interesting questions remain. Does the problem admit a polynomial kernel? Does there exist an algorithm for the problem with single-exponential dependence on k? We provide an affirmative answer to both questions below. In particular, we prove the following result.

Theorem 3. CLIQUE ELIMINATION *admits a kernel with at most $5k$ vertices for connected graphs.*

We would like to remark that the assumption that the input graph is connected is probably necessary, as CLIQUE ELIMINATION in general graphs admits a simple composition algorithm that takes the disjoint union of instances, so existence of a polynomial kernel in the general setting would imply that NP $\subseteq$ coNP/poly. We refer an interested reader to the work of Bodlaender et al. [2] for an introduction to the methods of proving implausibility of polynomial kernelization algorithms.

As a result of Theorem 3, an algorithm with single-exponential dependence on k can be obtained by kernelizing every connected component of the input graph separately, and then running a brute-force search on each kernel. This gives us a better running time than the aforementioned combination of meta-theorems.

Corollary 2. CLIQUE ELIMINATION *can be solved in* $\binom{5k}{k} n^{O(1)} \leq 12.21^k n^{O(1)}$ *time and polynomial space.*

The remainder of this section is devoted to the proof of Theorem 3. Before presenting the formal proof, we give some intuition behind our approach. Our kernelization algorithm is based on the observation that the max-leaf number of a graph, i.e., the maximum number of leaves a spanning tree of the graph can have, is a lower bound on the size of a complete graph that can be obtained as an elimination. Kleitman and West [10] showed that a connected graph G with minimum degree at least 3 admits a spanning tree with at least $|V(G)|/4 + 2$ leaves. Their result immediately leads to a linear kernel for CLIQUE ELIMINATION provided that the input graph G has minimum degree at least 3. Unfortunately, we are unable to get rid of all vertices of degree at most 2 in our setting. However, we can modify our input graph in polynomial time such that we either can solve the problem directly, or obtain a new graph G^* with no vertices of degree 1 and with no edge between any two vertices of degree 2. We then prove a modified version of the aforementioned result by Kleitman and West [10], namely that such graphs G^* admit a spanning tree with at least $|V(G^*)|/5 + 2$ leaves. This leads to Theorem 3.

We now proceed with the formal proof of Theorem 3. Following Observation 1, we will be looking for a set S that is a witness of cardinality k, i.e., every two non-adjacent vertices of S can be connected by a path all internal vertices of which are outside S.

We start by providing four reduction rules, i.e., polynomial-time algorithms that, given an instance (G, k) of CLIQUE ELIMINATION, output an equivalent instance (G', k'). Each time, we apply the rule with the smallest number among the applicable ones. We argue that if none of the rules is applicable, then the modified graph has no vertices of degree 1 and no edge between any two vertices of degree 2. Recognizing whether a rule can be applied, as well as the application of the rule itself, will trivially be polynomial-time operations. The total number of applications will be bounded by a polynomial in the input size.

Reduction Rule 1. *If $k \leq 3$ or $n \leq 3$, then resolve the instance in polynomial time via a brute-force algorithm, and output a trivial* **yes**- *or* **no**-*instance, depending on the result.*

The safeness of the above rule is obvious.

Reduction Rule 2. *If G contains a vertex v of degree 1, eliminate its sole neighbor v' to obtain a graph G'. Output the instance (G', k).*

Lemma 1. *Reduction Rule 2 is safe.*

Proof. We need to argue that if we can find a witness S, then we can also find a witness S' of the same size that does not contain v'. If $v' \notin S$, then we set $S' = S$. If $v' \in S$, then $v \notin S$. Otherwise, since v is adjacent only to v', $k \leq 2$ and we could have applied Reduction Rule 1. We now set $S' = (S \setminus \{v'\}) \cup \{v\}$ to obtain a witness set of the same cardinality that does not contain v'. $\square$

Reduction Rule 3. *If G contains a triangle v', v_1, v_2 such that v_1, v_2 are of degree 2, then eliminate v' to obtain a graph G'. Output the instance (G', k).*

Lemma 2. *Reduction Rule 3 is safe.*

Proof. Again, we need to argue that if we can find a witness S, then we can also find a witness S' of the same size that does not contain v'. If $v' \notin S$, then we set $S' = S$. Suppose that $v' \in S$. As Reduction Rule 1 was not applicable, we find that $k > 3$. Then neither v_1 nor v_2 belongs to S. We set $S' = (S \setminus \{v'\}) \cup \{v_1\}$ to obtain a witness set of the same cardinality that does not contain v'. $\square$

Reduction Rule 4. *If G has a path v_0, v_1, v_2, v_3 such that v_1, v_2 are of degree 2 and $v_0 \neq v_3$, then eliminate v_0 to obtain a graph G'. Output the instance (G', k).*

Lemma 3. *Reduction Rule 4 is safe.*

Proof. We need to argue that if we can find a witness S, then we can also find a witness S' of the same size that does not contain v_0. If $v_0 \notin S$, then we set $S' = S$. Suppose that $v_0 \in S$. As Reduction Rule 1 was not applicable, we find that $k > 3$. Hence, S contains at most one vertex from the set $\{v_1, v_2, v_3\}$, as otherwise one of them could be connected to at most two other vertices from S via paths avoiding other vertices from S. If $|S \cap \{v_1, v_2, v_3\}| = 0$, then we take $S' = (S \setminus \{v_0\}) \cup \{v_1\}$, while if $|S \cap \{v_1, v_2, v_3\}| = 1$, then we take $S' = (S \setminus \{v_0, v_1, v_2, v_3\}) \cup \{v_1, v_2\}$. It is easy to check that S' defined in this manner is a witness of the same cardinality that does not contain v_0. $\square$

If, after applying our four reduction rules exhaustively, we have not yet solved the problem, then we have obtained a graph G^* with no vertices of degree 1 and no edge between any two vertices of degree 2. If G^* has at most $5k - 11$ vertices, then we output the instance as the obtained kernel. Otherwise, i.e., if G^* has at least $5k - 10$ vertices, then we can safely return a trivial **yes**-instance due to the next result, which is our modified version of the aforementioned result by Kleitman and West [10]. This concludes the proof of Theorem 3.

Theorem 4. *Let G be a connected graph with minimum degree at least 2 such that no two vertices of degree 2 are adjacent. Then G admits a spanning tree with at least $|V(G)|/5 + 2$ leaves.*

Proof. We gradually grow a tree T in G keeping track of three parameters:

- n, the number of vertices in T;
- l, the number of leaves in T;
- m, the number of *dead* leaves in T, i.e., leaves that have no neighbor in $G \setminus T$.

The tree will be grown via a number of operations called *expansions*: by an expansion of a vertex $x \in V(T)$ we mean the adding of all the vertices $v \in V(G) \setminus V(T)$ with $xv \in E(G)$ and all the edges $xv \in E(G)$ with $v \notin V(T)$ to the tree T. We start with a tree T such that only leaves of T have neighbors in $G \setminus T$. Therefore, if we only use expansions to grow the tree, at each step of the growth process only the leaves of T are adjacent to $G \setminus T$. A leaf that is not dead, is called *alive*.

For a tree T, let us consider the potential $\phi(T)$ defined as $\phi(T) = 4l + m - n$. The goal is to

(a) find a starting tree T with $\phi(T) \geq 9$;
(b) provide a set of growing rules, such that there is always a rule applicable unless T is a spanning tree, and $\phi(T)$ does not decrease during the application of any rule;
(c) prove that during the whole process the potential increases by at least 1.

If goals (a), (b) and (c) are accomplished, then we can grow T using the rules until it becomes a spanning tree; in this situation we have $l = m$ and $n = |V(G)|$. As the potential increased by at least 1 during the whole process, we infer that $5l \geq |V(G)| + 10$, and hence $l \geq |V(G)|/5 + 2$, as claimed.

Goal (a) can be achieved by a careful case study; we omit the details due to page restrictions.

Having chosen the starting tree T, we can proceed with the growing rules. In order to grow the tree we always choose the rule that has the lowest number among the applicable ones, i.e., when applying a rule, we can always assume that the ones with lower numbers are not applicable. We would like to point out that the first three rules were already used in the original proof of Kleitman and West.

Growing Rule 1. If some leaf of T has at least two neighbors from $G \setminus T$, expand it. The potential $\phi(T)$ increases by at least $4 \cdot (d - 1) - d = 3d - 4 \geq 2$, where $d \geq 2$ is the number of the aforementioned neighbors from $G \setminus T$.

Growing Rule 2. If some vertex $v \in V(G \setminus T)$ is adjacent to at least two leaves of T, expand one of these leaves. Observe that, as Rule 1 was not applicable and only leaves of T are adjacent to $G \setminus T$, this expansion results in adding only v to T. Moreover, all the remaining leaves adjacent to v were alive but become dead, so the potential $\phi(T)$ increases by at least $1 - 1 = 0$.

Growing Rule 3. If there is a vertex $v \in V(G \setminus T)$ of degree at least 3 in G that is adjacent to a leaf of T, expand this leaf (which results in adding only v to T, as Rule 1 was not applicable) and then v. The potential increases by at least $4 \cdot (d - 2) - d = 3d - 8 \geq 1$, where $d \geq 3$ is the degree of v, as all the other neighbors of v are added to T as leaves, due to Rule 2 not being applicable.

Growing Rule 4. If there is a vertex $v \in V(G \setminus T)$ of degree 2 in G that is adjacent to a leaf of T, expand this leaf (which results only in adding v as a leaf, as Rule 1 was not applicable), then expand v, and then expand the second neighbor v' of v that became a leaf in T during the previous expansion. Note that v' could not be already in T, as otherwise Rule 2 would be triggered on vertex v. Since we assumed that no vertices of degree 2 are adjacent in G, the degree of v' is at least 3 and, as Rule 3 was not applicable, none of the neighbors of v' was in T. Denote by d the degree of v'; therefore, we have added to the tree T exactly $d + 1$ vertices (v, v' and $d - 1$ other neighbors of v') and increased the number of leaves by exactly $d - 2$. Hence, the increase of the potential is $4(d - 2) - (d + 1) = 3d - 9 \geq 0$, as $d \geq 3$.

It remains to argue that goal (c) is achieved. It is clear that if Growing Rule 1 or 3 is applied at least once, then the potential increases by at least 1. Suppose only Growing Rules 2 and 4 are applied during the whole process. Let x be a vertex of G that was added to T as a leaf during the very last rule application. Then x is a dead leaf. Since this was not taken into account when we determined a lower bound of 0 on the increase of the potential, the potential increases by

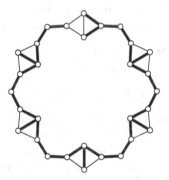

Fig. 1. A graph on 30 vertices for which the maximum possible number of leaves in a spanning tree is exactly 8; the bold (blue) edges indicate a spanning tree with 8 leaves. This example shows that the bound in Theorem 4 is tight.

at least 1. Thus, from the previously described analysis we conclude that using the presented method we are able to grow a tree with at least $|V(G)|/5 + 2$ leaves.   □

The bound $|V(G)|/5 + 2$ is best possible. A family of examples with tight inequality can be obtained by connecting a number of diamonds in the way as shown in Figure 1.

4   ELIMINATION on Trees

In this section, we study ELIMINATION when G or H is a tree. When H is a tree, we show that the problem can be solved in polynomial time. Then we show that when G is a tree, the problem is NP-complete.

For a tree H with at least two vertices, we denote by $L(H)$ the set of leaves of H. The remaining set of vertices is denoted by $I(H) = V(H) \setminus L(H)$ and called the *inner vertices*. For a graph G, by $C(G)$ we denote the set of cut-vertices of G. A connected graph is 2-*connected* if it does not contain a cut-vertex. A maximal 2-connected subgraph of G is called a *biconnected component* (*bicomp* for short), and we denote by $\mathcal{B}(G)$ the set of bicomps of G. Consider the bipartite graph $\mathcal{T}_G$ with the vertex set $C(G) \cup \mathcal{B}(G)$, where $(C(G), \mathcal{B}(G))$ is the bipartition, such that $c \in C(G)$ and $B \in \mathcal{B}(G)$ are adjacent if and only if $c \in V(B)$. This graph $\mathcal{T}_G$ is a tree if G is connected, and is called the *bicomp-tree* of G.

Let G and H be an instance of ELIMINATION where H is a tree. Since a graph G can be eliminated to a connected graph H if and only if at least one connected component of G can be eliminated to H, we assume without loss of generality that G is connected. Also it is easy to see that any graph G with at least one vertex can be eliminated to K_1, and K_2 is an elimination of a graph G whenever G has at least one edge. Hence, we can assume that H has at least three vertices. Therefore, $L(H) \neq \emptyset$ and $I(H) \neq \emptyset$.

Suppose that H is an elimination of G. Let $S = \{v_x \mid x \in V(H)\}$ be the witness, where v_x is the vertex of G that corresponds to the vertex x of H, and let $X = V(G) \setminus S$ be the corresponding solution yielding H. The witness S satisfies the structural properties given in the two following lemmas.

Lemma 4 (♠). *For any bicomp $B \in \mathcal{B}(G)$ it holds that $|V(B) \cap S| \leq 2$, and if $v_x, v_y \in V(B) \cap S$ for $x \neq y$, then $xy \in E(H)$.*

Lemma 5 (♠). *For any $x \in I(H)$, $v_x \in C(G)$.*

Now we choose an arbitrary inner vertex z of H and say that it is the *root* of H. The root defines the parent-child relation between any two adjacent vertices of H. For any two vertices $x, y \in V(H)$, we say that y is a *descendant* of x if x lies on the unique path in H from y to the root z. If y is a descendant of x and $xy \in E(H)$, then y is a *child* of x, and x is the *parent* of y. By definition, every vertex $x \in V(H)$ is a descendant of itself. For a vertex $x \in V(H)$, H_x denotes the subtree of H induced by the descendants of x, and for a vertex $x \in V(H)$ with a child y, H_{xy} is the subtree of H induced by x and the descendants of y.

Consider $r = v_z \in V(G)$. We choose r to be the *root* of the bicomp-tree $\mathcal{T}_G$ of G. By Lemma 5, r is a cut-vertex in G. The root r defines the parent-child relation on $\mathcal{T}_G$. Each bicomp B is a child of some inner vertex c in $\mathcal{T}_G$, and we say that the vertices of B are *children* of the corresponding cut-vertex c in G. A vertex $v \in V(G)$ is a *descendant* of a cut-vertex c if v is a child of some descendant of c in $\mathcal{T}_G$. For a cut-vertex c, we write G_c to denote the subgraph of G induced by the descendants of c. For a cut-vertex c and a bicomp B such that B is a child of c in $\mathcal{T}_G$, G_{cB} is the subgraph of G induced by the vertices of B and the descendants of all cut-vertices $c' \in V(B) \setminus \{c\}$.

Now consider two vertices x and y in H, such that neither is a descendant of the other, and let p be their lowest common ancestor. A crucial observation in our algorithm is that v_x and v_y are descendants of v_p in G, but they do not appear in the same subgraph $G_{v_p B}$ for some bicomp B that is a child of v_p. The following lemma formalizes this idea.

Lemma 6 ($\spadesuit$). *For any inner vertex $x \in V(H)$, if $y \in V(H)$ is a descendant of x in H, then v_y is a descendant of v_x in G. Moreover, if $y_1, \ldots, y_l$ are the children of x in H, then there are distinct children $B_1, \ldots, B_l$ of v_x in the bicomp-tree for which the following holds: for each $i \in \{1, \ldots, l\}$, if $y \in V(H_{xy_i})$, then $v_y \in G_{v_x B_i}$.*

We are now ready to describe our algorithm in the proof of the following theorem.

Theorem 5 ($\spadesuit$). ELIMINATION *can be solved in time $O(n^{9/2})$ when H is a tree.*

Proof. Let G and H be an instance of ELIMINATION where H is a tree. Clearly, if $|V(H)| > n$, then we have a no-instance of the problem. Hence, we assume that $|V(H)| \leq n$. Recall that it is sufficient to solve the problem for connected graphs G and trees H with at least three vertices. For the tree H, we choose an arbitrary inner vertex z and make it the root of H. For the graph G, we find the set of cut-vertices $C(G)$ and the set of bicomps $\mathcal{B}$, and construct the bicomp-tree $\mathcal{T}_G$. Then we construct a set $U \subseteq V(G)$ as follows: for each bicomp B that is a leaf of $\mathcal{T}_G$, we choose an arbitrary vertex $u \in V(B) \setminus C(G)$ and include it in U. It can be shown that H is an elimination of G if only if G can be eliminated to H with a witness $S \subseteq C(G) \cup U$. A formal proof of this statement requires some additional lemmas; we omit the details here due to page restrictions.

Suppose H is an elimination of G with a witness $S = \{v_x \mid x \in V(H)\}$. Since we chose z to be an inner vertex of H, the vertex v_z is a cut-vertex of G due to Lemma 5. Hence, by Lemma 6 there is a cut-vertex r in G such that if y is a descendant of x in H rooted at z, then v_y is a descendant of v_x in G rooted at r. We check all cut-vertices $r \in C(G)$, and for each r, we root G at r and try to find a witness that satisfies this condition. Clearly, H is an elimination of G if and only if we find such a witness for some r, and we have a no-instance of ELIMINATION otherwise.

From now on, we assume that the root vertex r of G is fixed, and we construct a dynamic programming algorithm. For each vertex $u \in C(G) \cup U$, the algorithm will create a set $R_u \subseteq V(H)$ such that:

- for any $u \in U$, $R_u = L(H)$;
- for any $u \in C(G)$, R_u is the set of all vertices x of H such that H_x is an elimination of G_u with the property that for any $y, y' \in V(H_x)$, if y' is a descendant of y in H_x, then $v_{y'}$ is a descendant of v_y in G_r, where $v_y, v_{y'}$ are the saved vertices in G_u corresponding to y, y'.

The algorithm returns **yes** if R_r contains z, and **no** otherwise.

Notice that the sets for $u \in U$ are already defined. The sets R_u for cut-vertices u are constructed as follows. Denote by $B_1, \ldots, B_k$ the bicomps of G that are children of the cut-vertex u in the bicomp-tree $\mathcal{T}_G$. Let D_u be the set of all vertices $w \in C(G) \cup U$ other than u that are descendants of u and are contained in some bicomp together with u. In other words, $D_u = (C(G) \cup U \setminus \{u\}) \cap \bigcup_{i=1}^{k} V(B_i)$. Suppose that the sets R_w have already been constructed for all $w \in D_u$. We then create R_u in two steps.

Step 1. All the vertices that are in R_w for some $w \in D_u$ are included in R_u.

Step 2. Let $T_i = \bigcup_{w \in D_u \cap V(B_i)} R_w$ for $i \in \{1, \ldots, k\}$. A vertex $x \in V(H)$ with children $y_1, \ldots, y_l$ is included in R_u if there is a set $\{i_1, \ldots, i_l\} \subseteq \{1, \ldots, k\}$ such that $y_j \in T_{i_j}$ for $j \in \{1, \ldots, l\}$.

In order to perform Step 2, whose correctness is guaranteed by Lemma 6, we need to solve a matching problem on an auxiliary graph. The full proof of correctness and the running time analysis of our algorithm will appear in the journal version of this paper. $\square$

Finally, we consider the case when G is a tree and H is an arbitrary graph. First, we make the following observation. A connected graph is called a *block graph* if each of its bicomps is a complete graph. Observe that if G is a block graph, then elimination of any vertex v results in another block graph, because this operation unites all maximal cliques that contain v into a single clique and then removes v. Since trees are block graphs, it gives us the following proposition.

Proposition 1. *If H is an elimination of a tree G, then H is a block graph.*

Despite the fact that graphs that are eliminations of trees have relatively simple structure, it turns out that ELIMINATION remains intractable when G is assumed to be a tree.

Theorem 6 (♠). ELIMINATION *is NP-complete, even if G is restricted to be a tree.*

Acknowledgements. We would like to thank Łukasz Kowalik for an inspiring discussion on the theorem of Kleitman and West.

References

1. van Bevern, R., Komusiewicz, C., Moser, H., Niedermeier, R.: Measuring Indifference: Unit Interval Vertex Deletion. In: Thilikos, D.M. (ed.) WG 2010. LNCS, vol. 6410, pp. 232–243. Springer, Heidelberg (2010)

2. Bodlaender, H.L., Downey, R.G., Fellows, M.R., Hermelin, D.: On problems without polynomial kernels. J. Comput. Syst. Sci. 75(8), 423–434 (2009)
3. Cai, L.: Fixed-parameter tractability of graph modification problems for hereditary properties. Inform. Process. Lett. 58(4), 171–176 (1996)
4. Courcelle, B.: The monadic second-order logic of graphs III: Tree-decompositions, minor and complexity issues. ITA 26, 257–286 (1992)
5. Downey, R.G., Fellows, M.R.: Parameterized Complexity. Springer (1999)
6. George, J.A., Liu, J.W.H.: Computer Solution of Large Sparse Positive Definite Systems. Prentice-Hall Inc. (1981)
7. Heggernes, P.: Minimal triangulations of graphs: A survey. Discrete Mathematics 306, 297–317 (2006)
8. Heggernes, P., van 't Hof, P., Jansen, B.M.P., Kratsch, S., Villanger, Y.: Parameterized Complexity of Vertex Deletion into Perfect Graph Classes. In: Owe, O., Steffen, M., Telle, J.A. (eds.) FCT 2011. LNCS, vol. 6914, pp. 240–251. Springer, Heidelberg (2011)
9. Kawarabayashi, K., Reed, B.A.: An (almost) linear time algorithm for odd cyles transversal. In: Charikar, M. (ed.) SODA 2010, pp. 365–378. SIAM (2010)
10. Kleitman, D.J., West, D.B.: Spanning trees with many leaves. SIAM J. Discrete Math. 4, 99–106 (1991)
11. Marx, D.: Chordal deletion is fixed-parameter tractable. Algorithmica 57(4), 747–768 (2010)
12. Marx, D., Schlotter, I.: Obtaining a planar graph by vertex deletion. Algorithmica 62(3-4), 807–822 (2012)
13. Natanzon, A., Shamir, R., Sharan, R.: Complexity classification of some edge modification problems. Discrete Appl. Math. 113, 109–128 (2001)
14. Parter, S.: The use of linear graphs in Gauss elimination. SIAM Review 3, 119–130 (1961)
15. Philip, G., Raman, V., Villanger, Y.: A Quartic Kernel for Pathwidth-One Vertex Deletion. In: Thilikos, D.M. (ed.) WG 2010. LNCS, vol. 6410, pp. 196–207. Springer, Heidelberg (2010)
16. Robertson, N., Seymour, P.D., Thomas, R.: Quickly excluding a planar graph. J. Comb. Theory B 62, 323–348 (1994)
17. Rose, D.J., Tarjan, R.E., Lueker, G.S.: Algorithmic aspects of vertex elimination on graphs. SIAM J. Comput. 5, 266–283 (1976)
18. Samdal, E.: Minimum Fill-in Five Point Finite Element Graphs. Master's thesis, Department of Informatics, University of Bergen, Norway (2003)
19. Yannakakis, M.: Edge-deletion problems. SIAM J. Comput. 10, 297–309 (1981)

On the Parameterized Complexity of Finding Separators with Non-Hereditary Properties[*]

Pinar Heggernes[1], Pim van 't Hof[1], Dániel Marx[2], Neeldhara Misra[3], and Yngve Villanger[1]

[1] Department of Informatics, University of Bergen, Norway
{pinar.heggernes,pim.vanthof,yngve.villanger}@ii.uib.no
[2] Computer and Automation Research Institute, Hungarian Academy of Sciences
(MTA SZTAKI), Budapest, Hungary
dmarx@cs.dme.hu
[3] Institute of Mathematical Sciences, Chennai, India
neeldhara@imsc.res.in

Abstract. We study the problem of finding small s–t separators that induce graphs having certain properties. It is known that finding a minimum clique s–t separator is polynomial-time solvable (Tarjan 1985), while for example the problems of finding a minimum s–t separator that is a connected graph or an independent set are fixed-parameter tractable (Marx, O'Sullivan and Razgon, manuscript). We extend these results the following way:

(1) Finding a minimum c-connected s–t separator is FPT for $c = 2$ and $W[1]$-hard for any $c \geq 3$.

(2) Finding a minimum s–t separator with diameter at most d is $W[1]$-hard for any $d \geq 2$.

(3) Finding a minimum r-regular s–t separator is $W[1]$-hard for any $r \geq 1$.

(4) For any decidable graph property, finding a minimum s–t separator with this property is FPT parameterized jointly by the size of the separator and the maximum degree.

We also show that finding a connected s–t separator of minimum size does not have a polynomial kernel, even when restricted to graphs of maximum degree at most 3, unless NP $\subseteq$ coNP/poly.

1 Introduction

One of the classic topics in combinatorial optimization and algorithmic graph theory deals with finding cuts and separators in graphs. Recently, the study of this type of problems from a parameterized complexity point of view has attracted a large amount of interest [5, 6, 11, 14–21]. Given a graph G and two vertices s and t of G, a subset of vertices $S \subseteq V(G) \setminus \{s, t\}$ is an s–t *separator* if s and t appear in different connected components of the graph $G - S$. In separation

[*] This work is supported by the Research Council of Norway and by the European Research Council (ERC) grant 280152.

M.C. Golumbic et al. (Eds.): WG 2012, LNCS 7551, pp. 332–343, 2012.
© Springer-Verlag Berlin Heidelberg 2012

problems, we are typically looking for *small* separators S. A natural extension of the problem is to demand $G[S]$, i.e., the subgraph induced by S, to satisfy a certain property. (For convenience, when the graph $G[S]$ has a certain property, we will say that the set S itself also has this property; for example, we say that a set $S \subseteq V(G)$ is 2-connected if $G[S]$ is 2-connected.) A classical result in this direction by Tarjan [22] shows that finding small *clique* separators is polynomial-time solvable. To our knowledge, this is the only known polynomial-time solvable problem of this type. Therefore, we explore here the problem from the viewpoint of parameterized complexity.

Parameterized complexity associates with every instance of a problem a non-negative integer k, called the *parameter*. As is common in the parameterized study of separator problems, the parameter k in this paper will always be the size of the separator we are looking for. We use n and m to denote the number of vertices and edges, respectively, in the input graph. A parameterized problem is *fixed-parameter tractable* (or FPT) if it can be solved in time $f(k) \cdot n^{O(1)}$ for some function f that only depends on k [9]. By showing that a parameterized problem is W[1]-hard, we can give strong evidence that it is unlikely to be FPT; we refer to [9] for more background on parameterized complexity.

For any graph class $\mathcal{G}$, let us consider the following parameterized problem.

$\mathcal{G}$-SEPARATOR
Input: A graph G, two vertices s and t of G, and an integer k.
Parameter: k.
Question: Does G have an s–t separator S of size at most k such that $G[S] \in \mathcal{G}$?

If $\mathcal{G}$ is the class of all complete graphs, then $\mathcal{G}$-SEPARATOR is polynomial-time solvable by the result of Tarjan [22]. Furthermore, Marx et al. [18, 19] showed that the problem is fixed-parameter tractable for many natural classes $\mathcal{G}$. We say that $\mathcal{G}$ is *hereditary* if, for every graph in $\mathcal{G}$, each of its induced subgraphs also belongs to $\mathcal{G}$.

Theorem 1 ([18, 19]). *For any decidable and hereditary graph class $\mathcal{G}$, the $\mathcal{G}$-SEPARATOR problem can be solved in time $f_{\mathcal{G}}(k) \cdot (n + m)$.*

For example, by letting $\mathcal{G}$ be the class of all graphs without edges, Theorem 1 shows that finding an independent set of size at most k separating s and t is FPT. The proof is based on a combinatorial statement called Treewidth Reduction Theorem, which shows (roughly speaking) that all the inclusionwise minimal s–t separators lie in a bounded-treewidth part of the graph and hence they can be found efficiently. Note that if $\mathcal{G}$ is hereditary, then we can always assume that the separator is inclusionwise minimal (otherwise we can remove vertices from it without leaving $\mathcal{G}$).

Theorem 1 naturally raises the question what the parameterized complexity of the $\mathcal{G}$-SEPARATOR problem is for graph classes $\mathcal{G}$ that are *not* hereditary.

Perhaps the most natural candidate is the class of *connected* graphs. The CON-NECTED SEPARATOR problem of deciding whether a graph G has a connected s–t separator of size at most k has been studied by Marx et al. [19]. Although it is not immediately clear how to apply the Treewidth Reduction Theorem to this problem, Marx et al. [19] managed to extend their framework from [18] to prove the following result.

Theorem 2 ([19]). *The* CONNECTED SEPARATOR *problem can be solved in time* $f(k) \cdot (n + m)$.

Our Results. Motivated by the results in [18, 19], we study the problem of finding small s–t separators satisfying different non-hereditary properties. Let us focus on the three tractable classes mentioned above (connected graphs, cliques, independent sets) and try to investigate further related classes.

As CONNECTED SEPARATOR is FPT, it is natural to explore what happens if we require higher-order connectivity. It turns out that, somewhat surprisingly, finding a c-connected s–t separator of size at most k remains FPT also for $c = 2$, but becomes $W[1]$-hard for any $c \geq 3$. In order to prove this, we show that the natural c-connected generalization of STEINER TREE is FPT for $c = 2$ and $W[1]$-hard for any $c \geq 3$. This result could be of independent interest.

We can generalize the class of cliques by considering the class of graphs with diameter at most d. We show that the problem of finding an s–t separator of size at most k that induces a graph with diameter d in G is $W[1]$-hard for any $d \geq 2$. This is in stark contrast with the case $d = 1$, as the problem of finding a *clique* separator of size at most k is known to be solvable in polynomial time [22].

Independent sets can be thought of as 0-regular graphs. This motivates exploring the problem of finding an r-regular s–t separator. We show that, unlike the $r = 0$ case which is FPT by Theorem 1, for any $r \geq 1$, it is $W[1]$-hard to decide if a graph G has an r-regular s–t separator of size at most k.

All the above results are on general graphs, i.e., graph G can be arbitrary. It comes as no surprise that the problem is much easier restricted to bounded-degree graphs. In particular, finding a small connected separator is FPT due to the fact that a bounded-degree graph contains only a bounded number of small connected sets. More interestingly, we show in Section 4 that for *every* (not necessarily hereditary) decidable graph class $\mathcal{G}$, the $\mathcal{G}$-SEPARATOR problem can be can be solved in time $h_{\mathcal{G}}(k, \Delta) \cdot m \log n$ on graphs of maximum degree at most Δ. We prove this by showing that the following problem can be solved in time $f(|V(H)|, \Delta) \cdot m \log n$ on graphs of maximum degree at most Δ: Given two graphs G and H and two vertices s and t of G, decide whether G has an s–t separator S such that $G[S]$ is isomorphic to H. This means that we can solve the $\mathcal{G}$-SEPARATOR problem by simply trying all members H of $\mathcal{G}$ having k vertices.

Finally, we investigate the existence of polynomial kernels for the problem of finding small s–t separators. A parameterized problem is said to admit a *kernel* if there is a polynomial-time algorithm that transforms each instance of the problem into an *equivalent* instance whose size and parameter value are bounded from above by $g(k)$ for some (possibly exponential) function g. It is known that a parameterized problem is FPT if and only if it is decidable and admits a kernel [9].

In the desirable case that $g(k)$ is a polynomial in k, we say that the problem admits a *polynomial kernel*. Many problems have been shown to admit polynomial kernels, including classes of problems that are covered by some kernelization meta-theorems [3, 12]. Recently developed methods for proving non-existence of polynomial kernels, up to some complexity theoretical assumptions [2, 4, 13], significantly contributed to the establishment of kernelization as an important and rapidly growing subfield of parameterized complexity.

Although the CONNECTED SEPARATOR problem is FPT by Theorem 2 and therefore admits a kernel [9], we show in Section 5 that this problem does not have a *polynomial* kernel, even when restricted to input graphs of maximum degree at most 3, unless NP $\subseteq$ coNP/poly. This means that techniques other than kernelization (e.g., treewidth reduction) seem to be essential for the efficient solution of the problem *even on bounded-degree graphs*.

2 Finding s–t Separators with Higher Connectivity

Theorem 2 states that the problem of finding a connected s–t separator of size at most k is FPT. In this section, we study the parameterized complexity of finding s–t separators of higher connectivity. A graph $G = (V, E)$ is *c-connected* if $|V| > c$ and $G - X$ is connected for every $X \subseteq V$ with $|X| < c$. Menger's Theorem provides an equivalent definition (see [8]): a graph is c-connected if any two of its vertices can be joined by c internally vertex-disjoint paths. For any integer $c \geq 1$, the c-CONNECTED SEPARATOR problem takes as input a graph G, two vertices s and t of G, and an integer k (the parameter), and asks whether there is an s–t separator of size at most k that induces a c-connected graph. Theorem 2 states that this problem is FPT when $c = 1$. Interestingly, it turns out that the problem remains FPT for $c = 2$, but becomes $W[1]$-hard for any $c \geq 3$.

The algorithm in [19] for finding a minimum connected s–t separator uses an FPT algorithm for STEINER TREE as a subroutine. For our purposes, we need to define the following natural c-connected generalization of the STEINER TREE problem. For any integer $c \geq 1$, the c-CONNECTED STEINER problem takes as input a graph G, a set $T \subseteq V(G)$ of terminals and an integer k (the parameter). The objective is to decide whether G has a c-connected subgraph H on at most k vertices such that H contains all the terminals. Such a graph H is called a *solution*. A solution H is *minimal* if no proper subgraph of H is a solution, and H is *minimum* if there is no solution H' with $|V(H')| < |V(H)|$. When $c = 1$, this problem is equivalent to the well-known STEINER TREE problem, which is known to be FPT when parameterized by k [10]. We show below that the c-CONNECTED STEINER problem remains FPT when $c = 2$, but becomes $W[1]$-hard for higher values of c.

A different way of generalizing STEINER TREE would be to require the weaker condition saying that H contains c internally vertex-disjoint paths between any two terminals. The following lemma shows that for $c = 2$ this is almost the same problem, as any minimal solution satisfying the weaker requirement satisfies the stronger requirement as well:

Lemma 1. $(\bigstar)^1$ *Let H be a graph and $T \subseteq V(H)$ a set of vertices such that there are two internally vertex-disjoint paths between any $t_1, t_2 \in T$. If H has no proper subgraph (containing T) having this property, then H is 2-connected.*

We note that for $c \geq 3$, the analog of Lemma 1 is not true. Thus the weaker requirement would result in a different problem, but we do not investigate it further in the current paper.

Our algorithm for 2-CONNECTED STEINER crucially depends on the following structural property of any minimal solution:

Lemma 2. *Let (G, T, k) be an instance of the 2-CONNECTED STEINER problem. If H is a minimal solution, then $H - T$ is a forest.*

Proof. Since the lemma trivially holds when $|T| \leq 1$, we assume that $|T| \geq 2$. Suppose H is a minimal solution. We show that every cycle in H contains at least one vertex of T, which implies that $H - T$ is a forest. For contradiction, let C be a cycle in H that contains none of the terminals. We will identify an edge e of C such that it remains true in $H - e$ that there are two internally vertex-disjoint paths between any two terminals. Then by Lemma 1, $H - e$ has a 2-connected subgraph which is a solution, contradicting the minimality of H.

We define a partition T_1, T_2 of the terminals as follows. A terminal $t \in T$ belongs to T_2 if there is another terminal $t' \in T$ such that for every pair P_1, P_2 of internally vertex-disjoint paths between t and t' in H, both P_1 and P_2 use at least one vertex of C, i.e., if t and t' belong to different connected components of $H - V(C)$. Note that in such a case t' is also in T_2. We define $T_1 = T \setminus T_2$.

Let $t \in T_1$. By definition, for any $t' \in T \setminus \{t\}$, there exist two internally vertex-disjoint paths in H between t and t' such that at least one of them does not use any vertex of C. Let H' be the graph obtained from H by deleting any edge of C. Then H' still contains two internally vertex-disjoint paths between t and any $t' \in T \setminus \{t\}$, as any path between t and t' that used the deleted edge can be rerouted on the cycle. Hence, if T_2 is empty, we can delete any edge from C and obtain a new solution, contradicting the minimality of H.

Now suppose $T_2 \neq \emptyset$. Let us define a *shortcut* of C to be a path P of length at least 2 between two vertices a and b of C, such that none of the internal vertices of P are in C. It follows from the definition of T_2 that for each $t \in T_2$, there are two distinct vertices a, b on C such that there are two internally vertex-disjoint paths P_a, P_b from t to a and b, respectively, whose internal vertices are not in C. In other words, for every $t \in T_2$, there is a shortcut of C that contains t. Let M be a shortest subpath of C such that there is a shortcut P^* of C between the endpoints a and b of M. Let $\overline{M}$ be the other path between a and b on the cycle C. Let a' be the neighbor of a on M (possibly $a' = b$). We claim that after removing the edge aa' from H, the obtained graph $H - aa'$ still contains two internally vertex-disjoint paths between each pair of terminals in T_2.

By the well-known properties of the 2-connected components of graphs, the relation "being in the same 2-connected component" (or equivalently, the relation

---

[1] Proofs marked with a star have been omitted due to page restrictions.

"there is a cycle containing both edges") defined on the edges of $H - aa'$ is an equivalence relation. Every edge of $\overline{M}$ is in the same equivalence class of this relation: $\overline{M}$ together with P^* forms a cycle containing all these edges. We claim that every $t \in T_2$ is also in this 2-connected component. As observed above, there is a shortcut P_t going through t. Let M_t be the subpath of the cycle C between the endpoints of P_t avoiding aa'. The paths P_t and M_t together form a cycle. This cycle contains at least one edge of $\overline{M}$, since M_t cannot be a proper subpath of M by the minimality of M. Thus the edges of this cycle are in the same 2-connected component as the edges of $\overline{M}$. We have shown that every $t \in T_2$ is in this 2-connected component of $H - aa'$. Consequently, there are two internally vertex-disjoint paths in $H - aa'$ between any two terminals $t_1, t_2 \in T$, yielding the desired contradiction to the assumption that H is a minimal solution.

We conclude that every cycle in H contains at least one vertex of T, which implies that $H - T$ is a forest.   □

Lemma 2 tells us that we have to find an appropriate forest that connects to the terminals in an appropriate way. Fixed-parameter tractability results for finding trees (or more generally, bounded-treewidth graphs) under various technical constraints can usually be obtained using standard application of dynamic programming. Here we need the following variant:

Lemma 3. (★) *Let F be a forest, G an undirected graph, and $c : V(F) \times V(G) \to \mathbb{Z}^+$ a cost function. In time $f(|V(F)|) \cdot n^{O(1)}$, one can find a mapping $\phi : V(F) \to V(G)$ such that $\phi(u)\phi(v) \in E(G)$ for every $uv \in E(F)$ and the total cost $\sum_{v \in V(F)} c(v, \phi(v))$ is minimized.*

The structural observation of Lemma 2 and the algorithm of Lemma 3 allow us to establish the fixed-parameter tractability of the 2-CONNECTED STEINER problem, which could be interesting in its own right. Furthermore, it will be used as a subroutine in our FPT-algorithm for finding a 2-connected s–t separator of size at most k.

Theorem 3. *The 2-CONNECTED STEINER problem is FPT.*

Proof. Let (G, T, k) be a yes-instance of the 2-CONNECTED STEINER problem and let H be a minimal solution. By Lemma 2, $H - T$ is a forest. We try all graphs H on at most k vertices that are candidates for being isomorphic to the solution H: that is, H is 2-connected, $T \subseteq V(H)$, and $H - T$ is a forest. The number of such graphs is a function of k only. For each such H, we define a cost function c such that for $x \in V(H - T)$ and $y \in V(G)$, we have $c(x, y) = 0$ if $N_H(x) \cap T \subseteq N_G(y) \cap T$ and $c(x, y) = \infty$ otherwise. In other words, we allow mapping x to y only if every terminal neighbor of x in H is also a neighbor of y in G. Let us use the algorithm of Lemma 3 to find a mapping ϕ of $H - T$ into G minimizing the cost. If the cost of ϕ is 0, then ϕ can be extended to a mapping of H into G, showing that H is a subgraph of G, which gives us a solution. Otherwise, we proceed with the next candidate H. If the algorithm finds no solution after processing all candidates, we can safely return "no".   □

In order to prove that 2-CONNECTED SEPARATOR is FPT, we will make use of the Treewidth Reduction Theorem due to Marx, O'Sullivan and Razgon [18, 19]. In fact, instead of using the Treewidth Reduction Theorem itself, we use a lemma (a slight reformulation of Lemma 2.8 in [18]) that forms its crucial ingredient. In order to state it, we need an additional definition. Let G be a graph and $C \subseteq V(G)$. The graph $torso(G, C)$ has vertex set C, and vertices $a, b \in C$ are connected by an edge if $ab \in E(G)$ or if there is a path in G connecting a and b whose internal vertices are not in C.

Lemma 4 ([18]). *Let s and t be two vertices of a graph G, let k be an integer, and let C be the union of all minimal s–t separators in G of size at most k. Then there is an $f(k) \cdot (n + m)$ time algorithm that returns a set $C' \supseteq C \cup \{s, t\}$, such that the treewidth of $torso(G, C')$ is at most $g(k)$.*

Note that even if G has a 2-connected s–t separator S of size at most k, G might not have a *minimal* s–t separator of size at most k that is 2-connected, since 2-connectivity is not a hereditary property. However, G does contain a minimal s–t separator that can be extended to a 2-connected set of size at most k. We call a set $S' \subseteq V(G)$ *k-biconnectable* if there is a 2-connected set $S \subseteq V(G)$ of size at most k such that $S' \subseteq S$.

Observation 4. *Let G be a graph. A set $S' \subseteq V(G)$ is k-biconnectable if and only if (G, S', k) is a yes-instance of the 2-CONNECTED STEINER problem.*

The set C' in Lemma 4 contains every minimal s–t separator S' that is k-biconnectable, but there is no guarantee that S' can be extended to a 2-connected set within C'. The next lemma shows that we can extend C' to a larger set C'' such that every k-biconnectable s–t separator $S' \subseteq C'$ can be extended to a 2-connected s–t separator $S \subseteq C''$ of size at most k.

Lemma 5. *Let s and t be two vertices of a graph G, and let k be an integer. There is a set $C'' \subseteq V(G)$ such that the treewidth of $torso(G, C'')$ is bounded by a constant depending only on k and the following holds: if G has a 2-connected s–t separator of size at most k, then G also has a 2-connected s–t separator S of size at most k such that $S \subseteq C''$. Moreover, such a set C'' can be found in time $h(k) \cdot n^{O(1)}$.*

Proof. Let $C' \subseteq V(G)$ be the set of Lemma 4 that contains every minimal s–t separator of G of size at most k, such that the treewidth of $torso(G, C')$ is bounded by a function of k. Let $K_1, \ldots, K_q$ be the connected components of $G - C'$, and let N_i be the neighborhood of K_i in C' for $1 \leq i \leq q$. By the definition of torso, each N_i forms a clique in $torso(G, C')$. Since each clique of a graph must appear in a single bag of any tree decomposition of that graph, we have $|N_i| \leq \text{tw}(torso(G, C')) + 1$, so the size of each N_i is bounded by a function of k only.

Our algorithm for constructing C'' iterates over all $i \in \{1, \ldots, q\}$, all non-empty subsets $X \subseteq N_i$, all graphs $F_{i,X}$ on at most $k - |X|$ vertices, and all possible ways in which the vertices of $F_{i,X}$ can be made adjacent to the vertices of X. For each of those choices, let $G_{i,X}$ be the graph obtained from $G[V(K_i) \cup X]$ and $F_{i,X}$ by adding edges between these two graphs in the way that we chose earlier. We then run the algorithm of Theorem 3 to check if there is a solution $H_{i,X}$ for the 2-CONNECTED STEINER problem with instance $(G_{i,X}, V(F_{i,X}) \cup X, k)$. If so, we take $H_{i,X}$ to be the minimum such solution; otherwise we let $H_{i,X} = \emptyset$. For each $H_{i,X}$, we mark all the vertices of $H_{i,X}$ that belong to K_i. Finally, we define C'' to be the set consisting of all the vertices of C' plus all the vertices that were marked during this entire process.

In order to prove the correctness of this algorithm, let us consider a 2-connected s–t separator S of size at most k in G such that $|S \setminus C''|$ is as small as possible. We need to show that $|S \setminus C''| = 0$. For contradiction, we assume that $|S \setminus C''| \geq 1$. Let K_i be a connected component of $G - C'$ such that K_i contains a vertex of $S \setminus C''$, let $S_i = S \setminus V(K_i)$, and let $X = S \cap N_i$. Note that $X \neq \emptyset$. Also note that S_i is a k-biconnectable set in the graph $G[V(K_i) \cup S_i]$. Hence, by Observation 4, $(G[V(K_i) \cup S_i], S_i, k)$ is a yes-instance of 2-CONNECTED STEINER. Since $X \neq \emptyset$, in some iteration of the algorithm, we considered a graph $G_{i,X}$ that is isomorphic to $G[V(K_i) \cup S_i]$ and hence found a minimum solution $H_{i,X}$ of 2-CONNECTED STEINER for exactly the instance $(G[V(K_i) \cup S_i], S_i, k)$. Let $S' = S_i \cup V(H_{i,X})$. By construction, S' is 2-connected. Note that $S \cap C'$ is an s–t separator, since otherwise there would be a minimal s–t separator of size at most k in G that contains a vertex outside C', contradicting Lemma 4. Since $S \cap C' \subseteq S'$, S' is an s–t separator. It is clear that $S' \subseteq C''$, which means that $|S' \setminus C''| = 0$. Hence $|S' \setminus C''| < |S \setminus C''|$, contradicting the minimality of S.

For each $i \in \{1, \ldots, q\}$, C'' contains at most $k|N_i|^k$ vertices of K_i, and hence the treewidth of torso$(K_i, C'' \cap V(K_i))$ is bounded by a constant depending only on k. It follows that the difference between the treewidth of torso(G, C'') and the treewidth of torso(G, C') is a constant depending on k (see also Lemma 2.9 in [19]), implying that the treewidth of torso(G, C'') is bounded by a function of k. Finding the set C' can be done in time $f(k) \cdot (m + n)$ by Lemma 4. For each choice of i and X, the possible number of different graphs $G_{i,X}$, and consequently the number of instances of 2-CONNECTED STEINER we have to solve, is bounded by some function of k. Since 2-CONNECTED STEINER is FPT by Theorem 3, the overall running time of the algorithm is $h(k) \cdot n^{O(1)}$ for some function h that depends only on k. $\square$

Theorem 5. *The* 2-CONNECTED SEPARATOR *problem is FPT.*

Proof. Let (G, s, t, k) be an instance of 2-CONNECTED SEPARATOR. We start by constructing the set $C'' \subseteq V(G)$ of Lemma 4. Let $G^* = \text{torso}(G, C'')$. We assign a color to each edge uv in G^*: we color uv black if uv is also an edge in G, and we color uv red otherwise. By Lemma 5, G contains a 2-connected s–t separator S of size at most k if and only if G^* contains an s–t separator S^* of size at most k such that deleting the red edges from $G^*[S^*]$ results in a 2-connected graph. The theorem now follows from Courcelle's Theorem [7] and the fact that this problem can be expressed in monadic second-order logic (see [19]). $\square$

We now show that the c-CONNECTED STEINER problem becomes hard when the connectivity of the solution is required to be at least 3.

Theorem 6. (★) c-CONNECTED STEINER *is* $W[1]$-*hard for any* $c \geq 3$.

Since we can transform an instance of c-CONNECTED STEINER into an instance of c-CONNECTED SEPARATOR by making two new vertices s and t adjacent to each of the terminals, Theorem 6 readily implies the following result.

Theorem 7. c-CONNECTED SEPARATOR *is* $W[1]$-*hard for any* $c \geq 3$.

3   More $W[1]$-Hardness Results on General Graphs

We say that a graph G is r-*regular* if the degree of every vertex in G is exactly r. For every $r \geq 0$, let r-REGULAR SEPARATOR denote the problem of deciding whether an input graph G has an s–t separator S of size at most k such that $G[S]$ is r-regular. Since the class of 0-regular graphs is hereditary, Theorem 1 implies that 0-REGULAR SEPARATOR, i.e., the problem of finding an s–t separator that is an independent set of size at most k, is FPT. We show that r-REGULAR SEPARATOR is $W[1]$-hard for every $r \geq 1$ when parameterized by k. Note that the class of r-regular graphs is not hereditary for any $r \geq 1$.

Theorem 8. (★) r-REGULAR SEPARATOR *is* $W[1]$-*hard for any* $r \geq 1$.

The *diameter* of a graph G is the maximum distance between any two vertices in G, where the *distance* between two vertices u and v is defined as the number of edges in a shortest path from u to v. The problem of finding an s–t separator that forms a *clique* is well-known to be solvable in polynomial time [22]. Since cliques induce subgraphs of diameter 1, it is natural to consider the problem of finding an s–t separator that induces a graph of diameter 2, or, more generally, of any fixed diameter $d \geq 2$. Note that for any $d \geq 2$, the class of graphs with diameter (at most) d is not hereditary; consider for example a chordless cycle on $2d + 1$ vertices. The class of graphs with diameter 1, however, *is* hereditary. d-DIAMETER SEPARATOR is the problem of deciding if an input graph G has an s–t separator S of size at most k such that $G[S]$ has diameter d.

Theorem 9. (★) d-DIAMETER SEPARATOR *is* $W[1]$-*hard for any* $d \geq 2$.

4   Finding s–t Separators in Graphs of Bounded Degree

Theorem 1 states that $\mathcal{G}$-SEPARATOR is FPT for any decidable and hereditary graph class $\mathcal{G}$. In the previous sections, we identified several non-hereditary graph classes $\mathcal{G}$ for which $\mathcal{G}$-SEPARATOR is $W[1]$-hard on general graphs. In this section, we prove that for any decidable (but not necessarily hereditary) graph class $\mathcal{G}$, the $\mathcal{G}$-SEPARATOR problem is FPT on graphs of bounded degree. We do this by showing that the following problem is FPT on graphs of bounded degree.

PATTERN SEPARATOR
Input: Two graphs G and H, and two vertices s and t of G.
Parameter: $k = |V(H)|$.
Question: Does G have an s–t separator S such that $G[S]$ is isomorphic to H?

We use a variant of the color coding technique of Alon, Yuster and Zwick [1] to reduce the PATTERN SEPARATOR problem on bounded degree graphs to the problem of finding an s–t separator of size at most k that has a certain hereditary property, which enables us to use Theorem 1.

For the remainder of this section, let G and H be two graphs, let s and t be two vertices of G, and let $H_1, \ldots, H_c$ be the connected components of H. We use n and m to denote the number of vertices and edges in G, respectively, and k to denote the number of vertices in H. Let ψ be a (not necessarily proper) coloring of G. A subset of vertices $V' \subseteq V(G)$ is *colorful* if ψ colors no two vertices of V' with the same color. For any subset C' of colors, we say that $V' \subseteq V(G)$ is C'-*colorful* if $|V'| = |C'|$ and every vertex in V' receives a different color from C'.

Definition 1. *Let $\psi : V(G) \rightarrow \{1, 2, \ldots, c, c+1\}$ be a $(c+1)$-coloring of G. We say that ψ is H-good if G has an s–t separator S satisfying the following properties:*

(i) each connected component of $G[S]$ is colored monochromatically with a color from $\{1, \ldots, c\}$;

(ii) no two connected components of $G[S]$ receive the same color;

(iii) the connected component of $G[S]$ with color i is isomorphic to H_i;

(iv) every vertex in $N_G(S)$ receives color $c + 1$.

It immediately follows from Definition 1 that (G, H, s, t) is a yes-instance of PATTERN SEPARATOR if and only if G has an H-good coloring. The main idea of our algorithm is that finding a separator S satisfying these requirements essentially boils down to finding a separator that is a colorful independent set, which is fixed-parameter tractable by the results of [18, 19]. The following lemma plays a crucial role in our FPT algorithm.

Lemma 6. (★) *Given a $(c+1)$-coloring ψ of G, we can decide in $g(k) \cdot (n+m)$ time whether ψ is H-good.*

Let (G, H, s, t) be an instance of PATTERN SEPARATOR, where the graph G has maximum degree at most Δ. Suppose (G, H, s, t) is a yes-instance, and let S be an s–t separator of G such that $G[S]$ is isomorphic to H. Since $|S| = |V(H)| = k$, and every vertex in S has at most Δ neighbors, $|N_G[S]| \leq (\Delta + 1)k$. Using the notion of a k-*perfect family of hash functions*, we can construct in time $((\Delta + 1)k)! \cdot 2^{O((\Delta+1)k)} \cdot \log n$ a family Φ of $(c+1)$-colorings of G such that (G, H, s, t) is a yes-instance if and only if Φ contains an H-good coloring, where the size of Φ is bounded by $((\Delta + 1)k)! \cdot 2^{O((\Delta+1)k)} \cdot \log n$ (see for example [1]). By Lemma 6, we can check for each coloring in Φ whether or not it is H-good in

$g(k) \cdot (n + m)$ time, for some function g that does not depend on n. This yields the following result.

Theorem 10. (★) PATTERN SEPARATOR *can be solved in* $f(k, \Delta) \cdot m \log n$ *time on graphs of maximum degree at most* Δ.

We can solve $\mathcal{G}$-SEPARATOR by using our algorithm for PATTERN SEPARATOR to try every member of $\mathcal{G}$ having size at most k:

Theorem 11. (★) *For any decidable class* $\mathcal{G}$, *the* $\mathcal{G}$-SEPARATOR *problem can be solved in time* $h_{\mathcal{G}}(k, \Delta) \cdot m \log n$ *on graphs of maximum degree at most* Δ.

5   No Polynomial Kernel for CONNECTED SEPARATOR

In this section, we show that the CONNECTED SEPARATOR problem does not admit a polynomial kernel, even when restricted to graphs with maximum degree at most 3, unless NP $\subseteq$ coNP/poly. The CONNECTED SEPARATOR problem is easily seen to be NP-complete by a simple polynomial-time reduction from STEINER TREE. The following result shows that the problem remains NP-complete on graphs of maximum degree at most 3.

Theorem 12. (★) *The* CONNECTED SEPARATOR *problem is NP-complete on graphs of maximum degree at most 3, in which the vertices s and t have degree 2.*

An *or-composition algorithm* for a parameterized problem $Q \subseteq \Sigma^* \times \mathbb{N}$ is an algorithm that receives as input a sequence $((x_1, k), \ldots, (x_r, k))$, with $(x_i, k) \in \Sigma^* \times \mathbb{N}^+$ for each $1 \leq i \leq r$, and outputs a pair (x', k'), such that

– the algorithm uses time polynomial in $\sum_{i=1}^{r} |x_i| + k$;
– k' is bounded by a polynomial in k; and
– $(x', k') \in Q$ if and only if there exists an $i \in \{1, \ldots, r\}$ with $(x_i, k) \in Q$.

A parameterized problem Q is said to be *or-compositional* if there exists an or-composition algorithm for Q.

Theorem 13. (★) *The* CONNECTED SEPARATOR *problem, restricted to graphs with maximum degree at most 3, is or-compositional.*

Combining results of Bodlaender et al. [2] and Fortnow and Santhanam [13] on the non-existence of polynomial kernels, together with Theorems 12 and 13, yields the following result.

Theorem 14. (★) *The* CONNECTED SEPARATOR *problem, restricted to graphs of maximum degree at most 3, has no polynomial kernel, unless NP $\subseteq$ coNP/poly.*

Acknowledgements. We would like to thank Pål Grønås Drange for fruitful discussions.

References

1. Alon, N., Yuster, R., Zwick, U.: Color-coding. J. ACM 42, 844–856 (1995)
2. Bodlaender, H.L., Downey, R.G., Fellows, M.R., Hermelin, D.: On problems without polynomial kernels. J. Comput. Syst. Sci. 75, 423–434 (2009)
3. Bodlaender, H.L., Fomin, F.V., Lokshtanov, D., Penninkx, E., Saurabh, S., Thilikos, D.M.: (Meta) Kernelization. In: FOCS 2009, pp. 629–638. IEEE Computer Society (2009)
4. Bodlaender, H.L., Thomassé, S., Yeo, A.: Kernel bounds for disjoint cycles and disjoint paths. Theor. Comput. Sci. 412(35), 4570–4578 (2011)
5. Bousquet, N., Daligault, J., Thomassé, S.: Multicut is FPT. In: Fortnow, L., Vadhan, S.P. (eds.) STOC 2011, pp. 459–468. ACM (2011)
6. Chen, J., Liu, Y., Lu, S.: An Improved Parameterized Algorithm for the Minimum Node Multiway Cut Problem. In: Dehne, F., Sack, J.-R., Zeh, N. (eds.) WADS 2007. LNCS, vol. 4619, pp. 495–506. Springer, Heidelberg (2007)
7. Courcelle, B.: Graph rewriting: an algebraic and logic approach. In: Van Leeuwen (ed.) Handbook of Theoretical Computer Science, Volume B: Formal Models and Semantics, pp. 193–242. Elsevier and MIT Press, Amsterdam (1990)
8. Diestel, R.: Graph Theory. Electronic Edition (2005)
9. Downey, R., Fellows, M.R.: Parameterized Complexity. Monographs in Computer Science. Springer, New York (1999)
10. Dreyfus, S., Wagner, R.: The Steiner problem in graphs. Networks 1, 195–207 (1971)
11. Feige, U., Mahdian, M.: Finding small balanced separators. In: Kleinberg, J.M. (ed.) STOC 2006, pp. 375–384. ACM (2006)
12. Fomin, F.V., Lokshtanov, D., Saurabh, S., Thilikos, D.M.: Bidimensionality and kernels. In: Charikar, M. (ed.) SODA 2010, pp. 503–510. SIAM (2010)
13. Fortnow, L., Santhanam, R.: Infeasibility of instance compression and succinct PCPs for NP. J. Comput. Syst. Sci. 77, 91–106 (2011)
14. Gottlob, G., Lee, S.T.: A logical approach to multicut problems. Inform. Process. Lett. 103(4), 136–141 (2007)
15. Guillemot, S.: FPT Algorithms for Path-Transversals and Cycle-Transversals Problems in Graphs. In: Grohe, M., Niedermeier, R. (eds.) IWPEC 2008. LNCS, vol. 5018, pp. 129–140. Springer, Heidelberg (2008)
16. Guo, J., Hüffner, F., Kenar, E., Niedermeier, R., Uhlmann, J.: Complexity and exact algorithms for vertex multicut in interval and bounded treewidth graphs. Eur. J. Oper. Res. 186(2), 542–553 (2008)
17. Marx, D.: Parameterized graph separation problems. Theor. Comput. Sci. 351(3), 394–406 (2006)
18. Marx, D., O'Sullivan, B., Razgon, I.: Treewidth reduction for constrained separation and bipartization problems. In: Marion, J.-Y., Schwentick, T. (eds.) STACS 2010, pp. 561–572 (2010)
19. Marx, D., O'Sullivan, B., Razgon, I.: Finding small separators in linear time via treewidth reduction. CoRR, arXiv:1110.4765 (2011)
20. Marx, D., Razgon, I.: Constant ratio fixed-parameter approximation of the edge multicut problem. Inform. Process. Lett. 109(20), 1161–1166 (2009)
21. Marx, D., Razgon, I.: Fixed-parameter tractability of multicut parameterized by the size of the cutset. In: Fortnow, L., Vadhan, S.P. (eds.) STOC 2011, pp. 469–478. ACM (2011)
22. Tarjan, R.E.: Decomposition by clique separators. Discrete Math. 55, 221–232 (1985)

Author Index